ACS SYMPOSIUM SERIES **346**

Polymers for High Technology
Electronics and Photonics

Murrae J. Bowden, EDITOR
Bell Communications Research

S. Richard Turner, EDITOR
Eastman Kodak Company

Developed from a symposium sponsored by
the Division of Polymeric Materials: Science and Engineering
at the 192nd Meeting
of the American Chemical Society,
Anaheim, California,
September 7–12, 1986

American Chemical Society, Washington, DC 1987

SEP /AE
CHEM

Library of Congress Cataloging-in-Publication Data

Polymers for high technology.
  (ACS symposium series, ISSN 0097-6156; 346)

  Includes bibliographies and indexes.

  1. Polymers and polymerization—Congresses.
2. Photoresists—Congresses. 3. Microlithography—
Materials—Congresses. 4. Microelectronics—
Materials—Congresses.

  I. Bowden, Murrae J., 1943-   . II. Turner,
S. Richard, 1942-   . III. American Chemical Society.
Division of Polymeric Materials: Science and
Engineering. IV. American Chemical Society. Meeting
(192nd: 1986: Anaheim, Calif.) V. Series.

TK7871.15.P6P627   1987   668.9   87-14573
ISBN 0-8412-1406-9

SP 9/22/87 UM

Foreword

The ACS SYMPOSIUM SERIES was founded in 1974 to provide a medium for publishing symposia quickly in book form. The format of the Series parallels that of the continuing ADVANCES IN CHEMISTRY SERIES except that, in order to save time, the papers are not typeset but are reproduced as they are submitted by the authors in camera-ready form. Papers are reviewed under the supervision of the Editors with the assistance of the Series Advisory Board and are selected to maintain the integrity of the symposia; however, verbatim reproductions of previously published papers are not accepted. Both reviews and reports of research are acceptable, because symposia may embrace both types of presentation.

Contents

v

Preface

THE ELECTRONICS REVOLUTION, WHICH BEGAN with the invention of the transistor at Bell Telephone Laboratories in 1948, has continued at a frenetic pace and shows no sign of abating. Polymers have played and continue to play an integral part in this revolution in a wide variety of applications. For example, the increasing complexity of microelectronic circuits has been due in no small measure to improvements in the lithographic art. Advances in the design and development of polymeric resists have been pivotal to lithography. Polymers play an enormously important role in the packaging and interconnection of electronic components and find wide use in other applications such as dielectrics. Optical technology, with its tremendous potential for applications in communications, memory, and information retrieval, has given impetus to research on such topics as organic materials (including polymers) for nonlinear optics and optical fiber coatings.

Definitive advances in polymeric materials pertaining to these many technological thrusts continue to be made, even in technologies we might consider to be relatively "mature", for example, resist materials for microlithography. Here, new materials and processes will be required in the not-too-distant future to meet the demands of new and evolving lithographic processes. In some of the relatively recent areas of research, such as polymers for nonlinear optics, molecular electronics, and conducting polymers, many fundamental scientific principles are still not fully understood. Real breakthroughs will be needed to convince skeptics in the solid-state community of the potential advantages offered by organic materials in applications currently limited by the properties of conventional semiconductor materials.

These two powerful forces—the ongoing electronic and photonic revolution, and the significant potential of polymers to contribute to the materials needs of that revolution—continue to stimulate chemists to explore the fundamental, chemically related principles underlying these technologies. Heightened awareness of research opportunities in these areas can in turn lead to further advances.

This book has been organized into eight sections, each representing a specific field. Each section contains an introduction written by the respective session chairperson of the symposium from which this book was developed. All session chairs are recognized experts in their fields. We are indebted to many people and organizations for making the symposium possible, especially the session chairs who assembled the technical presentations and served as coordinators of the reviewing process. We are particularly grateful to the Petroleum Research Foundation for providing a substantial grant that allowed several overseas speakers to attend the symposium. We also acknowledge the generous financial support of the IBM Corporation, Eastman Kodak Company, AZ Photoproducts Division of American Hoechst Corporation, Dynachem Division of Morton Thiokol Corporation, and the Division of Polymeric Materials: Science and Engineering. We are especially indebted to Lois Damick of Bell Communications Research who handled most of the administrative aspects in preparing this volume. Finally, we thank Robin Giroux and the production staff of the ACS Books Department for their efforts in getting this book published successfully.

MURRAE J. BOWDEN
Navesink Research and Engineering Center
Bell Communications Research
Red Bank, NJ 07701-7020

S. RICHARD TURNER
Corporate Research Laboratories
Eastman Kodak Company
Rochester, NY 14650

March 26, 1987

FUNDAMENTALS IN RADIATION CHEMISTRY
OF POLYMERS

FUNDAMENTALS IN RADIATION CHEMISTRY OF POLYMERS

The history of radiation chemistry effectively started in 1896 with the discovery of X-rays by Roentgen, and of natural radiation by Becquerel. However, it was not until the advent of nuclear fission on a large scale for power generation (and military purposes) in the 1950s that progress in radiation chemistry really began to be made. The nuclear industry has needed increasing knowledge of the radiation chemistry of materials. Also, there has been a strong incentive to discover new industrial processes which could use the radiation available from reactors. The manufacture of electron accelerators has had an important influence on fundamental research in radiation chemistry and has led to a variety of commercial radiation processes. Chemists have utilized these radiation sources to learn more about the production of reactive species and the mechanisms of chemical reactions initiated by radiation.

The term "radiation" can be used to describe all regions of the electromagnetic spectrum, from radiowaves with wavelengths of meters to gamma rays with wavelengths of nm. Particulate radiations, such as electrons and alpha particles, also have equivalent wavelengths which vary with their energies.

The importance of the radiation chemistry of polymers stems from the large changes in physical and mechanical properties that can be produced by small amounts of radiation. Only a few scissions or crosslinks per molecule, for example, can dramatically affect the strength or solubility of a polymer molecule.

Traditionally, radiation chemistry and photochemistry have been considered as distinct phenomena, being differentiated according to the energy of the photon or particle and the chemistry which follows the initial absorption event. Radiation chemistry derived from photon energies capable of ionizing the parent molecule, whereas photochemistry corresponded to processes resulting from excitation of specific groups in the molecule. However, increasing interest in deep UV radiation for microlithography, the manufacture of excimer lasers producing high intensities of radiation in this special region, and the use of very low energy electron beams in order to maximize absorption in thin films have led to blurring of this boundary.

Radiation must be absorbed before it can produce chemical changes. The initial or primary chemical species resulting from the absorption of high-energy radiation consist of excited states, ions and radicals. Complex sequences of chemical reactions then follow, leading to permanent chemical changes in the molecular structure of the parent molecule. It is these chemical changes that cause the properties of polymers to be modified forming the basis of a variety of applications, e.g., resists for microlithography and graft copolymerization. Such processes can also lead to deterioration of polymers and failure in mechanical applications. This sequence of events is illustrated in Figure 1.

The way in which energy is deposited in a material depends upon the energy of the incident radiation. UV radiation is absorbed selectively by chromophores, whereas high-energy radiation is absorbed according to the electron density of the material. However, energy and charge can migrate after the initial absorption event and the chemical reactions frequently end up being determined by relative bond strengths in the molecule for both types of radiation.

Fundamental studies of the radiation chemistry of polymers can be divided into investigations of (1) permanent chemical changes, including chain scission and crosslinking, structural changes in the polymer, and the formation of small, molecular products, and (2) transient intermediates in the sequence of chemical reactions which follow the absorption event, especially the formation and disappearance of the initial excited states, ions, and radicals. Pulse radiolysis using microsecond and nanosecond pulses of electrons with energies from 1-10 MeV has contributed greatly to knowledge of these intermediate species. Lasers have been used to provide pulses of UV radiation and for analysis. Studies of polymers in solution in various solvents have been compared with small molecules in the liquid phase and with polymers in the solid state. Studies such as these are providing us with an understanding of the relationships between the molecular structures of polymers and their radiation sensitivity/resistance.

It is important to recognize that fundamental research provides the foundation for technological developments. An understanding of the effects of radiation on polymers, for example, paved the way for the latter's utilization in the electronics industry today in the fabrication of integrated circuits, where polymers sensitive to UV light, electrons, X-rays, and ions are used as resists in lithographic processes. Such technological applications of the radiation chemistry of polymers are increasingly dependent on an inter-disciplinary approach to the development of new processes. Fundamental understanding of the radiation chemistry must be linked with a thorough knowledge of polymer chemistry. However, knowledge of chemistry alone is insufficient since polymers are used mainly in the solid state with all the implications of morphology on properties which is the province of materials science. Finally, any commercial process depends on the contribution of the engineer to convert a laboratory reaction up to an industrial scale. This cooperative approach is illustrated in Fig. 2 and is particularly important in utilizing radiation chemistry in the electronics field.

The following series of papers provides an overview of the fundamental chemistry of radiation-induced changes in polymers with special consideration of electronics applications. Such studies provide valuable insight into the chemical reactions that follow absorption of radiation, an understanding of which may provide the key to improving existing processes, perhaps even to developing new resist mechanisms.

James H. O'Donnell
Polymer and Radiation Group
Department of Chemistry
University of Queensland
Brisbane 4067, Australia

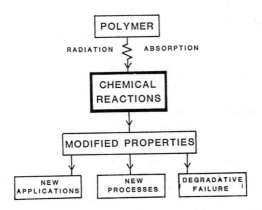

Fig. 1. The sequence of events from the absorption of radiation to its practical applications.

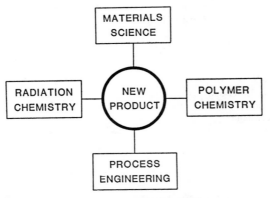

Fig. 2. Schematic representation of the cooperation necessary to utilize radiation chemistry in new technology.

Chapter 1

Development of Radiation Chemistry

G. Arthur Salmon

Cookridge Radiation Research Centre, Cookridge Hospital, University of Leeds, Leeds, LS16 6QB, United Kingdom

The scientific development of radiation chemistry is reviewed from the discovery in 1895 of x-rays and radioactivity by Roentgen and Becquerel through to the present.

The purpose of this article is to review the development of radiation chemistry which began with the discovery of x-rays by Roentgen(1) in 1895 and shortly afterwards of radioactivity by Becquerel(2), which in both cases involved the observation of chemical change in photographic plates and luminescence in certain phosphors. Clearly, in the space available, the review will be restricted and subjective, but will, it is hoped, give the general framework in which the subject has developed.

The Early Years

Very early studies of these radiations by the discoverers and by the Curies, Rutherford and others demonstrated that they were able to ionize the molecules of a gas upon which they acted. Indeed by 1900, the three kinds of rays, α, β and γ-rays, emitted by radioactive materials were characterised by their charges and their differing abilities to penetrate and ionize materials. Also, shortly after the discovery of radioactivity Pierre and Marie Curie(3) reported that radiation caused the coloration of glass and the formation of ozone from oxygen. Other chemical effects of radiation were quickly discovered. For example, Giesel(1900(4)) showed that radiation coloured alkali halides and decomposed water. Becquerel(1901(5)) showed that β- and γ-rays can induce many of the reactions that were known to be caused by absorption of light, such as the conversion of white to red phosphorus and the decomposition of hydriodic acid solutions. Jorissen and Woudstra(1912(6)) showed that the penetrating radiation from radium caused the coagulation of some colloidal solutions and Jorissen and Ringer(1906(7)) demonstrated that hydrogen and chlorine combine at room temperature under the action of these rays. Thus, during the first decade of this century the basic physical properties of ionizing radiations had been established as well as their ability to bring about chemical change.

0097–6156/87/0346–0005$06.00/0
© 1987 American Chemical Society

Absorption of Energy

Quantitative description of the chemical changes initiated by radiation requires an understanding of the processes by which the rays transfer energy to a system and knowledge of how much energy is transferred. For particulate radiations (α, β, e^- and e^+) the important interaction is inelastic collisions between the particles and the molecules of the medium resulting in their ionization and excitation.

$$M \wedge\!\wedge\!\wedge\!\rightarrow M^+ + e^-, M^* \qquad\qquad (1)$$

The rate of energy loss from the particle per unit length of track, or Linear Energy Transfer (LET), was known to follow the Bragg curve with a maximum LET close to the end of the particle's track. Bethe (1933(8)) derived theoretical expression for this quantity for electrons and other charged particles. For electrons this has the form:

$$-\frac{dE}{dx} = \frac{2\pi Ne^4 Z}{m_o v^2} \left[\ln \frac{m_o v^2 E}{2I(1-\beta^2)} -(2\sqrt{1-\beta^2} - 1 + \beta^2)\ln 2 \right.$$
$$\left. + 1 + \beta^2 + \frac{1}{8}(1-\sqrt{1-\beta^2})^2 \right] \qquad (2)$$

where v is the velocity of the electron, m_o its rest mass, β is v/c and I the mean excitation potential of the atoms of the stopping material.

For x- and γ-rays, the photons ionize the molecules of the medium by either the photoelectric effect(P. Lennard, 1902; A. Einstein, 1905), the Compton effect(9) or by pair production(10), depending on the photon energy and the atomic number of the absorbing material. Each of these mechanisms generates energetic electrons or positrons which are able to bring about further ionizations and excitations as discussed above.

Particle Tracks. Thus, irrespective of the particulate or photon nature of the primary radiation, the net effect is the formation of tracks consisting of ionized and excited molecules. These tracks, and their detailed structure can be revealed by the Cloud Chamber invented by Wilson in 1911(11). For fast electrons (low LET) the tracks mainly consist of spherical regions called spurs which contain from one to four ion-pairs which are separated in condensed phases by about 10^4 Å. For more highly ionizing particles such as α-particles the tracks are essentially cylindrical columns of ionized and excited molecules.

Thus, by the early thirties, the interaction of the various forms of radiation with matter to form ions and excited molecules in tracks was well understood and provided a firm foundation on which to base models for the chemical processes which follow this stage.

Gas Phase Radiation Chemistry in the 20's and 30's

In the 1920's and 30's most studies were confined to the gas phase and used α-particles from radon as the radiation source. The results were expressed as M/N values, the number of molecules converted per ion-pair formed in the gas. S.C. Lind and his co-workers played an important part in much of this work. In a number of cases, e.g., the polymerisation of acetylene, it was found that M/N values considerably exceeded unity. Lind(12) and Mund(13) proposed that M/N values greater than unity could be accounted for by the neutral molecules forming clusters around the ions and on neutralization of the cluster it was believed that the energy liberated could be used to cause chemical change in all the molecules forming the cluster.

A noteable landmark in the development of the subject was the publication in 1936 of two papers by Eyring, Hirschfelder and Taylor(14,15) in which they critically discussed the mechanisms of the radiation induced conversion of ortho to para hydrogen and the synthesis and decomposition of hydrogen bromide. In these papers they considered i) the nature of the initial ionization processes, laying stress on the information available from mass-spectrometric studes, ii) possible ion-molecule reactions iii) electron-attachment processes, iv) ion-neutralization to form neutral reactants and v) the role of ion-clustering. In general, they concluded that the reactants responsible for the bulk of the chemistry were neutral free radicals that participated in free radical chain reactions. They also showed that ion-molecule reactions were expected to be fast reactions and were important in determining the nature of the cations which were neutralized. Thus, these authors introduced many of the principles which form the basis of present day radiation chemical thinking.

Influence of The Second World War

The project to build the atomic bomb in the Second World War, the Manhattan project, had a very great influence on all nuclear research, and radiation chemistry was no exception due to the greater number of persons involved in studying the chemical effects of radiation. Radiation sources also became much more powerful and more readily available as non-natural radioisotopes, such as ^{60}Co, were produced in atomic reactors. Also at this time, radiation chemists concluded that it was preferable to express radiation-chemical yields in terms of the energy absorbed by a system, rather than as a M/N-value. This was because for condensed phase systems which were playing an increasingly important role in radiation chemistry, it is impossible to measure N, the number of ions formed, and values were based arbitrarily on gas phase W-values, i.e. the energy required to form an ion pair in the gas phase. Thus, the G-value was defined as the number of molecules converted per 100 eV absorbed by the system. This definition has held upto the present although recent recommendation by international bodies are for the definition to be modified to conform to the S.I. System of units. On this basis the G-value can be defined as the number of moles of product formed or reactant consumed per Joule absorbed and

$$G\text{-value/mol J}^{-1} = 1.037 \times 10^{-7} \times G\text{-value/molecules (100 eV)}^{-1}$$

Aqueous Solutions

The Free Radical Hypothesis. Although studies on aqueous solutions
had continued through the 30's, notably by H. Fricke and his
co-workers, detailed interpretation of data was prevented by the lack
of a clear hypothesis for the nature of the species responsible for
the chemistry. It was appreciated that for solutions the bulk of the
energy was absorbed by the solvent and this led to the concept of
indirect action(16,17) whereby activated species derived from the
solvent reacted with the solute, but the nature of the activated
species was unknown. However in 1944, Weiss(18) in England revived a
very early proposal by Debierne(19) that irradiation of water yielded
H-atoms and OH-radicals,

$$H_2O \rightsquigarrow H^{\bullet} + OH^{\bullet} \qquad\qquad (3)$$

and he indicated, for example, how this hypothesis explained the
formation of hydrogen peroxide when aerated water was irradiated and
also the oxidation and reduction of various metal ions in irradiated
water.
      This proposal was valuable in providing a concrete mechanism for
explaining the radiation chemistry of water, but it was based on the
invalid assumption that the same general principles apply to liquids
as to gases, and that dissociation processes are the important result
of irradiating materials. This ignores the profound influence that
solvent structure and polarity can have on the recombination of ions,
especially electrons. However, in 1953 Samuel and Magee(20) produced
a model for the primary event in water which suggested that the
electrons generated in the ionization event travel about 20A from the
positive ion while being thermalised. At this distance they would
still be within the strong Coulombic field of the cation and would be
pulled back to it leading to its neutralization and the formation of
$H^{\bullet}$ and $OH^{\bullet}$. Thus this model provided strong support for the Weiss
hypothesis which consequently had a very strong influence on the
thinking of radiation chemists until almost 1960 even though
Stein(21) and Platzman(22) had postulated in 1952 and 1953 that the
electron could be solvated and participate in reactions with solutes.
In fact, Platzman had considered the formation and fate of the
hydrated electron in some detail, but his ideas were presented at an
informal conference and did not make an immediate impact on the
radiation chemical community.
      The formation of the hydroxyl radical has been amply
demonstrated, but it is now accepted that the major route to its
formation is by the ion-molecule reaction (4).

$$H_2O^{+} + H_2O \rightarrow H_3O^{+} + {}^{\bullet}OH \qquad\qquad (4)$$

Allen(23) has indicated that M. Burton and J. Franck held the free
radical hypothesis of water radiolysis during their war-time work and
that they considered $^{\bullet}OH$ to be generated by reaction (4).

The Hydrated Electron. However, during the late 1950's results
became available on aqueous solutions which could not be reconciled
with the major reducing species being the hydrogen atom(24-28) and
also Czapski and Schwarz(29) and Dainton and co-workers(30,31) proved

that the reducing species carried unit negative charge. The invention of pulse radiolysis by Boag and Hart(32,33) and Keene(34) led to the observation in irradiated aqueous solutions of a broad absorption with $\lambda_{max} = 720$ nm, which was identified as being due to the hydrated electron, e_{aq}.

This species is essentially an electron stabilised by the surrounding water molecules. It has been the subject of detailed theoretical studies(35), but can be considered as an electron in a spherical potential well consisting of solvent molecules. Specific short-range solvation effects are thought to be important as well as long-range polarization forces.

With the discovery of the hydrated electron and the newly available technique of pulse radiolysis rapid advances were made in the understanding of the processes which occur in irradiated aqueous solutions. For an excellent review of the current state of knowledge see reference 36. The chemistry of the hydrated electron has been reviewed in detail by Hart and Anbar(37), but its reactions can be summarised as:

   i)   redox reactions with metal ions

$$eg. \quad M^{Z+} + e^-_{aq} \quad \rightarrow \quad M^{(Z-1)+} \tag{4}$$

   ii) electron attachment reactions

$$eg. \quad RCHO + e^-_{aq} \quad \rightarrow \quad RCHO^- \tag{5}$$

   iii) dissociative attachment

$$eg. \quad N_2O + e^-_{aq} \quad \rightarrow \quad N_2 + O^- \tag{6}$$

$$ClCH_2CO_2^- + e^-_{aq} \quad \rightarrow \quad Cl^- + {}^\cdot CH_2CO_2^- \tag{7}$$

and iv) reactions with Brønsted acids

$$eg. \quad H^+ + e^-_{aq} \quad \rightarrow \quad H^\cdot \tag{8}$$

Solvated Electrons in Organic Media. There was, of course, no reason to suppose that water was the only liquid capable of solvating the electron and over a very few years pulse radiolysis experiments demonstrated the existence of solvated electrons, e_s, in many liquids.

(Table I). $\underline{e_s^-}$ in Various Liquids

Liquid	λ_{max}/nm	(Ref)
water	720	(32–34)
alcohols	580–820	(38)
ammonia	1500	(39)
amines	>1700	(40,41)
dialkylamides	1500–1800	(41)
ethers	1900–2300	(42)
HMPA	~2300	(43)
hydrocarbons	>2000	(44)

The stability of the solvated electron in the various solvents varies greatly depending on the reactivity of the electron with the solvent. In liquid ammonia solvated electrons are stable for long periods in the blue solutions of alkali metals in ammonia(45). In water, e_{aq}^- decays slowly by reaction (9) with a rate constant of 16 $M^{-1}s^{-1}$(46)

$$e_{aq}^- + H_2O \rightarrow \ ^\bullet H + OH^- \tag{9}$$

while the analogous reaction in ethanol has a half-life of 9 μs at room temperature(47). In organic amides lifetimes of the order of microseconds have been observed(48) despite the presence of the carboxyl group.

Ionic processes in Non-Polar Media

Following the realisation that the reactions of the hydrated electron played an important role in the radiation chemistry of liquid water it was not long before evidence was sought, and found, that the electron and the counter cation could be involved in chemical reactions in non-polar liquids before they underwent neutralisation. Scholes and Simic (1964(49)) showed that on irradiation of solutions of nitrous oxide in hydrocarbons nitrogen was formed in the dissociative attachment reaction analogous to reaction (6). Similarly, Buchanan and Williams (1966(50)) attributed the formation of HD in γ-irradiated solutions of C_2H_5OD in cyclohexane to the transfer of a proton from the $C_6H_{12}^+$ ion to C_2H_5OD (reaction 10):

$$C_6H_{12}^+ + C_2H_5OD \rightarrow C_6H_{11}^{\ \bullet} + C_2H_5ODH^+ \tag{10}$$

$$e^- + C_2H_5ODH^+ \rightarrow C_2H_5OH + D^\bullet \tag{11}$$

$$D^\bullet + C_6H_{12} \rightarrow C_6H_{11}^{\ \bullet} + HD \tag{12}$$

The scavenging of electrons and cations in non-polar media can be explained if it is assumed that the thermalisation of the electron is a stochastic process giving rise to a distribution of ion-pair separations with a significant number of pairs with separations > 200 A. Most of the electrons drift back towards the cation under the influence of the Coulombic field and undergo recombination with their geminate partner. While drifting under the influence of their mutual field either of the partners may encounter a solute with which they can react. The dynamics of this recombination and scavenging process are, of course, determined by the diffusion coefficients of the two charged species and hence on their degree of localization.

Since reaction of the charges with solutes is in competition with their recombination in the Coulombic field, the dependence of the yield of product from the scavenging reaction is strongly dependent on the solute concentration. The empirical equation due to Warman, Asmus and Schuler(51) has proved to be particularly useful in treating scavenging data, where G(P) is the scavenging yield, G_{fi} the free-ion yield (see below) and G_{gi} the yield of ion-pairs which undergo geminate recombination in the absence of a scavenger.

$$G(P) = G_{fi} + G_{gi} \left\{ \frac{\alpha^{\frac{1}{2}}[S]^{\frac{1}{2}}}{1+\alpha^{\frac{1}{2}}[S]^{\frac{1}{2}}} \right\} \tag{13}$$

Due to Brownian motion, a small fraction of the ion-pairs escape recombination and become homogeneously distributed. This yield of ion-pairs is called either the <u>free-ion yield</u>, G_{fi} or the <u>escaped yield</u>, G_{esc} and is given by

$$G_{fi} = G_{ti} \int_o^\infty n(r) \exp(-r_c/r).dr \qquad (14)$$

where G_{ti} is the total ionization yield, $n(r)$ is the distribution of initial ion-pair separations and r_c, the Onsager critical distance, is $e^2/4\pi\epsilon_o\epsilon_r k_B T$. Expression (14) is only strictly valid for the case of a single ion-pair spur, but serves as a useful model description. The recent review by Warman(52) gives a detailed account of the factors governing the behaviour of electrons and cations in irradiated non-polar systems.

<u>Localized and Quasi-Free Electrons.</u> The mobility of electrons in non-polar media ranges from 5×10^{-3} cm^2 v^{-1} s^{-1} in liquid hydrogen to values as high as 2200 cm^2 v^{-1} s^{-1} in liquid xenon(see reference 35). For common hydrocarbons the value range from 0.013 cm^2 v^{-1} s^{-1} for trans-decalin(53) to 100 cm^2 v^{-1} s^{-1} for tetramethylsilane(54). These values are to be compared with that for e_{aq}^- of 2×10^{-3} cm^2 v^{-1} s^{-1} (quoted in reference 37).
    A simplified model(55(1972)) which explains the wide range of values, is that the electron in non-polar condensed media can exist in two states, a <u>localized</u> or trapped state whose mobility is comparable with that of small molecular ions and a <u>quasi-free</u> state the mobility of which is of the order of that in liquid xenon. In the <u>quasi-free</u> state the electron moves freely for much of its time in the uniform potential between molecules and only briefly interacts with the potential well surrounding molecules. Thus the localized and quasi-free states can be considered to exist in dynamic equilibrium and the mobility of the electron can be approximated by

$$\mu(e^-) = \mu_{d\ell}\exp(\Delta H_{d\ell}/k_B T) \qquad (15)$$

where $\mu(e^-)$ is the mobility of the electron, $\mu_{d\ell}$ the mobility in the <u>quasi-free</u> state, and $\Delta H_{d\ell}$ the enthalpy difference between the two states.
    Of course, electrons in non-polar media only display absorption spectra if they spend a major fraction of their existence in a localized state, as is the case with n-hexane.

<u>The Influence of the Polarity of the Medium</u>
The above discussion stresses the key role of solvent polarity and structure in determining the subsequent behaviour of the ionic species generated in the primary processes. Thus, in water with its high dielectric constant the bulk of the e_{aq}^- and H_3O^+ escape the Coulombic field and the spur processes depend only on their random diffusion.
    In non-polar media in the absence of electron or charge scavenging solutes, most electrons and ions will undergo fast geminate charge recombination leading to neutral free radicals, or in some media to long-lived excited states, which react to give the final products.

Ionic Processes in Solids

Concurrently with the view that reactions of electrons and positive ions play an important part in the radiation chemistry of liquids, it was being demonstrated, notably by Hamill(56) and his co-workers over the period 1962-1966, that charge trapping, migration and reaction were also important in the radiation chemistry of many solid systems, especially at low temperatures. As a result of these studies, γ-irradiation of low temperature solids has become perhaps the most versatile method of studying the spectroscopic properties of radical anions and cations.

Polymers

The history of the radiation chemistry of polymers is the subject of another lecture in this symposium by Professor Chapiro. However, since the main theme of the symposium is aimed at the use of electron beams in lithography for the manufacture of electronic devices, it is appropriate to refer briefly to the radiation chemistry of polymeric materials.

The first systematic study of the irradiation of polymers was undertaken by Dole and Rose during the period 1947-1949. These workers discovered that irradiation of polyethylene caused some degradation to low molecular weight products and the introduction of unsaturation in the polymer chains, but by far the most exciting discovery was that cross links were formed between polymer chains and this had a profound effect on the stress-strain curves and the cold-drawing properties of polyethylene(57-59). In 1952 Charlesby(60) published the first of many papers on the effects of radiation on polymers. The rapid development of the field upto 1960 is reviewed in books by Charlesby (61) and Chapiro (62).

The major studies of high polymer systems have concentrated mainly on relating the chemical changes of cross linking and chain degradation to the modification of the physical properties of the polymer. Most studies have utilised free radical mechanisms to explain cross linking and degradation of polymers. Since many polymers are polyolefines and non-polar, it is to be expected that most of the charges formed in the initial ionization event will undergo rapid recombination to yield radicals. However, the detailed mechanisms controlling the motion of the radicals which allow them to come together to from a cross link is still the subject of much un-certainly (63). Also for halogenated polymers dissociative electron capture processes analogous to reaction (7) are important in deter-mining the initial localisation of the free radicals. Recent developments in the field include the application of the picosecond pulse radiolysis technique to the study of the processes occurring in polymers(64) and the use of pulse radiolysis in conjunction with light scattering measurements to study their degradation (65).

The use of radiation to modify the physical properties of polymers has become a very important industry with products such as electrical cables with insulation capable of withstanding high temperatures and heat-shrinkable polyethylene. However, of direct relevance to this symposium was the recognition in the early 1970's that electron beam irradiation of polymer films could provide an important lithographic tool for the manufacture of microelectronic components. For consideration of the general principles of these processes see, for example, references (66) and (67). The products required in this field are complex requiring both microscopic resolution of elaborate patterns and proceedures for producing both positive and negative resists. Thus the emphasis has been on bringing about high efficiency degradation or cross linking of specialised polymers. One of the rationales behind the use of electron beams for this purpose is that the effective wavelength of electrons is so small that diffraction effects in the formation of patterns are negligible. However, it is probable that the resolution limit will prove to be determined by the thermalization length of the secondary electrons generated in the primary ionization events.

Current Trends

In the course of this lecture I have attempted to outline the development of radiation chemistry from its beginnings. The main historical theme has been the elucidation of the extent to which neutral free radicals as opposed to ionic species contributed to the overall chemistry. Our present understanding leads to the view that the interplay between free radical and ionic processes is very dependent on the system being considered, particularly its dielectric properties and the presence, or absence, of solutes which can react with electrons or cations.

Since the radiation chemistry of water and aqueous solutions is well understood and the yields of the primary species are well characterised, the radiolysis of aqueous solutions has become perhaps the most versatile means of studying a wide range of free radical and redox processes in solution, especially when combined with the nanosecond time-resolution which is routinely available with pulse radiolysis. It is outside the scope of this review to consider the wide range of processes which have been studied using these techniques and the reader is referred to the paper by Buxton(68) for further information.

The study of the very fast processes that follow on the absorption of radiation in organic systems is a very active field with pulse radiolysis with picosecond time resolution being one of the major tools. This technique, the latest version of which employs twin linear accelerators(69), has time resolution of about 20 ps. These methods are being used to investigate the fast recombination of charges, the formation of excited states and free radicals, mainly in hydrocarbon media, but have also recently been applied to the study of radiation effects in polymers(70).

In general, detailed knowledge of the radiation chemistry of organic liquids is restricted to the lower alcohols and some hydro-carbons and information on other systems in very sparse. This is one of the reasons why pulse radiolysis in organic solvents has not yet fulfilled its potential for application to the study of general chemical problems, as has been the case for aqueous systems.

Literature Cited

1. Roentgen, W.C. Sitzungosberichte der Physikalisch-Medizinischen Gesellschaft zu Würzbürg, 1895.
2. Becquerel, H. Compt. rend. 1896, 122, 420.
3. Curie, P.; Curie, M. Compt. rend. 1899, 129, 823.
4. Giesel, F. Verh. Deut. Phys. Ges. 1900, 2, 9.
5. Becquerel, H. Compt. rend. 1901, 133, 709.
6. Jorissen, W.P.; Woudstra, H.W. Zeit. Chem. u. Industrie d. Kolloide. 1912, 10, 280.
7. Jorissen, W.P.; Ringer, W.E. Berichte 1906, 39, 2093.

8. Bethe, H.A. Handbuch der Physik; Julius Springer: Berlin, 1933; vol. 23.
9. Compton A.H. Phys. Rev. 1923, 21, 483-502.
10. Blackett, P.M.S.; Occhialini, G.P.S. Proc. Roy. Soc. (London) 1933, A139, 699-727.
11. Wilson, C.T.R. Proc. Roy. Soc. (London) 1923, A104, 192.
12. Lind, S.C. The Chemical Effects of Alpha-Particles and Electrons; Chemical Catalog Co., Inc.: New York, 1928.
13. Mund, W. L'Action chimique des rayons alpha on phase gazeuze; Herman et Cie.: Paris, 1935.
14. Eyring, H.; Hirschfelder, J.O.; Taylor, H.S. J. Chem. Phys. 1936, 4, 479.
15. Eyring, H.; Hirschfelder, J.O.; Taylor, H.S. J. Chem. Phys. 1936, 4, 570.
16. Risse, O. Ergeb. Physiol. 1930, 30, 242.
17. Fricke, H. Cold Spring Harbor Symopsium 1934, 2, 241.
18. Weiss, J. Nature 1944, 153, 748.
19. Debierne, A Ann. Physique (Paris) 1914, 2, 241.
20. Samuel, A.H.; Magee, J.L. J. Chem. Phys. 1953, 21, 1080.
21. Stein, G. Disc. Faraday Soc. 1952, 12, 227.
22. Platzmann, R.L. Physical and Chemical Aspects of Basic Mechanisms in Radiobiology; U.S. National Research Council: 1933, No.305,22.
23. Allen, A.O. In The Chemical and Biological Action of Radiations; Haissinsky, M., Ed.; Academic Press: London, 1961; Vol. 5, p.14.
24. Hart. E.J. J. Am. Chem. Soc. 1954, 76, 4312.
25. Hart, E.J. Radiation Res. 1955, 2, 33.
26. Hayon, E.; Weiss, J. Proc. 2nd Int. Conf. Peaceful Uses of Atomic Energy, 1958, 29, 80.
27. Baxendale, J.H.; Hughes, G. Z. Phys. Chem. (Frankfurt), 1958, 14, 306.
28. Barr, N.F.; Allen, A.O. J. Phys. Chem. 1959, 63, 928.
29. Czapski, G.; Schwarz, H.A. J. Phys. Chem. 1962, 66, 471.
30. Collinson, E.; Dainton, F.S.; Smith, D.R.: Tazuke, S. Proc. Chem. Soc. 1962, 140.
31. Dainton, F.S.; Watt, W.S. Proc, Roy. Soc. (London), 1963, 275A, 447.
32. Boag, J.; Hart, E.J. J. Am. Chem. Soc. 1962, 84, 4090.
33. Boag, J.; Hart, E.J. Nature, 1963, 197, 45.
34. Keene, J.P. Nature, 1963, 197, 47.
35. Kestner, N.R. In Electron-Solvent and Anion-Solvent Interactions; Kevan, L. and Webster, B.C., Eds.; Elsevier: Amsterdam, 1976, Chapter 1.
36. Buxton, G.V. In The Study of Fast Processes and Transient Species by Electron Pulse Radiolysis; NATO Advanced Study Institute Series; Baxendale, J.H. and Busi, F., Eds.; Dordrecht: Holland, 1982, p.241.
37. Hart, E.J.; Anbnar, M. The Hydrated Electron; Wiley-Interscience: New York, 1970.
38. Taub, I.A.; Harter, D.A.; Sauer, M.C.; Dorfman, L.M. J. Chem. Phys. 1964, 41, 979,
39. Compton, D.M.J.; Bryant, J.F.; Cesena, R.A.; Gehman, B.L. In Pulse Radiolysis; Ebert, M.; Keene, J.P.; Swallow, A.J.; Baxendale, J.H., Eds.; Academic Press: London & New York, 1965, p.43.
40. Belloni, J.; de la Renaudiere, J.F. Nature Phys. Sci. 1971, 232, 173.

41. Gavlas, J.F.; Jou, F.Y.; Dorfman, L.M. J. Phys. Chem. 1974, 74, 2631.
42. Jou, F.Y.; Dorfman, L.M. J. Chem Phys. 1973, 58, 4715.
43. Shaede, E.A.; Dorfman, L.M.; Flynn, G.J.; Walker, D.C. Can. J. Chem. 1973, 51, 3905.
44. Richards, J.T.; Thomas, J.K. Chem. Phys. Lett. 1971, 10, 317.
45. Kraus, C.A. J. Am. Chem. Soc. 1908, 30, 1323.
46. Hart, E.J.; Gordon, S.; Fielden, E.M. J. Phys. Chem. 1966, 70, 150-6.
47. Baxendale, J.H.; Wardman, P. Chem. Comm. 1971, 429-30.
48. Guy, S.C.; Edwards, P.P.; Salmon, G.A. Chem. Comm. 1982, 1257-9.
49. Scholes, G.; Simic, M. Nature, 1964, 202, 895.
50. Buchanan, J.W.; Williams, F. J. Chem. Phys. 1966, 44, 4371.
51. Warman, J.M.; Asmus, K.-D.; Schuler, R.H. In A.C.S. Advances in Chemistry Series, 1968, 82, Vol. 2, 25.
52. Warman, J.M. In the Study of Fast Processes and Transient Species by Electron Pulse Radiolysis; NATO Advanced Study Institute Series; Baxendale, J.H. and Busi, F., Eds.; Dordrecht: Holland, 1082, p.433.
53. Warman, J.M.; Infelta, P.P.; de Haas, M.P.; Hummel, A. Can. J. Chem. 1977, 55, 2249.
54. Allen, A.O.; Gangwer, T.E.; Holroyd, R.A. J. Phys. Chem. 19755, 79,25.
55. Munday, R.M.; Schmidt, L.D.; Davis, H.T. J. Phys. Chem, 1972, 76, 442.
56. Hamill, W.H. In Radical Ions; Kaiser, E.T. and Kevan, L. Eds.; Wiley-Interscience: New York, 1968, p.321.
57. Dole, M.; Rose, D.; paper presented at the 114th Meeting of the American Chemical Society, Portland, Ore., 1948.
58. Dole, M.; Effect of Radiation on Colloidal and High Polymer Substances; in Report of Symposium IV; Technical Command, U.S. Army Chemical Center: Maryland, Sept. 18-20, 1959, p.120.
59. Rose, D.G. M.S. Thesis; Northwestern University; Evanston, Ill., 1949.
60. Charlesby, A. Proc. Roy. Soc. (London), 1952, A215, 187.
61. Charlesby, A. Atomic Radiation and Polymers; Pergamon Press: Oxford, 1960.
62. Chapiro, A. Radiation Chemistry of Polymeric Systems; Wiley-Interscience: New York, 1962.
63. Charlesby, A. Radiat. Phys. Chem, 1981, 18, 59-66.
64. Tagawa, S.; Schnabel. W.; Washio, M.; Tabata, Y. Radiat. Phys. Chem, 1981, 18, 1087-1095.
65. Schnabel, E.; Denk, O.; Grollmann, U.; Raap, A.; Washino, K. Radiat. Phys. Chem. 1983, 21, 225-231.
66. Ricker, T. Festkorperprobleme, 1976, XVI, 217-237.
67. Bowden, M.J. Radiat. Phys. Chem 1981, 18, 357-369.
68. Buxton, G.V. Proceedings of the Seventh International Congress of Radiation Research; Broerse, J.J., Barendsen, G.W., Kal, H.B., van der Kogel, A.J., Eds.; Martinus Nijhoff: Amsterdam, 1983, 119-128.
69. Tabata, Y.; Kobayashi, H.; Washio, M.; Tagawa, S.; Yoshida, Y. Radiat. Phys. Chem. 1985, 26, 473-479.
70. Tagawa, S. Radiat. Phys. Chem. 1986, 27, 455-459.

RECEIVED May 15, 1987

Chapter 2

Primary Action of Ionizing Radiation on Condensed Systems

J. K. Thomas[1] and G. Beck[2]

[1]Chemistry Department, University of Notre Dame, Notre Dame, IN 46556
[2]Hahn-Meitner Institut, Glienicker 100, 1-Berlin-39, Federal Republic of Germany

The initial effects of high energy radiation on
condensed systems are discussed. Evidence of short
lived intermediates, i.e. radical ions and excited
states is illustrated by fast pico-second and
nanosecond pulse radiolysis. A discusion of the
nature of early events leading to excited states is
discussed at length, and in particular comparisons
are made to corresponding experiments at low photon
energies, i.e. via laser flash photolysis.

Excitation of molecules by radiation enters different regimes which
depend on the energy of the quanta used. Excitation with low
energy quanta, as in photochemistry, leads to discrete excited
states as the quantum of energy is totally absorbed by the system;
wavelengths down to $\sim$ 2000 Å, i.e. energies of $\sim$ 6.0 eV are typical
in this work. High energy irradiation, or Radiation Chemistry
utilises x- or γ-rays and electrons of high energy, typically 200
keV to 20 meV. Energy loss is primarily via ompton scatter to the
electrons of the system, although the photoelectron effect becomes
increasingly important at lower energies. In photochemistry low
energy resonance processes excite the system or a part of the
system; while in radiation chemistry a statistical loss of energy
to the electrons of the system occurs, and the energy deposited in
a section of the system depends on the electron fraction of that
section (2). It is usual in photochemistry to excite a solute
directly, while in radiation chemistry the solvent is excited and
energy or free radical chemistry is then transferred to the
solute. In radiation chemistry, unlike photochemistry, the initial
high energy quantum is not completely lost on its first excitation,
but, a fraction of the energy is lost to ionise an electron
(secondary e$^-$) of the system. This process repeats as the
electrons, both primary and secondary, continue to excite the
medium causing ionization, until the last of secondary e$^-$ fail to
produce excitation. Thermalisation of the e$^-$ occurs to produce a
trapped electron of the medium, this may be a solvated electron.

0097–6156/87/0346–0016$06.00/0

Direct excitation to excited states has never been observed in radiation chemistry although it is predicted in the so called "optical approximation" (2,3). Spectroscopic evidence of direct excitation by very low energy electrons is forthcoming (4). Radiation chemistry is often called the chemistry of ionising radiation and the popular concept is that primary energy loss is to produce ionization, the subsequent chemistry then depending on the medium and its pertinent chemistry.

It is difficult to span the intervening energy gap between photo- and radiation chemistry, however, high powered pulsed lasers, utilising multiphoton absorption by the medium, do much to remedy this situation. For the most part, the work described falls into two categories, data with steady state irradiation i.e. light sources and ^{60}Co-γ rays, and pulsed experiments as with lasers and pulsed electron accelerators such as Van de Graaffs and Linacs. The interests of the meeting are in polymerised systems, and hence the majority of the paper will deal with hydrophobic systems. However, for completeness a general description of radiation chemistry in polar media is also included.

Radiolysis, general concepts

It is now agreed, that for the most part, the radiolysis of polar liquids such as alcohol, water, etc. leads to the formation of ions and subsequently radicals rather than excited states. However, the picture is changed as the polarity of the liquid decreases, and in the extreme case of arenes such as benzene only excited states both singlet and triplet are observed. In other liquids such as alkanes, dioxane, etc. both ions and excited states are observed, charge neutralisation of the ions also giving excited states.

Table I gives data for the yields of excited states and ions observed in the radiolysis of various liquid systems. The yield is stated in terms of the G value or number of molecules of product per 100 eV of energy absorbed by the system. An immediate generalization is possible: The radiolysis of nonpolar liquids, arenes, alkanes, etc. produces excited states and sometimes ions. Increasing the polarity of the liquid, e.g., benzene to benzonitrile, benzyl alcohol, phenol, leads to a decrease in the yield of excited states with a concomitant rise in the observed yield of ions. In very polar liquids such as water and alcohols only ions are observed.

A thesis that accommodates these data is one where radiolysis leads to significant yields of ions the recombination of which leads to excited states. Excited states may also be produced via direct excitation. It is an accepted fact that the recombination or neutralization of ions can lead to both excited triplet and singlet states, the process favoring the formation of triplet excited states (5). In polar media the overall energy change of the ion neutralization process may be too low to give excited states due to the increase solvent solvation energy of the ions in these media. In hydroxylic media ion molecule reactions may alter the ions and excited states are also eliminated. For example, in water the following processes take place:

$$H_2O \longrightarrow H_2O^+ + e^- \longrightarrow e^-_{aq} \text{solution}$$

$$\downarrow H_2O$$

$$H_3O^+ + OH^-$$

Solvation of both e^- and H_2O^+ leads to a situation where only solvated protons H_3O^+, hydrated electrons e^-_{aq} and hydroxyl radicals OH are formed. The relative yield of ions versus excited states depends on the relative rates of reaction of the primary ions with the solvent and the rates of ion neutralization. The rate of spin relaxation of the ions prior to neutralization determines the relative yield of triplets versus singlets (6).

The abundance of excited states produced directly can be calculated via the optical approximation (3). The optical approximation states that the energy lost to a particular electronic transition of a molecule is proportional to f/ε where f is the oscillator strength for that transition and ε is the energy. For the E_{1u} state of benzene $f \sim 1.0$ and ε is ~ 6.0 eV, the total oscillator strength for benzene being 42. Thus if we take an average energy of 14 eV for the electronic transitions of benzene then $\sim 5\%$ of the total energy lost is to the E_{1u} state. The yield per 100 eV is thus ~ 1.0.

Radiolysis of Alkanes

Pulse radiolysis (15, $\text{--}\rightarrow$ 18) of alkane solutions of arenes gives rise to excited singlet and triplet states, and anions and cations of the solute (19). The yields of excited states in alkanes are generally lower than those in arene solvents, while the reverse is often true of the ion yields.

More recent psec data (20) show that excited singlet states of the alkane also transfer singlet energy to the aromatic solute and that the rate of energy transfer is extremely rapid ($k \sim 10^{11}$ LM^{-1} s^{-1}).

Laser photolysis of biphenyl (ϕ_2) in benzene and C_6H_{12} may be compared to the pulse radiolysis of these systems to illustrate the above effects. Figure I shows data for single photon laser photolysis ($\lambda = 2650$ Å, energy 10 mj, pulse length 20 nsec), and 2 photon photolysis ($\lambda = 3471$ Å, energy 200 mj, pulse length 20 nsec) of biphenyl ϕ_2, in C_6H_{12} and benzene. The fluorescence decay of ϕ_2 is shown at $\lambda = 3300$ Å and the rate of appearance of the triplet ϕ_2^T at 3600 Å. No apparent growth of ϕ_2^T is seen in benzene in the two photon laser experiments. The laser experiments are analogous to the corresponding radiolysis studies, as the two photon energy is absorbed by benzene to give benzene excited states which then transfer energy to ϕ_2 (21-22). The lack of growth is due to almost precise overlap of the excited singlet ϕ_2^S and ϕ_2^T spectra as observed for naphthalene (22).

In the ϕ_2/C_6H_{12} system a significant growth of ϕ_2^T is seen following the 2650 Å laser pulse. The rate of growth of ϕ_2^T has a $\tau_{1/2} = 14.0$ nsec which is precisely the rate of decay of the fluorescence. Thus intersystem crossing, $\phi_2^S \rightarrow \phi_2^T$ is observed. The

Table I. Yields of Singlet and Triplet Excited States,
Ions and Total Yield in Radiolysis of Solvents

Solvent	G(T)	G(S)	G(ions)[a]	G(Total)[b]
O-xylene[7]	1.7	2.5	~0.1	4.3
M-xylene[7]	1.8	2.7	~0.1	4.5
P-xylene[7]	2.4	2.0	~0.1	4.5
Pseudo cumene[7]	1.8	1.6	~0.1	3.5
Mesitylene[7]	1.8	1.6	~0.1	3.5
Toluene[7]	2.4	2.1	~0.1	4.6
Benzene[7]	3.8	1.6	~0.1	5.5
Benzyl alcohol[8]	1.1	0.7	2.1	5.9
Benzonitrile[9]	1.4	1.2	1.4	4.0
Dimethylaniline[10]	3.1	0.9	<0.2	4.2
Phenol[11]	0	0	3.0	3.0
Cyclohexane[12]	0.7	0.7	1.6	3.0
Tetrahydrofuran[13]	0.1	0.04	0.66	0.8
Dioxane[14]	–	1.03	0.12	1.15
Methanol[15]	0	0	2.0	2.0
Water[15]	0	0	3.5	3.5

(a) Yield of anion or cation
(b) G(total = G(ions) + G(excited states).

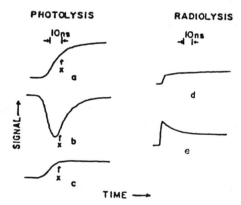

Figure 1. Laser photolysis and pulse radiolysis of biphenyl
(ϕ_2) in alkanes and arenes. Observation of development of
fluorescence or absorption (signal) versus time. (a)
Photolysis 15 nsec pulse of 2650 Å light of solution of 2 x
10^{-4} ϕ_2 in hexane, λ absorption at 3600 Å (biphenyl triplet).
(b) Photolysis of solution (a) with observation of development
and decay of biphenyl fluorescence at 3350 Å. (c) Photolysis
X marks the duration of pulse 15 nsec of pulse 3471 Å light of
0.1 M ϕ_2 in p-xylene. (d) Radiolysis of 0.1 m ϕ_2 in
cyclohexane λ = 3650 Å (triplet ϕ_2). (e) Radiolysis of 0.1 M
ϕ_2 in cyclohexane λ^1 = 4100 Å (anion of ϕ_2).

pulse radiolysis of this system shows that the anion ϕ_2^- is
produced rapidly and within the psec pulse, then decays
significantly over 10 nsec followed by a slower decay over 100 nsec
which matches that of the ϕ_2^T growth.

Radiolysis of alkane liquids fits the classical picture of
radiation chemistry, namely, large yields of ions are initially
produced which on recombination give rise to excited states and
other products. Much radiation chemistry of alkanes is also
interpreted in terms of free radicals. The exact connection
between the three reactive regimes of excited states, ions, and
free radicals is not always clearly established.

Radiolysis of low temperature alkane liquid also gives rise to
trapped radical ions (23). Thermal annealing of the irradiated
samples gives rise to luminescence characteristic of excited states
of the solvent and, or, solutes present (23,24). These data
conform exactly to those obtained in pulse radiolysis studies.

Radiolysis of Arene Systems

Radiolysis of arene liquids such as, benzene, toluene, p-
xylene, etc. gives rise to excited states of the liquid. Both
singlet and triplet excited states are formed, as illustrated by
radiolysis of various solutes in p-xylene (25). The symbols X and
S refer to the solvent and solute respectively, while the
superscripts S and T refer to singlet and triplet excited states
respectively. The initial radiation act gives X^S and X^T, followed
by energy transfer to the solute.

$$X^S + S \longrightarrow S^S \tag{1}$$

$$X^T + S \longrightarrow S^T \tag{2}$$

Intersystem crossing of the solute singlet occurs with quantum
yield ϕ_S, to give the solute triplet:

$$S^S \xrightarrow{\phi_S} S^T \tag{3}$$

The singlet excited state of p-xylene may also undergo intersystem
crossing to give the triplet with quantum yield ϕ_X:

$$X^S \xrightarrow{\phi_X} X^T \tag{4}$$

While the triplet state of p-xylene may be destroyed by processes
other than energy transfer to the solute

$$X^T \longrightarrow ? \tag{5}$$

The net result is that the final observed yield of the triplet
state of the solute S is

$$G(T) = G \frac{k_2[S]}{k_2[S] + k_5} + \phi_S G(x^S) \frac{k_1[S]}{k_1[S] + k_6} \quad (6)$$

where the subscript of k refers to the appropriate reaction and G is the total triplet yield of p-xylene which is the triplet produced directly $G(x^T)$ together with that produced by intersystem crossing via 4.

The yield G may be given by

$$G = G(x^T) + G(x^S)_{\phi X} \quad \frac{k_6}{k_6 + k_1(S)}$$

where k_6 is the total decomposition rate of the p-xylene via intersystem crossing, fluorescence, and radiationless decay. It can be seen from expression 6 that if ϕ_S is ~0 as for p-terphenyl then the singlet state only contributes to the triplet yield via reactions via 2 and 4, and at high solute concentrations $G(T) = G(x^T)$. If ϕ_S is unity as with benzophenone then at high concentration $G(T) = G(x^T) + G(XS)$, while intermediate values of ϕ_S give values of G_T lying between the limits.

The yields of triplet are shown in Figures 2a and b as calculated from "Equation 6", along with the experimental points.

It is concluded that $G(x^T) = 2.35$ and $G(x^S) = 2.05$ in p-xylene and $G(B^S) = 1.70$ in benzene radiolysis.

Quite similar kinetic data are obtained in two photon laser photolysis with, λ 3471 Å, of naphthalene in toluene. Fluorescence of the naphthalene excited singlet state and absorption spectra of the excited triplet state were observed, no excited species were observed in similar experiments in cyclohexane ([21]). This suggests that excitation of toluene is two photon into the S_3 state followed by rapid internal conversion to the first excited state S_1, which then leads to energy transfer to Naphthalene N.

$$S \xrightarrow{2h\nu} S_3^* \dashrightarrow S_1^* + N \dashrightarrow N_1^S + S$$

$$S_1^* \dashrightarrow S^T \quad\quad\quad\quad\quad\quad (7)$$

$$N_1^S \dashrightarrow N + \text{fluorescence} \quad\quad (8)$$

$$S^T + N \dashrightarrow N^T + S \quad\quad\quad (9)$$

All the rate parameters for the above processes are known and the yield of N^T as a function of naphthalene concentration can be calculated ([25]). The yield of naphthalene triplets, Y_T, is given by

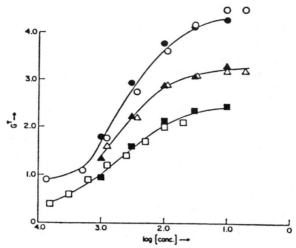

Figure 2a. Yield of solute triplets versus concentration
(M.liter) in p-xylene (radiolysis). Benzophenone (O);
1,1'binaphthyl (Δ); p-terpenyl (□). Filled points correspond
to calculated data. In these calculations, ϕ_S, is taken as
0.56, while ϕ_X is 1.0, 0.07 for benzophenone and p-terpenyl,
respectively, ϕ_X is taken as 0.5 for 1,1,'BN and k_2 is taken as
3×10^{10} M^{-1} sec^{-1} for benzophenone and p-terpenyl.
(Reproduced with permission from Ref. 25. Copyright 1973
Academic Press.)

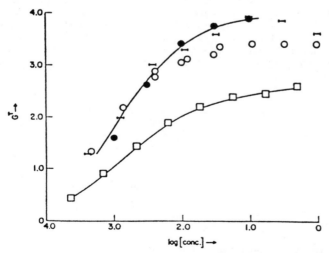

Figure 2b. Yield of solute triplets in p-xylene. Naphthalene
30°C (●); napthalene 100° C (O); pyrene (□); calculated for
naphthalene (⊶). Concentration units are M/liter. In these
calculations ϕ, is taken as 0.56, while ϕ_X is 0.8 for
naphthalene, and k_2 is taken as 3×10^{10} M^{-1} sec^{-1}.
(Reproduced with permission from Ref. 25. Copyright 1973
Academic Press.)

$$Y_T = \phi_N T_O (\frac{k_7 [N]}{k_7 [N] + \alpha}) + \phi_{S_1}^* T_O (\frac{\alpha}{\alpha + k_7 [N]}) (\frac{k_9 [N]}{k_9 [N] + \gamma})$$

where ϕ_N = 0.8 is the quantum yield for intersystem crossing for naphthalene singlets, k_7 and k are the rate constants for reactions "Equation 7 and 9", respectively, [N] is the naphthalene concentration and α and γ are the overall rate constants for decay of S_1^* and S_T^*, respectively. The quantum yield for intersystem crossing in toluene ϕ_{S_1} was taken as 0.7 (25).

T_O the yield of S_1^*, is calculated from ϕ_N and the yield Y_T at 0.1 mol dm^{-3} naphthalene, at which concentration all S_1^* reacts with N to give N_1^*. The calculated curve is shown in Figure 3. The agreement between calculation and experiment is good and supports the above mechanism. Rate data are interchangeable between two photon laser photolysis and radiolysis suggesting that only energy transfer between excited solvent and solute are important in the above studies.

Origin of Excited States in the Radiolysis of Aromatic Liquids

Scavenging Studies. The solvent excited states produced by radiolysis of aromatic liquids could be produced directly, or formed, via charge neutralisation of solvent ions. The low oscillator strengths of the first and second excited states of benzene, toluene and p-xylene preclude direct excitation into these states. However, the third excited state could be excited with a yield as high as unity.

Electron scavengers such as chlorobenzene (24), CCl_4 and $CHCl_3$ (21,26) reduce the initial yield of singlet states in the radiolysis of aromatic liquids. This suggests that at least some of the precursors arise from an initial event of ion - combination. In the case of chlorobenzene, it is also noted that this scavenger does not reduce the yield of excited states formed in the two photon laser photolysis of p-xylene, suggesting that the third excited state of p-xylene is unreactive with this scavenger. Some reduction in excited state yield by $CHCl_3$ and CH_3OH is observed in the two photon photolysis of toluene suggesting that these scavengers react with higher excited states as well as with electrons. Theory indicates that the yield of excited states I, in the presence of an electron scavenger (Q) is related to the initial yield I_O, via a relationship of the form

$$I_O/_I \quad \alpha \quad [Q]^n$$

where n is 0.5 (27) or (0.7) (28) which reflects on the kinetics of gemination in decay. Indeed, this is found for chlorobenzene scavenging of the precursors of excited states of p-xylene, (24) and in similar studies with $CHCl_3$ in toluene to be discussed shortly. However, in other systems cases, e.g. benzylchloride in siethylaniline (29) a simple Stern Volmer dependence of

$$I_O/_I \quad \alpha \quad [Q]$$

is found, for quenching of solvent emission indicating that the
excited state precursor decays exponentially, with time, unlike the
kinetics of geminate ion recombination. Typical data are given in
Figure 4, for $CHCl_3$ scavenging of toluene fluorescence in
radiolysis and in two photon laser photolysis, λ_2 = 343 nm. Stern
Volmer kinetics are obeyed in the case of two photon photolysis,
and ion-neutralisation kinetics in the radiolysis studies.
However, Stern Volmer kinetics are obeyed for radiolysis studies of
scavenging of toluene fluorescence with benzyl chloride. The
electron scavengers $CHCl_3$ and chlorobenzene reduce the fluorescence
yields of arenes in a manner that suggests:
     a) that in radiolysis the initial loss of energy leads to
ionization followed by neutralisation leading to excited states,
the scavenging of excited states varying with [Scavenger]^n; and
     b) that in the two photon laser photolysis $CHCl_3$, but not
chlorobenzene, reacts with the third excited state, the scavenging
varying as [Scavenger] n =1. Benzyl chloride always displays Stern
Volmer kinetics, even though it is an electron scavenger. These
data are difficult to fit in with the conventional picture
illustrated but other quenchers. If, unlike other quenchers
benzylchloride reacts efficiently, both with high excited states
and with electrons, then the data may be explained. It remains for
further studies to fully elucidate this point.

Pulsed Studies

Figure 5 shows the growth of fluorescence from biphenyl (ϕ_2) in
toluene irradiated by fine structure pulses of 30-ps duration. The
observed fluorescence is produced by energy transfer from the
excited solvent state T^S to biphenyl.

$$T^S + \phi_2 \xrightarrow{k_1} \phi_2^* + T$$
$$\downarrow$$
$$fluorescence$$

     The solid line is calculated utilising the response of the
electronic systems (accelator pulse length and rise time of the
detection electronics). Halflives of 26 and 34 psec were measured
for the rate of energy transfer from excited toluene to 1 and 0.5 M
biphenyl. The rate constant for energy transfer was then measured
as 5.3×10^{10} M^{-1} s^{-1} agreeing well with that measured by
photochemical techniques (30).
     The rise time of toluene fluorescence emission was within that
of the system, i.e., faster then 15 psec, which gives an upper
limit estimation for the time of ion recombination in toluene.
Figure 5 also contains similar studies in diethylaniline D.E. and,
shows that here the rise time of D.E. fluorescence is 24.4 psec and
slower than that in toluene. It was also noted that benzyl
chloride reduced the initial yield of D.E. fluorescence, the
kinetics being Stern Volmer.
     Decreasing temperature has a marked effect on energy transfer
processes in radiolysed systems, as noted earlier in solid polymer

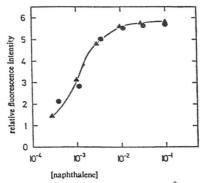

[naphthalene]

Figure 3. Yield of naphthalene triplet (N_T^*) in the laser photolysis of naphthalene in toluene; as a function of naphthalene concentration [N] in moles/L; ● . ▲ represents calculated yield as a function of naphthalene concentration, at [N] = 1.2 mol dm^{-3} with various concentrations of nitromethane. Concentrations in moles/L.

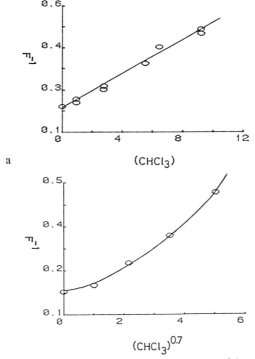

a   ($CHCl_3$)

($CHCl_3$)$^{0.7}$

Figure 4. Stern Volmer plots and ion quenching plots of scavenging precursors of fluorescence in the radiolysis and two photon laser photolysis of aromatic liquids. (a) $CHCl_3$ quenching of toluene fluorescence, F, two photon laser excitation. <u>Continued on next page</u>.

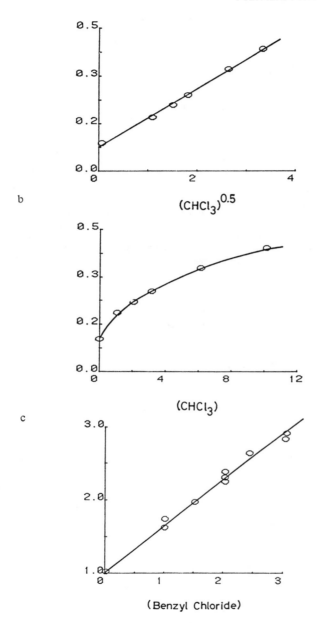

Figure 4.--<u>Continued</u>. (b) Similar to (a) but ionising
radiation used as excitation (radiolysis). (c) Benzyl
chloride quenching of toluene fluorescence (radiolysis).

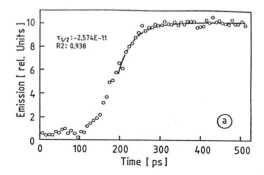

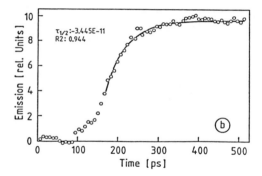

Figure 5.   Formation of fluorescence from biphenyl in toluene
at λ = 323 nm, compared to computer fit (solid line). R^2 =
regression parameter.   Concentration of biphenyl 1 M (a) and
0.5 M (b).

(31) where the yield of energy transferred to a solute in the solid polymer increased with decreasing temperature. As indicated earlier, trapped ions of solvent and solutes have been observed in the radiolysis of low temperature rigid glassy hydrocarbons. Ion recombination in these systems gives rise to excited states. At room temperature the rates of energy transfer to solutes such as biphenyl and diphenylanthracene are rapid and rate constants are in the range of 5 to 7 x 10^{10} M^{-1} s^{-1}, in agreement with values measured via low energy photochemical excitation (26). Solutes such as $CHCl_3$ and CCl_4 reduce the yield of singlet excited toluene via reaction of these solutes with an excited state precursor, presumably electrons which neutralise solvent cations producing excited states. Decreasing the temperature of the irradiated toluene systems to - 70° C leads to a marked increase in the rate of energy transfer from toluene to a solute. The nature of this effect and the effect of added e⁻ scavengers suggest that lowering the temperature leads to a change in the mechanism of energy transfer from direct energy transfer to the formation of solute ions followed by ion neutralization leading to solute excited states. The temperature effect as noted in Figures 6a and b. In Figure 6a the energy transfer between excited toluene and 9-10-diphenylanthracene DPA, can be quantitatively described by a simple bimolecular reaction with a k of 4.0 x 10^{10} M^- sec^{-1}. This is not true at lower temperatures, Figure 6b, where multiexponential behavior is required to fit the data. It can be seen that a single rate constant does not fit the experimental data. The calculated curve for k = 10 x 10^{10} M^{-1} s^{-1} is shown, which fits well at shorter times but not at longer time periods; conversely parameters that give, a good fit at longer time periods give a poor fit at shorter times. Two kinetic processes are required to fit the data, one with k_1 = 5 x 10^{10} M^{-1} s^{-1} and the other with k_2 = 20 x 10^{10} M^{-1} s^{-1}. The additive sum of the two rates fits the experimental data well.

Conclusion

The picture of radiolysis that emerges from various studies, two photon laser photolysis, low temperature radical ion trapping studies, and pulse radiolysis is:
    a)that the initial energy loss produces solvent ions.
    b)in polar liquids the reactions of these species and their radical products dominates the chemistry.
    c)in non-polar liquids excited states are formed; in alkanes a large portion of those arise from observed ion-recombination; in arenes interpretation of the excited state chemistry suggests that here also excited states are formed by ion-recombination.
    d)unusual processes exist in arene systems, i.e., precursor kinetics sometimes suggest direct reactions of higher excited states, and direct formation of excited states (as is photochemistry) has not been ruled out.

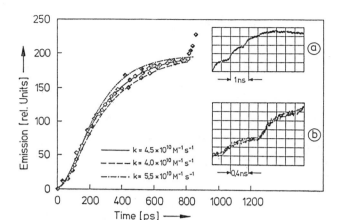

Figure 6a. Comparison of the observed growth of fluorescence at λ = 430 nm form a solution of 9 x 10^{-2} M, 9,10-diphenylanthracene (DPA) in toluene with computer simulation. The original data [insert (b) are indicated by squares. Insert (a) shows the total fluorescence, used for the determination of F_{∞}. (Reproduced with permission from Ref. 26. Copyright 1979 American Institute of Physics.)

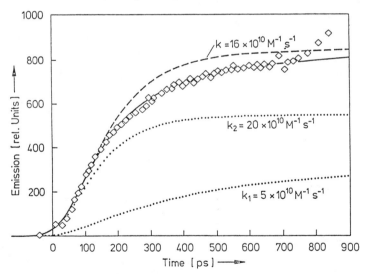

Figure 6b. Growth of fluorescence at λ = 430 nm from a solution of 5 x 10^{-2} M DPA in toluene at -50° C. The experimental data, indicated by open squares, are compared to computer fits with one rate constant or the sum of two kinetic processes with k_1 and k_2 (solid line) having the relative yields y_1 = 0.36 and y_2 = 0.64, respectively. (Reproduced with permission from Ref. 26. Copyright 1979 American Institute of Physics.

Literature Cited

1. The authors wish to thank the National Science Foundation, the Army Research Office, the Petroleum Research Foundation of the American Chemical Society, and the Hahn-Meitner Institut, Berlin for support of this work.
2. J. K. Thomas Chemistry of Excitation at Interfaces ACS monograph 181, 1984; Washington D. C.
3. R. L. Platzmann, in "Radiation Research" G. Silini, Ed., North Holland, Amsterdam, (1967).
4. E. N. Lassette, A. Skerbele, M. A. Dillon, K. J. Ross, J. Chem. Phys., _48_, 5066, (1968).
5. A. Weller & K. Zacharasse, J. Chem. Phys., _46_, 4984 (1967).
6. J. L. Magee & J. T. Huang, J. Phys. Chem., _76_, 3801 (1972).
7. T. Gangwer & K. K. Thomas, Rad. Res., _54_, 192 (1973).
8. A. Kira & J. K. Thomas, J. Phys. Chem., _60_, 2094 (1974).
9. A. Kira & J. K. Thomas, J. Phys. Chem., _60_, 766 (1974).
10. E. J. Land, J. T. Richards, and J. K. Thomas, J. Phys. Chem., _76_, 3805 (1972).
11. T. Platzner & J. K. Thomas, unpublished work.
12. G. Beck & J. K. Thomas, J. Phys. Chem., _51_, 3649 (1972).
13. O. Allen, NSRDS-NBS 57, NBS Reports (1976).
14. J. H. Baxendale & M. A. J. Rodgers, J. Phys. Chem., _72_, 3849 (1968).
15. J. T. Richards & J. K. Thomas, J. Chem. Phys., _48_, 5097 (1968).
16. G. A. Salmon, Int. J. Radiat. Phys. Chem., _13_, 1, (1976).
17. J. K. Thomas, Int. J. Radiat. Phys. Chem., _8_, 1, (1976).
18. R. Bensasson, & E. J. Land Trans. Farad. Soc., _67_, 1904, (1971).
19. J. K. Thomas, K. Johnson, K. Klippert, & R. Sowers, J. Chem. Phys., _48_, 1608 (1968).
20. G. Beck & J. K. Thomas, J. Phys. Chem., _76_, 3856 (1972).
21. G. Beck, J. T. Richards, & J. K. Thomas, Chem. Phys. Letts., _40_, 300 (1976).
22. G. Beck, & J. K. Thomas Trans. Farad. Soc I, _72_, 2610 (1976).
23. I. Platzner & J. K. Thomas, Int. J. Rad. Chem. & Phys. _7_, 573 (1975).
24. T. E. Gangwer & J. K. Thomas, Int. J. Rad. Chem. & Phys., _7_, 305 (1975).
25. T. E. Gangwer & J. K. Thomas, Rad. Res., _5_, 192 (1973).
26. G. Beck, A. Ding, & J. K. Thomas, J. Chem. Phys., _71_, 2611, (1979).
27. S. Rzad, P. Snfelta, J. Waman, R. H. Schuler, J. Chem. Phys., _52_, 3971, (1970).
28. H. T. Choi, J. A. Haglund, & S. Lipsky, J. Phys. Chem. _87_, 1583 (1983). Ibid, _88_, 5863, (1984), and _88_, 4247 (1984).
29. G. Beck & J. K. Thomas, J. Phys. Chem., _89_, 4062 (1985).
30. J. B. Birks, P. 613 "Photophysics of Aromatic Molecules" Wiley, Linden, 1970,
31. S. Siegel & T. Stewart, J. Chem. Phys., _55_, 1775 (1971).
32. G. Beck & J. K. Thomas, Rad. Res., _1_, 17 (1980).

RECEIVED June 17, 1987

Chapter 3

Historical Outline of Radiation Effects in Polymers

Adolphe Chapiro

Centre National de la Recherche Scientifique, 94320 Thiais, France

This historical development of the radiation technology of polymers is reviewed in this outline. The important applications of this technology are divided into two classes - large scale processes such as cross-linking of rubbers and plastics and specialized sophisticated processes such as microlithography. The initial fundamental studies that led to these applications are outlined and the slow process of commercialization is emphasized in this review.

The industrial use of radiation in the polymer field represents a well-established technology with multi-million dollars sales. However, if one looks back into the development of the basic techniques which underly today's commercial uses, one is surprised to find that the significant pioneering work, which was carried out ca. 40 years ago, was only accepted very slowly and reluctantly by industrial management in the 1950's and that the fast growth of this important technology took place only in recent years. There are many reasons for the slow start of radiation applications in industry, one of the most obvious being the link, established by public rumor as well as by some managing circles, between radiation and nuclear weapons. The lack of trained personnel, aware of the advantages of this new technology, also caused this delay, and it is significant that companies such as RAYCHEM or CRYOVAC, which enjoyed very early the expertise in the field, were very successful in their business right from the beginning.

The industrial applications in the radiation chemistry of polymers fall into several categories:

1. - Large scale operations :

   — Radiation cross-linking of rubbers and plastics;

   — Radiation sterilization of plastic medical supplies;

   — Radiation curing of monomer-polymer formulations.
The last topic is more recent but is rapidly increasing in importance.

2. - Sophisticated products :

   — specific polymers tailored for bio-medical applications;

   — polymers for electric and electronic applications, e.g., microlithography.

0097–6156/87/0346–0031$06.00/0
© 1987 American Chemical Society

This second group of materials involves small or medium-size production of polymers with very high added values.

In order to analyze the origins of this industry we shall examine in succession the elementary processes on which the technology is based.

Radiation Induced Polymerization

The first convincing observation that small molecules are condensed to solids under the influence of radiation is probably that of W. D. Coolidge who in 1925 subjected various organic substances to "cathode-rays", i.e., fast electrons, and noticed that solid deposits formed on the glass walls of his apparatus (1). In the same year, W. Mund and his co-workers made similar observations when bombarding gaseous hydrocarbons with α-particles (2). Other pioneers working in this field include S. C. Lind (3), J. C. McLennan (4) and C. S. Schoepfle (5).

In the 1930's, vinyl monomers were found to undergo chain polymerization when subjected to gamma-rays (6) or to neutrons generated by a cyclotron (7). However, the more extensive work was carried out in the late 1940's when it became clear that powerful radiation sources would become practical tools of interest to industry.

Systematic investigation of radiation-induced polymerization was conducted in Leeds by F. S. Dainton and his group (8) and in Paris by M. Magat and co-workers (9). The results of these studies convincingly demonstrated that radiation could initiate polymerization at any desired (low) temperature and led to a fundamental conclusion, viz., radiation-induced polymerizations occurred by conventional free radical mechanisms. This concept was extended to other radiation-chemical processes, and in the 1950's most radiation chemists used free radical theories for interpreting their experimental data.

However, in the early 1950's, V. L. Talroze et al. (10) and D. P. Stevenson and D. O. Schissler (11) showed that ionic species were responsible for a large number of ion-molecule reactions occurring in a mass-spectrometer in which the pressure was allowed to rise, thereby favoring bimolecular encounters. This unveiled an entirely new field of chemistry with "strange-looking" reactions such as:

$$CH_4^+ + CH_4 \rightarrow CH_5^+ + CH^{3\cdot} \qquad \text{(proton transfer)}$$

or:

$$CH_3^+ + CH_4 \rightarrow CH_2\,H_7^+ \qquad \text{(ionic condensation)}$$

and many others. Reactions like these occur with a very high cross-section in gases at low pressures.

Soon afterwards, in 1957, isobutene was found to undergo an ionic chain polymerization when irradiated in the liquid state (12,13), demonstrating that radiation could also initiate ionic reactions in condensed systems in which charge separation was much smaller than in a gas. These findings resulted in a dramatic change in the fundamental approach to radiation chemistry. Some of the reactions occurring in polymers were also reinterpreted since ionic processes had to be taken into account (a discussion relevant to these problems is presented in ref. 14).

Numerous monomers were irradiated, and it was shown that either free radical or ionic polymerizations would take place depending on experimental parameters. Several monomers were found to undergo a chain polymerization when irradiated in the crystalline state. This unexpected behavior led to a revision of basic concepts regarding reactivities in solids. Some interesting oriented polymer structures were formed in certain cases (see ref. 15).

The yields of radiation-induced polymerizations can be very high. No additives are required, which makes it possible to synthesize very pure polymers. The initiation step is temperature independent giving rise to an easily controlled process at any desired temperature. These features account for the commercial interest in radiation polymerization. The very high speeds attainable within the layers of monomers subjected to powerful electron beams explain the wide use of this technique in radiation curing of adhesives, inks and coatings. The corresponding formulations are "solvent-free" and involve pre-polymers and monomers as reactive diluents.

Radiolysis of Polymers

Early work in this field was conducted prior to the availability of powerful radiation sources. In 1929, E. B. Newton "vulcanized" rubber sheets with cathode-rays (16). Several studies were carried out during and immediately after world war II in order to determine the damage caused by radiation to insulators and other plastic materials intended for use in radiation fields (17, 18, 19). M. Dole reported research carried out by Rose on the effect of reactor radiation on thin films of polyethylene irradiated either in air or under vacuum (20). However, world-wide interest in the radiation chemistry of polymers arose after Arthur Charlesby showed in 1952 that polyethylene was converted by irradiation into a non-soluble and non-melting cross-linked material (21). It should be emphasized, that in 1952, the only cross-linking process practiced in industry was the "vulcanization" of rubber. The fact that polyethylene, a paraffinic (and therefore by definition a chemically "inert") polymer could react under simple irradiation and become converted into a new material with improved properties looked like a "miracle" to many outsiders and even to experts in the art. More miracles were therefore expected from radiation sources which were hastily acquired by industry in the 1950's.

The evidence that radiation would not produce miracles but was just another source of energy, available for activating chemical reactions, produced disappointments in industrial circles which had a long-lasting influence on subsequent developments.

Simultaneously with Charlesby's findings, work along similar lines was carried out in G. E.'s Research laboratories in Schenectady (22) and also in Research Institutes in the Soviet Union, although the latter only became known several years later (23). The results of this research demonstrated that in addition to polyethylene, many other polymers could be cross-linked by radiation. These include silicones, rubber, poly(vinyl chloride), polyacrylates and, to a lesser extent, polystyrene. In contrast, polymers such as polymethacrylates, polyisobutylene, polytetrafluoroethylene and cellulose underwent "degradation" by main-chain scission. These early findings were confirmed and extended to other compounds by numerous studies.

Large-scale production followed the early discovery of the improved thermal properties of cross-linked polyethylene in wire and cable insulators and various other electric and electronic applications. The "memory-effect", first described by Charlesby (24) led to the production of heat-shrinkable tubings. Today radiation cross-linking is a large-scale technology used for the production of improved wires and cables hot-water pipes, heat-shrinkable packaging films, automobile tire components and is also applied to the production of more sophisticated devices such as "solder sleeves", self-regulating heating cables, self-recovering switches and the like. Radiation degradation is also used commercially for reclaiming fluoropolymer wastes.

A good understanding of the mechanisms of radiation effects in polymers is required for operating another large scale radiation technology : the sterilization of plastic medical supplies which cannot be sterilized by heat (see ref. 25).

Electron microlithography is also based on radiation-induced cross-linking or degradation of specially designed polymers. In this technology, a very narrow beam of electrons is used to impinge a geometric pattern on the resist material. As a consequence of the resulting chemical transformation, the irradiated zones become easier or more difficult to dissolve. A remarkable precision of the pattern is obtained after selective dissolution in such "negative" or "positive" resists, giving well-defined line-widths in the sub-micron range.

Radiation-Induced Graft/Copolymerization

Two general methods of radiation grafting were developed in Paris in 1955 by M. Magat and co-workers (26,27). The first, referred to as "direct" or "simultaneous" grafting, derives from the study of radiation-induced polymerization. In the process, a polymer (A_n) is irradiated while in contact with a monomer (B). The radiolysis of A_n generates polymeric free radicals which initiate the polymerization of B. Since the initiating radicals become chemically attached to the ends of the growing chains, the reaction gives rise to the graft copolymer $A_n B_m$. Here radiation is merely used as a convenient tool for producing polymeric radicals. This method is easier to practice and often more economical than conventional chemical techniques.

The second method called "peroxidation" or "preirradiation" grafting is a two-step process. Polymer A_n is at first irradiated by itself in air. The resulting polymeric radicals react with oxygen to give peroxidic radicals which further react to form polymeric peroxides via a short chain reaction, analogous to conventional "auto-oxidation".

The peroxidized polymer is stable at room temperature, like an ordinary peroxide, and can be stored for considerable length of time (weeks, months...). In a second step the "activated" polymer is contacted with monomer B and the peroxides are broken down by heat or by a redox system, thereby generating the graft copolymer $A_n B_m$.

Both methods are very general. They apply to any polymer which undergoes radiolysis (i.e., "any" polymer) and the only limitation with respect to the monomer is that it polymerize by a free radical mechanism. Ionic grafting was also initiated by radiation (28,29) but the yields of this process are generally quite low.

The interest in radiation grafting lies in the fact that it is on route to the production of an almost unlimited range of new materials by an easy and clean method. If can be carried out in gels, in semi-solid substances and even in the bulk or on the surface of shaped products. At the time of these early findings, chemical grafting methods were scarce and were conducted in dilute solutions with very low yields of graft copolymers. This explains why these studies produced such a strong impact on radiation research.

Graft copolymers combine the properties of their polymeric constituents and as such are polymer alloys, which open a vast field of new polymeric species. This is why active research along these lines is performed in many academic and industrial research laboratories all over the world. However, only few applications have reached a commercial level today. They involve the production of specific polymeric adhesives, perm-selective membranes, bio-medical devices and the surface modification of certain products.

Radiation Curing of Coatings

This technology combines the basic aspects of radiation polymerization and radiation grafting. Formulations containing unsaturated pre-polymers and monomers are "cured", i.e., hardened, after exposure to fairly low doses of radiation : 10 to 50 Kilograys (1 to 5 megarads). Electron beams with the lowest available energies are used, selected for having the right penetrating power to cure coatings 10 to 50 microns thick. With the accelerators presently available in this range of energies (300 to 500 KeV), the curing dose can be delivered in a fraction of second, allowing considerable curing speeds. If the coating is applied to the surface of an organic substrate such as wood, paper or plastic, the monomer reacts with the polymeric radicals produced on the surface and becomes grafted to the substrate. The corresponding coatings are linked to the surface by covalent bonds and exhibit outstanding adhesion.

The pioneering work in this field was carried out in the late 1950's by a research team from the Ford Motor Co. and the resulting technology was practiced on a large scale by this Company for finishing the surface of plastic car components. Today, radiation curing is a rapidly expanding field and is used commercially for coating on wood, paper, fabrics, plastic films and metals, as well as for "drying" inks, adhesives and various other formulations, including magnetic pastes for tapes, discs and the like (see ref. 30 for a review of this field).

Conclusion

The history of the development of radiation uses in the polymer field is an interesting example of how a new technology is transferred from the laboratory to industry. It is concerned with classical techniques in polymer chemistry, suddenly confronted with a new tool, radiation, which at first is only available on a small scale, suitable for fundamental research but which, under the pressure of new and valuable results, is finally accepted by industry in spite of its frightening environment. Today we witness the development of this technology in many different branches of industry. It is, however, still in its infancy and further growth is expected to follow from presently known facets as well as from the discovery of new applications.

Literature Cited

[1]   Coolidge, W. D. *Science* 1925, *62*, 441.

[2]   Mund, W.; Koch, W. *Bull. Soc. Chim. Belge* 1925, *34*, 119.

[3]   Lind, S. C.; Bardwell, D. C. *J. Am. Chem. Soc.* 1926, *48*, 2335.

[4]   McLennan, J. C.; Glass, J. V. S. *Can. J. Research* 1930, *3*, 241.

[5]   Schoepfle, C. S.; Fellows, C. H. *Ind. Eng. Chem.* 1931, *23*, 1396.

[6]   Hopwood, F. L.; Phillips, J. T. *Proc. Phys. Soc. (London)* 1938, *50*, 438, *Nature* 1939, *143*, 640.

[7]   Joliot, F. *Fr. Pat.* 1940, *996*, 760.

[8]   Dainton, F. S. *Nature* 1947, *160*, 268; *J. Phys. Colloid, Chem.* 1948, *52*, 490.

[9]   Chapiro, A.; Cousin, C.; Landler, Y.; Magat, M. *Rec. Trav. Chim.* 1949, *68*, 1037.

[10] Talroze, V. L.; Lyubimova, A. K. *Doklady Akad. Nauk SSSR* 1952, *86*, 909.

[11] Stevenson, D. P.; Schissler, D. O. *J. Chem. Phys.* 1955, *23*, 1353; 1956, *24*, 926.

[12] Davidson, W. H. T.; Pinner, S. H.; Worral, R. *Chem. Ind. (London)* 1957, 1274.

[13] Grosmangin, J. Unpublished results quoted by M. Magat, *Collection Czechoslov. Chem. Commun.* 1957, *22*, 141.

[14] Chapiro A. "Radiation Chemistry of Polymeric Systems", *Interscience publishers, a division of John Wiley & Sons* 1962, New York.

[15] Chapiro A. "Radiation-Induced Reactions" in 'Encyclopedia of Polymer Science and Technology', *John Wiley & Sons* 1969, New York, Vol. *11*, 702-760.

[16] Newton, E. G. *U.S.P. 1*, 1929, 402 to the B.F. Goodrich Co.

[17] Davidson, W. L.; Geib, I. G. *J. Appl. Phys.,* 1948, *19*, 427.

[18] Burr, J. G.; Garrison, W. M. *A.E.C.D.* 1948, 2078.

[19] Sisman, O.; Bopp, C. D. *ORNL* 1951, 928.

[20] Dole, M. Report of Symposium IX, Chemistry and Physics of Radiation Dosimetry, *Army Chemical Center* 1950, 120, Maryland.

[21] Charlesby, A. *Proc. Roy. Soc.* 1952, *A215*, 187.

[22] Lawton, E. J.; Bueche, A. M.; Balwit, J. S. *Nature* 1953, *172*, 76.

[23] Karpov, V. L. "Sessiya Akad. Nauk SSSR po Mirnomu Ispolzovaniyu Atomnoi Energii" *Academy of Sciences of the USSR* 1955, 1 Moscow.

[24] Charlesby, A. *Nucleonics* 1954, *12* No. 5, 18.

[25] Gaughan E. R. L.; Goudie, A. J. *"Sterilization by Ionizing Radiation"* *International Conferences* 1974, *Vol. 2*, 1978, *Vol. 2*, Vienna, Austria, sponsored by *Johnson & Johnson*, Multiscience Publicatio Ltd. Montreal, Quebec, Canada.

[26] Chapiro, A.; Magat, M.; Sebban, J. *French Pat.* 1956, 1,130,099.

[27] Chapiro, A.; Magat, M.; Sebban, J. *French Pat.* 1956, 1,125,537, 1,130,100.

[28] Chapiro, A.; Jendrychowska-Bonamour, A. M. *Compt. Rend. Acad. Sci.* 1967, *265*, 484 (Paris) A.M.

[29] Kabanov V. Ya.; Sidorova, L. P.; Spitsyn, V. I. *Eur. Polym. J.* 1974, *10*, 1153.

[30] Nablo, S. "Radiation Safety Considerations in the Use of Self-Shielded Electron Processors" *in Rad. Phys. Chem.* 1983, *Vol. 22*, No 3-5, 369-377.

RECEIVED May 5, 1987

Chapter 4

Main Reactions of Chlorine- and Silicon-Containing Electron and Deep-UV (Excimer Laser) Negative Resists

Seiichi Tagawa

Research Center for Nuclear Science and Technology, University of Tokyo, Tokai-mura, Ibaraki 319-11, Japan

The main reactions taking place when chloromethylated polystyrene (CMS) and chloromethylated poly(diphenylsiloxane) (SNR) are irradiated with high energy electrons or deep UV (KrF excimer laser, 248 nm) radiation have been studied. The results are discussed in terms of short-lived reactive species generated using pulse radiolysis and laser (248 nm) photolysis techniques.

The increasing density of VLSI circuits has spurred development of sub-micron exposure techniques such as deep UV (excimer laser), X-ray, electron-beam, and ion-beam lithography. These techniques require resist materials specifically designed for the type of exposure, and the reader is referred to several excellent reviews of the various resist materials and processes reported over the past few years (1,2,3). Electron and deep UV (excimer laser) resists mainly utilize crosslinking reactions to achieve negative tone although other mechanisms have recently been developed (3). Much attention has been devoted to negative electron (4-11) and deep UV (7,12,13) resists containing chlorinated or chloromethylated phenyl side chains. Chloromethylation (4,5,6) and chlorination (7) of polystyrene greatly enhance the radiation sensitivity in a manner similar to iodination of polystyrene (8). This type of enhancement is also observed in other resists with chlorinated or chloromethylated phenyl side-chains such as chloromethylated poly-α-methylstyrene (9) and chlorinated polyvinyltoluene (10). The addition of aromatic groups also improves the dry-etching resistance of these resists.

Although a number of papers have been published on the lithographic characteristics of chlorine- and silicon-containing resists, details of the crosslinking reaction mechanism have been lacking. This is partly due to lack of experimental data on transient species produced during deep UV and electron beam irradiation of the resist at room temperature. Recently, pulse radiolysis (13,14) and laser photolysis (14,15) techniques have been applied to chlorine- and silicon-containing resists such as CMS (chloromethyl polystyrene), αMCMS (chloromethyl poly-α-methylstyrene), CPMS (chlorinated polymethylstyrene), SNR (chloromethylated poly(diphenylsiloxane)), polysilane, and poly(trimethylsilylstyrene)-co-CMS as well as laser photolysis and pulse radiolysis studies on polymers in chlorine-containing solvents (16). Experimental data on the time-resolved behavior of the many reactive intermediates produced by radiolysis or photolysis have begun to elucidate the reaction mechanisms of chlorine and silicon containing resists but some problems still remain.

EXPERIMENTAL

Samples were irradiated by a 10 ps single or 2 ns electron pulse from a 35 MeV linear accelerator for pulse radiolysis studies (17). The fast response optical detection systems of the pulse radiolysis system for absorption spectroscopy (18) is composed of a very fast response photodiode (R1328U, HTV.), a transient digitizer (R7912, Tektronix), a computer (PDP-11/34) and a display unit. The time resolution is about 70 ps which is determined by the rise time of the transient digitizer.

Details of the laser photolysis system have been reported elsewhere (19). Polystyrene or CMS in cyclohexane solution in a 1x1x2 cm quartz cell was excited by a 248-nm pulse from a KrF excimer laser (lambda Physik EGM 500). The width of the pulse was 15 ns (full width at half-maximum (FWHM)). Transient spectra were monitored point by point by using a conventional nanosecond laser photolysis method.

RESULTS AND DISCUSSION

Deep UV Resist Reactions of CMS. The laser photolysis studies on CMS and polystyrene solutions in cyclohexane were carried out at 248 nm using a KrF excimer laser. The intensity of monomer and excimer fluorescence of CMS becomes weaker with increasing chloromethylation ratio, but the lifetime of the excimer is essentially independent of the chloromethylation ratio being almost the same as the excimer lifetime of polystyrene (20 ns). The lifetime of the monomer fluorescence has not been investigated by nanosecond laser photolysis because of the short lifetime (about 1 ns) (20,21) of the singlet.

Fig. 1 shows the transient absorption spectra observed during photolysis of cyclohexane solutions of CMS. The absorption spectrum with the peak at 520 nm (spectrum a) is identical to that of the excimer of polystyrene (spectrum C) (22) and has a similar lifetime of 20 ns. The quenching rate of the absorption at 520 nm by O_2 is comparable to what one would expect for the CMS excimer.

The two sharp absorption peaks ((Fig. 1), spectrum b) are characteristic of a substituted benzyl type polymer radical ($\cdot P_1$) (14). The formation rate of these peaks is very rapid (within the limitation of the detection system) and is faster than the decay rates of the excimer fluorescence and the broad absorption between 300 and 400 nm ($T_n \leftarrow T_1$ absorption) described below. There was no clear correlation between the increase in the absorption peak attributed to $\cdot P_1$ and the decay of the excimer and triplet species of CMS suggesting that some other species is the main precursor of $\cdot P_1$.

The very broad absorption between 300 and 400 nm is comparable to the absorption spectrum of the triplet state of polystyrene (22). The lifetime of this intermediate is essentially independent of the chloromethylation ratio and is comparable to the lifetime of the triplet state of polystyrene (110 ns).

These results are consistent with the following mechanism: The 248 nm laser pulse first excites the phenyl rings of CMS (partially chloromethylated polystyrene) forming the lowest excited singlet state (which is the precursor of the benzyl type polymer radical ($\cdot P_1$)), together with excimer and excited state triplet species. The singlet undergoes dissociation fo the C-Cl bond forming $\cdot P_1$ and liberating a chlorine radical which reacts with the phenyl ring of CMS forming a complex whose absorption spectrum has been identified by studies of both laser photolysis and pulse radiolysis of polystyrene and poly-α-methylstyrene in chlorine containing solvents such as CCl_4, $CHCl_3$, and CH_2Cl_2 (16).

The reactions leading to crosslinking of CMS under deep UV irradiation are as follows:

$$CMS \xrightarrow{\ h\nu\ } CMS^* \tag{1}$$

$$CMS^* \longrightarrow -CH-CH_2 - (\cdot P_1) + Cl \tag{2}$$

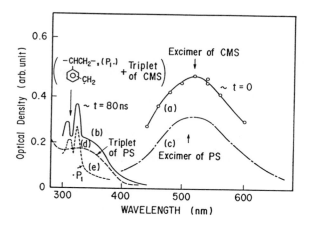

Figure 1. Transient absorption spectra observed for cyclohexane solutions of CMS (chloromethylated polystyrene) (a) immediately and (b) at 80 ns after a 248 nm laser pulse from KrF excimer laser. Absorption spectra of (c) excimer and (d) triplet state of polystyrene and (e) polymer radical, P_1, are also shown for reference. The excited singlet state of CMS is converted to $\cdot P_1$ and the excimer and the triplet states of CMS as follows.

$$CMS \xrightarrow{\text{248 nm}} {}^1CMS^*$$

$$^1CMS^* \longrightarrow \cdot P_1 + \cdot Cl$$

$$^1CMS^* \longrightarrow {}^3CMS^*$$

$$^1CMS^* \longrightarrow \text{excimer of CMS}$$

$$CMS + Cl \longrightarrow complex \tag{3}$$

$$complex \longrightarrow \cdot P_2 + HCl \tag{4}$$

$$\cdot P_1 + \cdot P_2 \longrightarrow P_1 - P_2 \tag{5}$$

$$\cdot P_1 + \cdot P_1 \longrightarrow P_1 - P_1 \tag{6}$$

$$\cdot P_2 + \cdot P_2 \longrightarrow P_2 - P_2 \tag{7}$$

The high sensitivity of CMS is due both to the high efficiency of production of the two polymer radicals ($\cdot P_1$ and $\cdot P_2$) at sites very close to each other and high efficiency of the crosslinking reaction (5).

Although the main reaction mechanisms of CMS can be inferred from time-resolved absorption and emission spectral data, some problems still remain.

(1)   In reaction (2), the C-Cl bond is broken directly but is is not clear whether it results from the excited singlet state, the excited triplet state or from vibrationally excited states (hot process);

(2)   The structure of the complex is not clearly determined yet. A possible structure is as follows

CT-complex radical   ion pair of cation radical and anion

(3)   The structure of $\cdot P_2$ has not been determined unambiguously. Possible structures include

mainly   partly

Electron Beam Resist Reactions of CMS. The lifetime of the excimer fluorescence of CMS observed in pulse radiolysis of CMS solutions in cyclohexane and tetrahydrofuran (THF) is almost independent of chloromethylation ratio from 0% to 24%. The intensity of the excimer fluorescence decreases with increasing degree chloromethylation indicating that the precursor of the excimer is scavenged by the chloromethylated part of CMS. In this case, an electron (quasi-free electron in cyclohexane and solvated electron in tetrahydrofuran, which are the precursors of the excimer), is scavenged by the chloromethyl group. The excited singlet state of cyclohexane and the monomer excited singlet state, which are also precursors of the excimer of CMS, are also quenched by the chloromethylated part of CMS.

The absorption due to the substituted benzyl type polymer radical, $\cdot P_1$ is observed in pulse radiolysis of CMS solutions in cyclohexane and THF. An electron in cyclohexane or THF reacts with CMS resulting in formation of $\cdot P_1$ and Cl^- by dissociative electron capture (reaction (10)).

Fig. 2 (a) shows the absorption spectra observed in CMS films following pulse radiolysis. There is no evidence of formation of excimer, triplet, or phenyl anion as observed in pulse radiolysis of polystyrene film (14) suggesting that these transient species or their precursors are scavenged by the chloromethyl group. Instead, absorption spectra with double peaks around 320 nm due to a substituted benzyl type polymer radical and the complex of phenyl rings and chlorine atoms are observed.

The transient absorption spectrum obtained from the pulse radiolysis of CMS solutions in benzene as shown in Fig. 2 (b) is very similar to the absorption spectrum obtained in a solid resist film as shown in Fig. 2 (a). However, the absorption in solution around 500 nm is mainly due to the complex of benzene with chlorine atoms whereas in the solid film, it is due to the complex of the phenyl ring of CMS with Cl. The absorption due to $\cdot P_1$ is observed in both solid CMS film and CMS solutions in benzene. The absorption around 320 nm may be due to the biradical of benzene. The very short lived species with absorption around 500 nm is mainly due to the benzene excimer with a small contribution from the CMS excimer. Benzene solutions of CMS have proven to be a very good model system for reactions occurring in solid films of CMS.

The main reactions leading to crosslinking of CMS under electron beam irradiation are as follows.

$$CMS \xrightarrow{\quad\quad} CMS^+ + e^- \tag{8}$$

$$CMS \xrightarrow{\quad\quad} CMS^* \tag{9}$$

$$CMS + e^- \xrightarrow{\quad\quad} -CH-CH_2-(\cdot P_1) + Cl^- \tag{10}$$

$$CMS^+ + Cl^- \xrightarrow{\quad\quad} complex \tag{11}$$

$$CMS^* \xrightarrow{\quad\quad} \cdot P_1 + \cdot Cl \tag{12}$$

$$CMS + \cdot Cl \xrightarrow{\quad\quad} complex \tag{13}$$

$$complex \xrightarrow{\quad\quad} \cdot P_2 + HCl \tag{14}$$

$$\cdot P_1 + \cdot P_2 \xrightarrow{\quad\quad} P_1 - P_2 \tag{15}$$

$$\cdot P_1 + \cdot P_1 \xrightarrow{\quad\quad} P_1 - P_1 \tag{16}$$

$$\cdot P_2 + \cdot P_2 \xrightarrow{\quad\quad} P_2 - P_2 \tag{17}$$

The high sensitivity of CMS as an electron beam resist is again due to the high efficiency for pair production of the two polymer radicals ($\cdot P_1$ and $\cdot P_2$) at sites close to each other.

Although the main reactions occurring during e-beam induced crosslinking of CMS are clarified by the time resolved data, several problems still remain.

(1) The problems in section III (pertinent to deep UV) also exist in the case of electron beam irradiation;

(2) Although both excited and ionic species play important roles in electron beam resist reactions, the ratio of the contribution of excited and ionic species, which depends on chloromethylation ratio, is not known.

Electron Beam Resist Reactions of SNR. The transient absorption spectrum observed in pulse radiolysis of SNR (partly chloromethylated diphenyl siloxane) solutions in benzene as shown in Fig. 3 is very similar to that of CMS (of Fig. 2).

The absorption around 320 nm due to the substituted benzyl type polymer radical and the absorption around 500 nm due to the complex of benzene and chlorine atom are formed within a 2 ns time scale of the electron pulse. The benzene biradical and the complex of SNR and chlorine atom may also be produced.

The scheme for the main electron beam resist reactions SNR is as follows:

$$SNR \xrightarrow{\hspace{1cm}\sim\hspace{1cm}} SNR^+ + e^- \tag{18}$$

$$SNR \xrightarrow{\hspace{1cm}\sim\hspace{1cm}} SNR^* \tag{19}$$

$$SNR + e^- \longrightarrow \quad -Si{-}O{-}\ (\cdot P_1^{'}) + Cl^{'} \tag{20}$$

$$SNR^+ + Cl^- \longrightarrow complex \tag{21}$$

$$SNR^* \longrightarrow \cdot P_1^{'} + \cdot Cl \tag{22}$$

$$SNR + Cl \longrightarrow complex \tag{23}$$

$$complex \longrightarrow \cdot P_2^{'} + \cdot Cl \tag{24}$$

$$\cdot P_1^{'} + \cdot P_2^{'} \longrightarrow P_1^{'} - P_2^{'} \tag{25}$$

$$\cdot P_1^{'} + \cdot P_1^{'} \longrightarrow P_1^{'} - P_1^{'} \tag{26}$$

$$\cdot P_2^{'} + \cdot P_2^{'} \longrightarrow P_2^{'} - P_2^{'} \tag{27}$$

The high sensitivity of SNR resist to electron beam and deep UV radiation (23,24) is again due to the high efficiency of the pair production of two polymer radicals ($\cdot P_1^{'}$ and $\cdot P_2^{'}$) at sites close to each other through reactions involving both excited and ionic species. The benzyl type polymer radical ($\cdot P_1^{'}$) and complexes are clearly identifiable from the time-resolved absorption spectrum data (5 b) shown in Fig. 3.

The major reaction pathways for SNR operating as a negative electron beam resist are general known, but several problems still remain.
(1) Details of reaction (22) are still obscure;
(2) The structure of the complex is not determined yet;
(3) The structure of P_2 has not been determined; possible structures include

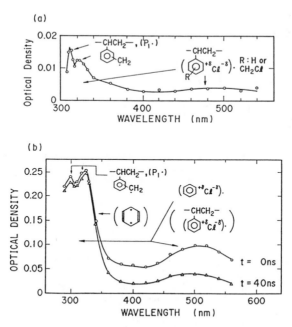

Figure 2. Transient absorption spectra observed in the pulse radiolysis of (a) CMS (chloromethylated polystyrene) solid film and (b) 200 mM CMS solutions in benzene immediately (o) and at 40 ns (Δ) after 2 ns pulses.

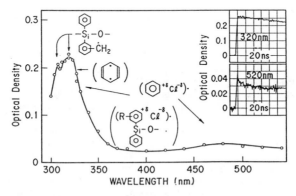

Figure 3. Transient absorption spectra observed in pulse radiolysis of SNR (partly chloromethylated poly(diphenylsiloxane) solutions in benzene.

mainly   partly

CONCLUSION

Time resolved absorption and emission spectral techniques can be used to probe the reactions occuring during radiolysis or photolysis of CMS and SNR. By identifying the intermediate species we have been able to develop some understanding of the crosslinking mechanism in these resists.

LITERATURE CITED

[1]   Willson, C. G. *"Introduction to Microlithography"*, Thompson, L. F.;Willson, C. G.; Bowden, M. J., Eds.; ADVANCES IN CHEMISTRY SERIES No. 219, *American Chemical Society:* 1983, Washington, D. C., p. 87.

[2]   Bowden, M. J. *"Materials for Microlithography"*,, Thompson, L. F.; Willson, C. G.; Frechet, Eds. *ADVANCES IN CHEMISTRY SERIES No. 266, American Chemical Society* 1984, Washington, D.C., p. 39.

[3]   Willson, C. G. paper in this ADVANCES IN CHEMISTRY SERIES.

[4]   a) Imamura, S. *J. Electrochem. Soc.* 1979, 126, 1628.
       b) Sugawara, S.; Kogure, K.; Harada, K.; Kakuchi, M.; Sukegawa, S.; Imamura, S.; Miyoshi *157th Meeting of the Electrochemical Society* 1980, Abstract No. 267, St. Louis, MO.
       c) Imamura, S.; Tamamura, T.; Harada, K.; Sugawara, S. *J. Appl. Polym. Sci.* 1982, 27, 937.

[5]   Imamura, S.; Tamamura, T.; Harada, K.; Sugawara, S. *J. Appl. Polym. Sci.* 1982, 27, 937.

[6]   Choong, H. S.; Kahn, F. J. *J. Vac. Sci. Technol.* 1981, 19, 1121.

[7]   a) Liutkus, J.; Hatzakis, M.; Shaw, J.; Paraszczak, J. *Proc. Reg. Conf. on "Photopolymers: Principles, Processes and Materials"* Nov. 8-10, 1982, Mid-Hudson Section, SPIE: Ellenville, New York, p. 223.
       b) Feit, E. D.; Stillwagon, L. E. *Polym. ENg. Sci.* 1980, 17, 1058.

[8]   Shiraishi, H.; Taniguchi, Y.; Horigome, S.; Nonogaki, S. *Polym. Eng. Sci.* 1980, 20, 1054.

[9]   Sukegawa, K.; Sugawara, S. *Japan J. Appl. Phys.* 1981, 201, 1583.

[10] Kamoshida, Y.; Koshiba, M.; Yoshimoto, H.; Harita, Y.; Harada, K. *J. Vac. Sci. Technol.* 1983, B1, 1156.

[11] Imamura, S.; Sugawara, S. *Japan J. Appl. Phys.* 1982, 21, 776.

[12] Harita, Y.; Kamoshida, Y.; Tsutsumi, K.; Koshiba, M.; Yoshimoto, H.; Harada, K.; *"Preprints of Paper Presented at 22nd Symposium of Unconventional Imaging Science and Technology SPSE"* Nov. 15-18, 1982, Arlington, VA, p. 34.

[13] a) Tagawa, S.; Hayashi, N.; Washio, M.; Tabata, Y.; Imamura, S. preparation for publication.
       b) Tagawa, S.; Shukujima, S.; Washio, M.; Tabata, Y.; Imamura, S. preparation for publication.

c) Tagawa, S. *Proc. Int'l Ion Engineering Congress-ISIAT83 & IPAT83* 1983, Kyoto, Japan.

d) Tabata, Y.; Tagawa, S.; Washio, M. *Materials for Microlithography Ed.* by Thompson, L. F.; Willson, C. G.; Frecht, J. M. *ACS Symp. Series* 1984, 255, 151.

[14]   Tagawa, *Radiat. Phys. Chem.* 1986, 27, 455.

[15]   Tagawa, S.; Nakashima, N.; Yoshihara, K. preparation for publication.

[16]   a) Tagawa, S.; Schnabel, W. *Makromol. Chem. Rapid Commun.* 1980, 1, 345

b) Tagawa, S.; Schnabel, W. *Polymer Photochemistry* 1983, 3, 203.

c) Tagawa, S.; Schnabel, W.; Washio, M.; Tabata, Y. *Radiat. Phys. Chem.* 1981, 18, 1087.

d) Washio, M.; Tagawa, S.; Tabata, Y. *Radiat. Phys. Chem.* 1983, 21, 239.

[17]   a) Tabata, Y.; Tanaka, J.; Tagawa, S.; Katsumura, Y.; Ueda, T.; Hasegawa, K. *J. Fac. Engng, Univ.* 1978, Tokyo 34B, 619.

b) Kobayashi, H.; Ueda, T.; Kobayashi, T.; Tagawa, S.; Tabata, Y. *Nucl. Instrum. Meth.* 1981, 179, 223.

c) Kobayashi, H.; Ueda, T.; Kobayashi, T.; Tagawa, S.; Tabata, Y. *J. Fac. Engng. Univ.* 1981, Tokyo, 36B, 85.

[18]   Kobayashi, H.; Ueda, T.; Kobayashi, T.; Washio, M.; Tagawa, S.; Tabata, Y. *Radiat. Phys. Chem.* 1983, 21, 13.

[19]   Nakashima, N.; Sumitani, M.; Ohmine, I.; Yoshihara, K. *J. Chem. Phys.* 1980, 72, 2226.

[20]   Chiggino, K. P.; Wright, R. D.; Phillips, D. *J. Polym. Sci. Polym. Lett. Ed.* 1987, 16, 1499.

[21]   Itagaki, H.; Horie, K.; Mita, I.; Washio, M.; Tagawa, S.; Tabata, Y. *J. Chem. Phys.* 1983, 79, 3996.

[22]   Tagawa, S.; Nakashima, N.; Yoshihara, K. *Macromolecules* 1984, 17, 1167.

[23]   Morita, M.; Tanaka, A.; Imamura, S.; Tamamura, T.; Kogure, O. *J. Appl. Phys.* 1983, 22, L659, Japan.

[24]   Morita, M.; Imamura, S.; Tanaka, A.; Tamamura, T. *J. Electrochem. Soc.* 1984, 131, 2402.

RECEIVED June 25, 1987

Chapter 5

Relations Between Photochemistry and Radiation Chemistry of Polymers

J. E. Guillet

Department of Chemistry, University of Toronto, Toronto, Ontario M5S 1A1, Canada

It is often assumed in studies of the radiation chemistry of polymers that because of the very high energies
of electron beams, x-rays and γ-rays that a complete
randomization of chemical processes and reactivity is
observed. However, concurrent experiments involving
both exposure to high-energy and deep-UV radiation on
the same polymer films show that even with very high energy photons such as γ-rays a considerable selectivity
of chemical reaction still occurs. This is due to the
fact that after the initial absorption step, followed by
the emission of a large number of Compton electrons
there is a cascading of the energy down the energy
scale, which tends to populate upper electronically excited states of the molecule. These states themselves
have short lifetimes and ultimately the energies appear
to be trapped in one of the lower excited states similar
to those which can be populated directly by UV. For example, studies of carbonyl containing polymers show that
the probability of a carbonyl group reacting is much
higher than would be predicted from its mass absorption
coefficient. Furthermore, the chemical products appear
to be identical to those which are formed from the direct irradiation of the carbonyl group with UV light.
However, when two or more competing reactions can occur
out of the excited state of the carbonyl it appears that
high energy radiation favors radical reaction over those
involving more extensive deformation or changes in the
shape of the absorbing molecule. This paper reviews experiments involving copolymers of a variety of monomers
with aromatic and aliphatic vinyl ketones. G values for
reaction of the carbonyl chromophore excited by electrons, γ-rays and soft x-rays will be compared with corresponding quantum yields of processes induced by the
direct absorption of UV light. The importance of cage
reactions in the dissociation of radical species will be
demonstrated.

0097–6156/87/0346–0046$06.00/0
© 1987 American Chemical Society

This paper is a review of work carried out in our laboratories at the University of Toronto in an attempt to understand
the relationship between polymer photochemistry, where excitation
is provided by UV or visible radiation, with processes which occur
when the polymer is excited by much higher energy radiation such
as γ-ray or elementary particles such as neutrons and electrons.

The earliest studies of radiation chemistry were on polymers
which were useful in the development of nuclear weapons and atomic
power in the early 1940's and 50's. The primary reason for such
studies was to determine the stability of polymeric materials for
technological applications in these industries. It was generally
assumed at that time, that because of the very large energies associated with the initial deposition step, very little selective
chemical reaction would occur and that a wide variety of random
bond-breaking processes would occur throughout the polymer liquid
or solid. Table I shows the energies associated with various
forms of radiation. The γ-ray photon from cobalt-60, for example,
has an energy 100,000 times greater than that of a typical UV photon. Much of the early work has been reviewed elsewhere (1-3).

Table I.   Relation Between Wavelength and Photon Energy

Wavelength λ (nm)	Type of radiation	Photon energy, ϵ_p		
		eV	kcal	kJ
1250	Infrared	1	23	96
125	Ultraviolet	10	230	960
12.5	Soft x-rays	100	2300	9600
1.25	Soft x-rays	1000	2.3×10^4	9.6×10^4
0.125	Soft x-rays	10,000	2.3×10^5	9.6×10^5
0.0125	X-rays	10^5	2.3×10^6	9.6×10^6
0.001	γ-rays (^{60}CO)	1.2×10^6	2.9×10^7	12.1×10^7

Source: Reproduced with permission from Ref. 10.   Copyright
1985 Cambridge University Press.

As the physics of the process became better understood, it
was realized that the absorption of a γ-ray photon or high-energy
particle by a polymer, although originally involving the formation
of a high-energy ion ultimately resulted in the production of
showers of Compton electrons of much lower energy. These were usually located in the tracks or spurs along the path of the original excitation or those of its successor particles or rays. Figure 1 shows a schematic diagram of the events following the absorption of a high-energy photon. It is for this reason that
there are many common features between the chemistry of x-rays and
γ-rays and processes initiated by electron beams, since the most
important energetic intermediates are likely to involve interactions of electrons with matter. Furthermore, it became apparent

that although the absorption of the high energy radiation occurred
more or less at random and was proportional to the density of the
absorbing medium, it was obvious by even a cursory reading of the
literature on radiation chemistry, and in our own studies, that
the chemistry did not occur at random but that the majority of reactions occurred at rather specific sites within the polymer
material.

In our early work we looked at the γ-radiolysis of simple
aliphatic ketones [4]. Some results are tabulated in Table II.
Although the carbonyl group of the ketone represented a minor part
of the total mass of the molecule, products arising from the excited state of the ketone carbonyl were four times greater than
were expected from its absorption cross-section. The photochemistry of ketones has been well studied. Bamford and Norrish [5]
showed that there are two main reactions: the Norrish type I reaction involving direct scission of the C-C bond adjacent to the
carbonyl group to form two free radicals, and the Norrish type II
which involves both the singlet and the triplet excited states and
is a rearrangement through the intermediate of a 1-4 biradical resulting in scission of the C-C bond in α-β position to form an
olefin and a lower ketone. Because these compounds are easy to

detect and measure by gas chromatography it is possible to look at
this radiolysis in some detail.

Studies were made both by γ-radiation and by UV photolysis
(Table III) [4]. There were substantial differences in the ratio
of G values for these two products. In UV photochemistry, excitation is to lower vibrationally excited states and the free radical
(Type I) yields are really quite low. Most of the reaction can be
attributed to the type II rearrangement rather than the radical
process. On the other hand, the G value for type I products in
γ-radiolysis is very much higher. It was suggested that in radiolysis, higher excited states were involved which eventually provided greater translational velocity to separate the free radical
pairs, thus giving rise to higher yields of chemical products.

Table II. Radiolysis Yields of Symmetrical Aliphatic Ketones[a]

R	Ketone	Phase	Products from ketone			G(A)[b]	G(B)[c]	$G(H_2)$	G(total products)	$\dfrac{G_I+G_{II}}{G_{total}}$	Fraction radiation absorbed by CO	$\left(\dfrac{G_I+G_{II}}{G_{total}}\right)\left(\dfrac{1}{f}\right)$
			Type I	Type II								
			G(alkane)	G(methyl ketone)	G (olefin)							
3	4-Heptanone	Liquid	1.42	0.49	0.49	0.15	0.32	0.47	2.87	0.84	0.25	3.4
4	5-Nonanone	Liquid	0.86	0.51	0.30	0.10	0.27	0.37	2.04	0.82	0.20	4.1
5	6-Undecanone	Liquid	0.56	0.45	0.24	0.17	0.40	0.57	1.82	0.69	0.16	4.2
6	7-Trideca-none	Liquid	0.36	0.28	0.17	0.07	0.48	0.65	1.44	0.55	0.14	3.9
7	8-Pentadeca-none	Solid	$<1 \times 10^{-4}$	$<1 \times 10^{-4}$	$<1 \times 10^{-4}$							
11	12-Tricosa-none	Solid	$<1 \times 10^{-4}$	$<1 \times 10^{-4}$	$<1 \times 10^{-4}$							

[a] Source: Reproduced with permission from Ref. 10. Copyright 1985 Cambridge University Press.
[b] A = sum of all low molecular weight products.
[c] B = sum of all high molecular weight products.

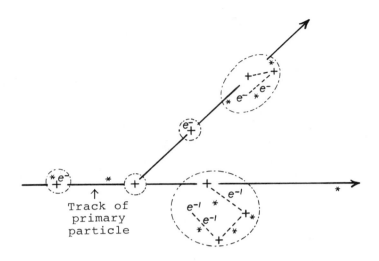

Figure 1. Schematic showing events following the absorption of a high-energy photon or particle. Those items circled are isolated ionizations and spurs. _ _ _ trace of low-energy secondary electron. + positive ion. e^{-1} = free electon. * = excited molecule.

Table III. Comparison of G Values for Photolysis (313 nm) and Radiolysis of Symmetrical Aliphatic Ketones at 35°C

Ketone	G(photolysis)			G (radiolysis)		
	$CO_{(I)}$	Methyl ketone$_{(II)}$	Ratio G_I/G_{II}	$\left(\dfrac{Alkane}{2}\right)_{(II)}$	Methyl ketone$_{(II)}$	Ratio G_I/G_{II}
4-Heptanone	0.44	3.8	0.12	0.71	0.49	1.45
5-Nonanone	0.025	2.8	0.0090	0.43	0.51	0.88
6-Undecanone	0.020	2.4	0.0083	0.28	0.45	0.62
7-Tridecanone	0.017	2.0	0.0085	0.18	0.26	0.69
8-Pentadecanone	0.015	1.7	0.0088	0.065	0.092	0.71
12-Tricosanone	0.012	1.5	0.0080	0.040	0.081	0.69

Source: Reproduced with permission from Ref. 10. Copyright 1985 Cambridge University Press.

Studies were also made of ketones which could be γ-irradiated, either in the liquid state or as a solid crystal. The G values for 8-pentadecanone and 12-tricosanone are very low in the crystal (Table II). Warming to slightly above the melting point gave the G values expected from other liquid members of the series. Apparently the pure crystal lattice does not permit enough motion to allow the radical fragments to separate and suggests that many reactions in semicrystalline materials like polyethylene probably take place primarily in the amorphous phase or at the interface between the crystalline and amorphous regions.

Ethylene-Carbon Monoxide Polymers

Among the first polymers studied in our program were copolymers of ethylene and carbon-monoxide in which the carbonyl groups were included primarily in the backbone of the chain [6]. It was observed that the presence of small amounts of ketone groups improved the efficiency of crosslinking by a substantial amount. For example, with only 1% of the carbonyl group, the G value for crosslinking of polyethylene was doubled. Furthermore, as shown in Table IV there was a concurrent reduction in the

Table IV. $G(H_2)$ Values for Polymers Irradiated in Film at 35°C under Vacuum to a 2-3 Mrad Dose

CO in polyethylene (wt-%)	$G(H_2)$ (ev)
0.00	3.29 ± 0.04
0.30	2.90 ± 0.07
0.55	2.56 ± 0.04
1.00	2.31 ± 0.03
3.50	1.45
9.10	1.37
0.50	2.37 ± 0.03
(mixture rule)	(2.80)
0.65	2.35 ± 0.04
(mixture rule)	(2.60)

Source: Reproduced with permission from Ref. 10.
Copyright 1985 Cambridge University Press.

amount of hydrogen produced from the CH_2 groups in the polymer chain. This is strong evidence that extensive energy migration was occurring in the solid polymer, and that energy originally deposited in the methylene group was being transferred efficiently to the carbonyl. We proposed that this occurred by a σ exciton band mechanism originally postulated by Partridge [7]. It was also observed that the G values for scission in this polymer were substantially lower than would be predicted from these studies

from model small alkanones. Similar effects are shown in conventional UV photochemistry. For example, Plooard and Guillet [8]
showed that whereas the quantum yield for photolysis of dimethyl
ketopimelate in solution was approximately 0.4, when the same
group was included in the backbone of a polymer chain by copolymerization with butylene glycol the total quantum yield dropped by
a factor of 20. This drastic reduction in photon efficiency is
attributed to the restrictions in mobility when a chromophore is
contained in the backbone of a polymer chain. This can reduce the
rate at which conformational changes can occur, as required in the
type II process and restrict the possibility of radical separation
required for the type I process to occur efficiently.

Studies of Polystyrene Copolymers

     Polystyrene is relatively resistant to high-energy radiation,
but does undergo crosslinking at high doses. However, one of the
major problems in radiation chemistry is the precise determination
of the G values for crosslinking. Recently, there have been efforts to obtain more precise measurements by the use of gel permeation chromatography (GPC). However, this method has serious disadvantages unless the polymers used are nearly monodisperse. In
many cases ultracentrifuge methods are much better than GPC because it is easier to resolve the individual peaks resulting from
one, two, three, or four crosslinks. Early data published by
Kells et al. [9] are shown in Figure 2. The methods which can be
used have been reviewed by Guillet [10]. When styrene is copolymerized with small amounts of methyl vinyl ketone the rates of
scission are increased [11]. With 4% or higher methyl vinyl ketone concentrations the polymer undergoes rather rapid chain scission, as shown in Figure 3, whereas at 2% MVK the rate of scission
and the rate of crosslinking balance out almost exactly so that no
net change in molecular weight occurs up to doses of nearly 200
megarads. In principle, by balancing the rate of crosslinking and
the rate of scission one should be able to obtain polymers whose
molecular weights are nearly independent of radiation dosage.
     Later studies of these same systems with higher amounts of
various ketone structures included by copolymerization have shown
that the polymers become quite photo and radiation sensitive and
are therefore of interest as potential candidates for resists used
in the manufacture of microcircuits. The large effect of relatively small amounts of ketone groups in these polymers is further
evidence that there is efficient exciton transfer of energy originally deposited in the aromatic rings to the ketone functional
group. This is a chemical analogy to the so-called scintillator
effect in such polymers.

Molecular Motion in Solid Amorphous Polymers

     From these early studies it became clear that both the type
and efficiency of chemical reactions occurring in polymeric solids
as a result of excitation by radiation will depend in no small degree on the amount of molecular motion which can occur in the
solid during the lifetime of the excitation, or the transient intermediates that result from energy deposition.

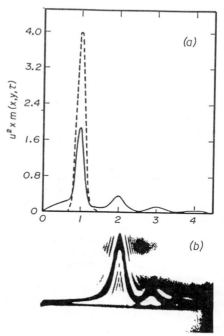

Figure 2. (a) Random scission and cross-linking of narrow molecular weight distribution (theoretical). (b) Schlieren photograph of material irradiated to 25.1 Mrad. (Reproduced with permission from Ref. 9. Copyright 1968, 1969 Wiley.)

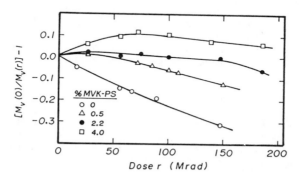

Figure 3. Changes in M_v versus radiation dose r for MVK–PS copolymers irradiated under vacuum in solid phase at 35 °C. (Reproduced with permission from Ref. 11. Copyright 1973 Wiley.)

If an amorphous polymer is cooled it will usually attempt to
crystallize, but because of the high internal viscosity of the
medium it is often precluded from packing into its lowest energy
conformation. At 0 K, the lack of thermal excitation prevents the
occurrence of most photochemical reactions. As the temperature is
increased, the specific volume of the polymer will also increase
as a result of forming "free volume", that is, space which is not
occupied by hard-shell dimensions of the atoms comprising the polymeric structure. The amount of free volume will depend to a certain extent on the previous thermal history. As free volume increases along with thermal excitation, various kinds of molecular
motions will be observed in the polymer which can be detected by
physical measurements.

A typical free volume temperature curve for an amorphous polymer such as polystyrene is shown in Figure 4 [10]. Theoccurrence
of various kinds of internal motion of the polymer solid can usually be detected by the observation of various low temperature
transitions by physical measurements including phosphorescence and
fluorescence emission as well as mechanical relaxations. The
largest increase in the rate of free volume formation occurs at
the glass transition, T_g. At this temperature, very large scale
motions occur of long segments of the polymer chain which may include from 30 to 50 carbon atoms.

Photochemical studies have shown that above this temperature,
polymeric chromophores behave as if they were in liquid solution
and similar quantum yields and efficiencies are observed [12].
Recent studies by our group [13] have demonstrated that the effects of these transitions on photochemical reactions are dependent on the amount of free volume required for a particular reaction to occur. Processes such as the Norrish type II which require extensive conformational changes in the solid state in order
to form the cyclic six-membered reaction intermediate are strongly
sensitive to free volume and therefore show greatly improved reaction efficiency above the glass transition temperature. On the
other hand, for example, the photo-Fries reaction which involves
the rearrangement of a caged radical species apparently involves
rather small free volume changes and shows no increase in quantum
yield at the glass transition, but decreases below the β-transi-
tion where motion of the phenyl rings is gradually frozen out
[14].

Effect of Ketone Structure

In our early photochemical studies we have shown that the
quantum efficiency for the type I processes in polymers containing
ketone groups is highly sensitive to the location of the ketone
group with respect to polymer chain, as shown in structures A and
B below

A

B

As discussed above, when the ketone group is in the polymer backbone, the excitation of the ketone produces two polymeric radicals
which must separate from each other within a short period of time
in order to produce chemical products. If, however, the ketone
group is in a side chain, as in structure B, then a polymeric radical is formed simultaneously with a small radical fragment. This
second fragment can diffuse rapidly through the polymer solid and
the quantum yields are increased by at least one order of magnitude [12,13]. Studies on electron beam irradiation of similar
polymers have confirmed this effect in high-energy radiation systems.

   If one wishes to prepare a positive photoresist it is important to obtain polymers which undergo efficient chain scission in
the solid phase. Recently we reported studies on a series of copolymers of styrene with a variety of ketone functional groups
which were introduced by copolymerization with substituted vinyl
ketone monomers. The copolymer structures are shown schematically
in Table V. Two processes are responsible for the reduction in
molecular weight in these polymers when irradiated with either UV
light or electron beams. These are shown schematically below.

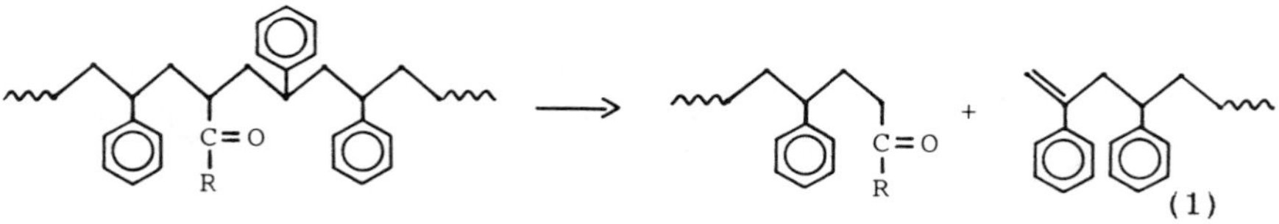

(1)

Norrish type I

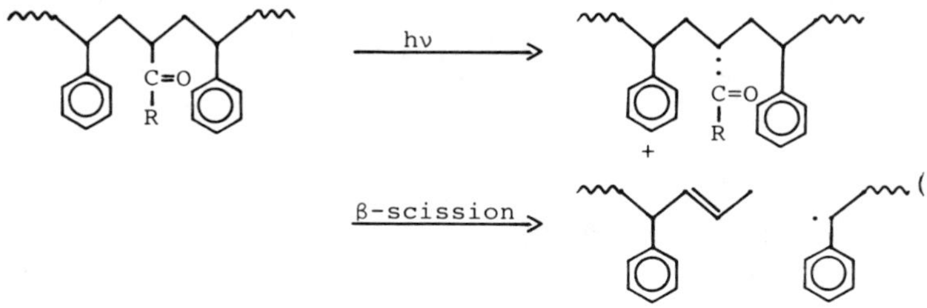

(2)

Norrish type I followed by β-scission

Table V. Structures of Vinyl Ketone Polymers

Copolymers	Copolymer	Structure number
Poly(ethylene-*co*-carbon monoxide) (PE–CO)	chain with C=O	I
Poly(ethylene-*co*-methyl vinyl ketone) (MP–MVK)	chain with C=O / CH$_3$	II
Poly(ethylene-*co*-methyl isopropenyl ketone) (PE–MIPK)	chain with CH$_3$ / C=O / CH$_3$	III
Poly(styrene-*co*-methyl vinyl ketone) (PS–MVK)	chain with phenyl groups, C=O / CH$_3$	IV
Poly(styrene-*co*-methyl isopropenyl ketone) (PS–MIPK)	chain with phenyl groups, CH$_3$ / C=O / CH$_3$	V
Poly(styrene-*co*-*tert*-butyl vinyl ketone) (PS–tBVK)	chain with phenyl groups, C=O / H$_3$C–C–CH$_3$ / CH$_3$	VI
Poly(styrene-*co*-phenyl vinyl ketone) (PS–PVK)	chain with phenyl groups, C=O / phenyl	VII
Poly(styrene-*co*-phenyl isopropenyl ketone) (PS–PIPK)	chain with phenyl groups, CH$_3$ / C=O / phenyl	VIII

Since the glass transition of these copolymers exceeds 60°C, restrictions on mobility during irradiation at room temperature will reduce the quantum yield of the type II process. However, we have recently shown that efficient chain scission can occur in the solid phase by formation of radicals by the type I process followed by β-scission of the polymer radical as is shown in eq. 2. The quantum yields are highly dependent on the structure of the ketone group included in the polymer. For example, the quantum yield for the type I process is 0.09 in MVK copolymers where the substituent on the ketone group is a methyl group but increases to 0.45 where the substituent is tertiary butyl [17]. There is a corresponding increase in the quantum yield of chain scission in the solid phase. Similar trends were observed when the same films were exposed to synchrotron radiation. The higher efficiency of the type I reaction in these structures is attributed to the formation of more stable radicals from the tertiary butyl as compared to the methyl ketone.   Adding an additional substituent to the carbon α to the carbonyl group creates still further stability in the radical formed by a type I process and still higher sensitivity to both light and γ-rays. For example, poly(*t*-butyl isopropenyl ketone) is one of the most sensitive polymers yet developed, both as a near UV and electron-beam resist [18].

Cage reactions

The concept of the so-called "cage reaction" has often been invoked in polymer chemistry to explain observed differences in reactivity between polymers and small molecules. It has been suggested that restrictions on mobility in the solid phase prevent the separation of reactive species, which causes them to recombine to form the original reactant, thus reducing the efficiency or quantum yield of the process in question.

The original concept of cage processes was due to Franck and Rabinowitsch [19], who invoked it to explain the reduction in quantum yields which occur on the photolysis of simple molecules like acetone and iodine when carried out in solution as compared to the gas phase. Extensive studies were carried out by Noyes and coworkers during the 1950s and 1960s [20] to develop a satisfactory theory, but it was not until the development of picosecond lasers, that the early stages of the processes of separation could be probed experimentally. It is now known that in the dissociation of two radical species, for example, recombination will occur in the primary cage in times from 20 to 100 ps. The situation is shown schematically in Figure 5. The radical pair will separate with a velocity dependent on the amount of excess energy partitioned into translational velocity along the axis defined by their centres of mass. This translational energy will be exchanged by momentum transfer with the solvent molecules surrounding them. If, as a result of these first collisions, they rebound to a position where they are within a distance σ of each other, where σ is the radius for reactive collision of the species, they will recombine to reform the initial species. In iodine atom dissociation this process occurs in 20-100 ps. If, on the other hand, the trajectory produces a pair of radicals or atoms separated by at least

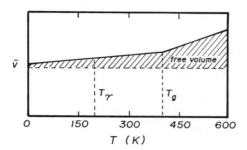

Figure 4. Specific volume and free volume of polymeric material.
(Reproduced with permission from Ref. 10. Copyright 1985 Cambridge
University Press.)

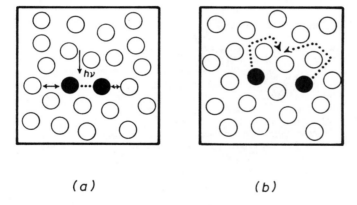

Figure 5. Scheme showing (a) primary and (b) secondary cage recombination.

one solvent molecule, then a random diffusion will occur. Because the local concentration is high there is still a good possibility (> 50%) that the two particles will diffuse to adjacent sites and recombine. This is called "secondary cage recombination", even though it has nothing to do with restriction of motion by a solvent cage, but is primarily an effect due to the non-uniform distribution of radical species. The probability of this reaction occuring depends both on the rate of diffusion of the radicals in the medium, the concentration of radicals from other dissociating molecules, and alternative reaction paths (such as H-atom abstraction from the solvent).

We have recently been concerned about how a polymer environment might affect these two processes. The first question is whether the environment of a solid polymer will increase the amount of primary cage recombination of a dissociating radical pair. Studies were made [21] of the quantum yield for photolysis of benzoyl peroxide in two solvents, benzene and toluene, and in solid polyethylene and polystyrene films:

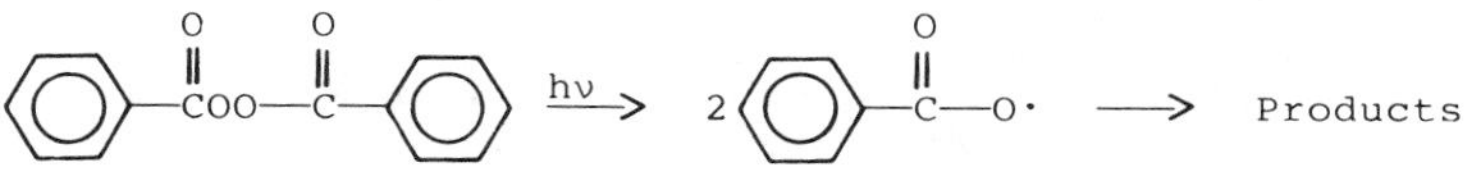

The quantum yields in both the solids and solution were uniformly high (~0.8), indicating that primary cage recombination represented at most 20% of the total and did not depend on either the bulk or microviscosity of the medium. Presumably the separating radicals exchange momentum only with small segments of the polymer chain and the motion which can occur in the short time scale of the primary cage recombination (20-100 ps) cannot be influenced by the polymeric nature of the solvent. It is interesting to note that there was no difference between the quantum yield in polyethylene, which is above its glass transition at 25°C and in polystyrene which is well below its T .

For this reason, any specific "polymer effects", if indeed they do occur, must be attributed to processes occurring outside the primary cage. Secondary cage recombination, for example, will be affected by the rate of diffusion in the polymer matrix. This might be expected to reduce the number of radicals which can escape the region associated their primary partners and become true "free" radicals.

Estimates of the probability of escape of radical pairs in conventional solvents have been made by product analysis of the decomposition of diacyl peroxides. For example, Braun et al. [22] estimated that 60 to 80% of the methyl radicals produced in the thermolysis of acetyl peroxide escape geminate cage recombination. However, Guillet and Gilmer [25] showed that for longer chain C_9 and C_{11} radicals the probability was much lower, ranging from 5% at 76°C to 16% at 262°C (Table VI).

Table VI. Extent of C_9 and C_{11} Radical
Escape from Cage [a]

T (°C)	% escape
76.5	4.0
96.5	5.4
152	8.3
183	12.4
262	16.1

[a]Solvent = Nujol Mineral Oil.

The lower escape probabilities in the latter work can be attributed to the much greater mass of the C_9 and C_{11} radicals and the
greater internal viscosity of the mineral oil solvent.

Any treatment of diffusion processes in polymers must include
estimates of the "internal viscosity", η_{int}, of the solid polymer
matrix. For example, for recombination of active free radicals
which occur at every collision, one can write the rate expression
in the form:

$$\text{Rate} = k_{diff}[R \cdot]^2 \qquad (4)$$

The simplified form of the Einstein-Smoluchowski equation gives
for radicals 1 and 2

$$k_{diff} = \frac{4\pi\rho'N}{1000}(D_1+D_2) \qquad (5)$$

where ρ' is some collision radius for reaction, N is Avogadro's
number and D_1 and D_2 are the diffusion constants for species 1 and
2, respectively.

We consider three limiting cases, depending on the relative
size of the radical species.

Case 1. Two small radicals of equal size, in which case Equation
1 reduces to the Debye equation:

$$k_{diff} \doteq \frac{8\ RT}{3000\ \eta} = 2.2 \times 10^5(T/\eta) \qquad (6)$$

where η is the viscosity of the medium.

Case 2. One small radical and one large polymer radical. In this
case $D_2 = 0$ and

$$k_{diff} = 1.1 \times 10^5(T/\eta) \qquad (7)$$

The rate is only one-half that for two small radicals.

Case 3. Two polymer radicals.

$$k_{diff} = \text{const} \times (D_1 + D_2) \qquad (8)$$

It is not possible to estimate the value of k_{diff} in this case, but it will remain proportional to T/η. In concentrated polymer solutions and in solid polymers, diffusion will almost certainly involve movement by reptation and will be very slow. This explains the results reported earlier for low quantum yields for photolysis of backbone carbonyls.

When considering the rates of chemical processes in polymers which require diffusion of reagents and products, it is necessary to estimate the internal viscosity η_{int} to substitute in expressions such as Equations 6 and 7. We have made estimates for η in solid polyethylene from luminescence quenching of naphthalene fluorescence in ethylene —CO copolymers [24]. (Table VII)

Table VII.   Internal Viscosities and k_{diff} for Reaction in Various Media

Medium	T (°C)	η (poise)	k_{diff} (L mol^{-1} s^{-1})
Hexane	20	3×10^{-3}	2×10^{10}
Ethylene glycol	20	0.2	4×10^{8}
Polyethylene	80	1.0	1.0×10^{8}
Glycerin	20	11	6×10^{6}

It is noteworthy that the internal viscosity in amorphous polyethylene at 80°C is only 1 poise. Extrapolation of this value to 20°C would give a value similar to that of glycerine (~10 poise). The macroscopic viscosity of solid polyethylene, based on creep measurements is of the order of 10^{10} poise. As a result of this, low internal viscosity in polyethylene, small molecules diffuse rapidly thorugh it and relatively rapid bimolecular reactions can occur. This type of molecular mobility in solid polymers makes possible a wide variety of photo- and radiochemical reactions. Furthermore, it makes possible a rational explanation for effects which were hitherto obscure. For example, recent studies carried out in our laboratory in collaboration with Miller's group at IBM (San Jose), have shown that a variety of substituted polysilanes show high quantum yields for photolysis in solution but very low values in the solid state under near UV irradiation. On the other hand, these solid polymers show quite respectable G values as electron beam resists. Presumably some of the extra energy deposited by the electron beam is converted to translational momentum in the separating radical fragments, allow them to escape cage recombination. This explanation seems to fit most of the radical processes we have studied so far.

In conclusion, studies of the photo and radiation chemistry
of ketone-containing polymers show many similar features, suggesting that at least some of the excited intermediates induced by
high-energy radiation are similar to those induced by direct UV
absorption. As a general rule, the efficiency of radical processes appears to be greater in radiation-induced processes, possibly
because of the higher energy of precursor states. There is strong
evidence that extensive energy migration occurs in both solids and
liquids, and must be considered as contributing to the overall
mechanism of photo and radiation processes in polymers.

Literature Cited

1. Charlesby, A. Atomic Radiation and Polymers; Pergamon Press:
    Oxford, 1960.
2. Bovey, F. A. The Effects of Ionizing Radiation on Natual and
    Synthetic High Polymers; Interscience: New York, 1958.
3. Chapiro, A. Radiation Chemistry of Polymer Systems;
    Interscience: New York, 1962.
4. Slivinskas, J. A.; Guillet, J. E. J. Polym. Sci., Polym.
    Chem. Ed., 1973, 11, 3043.
5. Bamford, C. H.; Norrish, R. G. W. J. Chem. Soc. 1935, 1504;
    1938, 1521; 1938, 1544.
6. Slivinskas, J. A.; Guillet, J. E. J. Polym. Sci., Polym.
    Chem. Ed. 1974, 12, 1469.
7. Partridge, R. H. J. Chem. Phys. 1970, 52, 2485, 2491, 2501.
8. Plooard, P. I.; Guillet, J. E. Macromolecules 1972, 5,
    405.
9. Kells, D. I. C.; Koike, M.; Guillet, J. E. J. Polym.   Sci.,
    Part A-1 1968, 6, 595. Kells, D. I. C.; Guillet, J. E.   J.
    Polym. Sci., Part A-2 1969, 7, 1895.
10. Guillet, J. E. Polymer Photophysics and Photochemistry;
    Cambridge University Press: Cambridge, 1985.
11 Slivinskas, J. A.; Guillet, J. E. J. Polym. Sci., Polym.
    Chem. Ed. 1973, 11, 3057.
12. Dan E.; Guillet, J. E. Macromolecules 1973, 6, 230.
13. Guillet, J. E.; Li, S.-K. L.; Ng. H. C. In Materials for
    Microlithography; Thompson, L. F.; Willson, C. G.;
    Frechet, J. M. J., Ed.; ACS Symposium Series No. 266,
    American Chemical Society: Washington, D. C., 1984; p
    165.
14. Li, S.-K.; Guillet, J. E. Macromolecules 1977, 10, 840.
15. Sitek, F.; Guillet, J. E.; Heskins, M. J. Polym. Sci.,
    Symposium 1976, No. 57, 343.
16. Li, S. K. L.; Guillet, J. E. J. Polym. Sci., Polym. Chem.
    Ed. 1980, 18, 2221.
17. Guillet, J. E.; Li, S.-K. L.; MacDonald, S. A.; Willson,
    C. G. In Materials for Microlithography; Thompson, L. F.;
    Willson, C. G.; Frechet, J. M. J., Ed.; ACS Symposium
    Series No. 266, American Chemical Society: Washington, D.
    C., 1984; p 389.

18. MacDonald, S. A.; Ito, H.; Willson, C. G.; Moore, J. W.; Gharapetian, H. M.; Guillet, J. E. In <u>Materials for Microlithography</u>; Thompson, L. F.; Willson, C. G.; Frechet, J. M. J., Ed.; ACS Symposium Series No. 266, American Chemical Society: Washington, D. C., 1984; p 179.
19. Franck, J.; Rabinowitsch, E. <u>Trans. Faraday Soc.</u> 1934, <u>30</u>, 120.
20. Noyes, R. M. <u>Progr. React. Kinet.</u> 1961, <u>1</u>, 128.
21. Moore, J. W.; Guillet, J. E. Unpublished work.
22. Braun, W.; Rajbenback, L.; Eirich, F. R. <u>J. Phys. Chem.</u> 62, <u>66</u>, 1951.
23. Guillet, J. E.; Gilmer, J. C. <u>Can. J. Chem.</u> 1969, <u>47</u>, 4405.
24. Heskins, M.; Guillet, J. E. In <u>Photochemistry of Macromolecules</u>; Reinisch, R. F., Ed.; Plenum: New York, p 39.

RECEIVED April 30, 1987

RESIST MATERIALS FOR ELECTRON AND X-RAY LITHOGRAPHY

RESIST MATERIALS FOR ELECTRON AND X-RAY LITHOGRAPHY

Up until 1970, there was little cause for concern in the integrated circuit industry from the standpoint of optical fabrication techniques. The smallest features used in IC design were $6\mu m$ or greater, which were comfortably within the resolution capabilities of optical lithography. People began to recognize, however, that if the scale of integration were to continue to increase at the same rate heretofore, device features would approach $1\mu m$ or smaller within 15 years, and many believed that conventional photolithography lacked the resolution to meet these requirements. This reasoning led several companies to begin working on alternative lithographic technologies, most notably electron-beam lithography.

The concept of e-beam writing developed from scanning electron microscopy which, by the early 1970s, had been developed to a high degree of refinement. It was recognized that by accurately deflecting and modulating (turn off and on) a finely focused beam of electrons with the aid of a digital computer, a scanning electron-beam machine could be made to "write" high-resolution patterns on a resist-coated substrate. However, the only resist materials available at that time were conventional photolithographic resists which lacked adequate sensitivity to electron-beam radiation. Moreover, they were sensitive to visible/ultraviolet light which represented an unnecessary complication for e-beam lithography. Clearly the commercial realization of the latter would require new resists with sensitivity and chemical and physical properties specifically tailored to the needs of this new technology.

The 1970s witnessed an unprecedented spate of activity in the design of resist materials, not only for e-beam lithography but also for the other forms of lithography which were under development such as short-wavelength (deep UV) photolithography, X-ray and ion-beam lithography. Each technology required resist materials specifically designed and optimized for the particular radiation being contemplated. In some respects, much of the impetus that lay behind such efforts has abated, primarily because the early predictions of the demise of photolithography at resolutions of $1\mu m$ have not proved correct. The current resolution limitation is now thought to be near $0.5\mu m$ which means photolithography will continue to serve the IC industry well into the 1990s. Nevertheless research has continued on alternative lithographic processes, including both hardware and associated resist materials in the belief that they will eventually find a niche in manufacturing beyond today's limited applications. Particularly important have been the advances in processing which have continued to make new demands on resist performance, irrespective of the type of exposure.

The papers in this session address some of the current thinking on resist design and process requirements. Four of the seven papers for example, deal with polymers containing elements which form refractory oxides in an oxygen plasma environment. Such resists have application in multilayer processing techniques which are becoming increasingly important in present and future device fabrication processes.

The future of microlithography is clearly bright and contains many challenges in the areas of resist research and processing. By the end of this century, it is almost certain that some new form of microlithographic patterning will be commonplace in manufacturing, and that a new resist and resist process will be required. The next decade promises to be very exciting for the resist and process engineer. The work reported in this book will definitely be an important part of the scientific foundation on which these new technologies will be based.

Larry F. Thompson
AT&T Bell Laboratories
Murray Hill NJ USA 07974

Chapter 6

Characteristics of a Two-Layer Resist System Using Silicone-Based Negative Resist for Electron-Beam Lithography

Toshiaki Tamamura and Akinobu Tanaka

NTT Electrical Communications Laboratories, Tokai, Ibaraki 319-11, Japan

Two different molecular weight grades of the sili-
cone-based negative resist (SNR) have been evaluated
as the top resist of a 2-layer resist system for
practical 0.5 µm electron beam (E-beam) lithography
and nanometer E-beam lithography. The high molecular
weight grade (Mw = 50,000) obtained by fractionation
of a broad molecular weight distribution polymer
showed 0.5 µm resolution with sensitivity less than
10 µC/cm^2. Low molecular weight SNR (Mw = 2,800)
facilitated fabrication of nanometer sized features
in thick planarizing substrates by O_2 RIE pattern
transfer. Using SNR (80 nm) on top of a 250 nm CVD
carbon film, resolution of 40 nm lines with a 0.1 µm
in pitch on a solid substrate has been demonstrated.

The role of electron beam (E-beam) lithography as a mask making
technology for conventional photolighography has been firmly established. However, E-beam lithography has not found wide acceptance
for direct device fabrication (except perhaps for Ga-As devices),
primarily because of substantial progress in the resolution
capability of optical lithography. However, E-beam lithography may
still be a promising candidate for practical lithography below 0.5
µm which many believe to be the limit of conventional photolighography. The increasing importance of custom LSI, such as application specified integrated circuits (ASIC), could further promote
the use of direct E-beam lithography in the near future. Another
important application of E-beam lithography is in nanometer lithography where it is being used to fabricate devices with very small
structures (1), and to demonstrate new physical phenomena.
    For a variety of reasons (2), two-layer resist processes using
silicon-containing resists for the top imaging layer are regarded
as the most practical processing technology for future device
fabrication. Although extensive studies have been carried out on
the development of E-beam resist materials, no material with the
universal appeal comparable to the Novolak-type resists in
photolithography has emerged. We have developed a novel silcone-
based negative resist (SNR), which serves as a top imaging resist

for both E-beam and deep UV lithography (3,4). The structure of
SNR is shown in Fig. 1. SNR is characterized by a very slow etching rate in an oxygen reactive ion etching (O_2RIE) environment
which is attributed to the presence of silicon in the siloxane main
chain of the polymer. The diphenyl-substituted structure increases
the glass transition temperature of the resist, which improves its
handling and resolution; the chloromethyl groups function to
increase resist sensitivity.

In this paper we examine the effect of molecular weight on the
lithographic performance of SNR with the objective of developing
materials suitable for practical E-beam lithography and nanometer
E-beam lithography using 2-level processing techniques.

Experimental

SNR was synthesized by reaction of diphenylsilanediol with chloro-
methylmethylether using $SnCl_4$ as a catalyst. The molecular parameters of SNR were determined by gel permeation chromatography
using a Model HLC-802 A (Toyo Soda) instrument and polystyrene
standards. Either hard baked (at 200°C 30 min.) Novolak-type
positive photoresist: Microposit 1400 (Shipley), or plasma-CVD
carbon film was used as the bottom resist in 2-layer application.
Carbon films were deposited by an electron cyclotron resonance
(ECR)-type plasma CVD process using acetylene as a carbon source
(5). SNR was spin-coated onto silicon wafers that had been precoated with the bottom resist, and exposed with a computer-con-
trolled E-beam exposure system (Elionix ELS-5000) operated at 20
keV. After development of the SNR, the bottom resist was patterned
by O_2RIE using a parallel plate RIE instrument (ANELVA DEM-451) at
an oxygen pressure of 10 mTorr. The transferred resist patterns
were then evaluated with a scanning electron microscope (JSM-840,
JEOL). The reactive ion-etching resistance of the resists to
various etchant gases was determined by measuring residual film
thickness with Talystep (Taylor-Hobson).

Resist Properties

In our earlier work, SNR was prepared by the chloromethylation of
oligomeric diphenylsiloxane (Petrarch Systems, average Mw = 1400)
containing OH end groups. We found that the polymerization of the
oligomers occurs during the chloromethylation reaction (6), and
high molecular weight polydiphenylsiloxane was prepared. One
problem with this approach was a difficulty in sufficient control
of the final molecular parameters, particularly in the case of very
high Mw polymers. Consequently, we have developed an improved
synthetic procedure which uses diphenylsilanediol instead of
diphenylsiloxane oligomers.

In contrast to the narrow molecular weight distribution
(Mw/Mn = 1.2 - 1.5) of the low molecular weight resins, the high
molecular weight polymers exhibited fairly broad molecular weight
distriblutions. Thus, in order to obtain SNR with high molecular
weight and narrow molecular weight distribution, it was necessary
to fractionate the material. The low molecular weight polymer was

removed by fractionation using toluene as a good solvent and hexane as a non-solvent. The molecular and lithographic parameters of SNR samples in this study are summarized in Table 1.

Table 1. Molecular and Lithographic Parameters of SNR

	Mw	Mw/Mn	D_0	$D_{0.8}$	γ	O_2 RIE rate
				($\mu C/cm^2$)		(nm/min.) (ratio*)
SNR-H	51,000	2.3	1.4	3.8	2.0	2.5	37
SNR-M	35,000	3.2	3.4	12.5	1.9	2.5	37
SNR-L	2,800	1.3	75	150	2.4	5.5	16

*The ratio of O_2 RIE rates against Microposit 1400 baked at 200°C 30 min.

Sensitivity curves are shown in Fig. 2. The contrast γ of SNR-H and SNR-M was not definable by the method recommended in the literature (7). In this study, γ was obtained by the equation shown in Fig. 2.

The SNR-M sample corresponds to the as prepared material and was characterized by a relatively large Mw/Mn. Its sensitivity curve (Fig. 2) initially rose steeply, but decreased at higher doses. Although $D_{0.8}$ was only 12.5 $\mu C/cm^2$, the dose required to retain 100% of the initial thickness was about 50 $\mu C/cm^2$. This decrease in contrast results from the presence of a fairly large fraction of the low molecular weight polymers. On the contrary, SNR-H, obtained from SNR-M by fractionation, showed a more typical negative-type sensitivity curve. The higher molecular weight and narrower distribution of the fractionated material resulted in a $D_{0.8}$ of 3.8 $\mu C/cm^2$ which was almost 4 times higher than that of the original polymer. SNR with low molecular weight was obtained by stopping the reaction at a lower degree of conversion. SNR-L has a Mw of only 2800 which is about the lowest molecular weight that will still form a uniform film by spin coating. This low-molecular-weight sample had a higher contrast than the others, but its O_2 RIE rate was significantly higher. Its etch rate ratio relative to Microposit 1400 was only 16 compared to 37 for the other samples. This appears to reflect a lower rate of formation of the SiO_2 layer during O_2 RIE. This reason is not well understood, but we believe that the polymer structure of high molecular weight material is slightly different from that of low molecular weight polymer. Whereas the low molecular weight polymer has a linear structure as shown in Fig. 1, the high molecular polymer may contain a certain amount of crosslinked structure, which may facilitate the formation of the SiO_2 layer. This speculation also explains the very broad molecular weight distribution of the as prepared material.

High Glass-transition Temperature ⇐
High Oxygen RIE Resiistance ⇐
High reactivity to E-beam ⇐

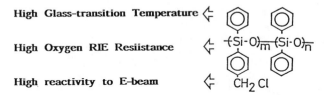

Figure 1.   Structure of SNR

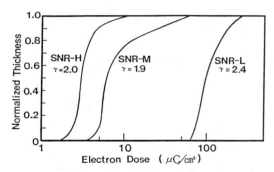

Figure 2.   Sensitivity curves of SNR to 20 kV E-beam on 1.0 μm
thick Microposit 1400 on Si substrate.   The contrat
γ was obtained by the following equation:
$$\gamma = 1.0/\log (D_1 /D_0)$$

Patterning of SNR/Microposit 2-layer resist. The resolution of SNR
-H and SNR-M/Microposit 1400 2-layer resist systems was evaluated
from analysis of the patterns transferred into the underlying planarizing layer by O_2 RIE. The initial thickness of SNR and Microposit layers was 0.18 and 1.0 μm, respectively. Two test patterns
were evaluated, viz., 0.5 μm line & space (1:1) patterns and 0.5 μm
gaps with 3 μm wide lines. The respective micrographs corresponding to optimum exposure are shown in Fig. 3. In all cases the fine
line-&-space patterns required a higher dose than the large patterns which was due to the proximity effect. SNR-H resolved the
two test patterns at 5.8 μC/cm^2, and 8.6 μC/cm^2, respectively.
Both patterns showed irregular pattern edges following solvent
development which we attributed to swelling effects. The patterns
shown in Fig. 3 were obtained by significant overetching to remove
the resist scum between the patterns. The high O_2 RIE resistance of
SNR coupled with the anisotropic character of O_2 RIE pattern transfer permits such overetching without degradation of pattern quality. The pattern edges were smoother with SNR-M as shown in Fig.
3, but the dose had to be increased to 12 and 21 μC/cm^2 for 0.5 μm
gap patterns and 0.5 μm line & space patterns, respectively. However, in all cases the gaps obtained were slightly wider than the
designed rule, perhaps due to shrinking of the SNR patterns following development. It is apparent then that the SNR-H/Novolak-type
two-layer resist system can be used for 0.5 μm E-beam lithography
with a practicl sensitivity less than 10 μC/cm^2.
        Nanometer patterning using SNR-L was carried out by using a
thinner top resist layer (80 nm) on a 0.3 μm thick Microposit
layer. Resolution was taken as the minimum linewidth and pitch of
well-resolved patterns. As the lines became finer, the resolution
became dependent not only on the development properties of SNR, but
also on the O_2 RIE process characteristics. SEM micrographs of 150
nm pitch gratings after O_2 RIE with three different doses are shown
in Fig. 4. As expected, the line width increases with increasing
dose. At a line dose of 1.8 nC/cm, the lines are 70 nm wide (Fig.
4-a) and are well resolved when compared with the spaces. At a
lower dose, the line width was somewhat narrower, but the lines
were collapsed. This problem became particularly severe for line
dose of 1.35 nC/cm (line width = 40 nm, Fig. 4-c). Collapse of the
patterns does not appear to result from swelling of SNR lines during development. If this were the case, one would expect the bottom layer between the lines not to be etched during RIE pattern
transfer, but as can be seen in Fig. 4, the resist under the
bridged lines has been completely removed. Moreover, in both cases
the first and last grating lines tend to bend, and thus cannot act
as the resist mask for the subsequent RIE patterning of the substrate layer.
        The deformation of the nanometer patterns, which has also been
observed in the case of other polymeric planarizing materials, is
caused by a lack of mechanical integrity of the bottom resist during O_2 RIE processing. Because of this mechanical instability, the
minimum linewidth and pitch were limited to about 70 nm and 120 nm,
respectively. Forced cooling of the substrate may prevent this
problem from occurring.

1-a

5.8 $\mu C/cm^2$

1-b

8.6 $\mu C/cm^2$

2-a

12 $\mu C/cm^2$

2-b

21 $\mu C/cm^2$

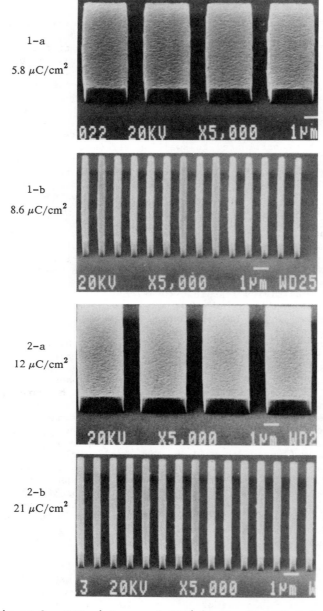

Figure 3. SEM pictures of SNR/microposit 1400 2-layer resist patterns (The thickness of Microposit 1400 is 1.0 μm) 1-a: 0.5 μm gaps/3.0 μm lines with SNR-H, 1-b: 0.5 μm line and space with SNR-H, 2-a: 0.5 μm gaps/3.0 μm lines with SNR-M, 2-b: 0.5 μm line and space with SNR-M.

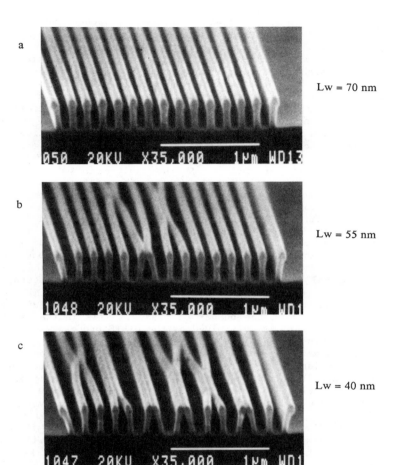

Figure 4.   SEM pictures of 150 nm pitch SNR-L/Microposit 1400
2-layer resist patterns on Si substrate. (The
thickness of Microposit 1400 is 300 nm). a:
Resolved lines with a line dose of 1.8×10^{-9} C/cm,
b: Collapsed lines with a line dose of $1.5 \times$
10^{-9} C/cm., c: Collapsed lines with a line dose of
1.35×10^{-9} C/cm.

<u>SNR-L/CVD Carbon Resist System</u>. We have proposed another approach
to improve the nanometer-scale resolution of the SNR 2-layer resist
system, viz., the application of carbon films as the bottom layer
material. (5)   Carbon films prepard by plasma CVD are hard and
thermally stable. Figure 5 shows about 40 nm-wide SNR/carbon patterns with a 150 nm pitch on a Si substrate. The narrow lines are
well resolved with a steep profile, and the lines at both edges
of the pattern have not bent or fallen down. This excellent stability of nanometer-scale carbon patterns facilitates the evaluation of the resolution limit SNR in 2-layer resist application.

In the case of a positive electron resist such as PMMA, the
finest lines are normally written using a single E-beam scan.
However, similar exposure of SNR-L resulted in wavy lines, which
reduced the dose margin for the fabrication of fine gratings. This
problem is caused by the swelling of SNR during development. Thus,
the resolution limit was investigated by changing the number of
passes to form a line with a positional increment of 2.5 nm for
each pass. Figure 6 shows the dose dependence of linewidth for
nominal 150 nm-pitch gratings exposed using 1, 4, 8 and 16 passes.
With the increase of passes, the dose margin increased and the
problem of wavy lines was alleviated. 8 passes, which correspond
to a nominal linewidth of 20 nm provided the best pattern quality
for a pitch of 150 nm. Patterns corresponding to this exposure
sequence are shown in Fig. 5. The minimum pattern width obtained
does not depend on the number of passes. This indicates that the
minimum linewidth is determined by the O_2RIE process rather than by
the resolution of SNR itself.

The maximum resolution obtained in this system has been 35-40
nm in linewidth on a 100 nm pitch. Although these values are somewhat higher than those obtained in nanometer patterning of PMMA on
a solid substrate (8), the SNR/carbon resist patterns are much more
practical for fabrication of ultrasmall structures by dry-etching
techniques. This is because carbon films are 6 - 10 times more
resistant than PMMA during RIE using using various fluorocarbon
gases. This negative type E-beam nanolithography will be very
useful in the fabrication of very narrow lines by RIE pattern
transfer and may replace positive resists and the lift-off process
in such applications.

Conclusion

A two-layer resist system was evaluated for E-beam lithography
based on the silicon containing negative resists SNR as the imaging
layer. Two grades of SNR were evaluated, one a fractionated high
molecular weight, narrow molecular-weight-distributon sample, the
other a low molecular weight sample. 0.5 µm features could be
resolved in the high molecular weight sample at a practical sensitivity of 5 µC/cm^2 to 20 kV E-beam. Nanometer-scale patterns were
delineated in the low molecular weight sample, albeit at much
higher dose than the high molecular weight grade. Using mechanically strong carbon films as the bottom layer, 40 nm wide lines
with 100 nm pitch could be transferred by O_2RIE.

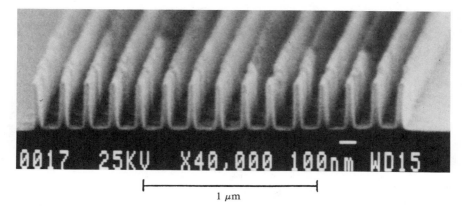

Figure 5.   SEM picture of 40 nm wide, 150 nm pitch SNR-
L/carbon 2-layer resist patterns on Si substrate.
(The carbon thickness is 250 nm).

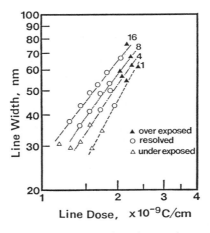

Figure 6.   Dose dependence of linewidth with a pitch of 150 nm
SNR-L/carbon 2-layer resist patterns exposed with
the different numbers of passes with an increment
of 2.5 nm.

Literature Cited

1. Howard, R. E.; Prober, D. E.; VLSI Electronics, Microstructure
    Science 5, Academic Press Inc., New York 1982, Chapter 4.
2. Lin, B. J.; In "Introduction to Microlithography," Chapter 6,
    ACS Symposium Series 219, American Chemical Society:
    Washington D. C., 1983.
3. Morita, M.; Imamura, S.; Tanaka, A.; Tamamura, T.; J.
    Electrochem. Soc. 1985, 131, 2402.
4. Tanaka, A; Morita, M.; Imamura, S.; Tamamura, T.; Kogure, O.;
    In "Materials for Microlithography," ACS Symposium Series 266,
    American Chemical Society: Washington D. C., 1984, 311.
5. Kakuchi, M.; Hikita, M.; Tamamura, T.; Appl. Phys. Lett. 1986,
    48, 835.
6. Imamura, S.; Morita, M; Tamamura, T.; Kogure, O.;
    Macromolecules, 1984, 17, 1412.
7. Taylor G. N.; Solid State Technol., 1984, 27, no. 6, 105.
8. Howard, R. E.; Hu, E. L; Jackel, L. D.; IEEE Trans. Electron
    Devices, 1980, ED 28, 592.

RECEIVED May 5, 1987

Chapter 7

Phenolic Resin-Based Negative Electron-Beam Resists

H. Shiraishi, N. Hayashi, T. Ueno, O. Suga, F. Murai, and S. Nonogaki

Central Research Laboratory, Hitachi, Ltd., Kokubunji, Tokyo 185, Japan

Phenolic resin-based negative resists such as MRS (deep UV negative resist) do not swell aqueous alkaline developers. An attempt has been made to clarify the non-swelling dissolution mechanism of the resist containing poly-p-vinylphenol as a matrix following exposure to electron beam irradiation. The following results have been obtained: (1) electron beam exposure causes an increase in molecular weight of the poly-p-vinylphenol matrix, (2) the dissolution rate of poly-p-vinylphenol in the developer decreases with increasing molecular weight, and (3) there is no fractional dissolution during the development of the resist. It is concluded that non-swelling development, non-fractional dissolution in the proceeds by an etching-type mechanism rather than the fractional dissolution mechanism that occurs with most organic-laser developers.

Many reports have been published on negative electron-beam resists. Most of these resists utilize radiation-induced gel-formation as the insolubilzation reaction. However, a major problem with these resists, is that their resolution is limited by swelling which is induced by the developer during development.

Novolac- or phenolic resin-based resists usually show no pattern deformation induced by swelling during development in aqueous alkaline solution. Examples of such resists are naphthoquinonediazide/novolac positive photoresists, novolac-based positive electron-beam resist (NPR) (1), and azide/phenolic negative deep-UV resist (MRS) (2). Iwayanagi et al.(2) reported that the development of MRS proceeds in the same manner as the etching process. This resist, consisting of a deep-UV sensitive azide and phenolic resis matrix, is also sensitive to electron-beams. This paper deals with the development mechanism of non-swelling MRS and its electron-beam exposure characteristics.

0097–6156/87/0346–0077$06.00/0
© 1987 American Chemical Society

Experimental

Materials. The resist used in this study consisted of poly(p-vinylphenol) as the phenolic resin matrix and 3,3'-diazidodiphenyl-sulfone as the sensitizer. The composition was essentially the same as that of RD-2000N (trade name of Hitachi Chemical Co.'s MRS). Two kinds of poly(p-vinylphenol) were used. One was commercially available from Maruzen Oil Co., under the name of RESIN M, and the other was synthesized in our laboratory. The latter polymer, mostly with high molecular weight, was prepared by hydrolyzing poly(p-acetoxystyrene) with a hydrazine(3). The poly(p-acetoxystlyrene) was obtained by polymerization of the monomer using 2,2'-azobisisobutyronitrile as an initiator. The alkaline developers used were aqueous tetramethylammoniumhydroxide (TMA) solutions. Isoamylacetate was used as the organic solvent developer.

Characteristics. Electron-beam exposure experiments were carried out by using a prototlype HL-600 Hitachi Electron-Beam Lithography System which is a vector scanning type variable-shaped electron-beam machine. The acceleration voltage was 30 kV. Resist films were formed on silicon wafers by spin-coating and prebaked at 80 C for 20 min before exposure.

The molecular weight distributions were measured by gel permeation chromatography (GPC) with a Hitachi 635 liquid chromatography slystem equipped with Gelpack Al50, Al40, and Al20 GPC columns (Hitachi Chemical Co.). The GPC solvent was tetrahydrofuran.

Film thickness was determined with an Alpha Step 200 (Tencor) profilometer.

The dissolution rate of the sample films during development was measured by laser-interferometry(4). A 5 mW He-Ne laser was used as the monitering light source and a silicon photodiode connected to a chart recorder was used as the signal detector.

Results and Discussion

In the resists containing phenolic resin and bisazide, electron-beam exposure resulted in the production of primarly amines and increased the molecular weight of the phenolic resin (5). A typical example is shown in Fig. 1. In this figure, peaks 1 and 2 are bisazide and the poly(p-vinylphenol) matrix, respectively. Peak 3 is a primary amine produced from the decomposition of bisazide. The primary amine does not affect the solubility of the resist (6). Since high-molecular-weight phenolic resins are less alkaline-soluble than low-molecular-weight resins (7), it follows that the reduced solubility must be due to the increase in molecular weight. The change in molecular weight distribution of phenolic resin contained in MRS following electron-beam exposure is shown in Fig. 2. The molecular weight of the resin increased as exposure proceeded. In a previous paper(8), we concluded that the photo-induced decomposition of azide in the phenolic resin matrix causes an increase in the resin's molecular weight due to cross-linking. This scheme is the same mechanism as that in the insolubilization of cross-linking negative resist. Therefore, as

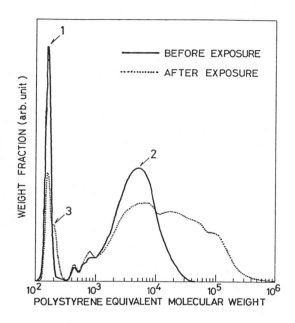

Fig. 1. Molecular weight distributions of MRS before and after electron-beam exposure. Dose: 50 μC/cm². Peaks 1, 2, 3 are azide, poly(p-vinylphenol), and primary amine, respectively.

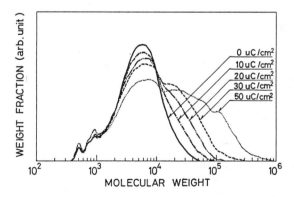

Fig. 2. Molecular weight distributions of poly(p-vinylphenol) matrix in electron-beam-exposed MRS films as a function of electron dose.

predicted in such cases, high sensitivity can be expected when a higher molecular weight phenolic resin is used as the matrix.

Figure 3 shows the electon-beam exposure characteristics of two MRS samples using high molecular weight poly(p-vinylphenol) (Mw=7.4x10^4) and standard molecular weight poly(p-vinylphenol) (Mw=6.7x10^3) as matrices. Azide content in each resist was 16.7% by weight. The developers were aqueous alkaline solutions. The development of each resist was stopped immediately after the unexposed areas were dessolved by the developer. The increase in sensitivity shown in Fig. 3 is smaller than expected given the assumption that the sensitivity is proportional to the weight average molecular weight of the resist. Some pattern deformation due to swelling was seen in the case of high molecular weight MRS. In convlentional negative electron-beam resists, the gel-point dose is inversely proportional to the weight average molecular weight (9). In order to determine the gel-points of these MRS samples, the resist films were developed with organic solvent. The exposure characteristics thus obtained are shown in Fig. 4. Gel-points were 2.3 μC/cm^2 for the high molecular weight MRS, and 70 μC/cm^2 for the standard molecular weight MRS.

As shown in Fig. 3 and 4, the exposure characteristics of high molecular weight MRS are similar when developed in different developers. In addition, patterns in high molecular weight MRS developed with organic solvent are deformed as a result of swelling. To elucidate this phenomenon, the relationship between dissolution rate in alkaline developer and the molecular weight of the polymer was investigated over a wide range of molecular weight. The results are summarized in Fig. 5. The dissolution rate of poly(p-vinylphenol) in the alkaline solution decreased rapidly with increasing molecular weight up to 10^4. However, the decrease tended to saturate after the molecular weight exceeded 10^4. This is the reason the high molecular weight MRS was not as sensitive as expected when developed in alkaline developer.

On the contrary, the standard molecular weight MRS exhibits a large difference in exposure characteristics between Fig. 3 and 4. In order to investigate this difference, the molecular weight of poly(p-vinylphenol) in the exposed MRS film was measured as a function of electron-beam dose. The results are shown in Fig. 6. The molecular weight of the matrix-resin increased after the electron-beam dose exceeded 5 μC/cm^2. Within the dose range of 5 to 20 μC/cm^2, the molecular weight ranged from 7x10^3 to 1x10$\{$. This molecular weight range corresponds to the rapidly changing range of the dissolution rate in Fig. 5. This dissolution rate change gave the standard molecular weight MRS much higher sensitivity in alkaline developer comared with the gel-point dose.

Figure 7 shows SEM photographs of the standard MRS fine patterns delineated by electron-beams. As shown in the figure, no swelling occured during the development of MRS having a standard molecular weight matrix, because the dissolution process is an etching-like process, i.e, the resist film dissolves from the surface gradually without any fractional dissolution. We have previously reported a similar non-fractional development process (10). In order to confirm the non-fractional dissolution, molecular weight distributions were measured for electron-beam-exposed MRS

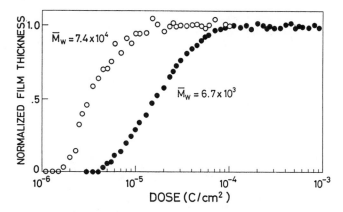

Fig. 3. Exposure characteristics of MRS obtained by aqueous alkaline development. o: molecular weight MRS (Mw=7.4x10⁴). ●: standard molecular weight MRS (Mw=6.7x10³).

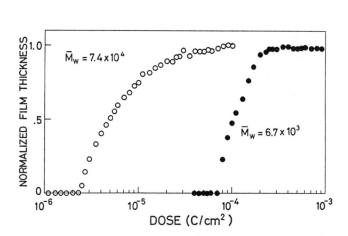

Fig. 4. Exposure characteristics of MRS obtained by organic solvent development. o: high molecular weight MRS (Mw=7.4x10⁴). ●: standard molecular weight MRS (Mw=6.7x10³).

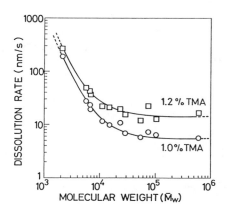

Fig. 5. Dissolution rate of poly(p-vinylphenol) as a function of
molecular weight. o: 1.0% tetramethylammonium hydroxide(TMA)
aqueous solution. □ : 1.2% TMA aqueous solution.

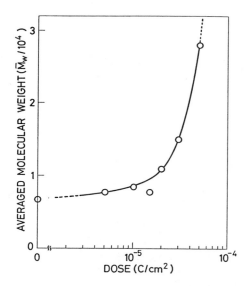

Fig. 6. Molecular weight of poly(p-vinyl phenol) in exposed MRS as
a function of electron-beam dose.

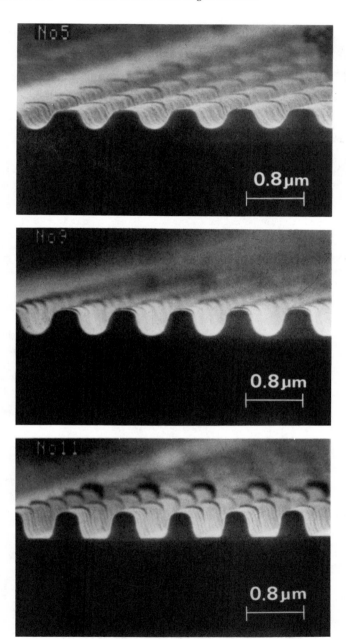

Fig. 7. SEM photographs of MRS patterns during alkaline development. Patterns: 0.4 μm line and space. Dose: 40 μC/cm^2. Developer: 0.72 wt.% TMA solution.

films before and after development. The results are shown in Fig. 8. The exposure dose was 15 $\mu C/cm^2$, and development was stopped so as to leave 50% of the initial film thickness remaining in the exposed areas. The molecular weight distribution of MRS before development was almost the same as that after half-development. It is concluded that there is no evidence of fractional dissolution in which lower molecular weight components are dissolved preferentially.

Conclusion

Electron-beam exposure characteristics of the negative deep-UV resist (MRS) were investigated when the resist was used as a phenolic resin-based negative electron beam resist. In particular, the development process was studied in an aqueous alkaline solution by measuring the dissolution rate of phenolic resin and the molecular weight distributions of electron-beam-exposed resist films. The molecular weight of the resin matrix increased continously as exposure proceeded. The dissolution rate of the resin matrix decreased rapidly with increasing molecular weight in the molecular weight range below 10^4, and the decrease tended to saturate when molecular weight exceeded 10^4. When a standard molecular weight phenolic resin matrix and alkaline developer were used, no fractional dissolution was seen to occur during the development of the resist. This development process corresponded to an etching-like development of the resist.

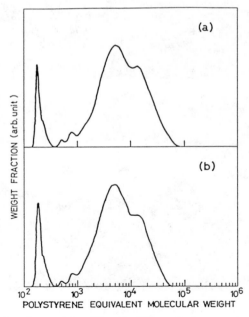

Fig. 8. Molecular weight distributions of electron-beam-exposed MRS resist. Dose: 15 $\mu C/cm^2$. (a) before development, (b) after half development (50% remaining).

REFERENCES

(1) M. J. Bowden, L. F. Thompson, S. R. Fahrenholtz, and E. M. Doerries, J. Electrochem. Soc., 128, 1304-1313 (1981).

(2) T. Iwayanagi, T. Kohashi, S. Nonogaki, T. Matsuzawa, K. Douta, and H. Yanazawa, IEEE trans. Electron Devices, ED-28, 1306-1310 (1981).

(3) R. Arshady, G. W. Kenner, and A. Ledwith, J. Polymer Sci.: Polymer Chem. Ed., 12, 2017-2025 (1974).

(4) C. G. Willson, ACS Symp. Series 219 "Introduction to Micro-lithography", Ed. by L. G. Thompson, C. G. Willson, and M. J. Bowden, Chapt. 3, (1983).

(5) H. Shiraishi, T. Ueno, O. Suga, and S. Nonogaki, ACS Symp. Series 266 "Materials for Microlithography", Ed. by L. F. Thompson, C. G. Willson, and J. M. Frecht, 423-434 (1984).

(6) T. Iwayanagi, M. Hashimoto, S. Nonogaki, S. Koibuchi, and D. Makino, Polymer Eng. Sci., 23, 935-940 (1983).

(7) S. Nonogaki, M. Hashimoto, T. Iwayanagi, and H. Shiraishi, Proc. SPIE 539, Advances in Resist Technology and Processing II, 189-193 (1985).

(8) M. Hashimoto, T. Iwayanagi, H. Shiraishi, and S. Nonogaki, Technical Papers, SPE Regional Technical Conference "Photopolymers: Principles, Processes and Materials", pp. 11-33, Ellenville, New York, Oct. 28-30 (1985).

(9) H. Y. Ku and L. C. Scala, J. Electrochem. Soc., 116, 980-985, (1969).

(10) Y. Hatano, H. Shiraishi, Y. Taniguchi, S. Horigome, S. Nonogaki, and K. Naraoka, Proc. Symp. Electron and Ion Beam Sci. Technol. 8th Int. Conf., 332-340 (1978).

RECEIVED May 5, 1987

Chapter 8

Electron-Beam Sensitivity of Cross-Linked Acrylate Resists

Nigel R. Farrar and Geraint Owen

Hewlett–Packard Laboratories, Palo Alto, CA 94303–0867

Acrylate resists such as PMMA have excellent resolution
and contrast but poor sensitivity. Improved contrast
or sensitivity can be achieved by forming radiation
sensitive crosslinks in copolymers of PMMA by thermal
treatment on the wafer. In this work, we have evaluated
the performance of a commercial resist which is a mix
of two copolymers, and have explored methods for optimizing its use. We have examined the effect of altering
the crosslink density by controlling softbake temperature, using a flood exposure in addition to the
patterning exposure, diluting the resist with PMMA and
changing the copolymer mix ratio. All of these techniques lead to improvements in sensitivity at the expense
of contrast, with the most promising results being shown
by the mix variations. The optimum crosslink density
for acceptable contrast with maximum sensitivity has
been determined. However, the most satisfactory method
for achieving the reduced crosslink density is by controlling the chemical structure of the resist, since all
the methods explored in this work involve additional
process complexity.

Acrylate resists such as polymethylmethacrylate (PMMA) have been
used extensively in electron beam lithography because of their excellent resolution and contrast, despite their limited dry etch resistance and low sensitivity (1). Copolymers of PMMA, containing
chemical groups more sensitive to radiation induced degradation,
have also been studied and have shown up to a four-fold improvement
in sensitivity (2). One approach has been to form a crosslinked gel,
in-situ on the wafer, which contains radiation sensitive crosslinks
and leads to improved sensitivity and improved contrast during
development (3-7).
    Various crosslinked acrylate resists have been reported in the
literature, mainly based on methacrylic acid anhydride (MANH) crosslink units. In the simplest case, the anhydride crosslinks may be
formed by baking methacrylic acid (MAA) homopolymer, although this

0097–6156/87/0346–0086$06.00/0
© 1987 American Chemical Society

requires a temperature of 230°C and is not suitable for all situations (8). One approach has been to carry out the dehydration reaction in the solid phase (9). The resulting material contains both inter- and intra-molecular anhydride. The crosslinked material can be filtered out of solution leaving a sensitive terpolymer which is completely soluble. This material is difficult to prepare and does not offer the advantage of an insoluble unexposed phase to generate high contrast.

MAA-methacryloyl chloride (MAC1) reactions have been shown to form anhydride crosslinks at lower temperatures, although the starting material is more complex as it must comprise two copolymers (3-4), or a copolymer and terpolymer (5-6). Both materials make use of the greater sensitivity of the acid anhydride group but differ in the contrast mechanism involved. Roberts (3) shows that anhydride is destroyed during exposure and correlates sensitivity with anhydride content. Kitakohji et. al. (5) also associate the anhydride group with the material sensitivity but retain excess acid as a means of improving contrast. MMA-MAA and MMA-MAC1 copolymers have also been used separately to form crosslinked resists by reaction with a difunctional monomer (10-12).

Of the reported materials, the only commercially available resists are one of the MMA-MAA/MMA-MAC1 compositions and a t-butyl methacrylate copolymer material (13). In this work, we have chosen to examine the copolymer mixture, which crosslinks at a temperature of 160°C compared to 250°C for the single copolymer. We have carried out experiments to optimize its use for direct write electron beam lithography.

Theory

The absorbed energy density, E_g, required to destroy the gel completely in pre-crosslinked resists can be predicted as follows. Assume that the density of the material is $\rho kg/m^3$, the monomer molecular weight is M_0 and that Avogadro's number is $N_a/(kg.mole)$. The number of monomer units per m^3 is $N_a.\rho/M_0$. If the crosslink density (i.e. the fraction of monomer units which are crosslinked) is d_0, then the number of crosslinks per unit volume is

$$N_a.\rho.d_0/2M_0$$

The factor of two appears since one crosslink joins two monomer units. At the point at which the gel is destroyed, the crosslink density is $1/y_w$ where y_w is the weight average degree of polymerization of the resist (14). It is assumed that irradiation destroys only crosslinks, not main chain bonds, and that y_w remains constant. If the crosslink density in the unexposed resist is d_0, then the number of crosslinks per unit volume which must be broken to destroy the gel is

$$\rho.N_a.(d_0-1/y_w)/2.M_0$$

If the energy required to destroy a crosslink is ε_x, then the absorbed energy per unit volume, E_g, required to destroy the gel is given by

$$E_g = \rho.N_a.\varepsilon_x.(d_0-1/y)/2.M_0 \tag{1}$$

This equation is of the same form as one proposed by Suzuki and Ohnishi (12).

Thus, sensitivity depends strongly on the crosslink density, c_0, which is controlled by the fraction of crosslinkable units in the material and the extent of the crosslinking reaction during baking. In the present work, the number of crosslinkable sites on each copolymer was fixed so different methods of changing the cross-link density were explored. Three approaches were used: 1) the extent of the crosslink reaction was controlled by varying the bake conditions, 2) the crosslink reaction was carried out to completion and then the crosslink density was modified by a subsequent process step, 3) the total number of crosslink sites was altered and the reaction was allowed to proceed to completion.

Experimental

The resist used in most of these experiments was Isofine E-B Positive Resist PM-15 purchased from MicroImage Incorporated, Orange, Connecticut. The two copolymers which comprise this material are MMA-MAA and MMA-MAC1, with an MMA content of 90% in both cases, and are mixed in equal proportions. The copolymers were also provided as separate solutions by MicroImage Technology Ltd., Riddings, Derbyshire, England. Both copolymers were reported, by the manufacturer, to have a number average molecular weight of about 25,000.

Resist films of approximately 0.5μm thickness were spun on silicon wafers and crosslinked by baking either in an oven or on a hotplate. Incremental exposures were made by a JEOL JBX6A2 electron beam machine at 20 keV. The UV flood exposures were carried out under nitrogen using a 185nm UV lamp. UV dosimetry was carried out on the basis of exposure time which had previously been correlated with the equivalent electron beam exposure by measuring dissolution rates.

Clearing doses were determined by immersing the wafers in a strong solvent (acetone) for two minutes. Standard dip development was carried out by immersion in MIBK (methyl-isobutyl-ketone) at 21°C for time increments from one minute to ten minutes. The dissolution rate was calculated from thickness loss, measured using a Nanospec, and development time. SEM examination of test structures was used to evaluate the resolution of the resist under the different processing conditions. Unless specifically mentioned, each experiment showed that 0.5μm features could be resolved in the resist.

Results and Discussion

(a) Performance of As-received Material. The commercial PM-15 resist was processed according to the manufacturer's specifications with a 30 minute, 160°C softbake and development in MIBK. It was found that the dose required to destroy the gel in the exposed regions was 50-85μC/cm². The dose is feature dependent due to the proximity effect, caused by electron backscattering from the substrate. An isolated exposed line requires 85μC/cm² to clear because it receives virtually no backscattered energy while a large pad requires only 50µC/cm^2 because it receives substantial additional deposited energy from backscattered electrons. For crosslinked positive resists, the isolated lines are the most difficult features to clear and require the greatest dose. Therefore, the sensitivity of the resist will be given as the clearing dose for an isolated exposed line.

At the clearing dose, the value of relative dissolution rate, S_{rg}, is very large, indicating excellent contrast. The S_{rg} parameter (15) is a measure of the rate at which dissolution rate increases with dose. Its value at any dose is defined as the dissolution rate at that dose divided by the rate at 40% of that dose, which is equivalent to the relative energy density between an isolated unexposed region and the large adjacent exposed pads. This represents the worst case situation for resist contrast in electron beam lithography. The value of S_{rg} can be related to the feature profile of the resist and previous work has indicated that a value of S_{rg}=8.7 corresponds to the good line profile characteristic of PMMA exposed at 80µC/cm^2 (15). For PM-15, values well over 100 were obtained, indicating a level of contrast much greater than required.

It was clear that the initial crosslink density of the resist was much higher than necessary to generate acceptable contrast and was leading to reduced sensitivity. The subsequent experiments were designed to improve sensitivity by reducing the initial crosslink density. Although this was expected to degrade contrast, it was felt that an adequate level of contrast could be maintained while the other resist parameters were optimized.

(b) Effect of Bake Temperature. The PM-15 resist was baked at temperatures between 110°C and 160°C in order to vary the extent of the crosslinking reaction. The bake time was held constant at 30 minutes at each temperature. Above 160°C the crosslinking reaction appeared to have reached completion and the resist performance was not sensitive to changes in bake temperature and time in this region. However, at lower temperatures than 160°C, the dissolution rate of material exposed at a given dose decreased with increasing bake temperature, as shown in figure 1. Also, the gradient of the curves, which determines the S_{rg} parameter, decreased with decreasing bake temperature. The minimum bake temperature required for the resist to have an equivalent contrast to PMMA was between 120°C and 130°C, as seen when comparing the dissolution rate curves to the S_{rg}=8.7 line in figure 1. However, the resist thickness loss in unexposed regions became greater at lower temperatures, see figure 2, and the optimum bake temperature was found to be 130°C. At this temperature, the clearing dose was 20-35µC/cm^2 (depending on feature), as shown in figure 3, and the sensitivity was 35µC/cm^2.

Since the crosslinking reaction at these temperatures had not proceeded to completion the resist characteristics were potentially more susceptible to bake fluctuations. Although our results were quite consistent, a process in which the material was baked at 160°C was felt to be more desirable.

(c) Changing the Crosslink Density by Flood Irradiation. Wafers were prepared as in section (a) and then flood exposed by deep UV or electron beam irradiation to reduce the crosslink density before the

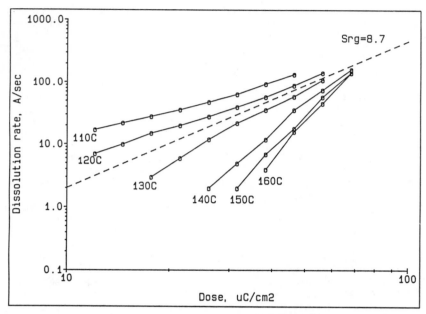

Figure 1.   Dissolution Rate vs Dose for PM-15 Resist Baked at
Various Temperatures.

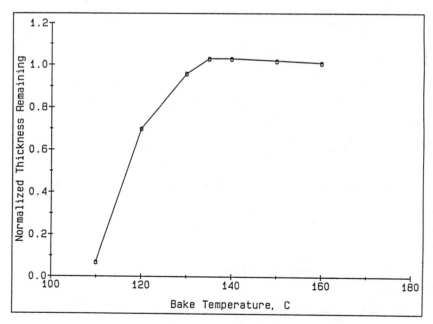

Figure 2.   Thickness Remaining vs Bake Temperature for Unexposed
Resist.

patterning exposure. This procedure reduces the patterning dose required but degrades the contrast of the resist. There are optimum values of the flood exposure, Q_f, and patterning exposure, Q_p, such that Q_f is as large as possible and Q_p as small as possible without degrading the contrast to an unacceptable level.

The type of feature requiring the greatest exposure to destroy the gel is an isolated line, because it receives less backscattered energy than any other type of feature. From the results of section (a) it is known that the required dose, Q_c, is $85\mu C/cm^2$ in the absence of flood exposure. For a flood exposure dose, Q_f, the new patterning dose, Q_p, required to clear the line is

$$Q_p = Q_c - Q_f(1+n_e)$$

where n_e is the backscattered energy coefficient for the substrate. For silicon and 20keV electrons, $n_e = 0.73$. Hence

$$Q_p = 85 - 1.73Q_f$$

and is plotted as line A in figure 4. For the exposed regions to clear during development Q_f and Q_p must be chosen such that Q_p lies above this line.

An isolated unexposed space is the type of feature with the worst possible contrast in electron beam lithography since it receives more backscattered energy than any other type of feature (16). For such a feature the ratio of the energy deposited in the adjacent exposed region to that deposited in the nominally unexposed region, after both flood and patterning exposures, is

$$\frac{(1+n_e).(Q_f+Q_p)}{(1+n_e).Q_f + n_e.Q_p} \qquad (2)$$

From the experimental data of section (a), it is known that this ratio must have a minimum value of 1.63, equivalent to $S_{rg}=8.7$. Equation 2 thus reduces to

$$Q_p = 1.97Q_f$$

and gives the relationship between the flood and patterning exposures for acceptable contrast. This is plotted in figure 4 as line B. Thus, the flood and patterning exposure doses must be chosen such that Q_p lies above both lines A and B.

The allowed working region is shown in figure 4 and indicates that the minimum possible patterning exposure dose is $45\mu C/cm^2$, at which the flood exposure required is $23\mu C/cm^2$. This prediction was confirmed by experiment, with a $45\mu C/cm^2$ patterning exposure required when a $25\mu C/cm^2$ (or equivalent UV dose) flood exposure was used.

However, during these experiments, it was observed that, for a given total dose (flood + patterning), the dissolution rate depended on the flood/patterning exposure dose ratio. This effect appears to occur in several acrylate resists and is illustrated for PMMA resist in figure 5. Shiraishi et. al. (17) observed a similar effect in a two component resist and explained the result on the basis of two

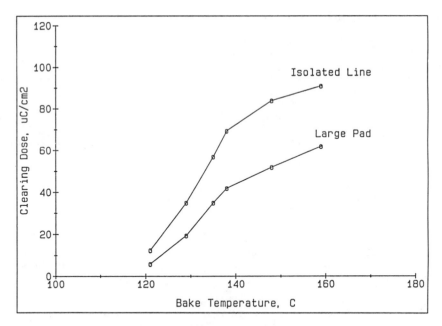

Figure 3. Clearing dose vs. bake temperature for PM-15 resist.

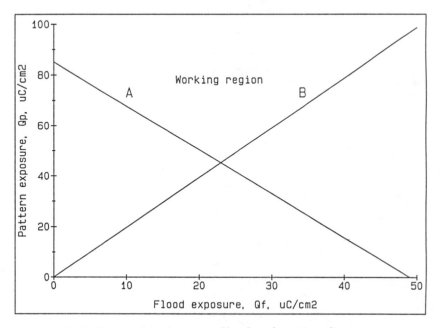

Figure 4. Relationship between flood and patterning exposures.

competing reactions taking place after exposure. They postulated
that the effect of irradiation is to produce free radicals and that
their concentration decreases with time due to both a first order
and second order reaction. These reactions are associated with free
radical stabilization and free radical recombination respectively.
This concept may be applied to bond scission resists since permanent
bond scission is a first order process and recombination is a second
order process. After all the radicals have reacted, several hours
after exposure, the number of bond scissions per unit volume, N_1, is

$$N_1 = K.\ln(1 + k.C_0)$$

where C_0 is the initial concentration of free radicals and K and k
are related to the rate constants of the first and second order
reactions respectively.

This result may be extended to the case of two or more exposures several hours apart. If the initial free radical concentrations are αC_0 and $(1-\alpha)C_0$, for two exposures, then the number of
bond scissions, N_2, several hours after both exposures is

$$N_2 = k.\ln\{1 + k.C_0 + \alpha(1 - \alpha)C_0{}^2\}$$

It is clear that $N_2 > N_1$ for all values of α between 0 and 1. It is
evident that N_2 reaches a maximum when $\alpha = 1/2$ and that the relationship is symmetrical about this value. This behavior describes the
graph of figure 5 quite well since the dissolution rate of PMMA
depends on the total number of bond scissions. The effect is also
dependent on the time intervals and wafer history between the two
exposures.

The double exposure complications introduced by the first and
second order reaction mechanisms in acrylate resists led to the
conclusion that the flood and patterning exposure process was not a
practical solution to the problem of high crosslink density in PM-15
resist.

(d) Diluting the Resist with PMMA. Solutions of PM-15 resist and
PMMA resist (M_w = 496,000) were mixed in equal proportions in order
to reduce the total number of crosslinking sites and hence reduce
the crosslink density after a 160°C bake. The maximum clearing dose
was reduced to $55\mu C/cm^2$ and contrast was preserved with $S_{rg} > 8.7$.
However, from SEM pictures of the test resolution patterns, it was
clear that the structure of the resist film was not uniform and
that there were regions of differing solubility. This may have been
due to non-uniform crosslinking or incompatibility of the polymers
in the mixture.

(e) Mixing the Copolymers in Unequal Proportions. In order to overcome the incompatibility problem observed in the PMMA mixtures, the
total number of crosslink sites was reduced by mixing polymers with
known compatibility. The MMA-MAC1 and MMA-MAA copolymers were
mixed in different ratios (1:2, 1:4, and 1:10) and baked at 160°C.
For unequal amounts of chloride and acid groups, the crosslink
density has a lower value than that for PM-15, in which the mix
ratio is 1:1, and may be calculated as follows. Assume that the

ratio of chloride to acid copolymer molecules is 1:n, and that n>1.
Assume that the chloride and acid copolymers both have a number
average degree of polymerization, y_n, and that they both have a
fraction, f, of crosslinkable units (f=0.1 for PM-15). For a
collection of N polymer molecules, there are $N/(n+1)$ chloride molecules and $Nn/(n+1)$ acid molecules. Since there are fewer of them,
it is the chloride molecules which control the crosslink density
because when all the available chloride units have reacted no
further crosslinking is possible. The total number of chloride units
is $y_n.f.N/(n+1)$ and, when all of these have reacted, the total
number of crosslinked units in the N molecules will be

$$2.y_n.f.N/(n+1)$$

since each chloride unit will be crosslinked to an acid unit. The
total number of monomer units is $y_n.N$ and so the crosslink density,
d_0, will be

$$d_0 = 2.f/(n+1) \tag{3}$$

The points in figure 6 show the values of d_0 for the different mixes
plotted against the clearing dose for a large exposed pad. Equation
3 was substituted into equation 1 and ε_x was used as a fitting
parameter to plot the lines through the data shown in figure 6.
E_g was converted to dose assuming that $1\mu C/cm^2$ is equivalent to
3.6×10^7 J/m^3. The dependence of clearing dose on crosslink density
agreed very well with the theory. The two sets of data for "new
material" and "aged material" correspond to resist films which were
spun from solution immediately after mixing and 137 days later, respectively. The properties of the spun film change with time and the
material appears to become more sensitive. This effect is discussed
further at the end of this section.

     As with the results in section (b), reducing the crosslink
density reduces the gel content in the unexposed regions and leads
to greater thickness loss during development. An estimate of this
effect may be made on the basis of a nearest-neighbor crosslinking
analysis. Assuming that i) all the chloride molecules are incorporated into the gel, ii) the polymer molecules are represented as
close-packed spheres and hence have twelve immediate neighbors, and
iii) neighboring acid and chloride molecules are crosslinked by at
least one bond, it follows that any acid molecule that has a
chloride molecule neighbor is a part of the gel. Therefore, an acid
molecule is only a part of the sol if it is completely surrounded
by other acid molecules. The probability that a molecule is an
acid molecule is $n/(n+1)$ and the probability that its neighbors are
all acid molecules is $[n/(n+1)]^{12}$. For N polymer molecules the
number of acid molecules completely surrounded by other acid molecules is

$$[N.n/(n+1)].\{[n/(n+1)]^{12}\}$$

These molecules constitute the sol, so the sol fraction is
$[n/(n+1)]^{13}$. Therefore the gel fraction is

$$1 - [n/(n+1)]^{13}$$

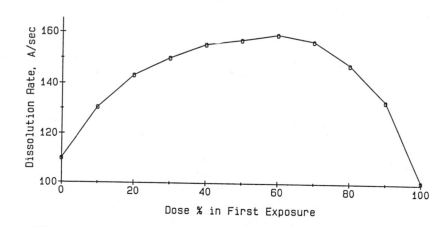

Figure 5.   Dissolution Rate vs Dose Ratio for Double Exposed PMMA.

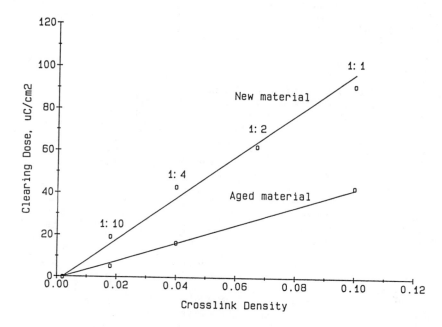

Figure 6.   Clearing Dose vs Crosslink Density for Unevenly Mixed Copolymers.

This curve is plotted as the line in figure 7 and shows good agreement with the experimental points.

For mixtures with a useful gain in sensitivity over standard PM-15 resist, dissolution rate was measured as a function of dose and is shown in figure 8. By comparison with the $S_{rg}=8.7$ line, it is seen that the contrast is adequate for both the 1:4 and 1:10 mixes at about $20 \mu C/cm^2$. This dose is higher than the clearing dose for the 1:10 mixture but necessary for acceptable contrast. However, the resolution and thickness remaining for this mixture are inadequate, but may have been degraded by using an inappropriate developer. Because the polarity of the material increases as the excess MAA content increases, further work is required to optimize the development system for each mixture. The 1:4 mixture meets the contrast and resolution requirements but could probably be further improved in terms of sensitivity. It appears that a mixture between the 1:4 and 1:10 ratios would offer the optimum crosslink density to maximize sensitivity and retain sufficient contrast.

A concern that remains for these materials is that the performance of films cast from the mixed solutions changes over time, with a decrease in the clearing dose of about 30% after 100 days from the preparation of the resist, as shown in figure 9. However, the properties appeared to approach a stable level. This effect had obviously occurred in the commercial premixed material also, since the clearing dose had stabilized close to that of the aged 1:1 solution, and was much lower than the clearing dose for the freshly mixed solution. Hydrolysis of the chloride groups by atmospheric moisture could explain a variation in performance but not the apparent stabilization at longer times.

Conclusions

The commercially available PM-15 resist containing 10% crosslinkable groups shows no improvement in sensitivity over PMMA although, at the operating dose, the contrast is superior. Reducing the crosslink density by changing the bake conditions leads to a sensitivity improvement of 60% by sacrificing the previously high contrast. Other methods of reducing the crosslink density have also led to sensitivity improvements, of which the most promising is the uneven copolymer mixtures with a potential 60-70% reduction in operating dose. However, the solutions must be aged after mixing in order to obtain reproducible performance. This technique also requires work on a developer system. Flood exposure yields a 50% improvement in sensitivity but creates additional processing difficulties due to double exposure effects.

A comparison of the results is shown in Table I. It is clear that a reduced crosslink density is desirable for improved sensitivity. This must be optimized to ensure acceptable contrast and minimal thickness loss in unexposed regions. Although all the methods explored in this work gave improved sensitivity and acceptable contrast, it appears that the simplest and most reproducible method for achieving lower crosslink density is to modify the

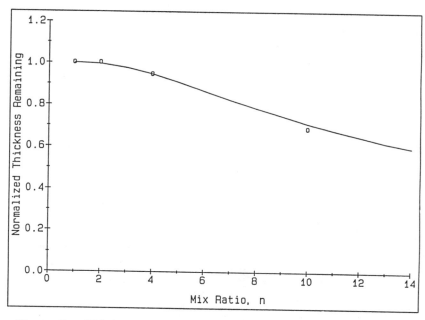

Figure 7. Thickness Remaining vs Mix Ratio for Unevenly Mixed Copolymers.

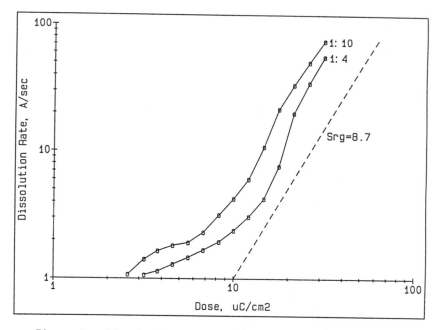

Figure 8. Dissolution Rate vs Dose for Unevenly Mixed Co-polymers.

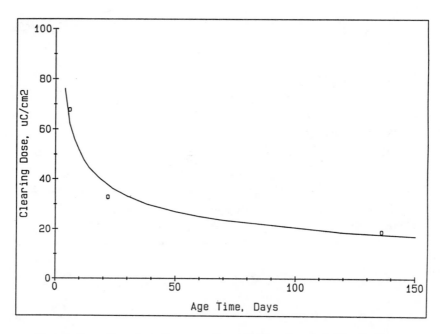

Figure 9.   Clearing Dose vs Age of Mix for 1:4 Mix Ratio.

structure of the resist by reducing the number of crosslinkable groups in both copolymers. The crosslinking reaction can then be carried to completion, which reduces its sensitivity to bake temperature fluctuations. Also, the gel structure should be more homogeneous and the polarity more uniform than the unevenly mixed copolymers. If such resists were available, as copolymer mixtures or a single terpolymer, they would be very attractive candidates for $20\mu C/cm^2$ electron beam lithography.

Table I. Summary of Results

Material	Clearing dose, $\mu C/cm^2$ (Isolated line)	S_{rg} Contrast (at clearing dose)	Resolution
As received (160°C bake)	85	200	<0.5µm
As received (130°C bake)	35	9	<0.5µm
PMMA mixture	55	30	poor
1:4 mixture	30	15	<0.5µm
1:10 mixture	15	6	poor
Flood exposed ($25\mu C/cm^2$)	45	9	<0.5µm

Acknowledgments

Thanks are due to Marsha Long for the results in section (e). Thanks are also due to Judith Seeger and Marge McDermott for electron beam exposures.

Legend of Symbols

Symbol	Units	Value	Definition
N_a	1/kg.mole	6×10^{26}	Avogadro's number
ρ	kg/m^3	1200	Density of the resist
M_0		100	Monomer molecular weight
y_w		500	Weight average degree of polymerization
f		0.1	Fraction of crosslinkable units
n_e		0.73	Backscattered energy coefficient Si, 20keV

Literature Cited

1) Hatzakis, M. J. Electrochem. Soc. 1969, 116, 1033.
2) Moreau, W. M. Proc. SPIE Mtg. on Submicron Lithography 1982, 333, 2.
3) Roberts, E. D. ACS Preprints for Organic Coatings Mtg. 1973, 33, 359.
4) Roberts, E. D. Applied Polymer Symposium 1974, 23, 87.
5) Kitakohji, T. et al. J. Electrochem. Soc. 1979, 126(11), 1881.
6) Kitamura, K. et al. Fujitsu Sci. Tech. J. 1980, 16(1), 89.

7) Harada, K. et al. <u>IEEE Trans. on Electron Dev</u>. 1982, <u>ED-29(4)</u>,
    518.
8) Hatzakis, M. J. <u>Vac. Sci. Tech</u>. 1979, <u>16(6)</u>, 1984.
9) Moreau, W. M. et al. <u>J. Vac. Sci. Tech</u>. 1979, <u>16(6)</u>, 1989.
10) Asmussen, F. et al. <u>J. Electrochem. Soc</u>. 1983, <u>130(1)</u>, 180.
11) Suzuki, M. et al. <u>J. Electrochem. Soc</u>. 1985, <u>132(6)</u>, 1390.
12) Suzuki, M.; Ohnishi, Y. <u>J. Electrochem. Soc</u>. 1982, <u>129(2)</u>, 402.
13) Saeki, H.; Kohda, M. <u>Proc. 11th Symp. Sem. Int. Cir. Tech</u>. 1979.
14) Flory, P. J. <u>Principles of Polymer Chemistry</u>; Cornell: Ithaca,
    1953, Ch. 9
15) Rissman, P.; Owen, G. <u>J. Vac. Sci. Tech</u>. 1985, <u>B3(1)</u>, 159.
16) Owen, G.; Rissman, P. <u>J. Applied Phys</u>. 1983, <u>54</u>, 3573.
17) Shiraishi, H. et al. <u>ACS Symposium 242</u>, 1984, 167.
18) Owen, G. <u>Rep. Prog. Phys</u>. 1985, <u>48</u>, 795.

RECEIVED May 1, 1987

Chapter 9

A "One-Layer" Multilayer Resist

R. D. Allen, S. A. MacDonald, and C. G. Willson

Almaden Research Center, IBM, San Jose, CA 95120-6099

Multilayer resist technology offers a number of advantages in the generation of relief images but carries the burden of process complexity. We wish to report a novel process that greatly simplifies the optical MLR sequence. This concept is based on selective surface modification of the resist with a reactive dye which masks selected areas toward later flood exposure and solvent development.

The use of multilayer resist (MLR) schemes (1) in the printing of complex sub-micron structures has experienced a continuing evolution during the past decade. Early multilayer processes were highly complex, consisting of three or more functional layers, each of which performed one specific task (2,3). The utility of these complex multilayer schemes spurred a widespread interest in process simplification. Advances in synthetic polymer chemistry resulted in new materials which combined the function of two or more layers employed in the original complex schemes. This was accomplished through advances in the synthesis of organometallic polymers, specifically structures containing silicon (4-7) and tin (5,8). These organometallic polymers are designed to be both radiation sensitive and oxygen etch resistant. This combination of properties has lead to the current state of the art — the two layer MLR. We believe that this methodology can be further simplified. The goal of this work is to demonstrate the design of a "one layer" MLR process.

Multilayer resist technology, in its most simple form, involves the imaging of a top, thin resist layer, followed by pattern transfer through the bottom "planarization" layer (Scheme I). Multilayer resist technology offers a number of advantages in the generation of relief images. Among these advantages are: insensitivity to topographic features, the ability to generate high aspect ratio images, and the opportunity to eliminate image distortion resulting from incident radiation reflecting off the substrate. The disadvantages of MLR technology are associated with the complexity introduced by the extra steps required to process the additional resist layers.

Currently two methods are utilized for transfer of the relief image from the top resist layer to the bottom planarizing layer; oxygen reactive ion etching (RIE), and optical flood exposure (1). Oxygen RIE systems have not found widespread acceptance due to the high cost and low throughput of RIE equipment. Optical transfer processes, while less cost- and time-intensive, have been slow in acceptance due to interfacial mixing problems, the added processing steps required to coat and bake two resist layers, and the practical problem of dealing with two resists and two developers on a manufacturing line.

The lithographic design strategy in this work is to develop a one layer MLR process in which optical pattern transfer techniques are employed. Taylor and coworkers (9) first discussed the benefits of near-surface functionalization as an imaging strategy, taking advantage of limited permeation of inorganic RIE-resistant reagents. These workers also alluded to the possibilities of surface functionalization by very low voltage e-beam (≤ 5 keV) and photolithography using very strong absorbers to surface expose materials in a one layer MLR-type scheme. A very recent communication by Coopmans and Roland (10) discusses the advantages of near-surface functionalization of an as yet undisclosed resist by moderate doses (50-100 mj/cm^2) of near UV light. After selectively silylating the exposed areas, pattern transfer by oxygen RIE is performed. Optical pattern transfer of two layer MLR systems was accomplished in B. J. Lin's well known work on the "Deep UV-PCM" (11), in which AZ1350 photoresist is imaged on top of a deep UV resist (in this case PMMA). Lin took advantage of the absorbance characteristics of novolak resins, i.e., their extreme opacity in the DUV. The imaged photoresist thus becomes a "portable conformable mask" in a subsequent DUV flood exposure. This concept inspired our work in the optical one layer MLR.

We have recently developed a methodology which combines the performance advantages of MLR systems with the processing ease associated with single layer resists. This process entails the imagewise surface irradiation of a resist to generate reactive functionalities near the air-resist interface. This exposed surface is then reacted with a "dye" which acts as a contact mask for a following flood exposure which transfers the image. The relief image is then developed in the usual manner. The specific system to be discussed employes a protected phenolic resin as the resist layer. Imagewise surface deprotection (acid catalyzed) provides phenolic hydroxyls which react with reagents containing isocyanate functionalities. The surface carbamates (urethanes) formed in this process act to mask the resist toward DUV irradiation.

Experimental

Resist films (1-2 microns) containing t-butoxycarbonate protected phenolic resin and appropriate onium salt photoinitiators were spin coated on quartz wafers (UV analysis), sodium chloride plates (IR) and silicon wafers (imaging). Near-surface irradiation with deep UV light was performed with a mercury-arc lamp through 200 and 220 nm bandpass filters (Oriel).

Urethane forming reactions were run in the gas phase, by placing the appropriate wafers in a temperature controlled vacuum oven connected with a mechanical pump and fitted with a septum-capped injection port through which reagents were introduced. Triethyl amine was injected into the evacuated oven at a temperature exceeding 100°C. After several minutes, excess amine was removed and phenyl isocyanate introduced. (Experience dictates that this two step injection procedure together with removal of excess amine prevents major residue buildup inside the vacuum chamber.) The isocyanate is allowed to react for several minutes, followed by re-evacuation of the chamber. After the wafers are removed from the oven, baking on a hot plate is generally performed to remove any volatile reagents which may have diffused into the film during the earlier reaction sequence. The wafers are then flood exposed with 254 nm filtered light, post baked, and spray developed with alcoholic solvents.

Instrumental analysis was performed with an IBM IR/32 (FTIR), Hewlett Packard 8450A UV/Vis spectrophotometer, and a Tencor Alpha Step 200 (film thickness). Photomicrographs were taken with a Reichert-Jung Polyvar-met optical microscope and electron micrographs with a Philips SEM505.

Results and Discussion

The basic principals of the one layer MLR process are as follows:
1. Coat the substrate with a single layer of an appropriate resist.
2. Expose the resist with radiation (UV light, e-beam, etc.) in a fashion such that radiation does not reach the substrate, and confine the radiation induced chemical changes near the air-resist interface. As first introduced by Taylor and co-workers in 1984, this can be accomplished with UV light by exposing the resist at a wavelength where the film is strongly absorbing, or by using e-beam irradiation modified by the use of a retarding potential (9).
3. The exposed resist is treated with a reagent which reacts preferentially with either the exposed, or the unexposed regions of the film. This reagent must be a "dye" such that after the film has been treated, the treated areas absorb strongly at some wavelength where the original resist is both sensitive and significantly transparent.
4. The dyed film is flood exposed at a wavelength where the dyed portions are strongly absorbing and the original resist is both sensitive and significantly transparent.
5. The relief image is developed in the usual manner with an appropriate solvent.

The overall sequence of this process is shown in Scheme II. The net effect of this process is to first generate a mask in the surface of the resist and then to contact print that mask down into the underlying material. The resist can be any structure that will undergo a radiation induced transformation that results in the generation of a functional reactivity difference between the exposed and the unexposed areas. Chemically amplified functional group deprotection reactions are ideally applied to the one layer MLR process, due to the high sensitivity and large functional reactivity differences generated with such resist systems (5,12). For illustrative purposes, we have shown a negative-tone process. The concepts described can conceivably be used to produce a positive-tone image, by selecting a dye reagent that will react with the surface of the unexposed film. In addition, a positive-tone image may also be obtained by development of the dyed and underlying unexposed regions with an appropriate solvent.

We first demonstrated the workability of this concept with a t-butoxycarbonyl protected polyvinyl phenol/onium salt resist. As this resist system is strongly absorbing below 240 nm, imagewise exposure with 220 nm filtered light should confine the irradiation to the air-resist interface. Very low doses of 220 nm filtered light (e.g., 0.5 mj/cm^2) clearly show the presence of both carbonate protecting group (1755 cm^{-1}) and unprotected phenol (3200-3500 cm^{-1}) in the infrared spectrum after post-exposure baking (Figure 1). The surface phenolics are capable of a variety of reactions. We have chosen to react the phenolic with aromatic isocyanates,

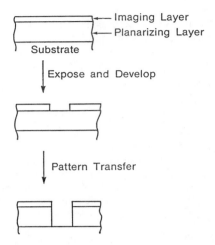

Scheme I. The two layer MLR process.

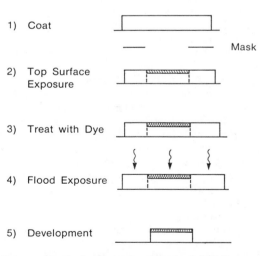

Scheme II. Optical transfer: one layer MLR.

principally due to the high vapor pressure of these reagents (reactions are often run in the gas phase), their high reactivity with phenolics, and the high optical density (at 254 nm) of the aromatic carbamates formed in the reaction. For example, phenyl isocyanate is successfully reacted with poly(p-hydroxystyrene) derived from acid catalyzed deprotection of poly(p-t-butoxycarbonyloxy-styrene) as shown in Scheme III. The IR spectra of the surface phenolic and derived urethane are shown in Figure 2. A material with very high optical density at wavelengths shorter than 260 nm results from this carbamate formation. The UV spectra of the starting polymer, the deprotected phenolic, and the carbamate modified polystyrene are shown in Figure 3. The extreme absorptivity at 254 nm demonstrates that these carbamate modified materials are excellent "mask" candidates for our one layer MLR process.

The principles of the process chemistry have been demonstrated by successfully imaging relatively large structures using the one layer MLR process with the materials mentioned above (13). Optical micrographs demonstrate the capability of making fine line 1.25 μm (nominal) contact printed structures (Figure 4). Further characterization by SEM reveals a curious "footing" phenomenon (Figures 5 and 6) which may be due to the inadequate masking of the edges of the "surface mask."

Conclusions

Acid catalyzed resist chemistry involving functional group deprotection (onium salt mediated chemical amplification) is ideally suited for one layer MLR methodology. Low doses (<1 mj/cm^2) of deep UV light (200-220 nm) have been used to selectively deprotect the upper regions of a protected phenolic resin. The functional reactivity change has been utilized to imagewise react dye molecules into the upper regions of the film. Phenyl isocyanate is reacted with the phenolic, forming carbamates which show high absorptivity to 260 nm. Aromatic isocyanates thus are useful dye

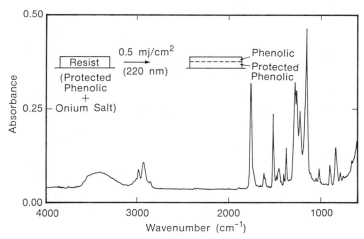

Figure 1. Infrared spectrum of surface irradiated carbonate protected polyvinyl phenol/onium salt resist.

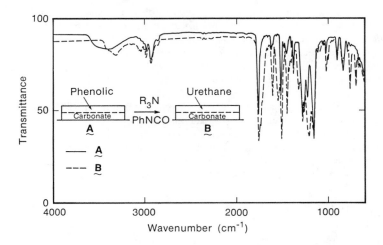

Figure 2. Infrared spectra of surface irradiated resist A and the dye modified masked resist B.

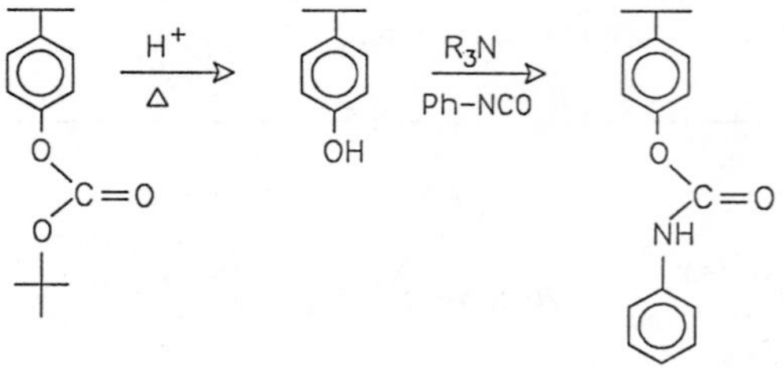

Scheme III. Reaction sequence employed in the one layer MLR process.

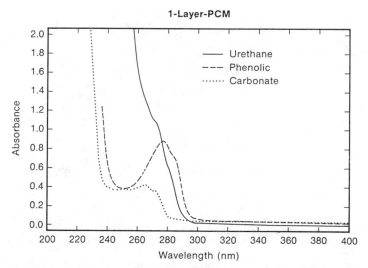

Figure 3. UV spectra of the carbonate protected phenolic, after surface irradiation with 500 μJ/220 nm light, and after reaction with phenyl isocyanate. Note the extreme absorptivity at the flood exposure wavelength of 254 nm.

Figure 4. Optical micrograph of developed contact print formed by the one layer MLR process.

Figure 5. SEM micrograph of contact printed images formed by the one layer MLR process.

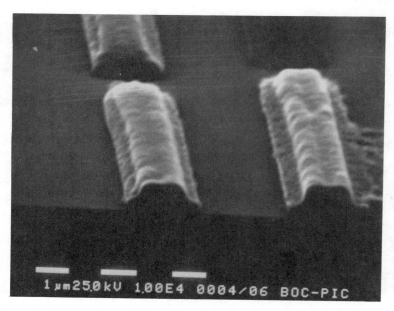

Figure 6. SEM micrograph of contact printed images formed by the one layer MLR process.

candidates due to their additional qualities of suitable vapor pressure and reactivity. Flood exposure at 254 nm followed by development with polar solvents leads to negative image formation. Fine lines are resolved, but wall profiles show "footing" problems. The reasons for this unusual behavior are currently under investigation.

Literature Cited

1.   Lin, B. J. "Multi-Layer Resist Systems," in Introduction to Microlithography; Thompson, L. F.; Willson, C. G.; Bowden, M. J., Eds.; ACS Symposium Series No. 219; 1983; Chapter 6.

2.   (a). Franco, J. R.; Havas, J. R.; Rompala, L. J. U.S. Patent 4 004 044, 1975; (b). Havas, J. R. Electrochem Soc. (Extended Abstracts) 1976, 76(2), 743.

3.   Moran, J. M.; Maydan, D. J. Vac. Sci. Technol. 1979, 16, 1620.

4.   Shaw, J. M.; Hatzakis, M.; Paraszczak, J.; Liutkis, J.; Babich, E. Polym. Eng. Sci. 1983, 23, 1054.

5.   MacDonald, S. A.; Ito, H.; Willson, C. G. Microelectronic Engineering (North-Holland) 1983, 1, 269.

6.   Suzuki, M.; Saigo, K.; Gokan, H.; Ohnishi, Y. J. Electrochem. Soc. 1983, 130, 1962.

7.   Reichmanis, E.; Smolinsky, G. SPIE Proceedings, Advances in Resist Technology, 1984, 469, p 38.

8.   Labadie, J. W.; MacDonald, S. A.; Willson, C. G. J. Imaging Sci. 1986, 30(4), 169.

9.   Taylor, G. N.; Stillwagon, L. E.; Venkatesan, T. J. Electrochem. Soc. 1984, 131(7), 1658.

10. Coopmans, F.; Roland, B. SPIE Proceedings, Advances in Resist Technology and Processing III, 1986, 631, p 34.

11. (a). Lin, B. J. SPIE Proceedings, 1979, 174, p 114; (b). Lin, B. J. J. Electrochem. Soc. 1980, 127, 202.

12. (a). Ito, H.; Willson, C. G. "Applications of Photoinitiators to the Design of Resists for Semiconductor Manufacturing," in Polymers in Electronics; Davidson, T., Ed.; ACS Symposium Series No. 242; 1984; Chapter 2; (b). Willson, C. G.; Ito, H.; Fréchet, J. M. J.; Tessier, T. G.; Houlihan, F. M. J. Electrochem. Soc. 1986, 133(1), 181.

13. Allen, R. D.; MacDonald, S. A.; Willson, C. G. ACS PMSE Preprints, Anaheim Meeting , 1986, 55, p 293, Figure 3.

RECEIVED April 8, 1987

Chapter 10

New Silicon-Containing Electron-Beam Resist Systems

E. Reichmanis, A. E. Novembre, R. G. Tarascon, and A. Shugard

AT&T Bell Laboratories, Murray Hill, NJ 07974

The incorporation of silicon into resist systems has
been shown to effectively instill oxygen etching
resistance while maintaining lighographic utility.
Such materials may be used as the imaging layer for
two-level processes involving RIE pattern transfer
through a planarizing layer of organic polymer. We
have used the trimethylsilylmethyl appendage to
effect oxygen etching resistance in both positive and
negative e-beam resist systems. Compatible silyl
novolac/polyolefin sulfone blends afford sensitive,
high resolution resists, while the inherently positive acting trimethylsilylmethyl methacrylate can be
copoloymerized with chlorinated styrenes to yield
negative resists capable of submicron resolution.
The synthesis and radiation chemistry of these materials is discussed, in addition to a brief analysis
of their lithographic properties.

Multi-layer resist systems have secured an important role in the
fabrication of devices with geometries of 1.0 μm or less (1).
However, the complexity associated with these processes must be
simplified. Since the introduction of tri-layer systems (2,3), an
obvious simplification is to combine the properties of both the
upper resist imaging layer and the oxygen reactive ion etching
(RIE) resistant masking layer (typically SiO$_2$) into one. One mechanism to accomplish this is to incorporate metal atoms into polymeric materials that function as resists (4-7). Organosilicon
species have been shown to provide excellent oxygen RIE resistance
and are typically copolymerized with other monomers that effect the
radiation sensitivity necessary for imaging (8).

Unfortunately, the incorporation of silicon into polymeric
resists can alter the desirable materials characteristics. A
decrease in the glass transition temperature often accompanies the
inclusion of silicon into a resin, and most silicon substituents
will drastically change the solubility properties of the parent
polymer. For example, polymerization of propylpentamethyldisiloxyl
methacrylate affords rubbery, low T_g polymers that are

inappropriate for resist applications (9). While other components may be incorporated into the system to raise the T_g, the synthesis could become complex if enhanced radiation sensitivity is also desired. Silicon substituents are also typically hydrophobic in nature such that if a base soluble resin is desired, there will often be a delicate balance between the amount of silicon that can be incorporated into the system to effect RIE resistance and yet allow aqueous base solubility.

We have successfully employed the trimethylsilylmethyl appendage to effect oxygen RIE resistance in both positive and negative acting electron-beam resist systems (10,11). The relatively compact nature of this substituent allows the preparation of glassy polymers useful for lithographic applications. The preparation and characterization of select trimethylsilylmethyl substituted resists will be presented in addition to a study of their radiation chemistry and lithographic properties.

Experimental

Materials. Trimethylsilylmethyl methacrylate (SI) and chloromethylstyrene (CMS) (mixed m,p isomers) were obtained from Petrarch Systems Inc. and Dow Chemical Co. Inc., respectively. Both monomers were purified by distillation at reduced pressure. Copolymers were prepared by free-radical solution polymerization at 85°C in toluene. Reactions were initiated using benzoyl peroxide.

Trimethylsilylmethylphenol and o-cresol were obtained from Petrarch Systems, Inc. and Aldrich Chem. Co. Inc., respectively. Silylated novolac (SI-novolac) resins were prepared by condensation polymerization of p-trimethylsilylmethyl phenol, o-cresol and formaldehyde. Poly(2-methyl-1-pentene-sulfone) (PMPS) was prepared as described in the literature (12).

Characterization. P(SI-CMS), polystyrene equivalent molecular weight was determined by high pressure size exclusion chromatography (HPSEC) using a Water's Model 510 pump, 401 differential refractometer, and duPont bimodal silanized columns; SI-novolac number average molecular weight was determined by vapor phase membrane osmometry (Wescan Instruments, Inc., Model 232A). Glass transition temperatures were determined using a Perkin-Elmer DSC-2 differential scanning calorimeter. The temperature heating rate was 10°C/min with sample masses ranging from 10 to 20 mg.

Reactive Ion Etching. Etching experiments were carried out in an Applied Materials Model 8110 Hex reactor. Alternatively, a Cook Vacuum Products Inc. Model C71 RF/DC Sputtering Module was employed. Film thickness measurements were taken before and after etching to determine etching rates, and the rates were typically compared to that of the novolac-diazoquinone photoresist, HPR-206, baked at 210°C for 1 hour. Measurements were obtained on a Dektak Model IIA profilometer.

P(SI-CMS) resist films, 0.5 to 1.0 μm thick, were spun from 10-12 w/v % solutions of the polymers in 2-methoxyethyl acetate. Silylated novolac-PMPS (SI-NPR) resist solutions were prepared by dissolving the SI-novolac (10 wt%) and PMPS (0.9 wt%) into isoamylacetate. All resist solutions were consecutively filtered through a 0.5 and 0.2 µm Teflon filter stack (Millipore, Inc.). Resist solutions were spun onto four inch silicon substrates that were either bare, or precoated with 1.5 µm of Hunt photoresist (HPR-206) baked at 210°C for 1 hour in air. Prior to exposure, the resist coated substrates were given a 30 to 60 min. prebake at 80 to 100°C in air. The negative resist was post-develop-baked at 120-150°C.

Lithographic Characterization. Electron-beam exposures were conducted on an EBES system operating at 20 kV, with a beam address and spot size both equal to 0.25 µm. Electron response parameters were evaluated using linewidth control patterns. P(SI-CMS) was spray developed after exposure using an APT Model 915 resist processor in toluene-methanol (1:1) for 30 sec followed by a methanol rinse for 45 sec. Aqueous solutions of tetramethylammonium hydroxide (TMAH, 25% in water, Fluka Inc.) were used for the novolac resist development. Exposed films were dip-developed for 20 sec. in TMAH-water (1:2.5) solutions.

Film thicknesses remaining after exposure and development were measured optically using a Nanometrics Nanospec/AFT micro area thickness gauge, and a Dektak IIA profilometer. Scanning electron micrographs (SEM) of processed patterns were taken using either a JEOL IC 35CFS or Cambridge Stereoscan 100 scanning electron microscope.

Results and Discussion

Materials Synthesis and Characterization. In addition to the requirements of etching resistance, sensitivity, solubility and high glass transition temperature (T_g), one of the criteria used in designing both a negative and positive electron-beam resist system was synthetic simplicity. The trimethylsilylmethyl appendage allows the incorporation of silicon into polymeric resists without adverse synthetic complications. Standard free radical or condensation polymerization techniques can be employed with appropriately substituted monomers that are readily available.

While poly(trimethylsilylmethyl methacrylate) is inherently a positive acting resist system (13), the silylated monomer can be readily copolymerized with such moieties as chloromethyl styrene to generate crosslinkable polymers. A series of P(SI-CMS) polymers (Figure 1) were prepared and the resulting materials properties are listed in Table I. Using the Fineman-Ross treatment (14), the reactivity ratios for SI and CMS were 0.49 and 0.54, respectively. A slightly greater mole percentage of CMS was therefore observed in the copolymer with respect to the monomer feed composition. Due to the compact nature of the silylated ester group, only a limited reduction in T_g is observed for these systems. The silylated homopolymer exhibits a T_g of 68°C and copolymerization with chloromethyl styrene effects an increase up to 78°C for the polymer containing 54 mole% SI. These values of T_g are typical of many negative resist systems and should not affect image stability during lithographic processing.

The trimethylsilylmethyl unit may also be incorporated into phenolic resins that are components of solution inhibition positive

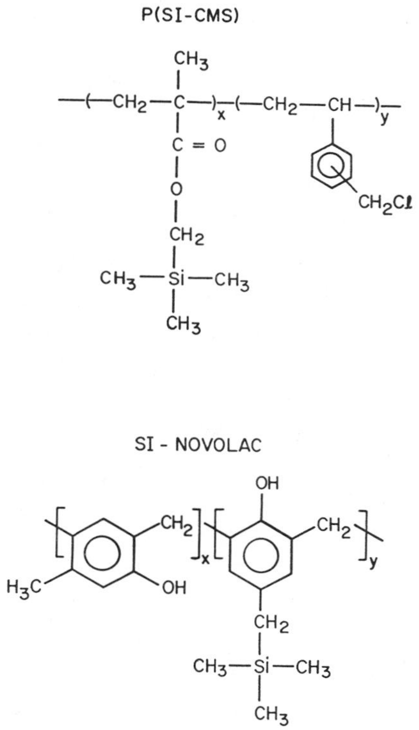

P(SI-CMS)

SI - NOVOLAC

Figure 1.   Schematic representation of P(SI-CMS) and silylated novolac.

resists. A schematic representation of the resin is shown in
Figure 1. The methylene spacer between the aromatic ring and silyl
species in trimethylsilylmethyl phenol eliminates the problem of
Si-C bond cleavage observed in other systems (15). While
condensation polymers of the silylated phenol and formaldehyde were
insoluble in aqueous base because of the hydrophobic nature of the
silyl moiety, incorporation of o-cresol afforded alkaline soluble
resins with up to ~ 10 wt % silicon. Polymer molecular parameters
for these systems are also given in Table I. While the T_g's are
lower than desired, no adverse effects on lithographic imaging were
observed.

Table I. Polymer Molecular Parameters

Polymer	Composition	wt.% Si	$\overline{M}_w$ (x 10^{-5})	M_w/M_n	$\overline{M}_n$	T_g (°C)
P(SI–CMS)	0.54:0.46[a]	6.7	0.85	2.3	–	78
	0.66:0.34[a]	10.3	1.16	1.9	–	76
	0.91:0.09[a]	14.6	1.87	2.1	–	70
	1.0:0[a]	16.3	2.17	2.2	–	68
SI–Novolac	0.3:0.7[b]	9.7	–	broad	750	26
	0.6:0.4[b]	5.4	–	broad	404	26
	0.8:0.2[b]	2.5	–	–	–	–

[a] Given as the ratio of SI:CMS in the polymer, as determined by
   elemental analysis.

[b] Given as the ratio of o-cresol to trimethylsilylmethylphenol in
   the feed.

Oxygen RIE Behavior. Oxygen RIE treatment of the silylated polymers prepared above leads to the generation of a surface layer of
SiO_x, presumed to be SiO_2. Auger sputter depth profiles (Figure
2) of the silylated novolac reveals that a 30-50Å surface layer of
oxidized silicon is in fact formed. The thickness of this layer is
an estimate based on the sputtering rate of SiO_2 (20Å per min)
under similar conditions. The composition of that layer is primarily silicon and oxygen, with some carbon also present. Removal of
resist via sputtering, effects a decrease in the oxygen and oxidized silicon signals with a concomitant increase in carbon intensity. Untreated SI-novolac exhibits silicon signals at 84 and 1614
EV; oxygen RIE treatment effects a shift to 77 and 1609 EV. The
signals at 77 and 1609 EV are typical of oxidized silicon (SIO_x).
     The etching properties of the materials described earlier were
evaluated as a function of silicon content, and the results are

shown in Figure 3.   The relation is non-linear for both resist systems and becomes asymptotic at high percentages of silicon. Examination of the plots of etching rate ratios vs. wt. % silicon indicates that small incremental increases in silicon content, in the range of 10-16%, will effect large increases in etching ratios between a silicon resist and planarizing layer.   Note the increase in selectivity from 12 to 20 for the P(SI-CMS) copolymer containing 14.6 wt % Si and silyl methacrylate homopolymer that contains 16.3 wt% Si, respectively.

Etching ratios for silylated polymers are typically highly dependent on the etching conditions employed.   Changes in oxygen pressure, RF power, and DC bias will effect changes in both etching rate ratios between resist and planarizing layer, and etching anisotropy.   For example, a decrease in bias voltage from -550V to -200V leads to an improvement in selectivity from 4.5 to 17 for a P(SI-CMS) polymer containing 10 wt % silicon.   However at -200V, the etching process contains a significant isotropic component and an unacceptable linewidth loss will be observed.   While etching at -350V will afford highly anisotropic profiles, the selectivity of 6.5 observed for the 10 wt % Si polymer is inadequate for sub-micron pattern transfer, and the higher silicon content material must be employed.   It is interesting to note that the wt. % silicon in a polymer is not the only factor in determining oxygen RIE selectivities.   The novolac resin containing only 10 wt% silicon etches at approximately the same rate as the 14% Si P(SI-CMS) poly-mer.   Clearly, such factors as polymer structure and composition play an added role in determining RIE chemistry (16).   Oxygen etch-ing rates as a function of bias voltage for two P(SI-CMS) polymers and the silyl novolac containing 10 wt% Si are given in Table II.

Table II.   Oxygen RIE selectivity as a function of bias voltage

Resist	wt % Si	Etching Rate Ratio (HPR:Resin)		
		-500V	-350V	-200V
HPR-206	0	1	1	1
P(SI-CMS)	10.3	4.5	6.5	17
P(SI-CMS)	14.6	11	12	17
SI-Novolac	9.7	3.5	11	>11

Lithographic Characteristics.   The exposure response curves for P(SI-CMS) and SI-novolac containing PMPS (SI-NPR) are shown in Figure 4, and their lithographic characteristics are summarized in Table III.   The sensitivities are 2 $\mu C/cm^2$ ($D_g^{0.5}$) and 8 $\mu C/cm^2$ for P(SI-CMS) and SI-NPR, respectively.   In the case of SI-NPR, PMPS affords radiation sensitivity via spontaneous unzipping after expo-sure (17).   The mechanism is equivalent to that of NPR and the lithographic characteristics are quite similar.   No evidence of incompatibility was observed.

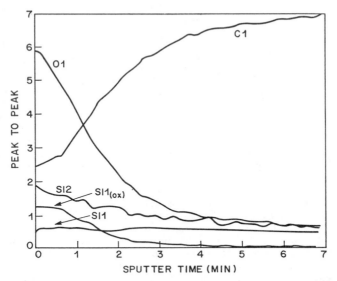

Figure 2.   AES sputter depth profiles for C, O, and Si
            (unoxidized (SI1 and SI2) and oxidized (SI1(ox)) of
            the surface of SI-novolac after O_2 RIE.

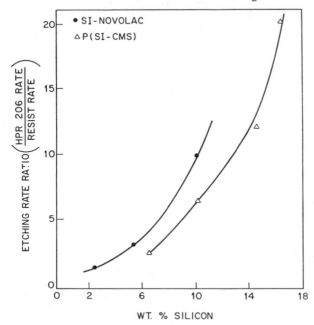

Figure 3.   Plot of etching rate ratios as a function of
            silicon content for P(SI-CMS) and silylated novolac
            (-350V, 4mtorr O_2).

Table III. Summary of Lithographic Characteristics

Resist	Sensitivity	γ	Resolution[a]
P(SI–CMS)	$2 \ \mu C/cm^2$	1.8	0.75 μm
SI–NPR	$8 \ \mu C/cm^2$	1.1	0.5 μm

[a]Minimum demonstrated coded line/space resolution.

As the dose requirements for crosslinking chloromethylated polystyrenes are significantly below those for methacrylate chain scission, P(SI–CMS) polymers are sensitive, negative acting resists. Ledwith, et al. have shown that only small amounts of CMS are required to effect over an order of magnitude enhancement in poly(methylstyrene) sensitivity (18), and the sensitivity observed for our copolymer is in the range of that observed for the methylstyrene-CMS copolymers of equivalent molecular weight. An approximation of the radiation chemical yields for P(SI–CMS) were determined as per the method described by Novembre and Bowmer (19). The calculated values of G(x) and G(s) are 1.26 and 0.5, respectively, and the Charlesby-Pinner (20) plot used to determine those values is shown in Figure 5. The relatively high value of G(s) is clearly attributable to the presence of an alkyl methacrylate. The efficiency of C-Cl bond cleavage leading to crosslinking of the polymer network is evidenced by the high value of G(x).

Submicron patterns have been generated in both resists and effectively pattern transferred through thick planarizing photoresist with little linewidth loss. Figure 6 depicts coded 1.0, 0.75 and 0.5 μm line-space patterns printed in SI–NPR prior to oxygen RIE treatment. Similar patterns have been obtained with P(SI–CMS) and Figure 7 shows the results of oxygen RIE pattern transfer of coded 1.0 and 0.75 μm images obtained with the negative resist.

Conclusion

Both negative and positive acting, oxygen RIE resistnt e-beam resist systems have been prepared through the incorporation of the trimethylsilylmethyl functionality into standard resist chemistry. Resins containing >10 wt% silicon display an RIE resistance more than 10 times greater than conventional photoresists and allow submicron pattern transfer with minimum linewidth loss.

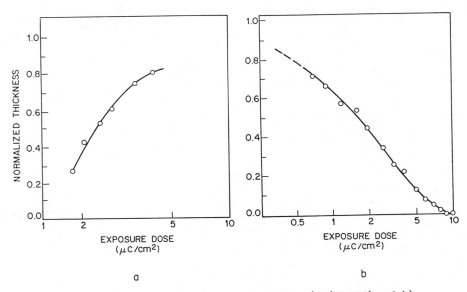

Figure 4.   Exposure response curves for a) P(SI-CMS) and b)
            SI-NPR.

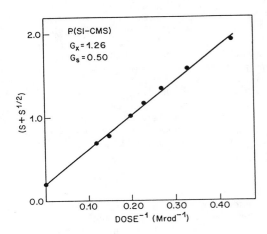

Figure 5.   Charlesby-Pinner plot of P(SI-CMS).

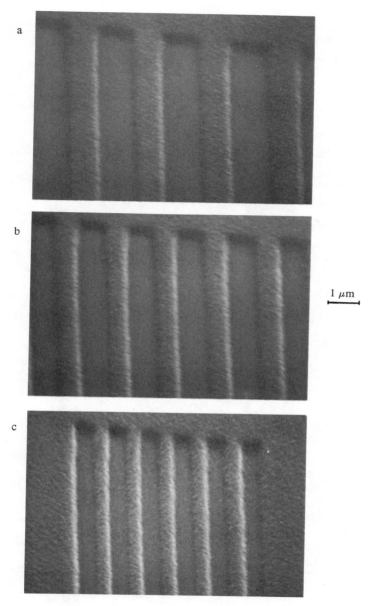

1 μm

Figure 6.   SEM micrographs depicting coded (a) 1.0, (b) 0.75 and (c) 0.5 μm line-space patterns printed in SI-NPR.

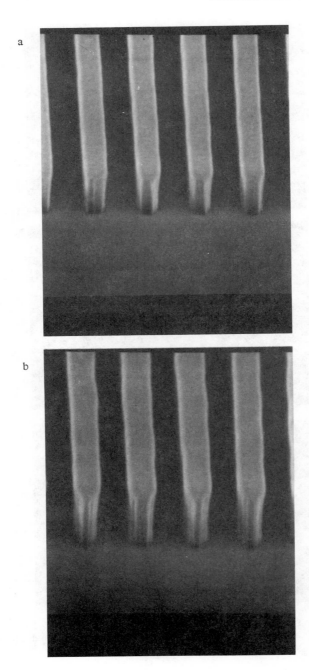

Figure 7. SEM micrographs depicting coded 1.0 (a) and 0.75 (b) μm line-space patterns printed in P(SI-CMS) followed by oxygen RIE pattern transfer.

Acknowledgments

The authors wish to thank T. G. Melone and C. Lochstampfor for e-beam irradiations, A. Kornblit for sample oxygen RIE determinations, and H. Luftman for Auger analyses.

Literature Cited

1. Lin, B. J. In "Intro. to Microlithography," Thompson, L. F.; Willson, C. G.; Bowden, M. J., eds.; ACS Symposium Series, 219, (1983).
2. Havas, J. Electrochem Soc. Extended Abstracts 1976, 76(2), 743-44.
3. Moran, J. M.; Maydan D. J. Vac. Sci. Technol., 1979, 16, 1620.
4. Hatzakis, M.; Paraszczak, J.; Shaw, J. In "Microcircuit Engineering 81," Oosenbrug, A., ed., Swiss Fed. Inst. of Technol., Lausanne, pp 386-96.
5. Ohnishi, Y.; Suzuki, M.; Saigo, K.; Saotome, Y.; Gokan, H. Proc. SPIE 1985, 539, 62.
6. MacDonald, S. A.; Steinmann, A. S.; Ito, H.; Hatzakis, M.; Lee, W.; Hiraoka, H.; Willson, C. G. "Int'l. Symp. Electron, Ion, Photon Beams," Los Angeles, CA, May 31-June 3, 1983.
7. Taylor, G. N.; Wolf, T. M. Polym. Engrg. Sci. 1980, 20, 1087.
8. Reichmanis, E.; Smolinsky, G.; Wilkins, C. W., Jr. Solid State Technol. Aug. 1985, 130.
9. Reichmanis, E.; Smolinsky, G. Proc. SPIE 1984, 469, 38.
10. Novembre, A. E.; Reichmanis, E.; Davis, M. Proc. SPIE 1986, 631, 14.
11. Tarascon, R. G.; Shugard, A.; Reichmanis, E. Proc. SPIE, 1986, 631, 40.
12. Bowden M. J.; Thompson, L. F. J. Appl. Polym. Sci. 1973, 17, 3211.
13. Reichmanis, E.; Smolinsky, G. J. Electrochem. Soc. 1985, 132(5), 1178.
14. Fineman, M.; Ross, S. D. J. Polym. Sci., 1950, 5, 259.
15. Wilkins, C. W., Jr.; Reichmanis, E.; Wolf, T. M.; Smith, B. C. J. Vac. Sci. Technol. 1984, B3, 306.
16. Ohnishi, Y.; Watanabe, F. J. Vac. Sci. Technol. 1986, B4, 422.
17. Bowden, M. J.; Thompson, L. F.; Fahrenholtz, S. R.; Doerries, E. M. J. Electrochem. Soc. 1981, 128 (6), 1304.
18. Ledwith, A.; Mills, M.; Hendy, P.; Brown, A.; Clements, S.; Moody, R. J. Vac. Sci. Technol. 1985, B3, 339.
19. Novembre, A. E.; Bowmer, T. N. In "Materials for Microlithography," Thompson, L. F.; Willson, C. G.; Frechet, J. M. J.; eds.; ACS symposium Series, 266, 1984, 241.
20. Charlesby, A. "Atomic Radiation and Polymers," Permagon Press, New York, 1960.

RECEIVED April 8, 1987

Chapter 11

Lithographic Evaluation of Poly(methyl methacrylate)-*graft*-poly(dimethylsiloxane) Copolymers

Murrae J. Bowden, A. S. Gozdz, C. Klausner, J. E. McGrath[1], and S. Smith[1]

Bell Communications Research, Red Bank, NJ 07701–7020

Poly(methyl methacrylate)-g-poly(dimethylsiloxane) graft copolymers have been shown to exhibit positive deep UV resist behavior and to function as a suitable etch mask for O_2 RIE pattern transfer in two-layer resist applications. Phase separation is restricted to a regime (100-200 Å) significantly below the size of normal lithographic features. There is some degree of phase miscibility, the extent of which depends on the precise molecular architecture and relative amount of the two components. Since poly(dimethylsiloxane) is considerably more sensitive to electron beam irradiation than poly(methyl methacrylate) and is negative acting, its presence in the continuous PMMA phase dictates the tone of the response in electron-beam lithography. Equal line and space patterns are difficult to generate because of the competing radiation chemistries of the block components.

There has been considerable interest in recent years in multi-layer resist processes as a means of overcoming the resolution limitations associated with single-layer lithography over topographical features on the surface of a wafer. In the multi-layer approach, a resist film, typically 1-2 μm thick, is first applied to the substrate in order to planarize the surface. A second resist layer is then coated on top of the planarized surface either directly (two-layer) or following prior deposition of an intermediate layer such as SiO_2 (three-layer). The top resist serves as the imaging layer and subsequent mask for pattern transfer into the underlying layer.

Several pattern transfer techniques have been developed, each of which places certain constraints on the design of the imaging resist. For example, in the case of transfer via oxygen reactive-ion etching (O_2 RIE) involving a two-layer resist structure, an etch rate ratio greater than 10:1 between bottom and top resist layers is required in order to achieve acceptable quality in the transferred image (1). The common resist materials are carbon-based polymers and their etch rates relative to those of Novolac-based materials which are commonly used for the planarizing layer, are less than 1.0. Thus, they are not suitable for RIE pattern transfer.

[1]Current address: Chemistry Department, Virginia Polytechnic Institute and State University, Blacksburg, VA 24060

Several years ago, Taylor et al. (2,3) showed that the oxygen plasma-etch resistance of carbon-based polymers could be markedly enhanced by incorporating certain atoms into the polymer chain. Particularly effective were those elements such as silicon or titanium that form a refractory oxide during oxygen RIE. The oxide is formed at the surface of the resist and greatly retards the subsequent etching rate of the remaining resist.

This approach has been applied to both positive and negative resists, particularly in the case of silicon because of the wide variety and availability of organosilicon compounds. Resists have been designed with the silicon moiety present as a substituent either on the side chain, e.g., as in the positive resist poly(3-butenyltrimethylsilane sulfone) (4), or in the main chain itself as in substituted polysiloxanes (5,6) or polysilanes (7,8). Since the objective is also to maintain high sensitivity, the overall design approach has generally been to incorporate both functions, viz., sensitivity and RIE resistance, through two complementary moieties. For example, one might design a random copolymer composed of two different monomer units, one of which provides radiation sensitivity while the other contains the refractory oxide-forming element to impart oxygen RIE resistance. This principle is exemplified by the negative resist poly(trimethylsilylstyrene-*co*-chloromethylstyrene) (9) and is by far the simplest design approach. Alternatively, one could prepare homopolymers from the respective monomers and simply mix them together as in the case of NPR resist (10,11). This approach has had only limited success because polymer mixtures or blends are generally incompatible and the resulting macroscopic phase separation results in poor lithographic performance.

Hartney *et al.* (12,13) proposed the use of block copolymers as a means to circumvent this problem of macroscopic phase separation. The chain segments in block copolymers are confined to microscopic domains, typically 100-200 Å in size, which are composed of similar segments with the covalent bond between segments being located at the interface between domains. The precise equilibrium morphology depends on the relative size of the individual blocks as well as the casting conditions (14). The key point is that the size of the phase-separated domains is well below that of typical lithographic features and thus does not limit image quality.

Hartney *et al.* (12,13) demonstrated the lithographic utility of block copolymers using poly(*p*-methylstyrene - dimethylsiloxane) block copolymers. The *p*-methyl group was subsequently chlorinated to enhance the sensitivity of the resist. Through control of block lengths, they were able to obtain a morphology resembling either poly(dimethylsiloxane) (PDMS) rubber spheres in a continuous glassy chlorinated poly(methylstyrene) (CPMS) matrix or a lamellae structure, and showed that good electron beam sensitivity, resolution, and reactive-ion etch resistance could be obtained. Since both components of the CPMS/PDMS block copolymer crosslink when irradiated with high-energy electrons or deep UV, it is not surprising that the resist was negative acting.

It should be possible to extend this principle to the design of a positive resist by making a degrading polymer the continuous glassy phase. Unfortunately, block copolymers with PDMS are prepared by living anionic polymerization techniques and there are no really suitable degrading polymers that can be conveniently formed into a block copolymer with PDMS. Methyl methacrylate can be polymerized anionically, but control of the molecular parameters is difficult because of the protic impurities present in most commercially available grades of monomer and the inherent side reactions associated with the ester functionality during anionic polymerization (15).

Copolymerization involving macromonomers offers an alternative convenient route to the synthesis of block and graft copolymers with well-defined phase-separated morphology. The anionic organolithium-initiated ring-opening polymerization of hexamethylcyclotrisiloxane allows the preparation of well-defined polymers with controlled molecular weight, narrow molecular weight distribution and functional termination (16). By terminating the living siloxanolate anion with a chlorosilane functional methacrylate monomer, Smith and McGrath (17) were able subsequently to copolymerize this macromonomer with methyl methacrylate in solution to produce a polymer consisting of a PMMA backbone with pendant PDMS grafts at varying intervals along the chain. Smith and McGrath reported average PDMS domain sizes in cast films of 12-21 nm (depending on the length of the PDMS "graft") which are of comparable size to the domains in the block copolymers of Hartney *et al.* (12,13). Since PMMA represents the continuous phase and degrades under high energy electron or deep UV radiation, we reasoned that such graft copolymers should function as positive resists in lithographic application. This paper reports the lithographic behavior of these materials under electron beam and deep UV irradiation.

EXPERIMENTAL

Poly(methyl methacrylate)-*g*-poly(dimethylsiloxane) copolymers were prepared by copolymerizing methyl methacrylate with a poly(dimethylsiloxane) macromonomer. The latter was prepared by anionic polymerization of hexamethylcyclotrisiloxane (D_3) initiated by *s*-butyl lithium, and terminated with 3-methacryloxypropyl-dimethylchlorosilane. Details of the synthesis are given elsewhere (17). The macromonomer was precipitated in methanol and dried under vacuum. Free radical copolymerizations were carried out in toluene at 60°C initiated with 0.1% azobisisobutyronitrile. The copolymers were extracted extensively with hexanes or isopropanol (depending on polymer solubility) to remove any PDMS homopolymer.

The polymers were dissolved in chlorobenzene, filtered, and films approximately 3000 Å-thick spin-coated onto silicon wafers and baked for 1 hr at 160°C. They were then exposed to an electron beam (AT&T Bell Laboratories EBES machine) at 20 keV. Doses ranged from 1 to 100 $\mu C/cm^2$. Alternatively, the samples were exposed to deep UV radiation at 260 nm using an Optical Associates Inc. (OAI) deep-UV exposure tool. The exposed patterns were developed by dipping in mixtures of methyl isobutyl ketone and isopropanol of varying strength whose composition depended on the structure (graft molecular weight and grafting ratio) of the copolymers.

Reactive-ion etch resistance was evaluated using a Cook Vacuum Products RIE system, Model C71-3 operating at 13.56 MHz. The etch resistance was determined for a two-layer configuration in which the resist was spun on 1.5 μm of a polyester planarizing layer (PC-1) supplied by Futurrex Inc. The latter etches approx. 30-50% faster than HPR-204 (~1500 Å/min compared to 1000 Å/min for HPR). The etching conditions were: Power: 0.16 W/cm^2, Bias: −350 V: Pressure, 20 mTorr O_2; Flow rate: 15 sccm O_2.

Differential Scanning Calorimeter (DSC) thermograms were obtained on a Perkin Elmer DSC 4 run at 10°C/min. Dynamic Mechanical Thermal Analysis (DMTA) spectra were obtained on a Polymer Labs DMTA at a frequency of 1 Hz with a temperature range from -150°C to +150°C at a scan rate of 5°C/min.

ESCA measurements were carried out on a Kratos XSAM-800 instrument with a Mg anode (200 watts) at a vacuum of 10^{-9} torr. TEM analysis was done on a Phillips IL 520T STEM at 100 kV in the TEM mode.

RESULTS AND DISCUSSION

The samples investigated in this study are shown in Table I. They cover a range of graft molecular weights with M_n varying from 1,000 to 20,000 (1-20K) and grafting ratios (characterized by the weight fraction of PDMS which varied from 5 to 45%). These two parameters dictate the properties of the copolymers. Figure 1 shows a plot of the glass transition temperature of the continuous (PMMA) phase as a function of graft molecular weight for a constant PDMS composition (25-31%) while Figure 2 shows the effect of composition at a constant graft MW of 10K. The T_g of PMMA is shifted to lower temperatures as the molecular weight of the PDMS graft decreases from 20K to 1K. Likewise, for a constant graft MW of 10K, the T_g decreases with increasing PDMS content. Similar results were observed by dynamic mechanical thermal analysis (17).

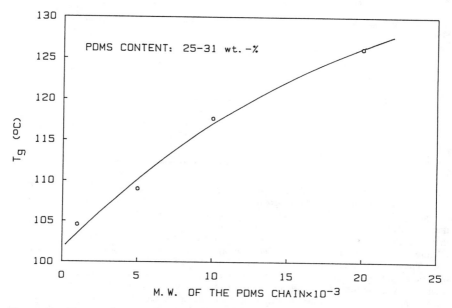

Figure 1. Glass transition temperature of PMMA in PMMA-*g*-PDMS copolymers as a function of molecular weight of the PDMS macromonomer at constant PDMS content.

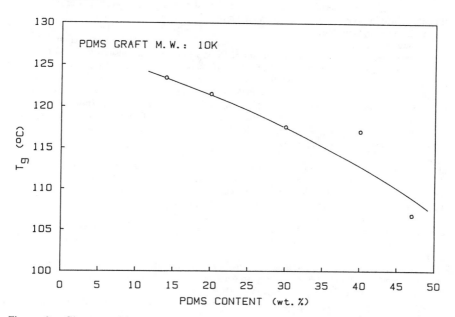

Figure 2. Glass transition temperature of PMMA in PMMA-*g*-PDMS copolymers as a function of PDMS content at constantgraft molecular weight (10K).

Table I. Characteristics of PMMA-*g*-PDMS Copolymers

Sample	M_n of PDMS	PDMS content (wt.-%)
7Q	1	32
7G	5	31
7L	10	30
11O	20	25
7J	10	14
7K	10	20
7L	10	30
11I	10	40
11J	10	47

As seen in Figure 1, the T_g of the graft copolymers only approaches that of PMMA for PDMS side chains on the order of 20K or greater. If complete phase separation were maintained on decreasing the length of the side chain, T_g would be expected to remain constant and equal to that of PMMA. Thus, the steady drop in T_g with decreasing length of the side chain implies that a degree of phase mixing exists. Similarly, for a constant 10K side chain, T_g decreases (Figure 2) as the total PDMS content increases again suggesting that phase mixing is occurring.

These results imply that the molecular weight of the PDMS graft needs to be on the order of 20K or greater in order to ensure complete phase separation. This would also seem to be desirable from the lithographic standpoint since this maximizes the T_g of the glassy continuous phase which is believed to be important in maintaining pattern acuity during development. There are however practical considerations which must be taken into account. Uniform samples with graft MW of this magnitude are difficult to prepare. A total molecular weight of 200,000 and side chain of 25K would correspond on average to only two grafted chains per main chain for a 25% PDMS composition. Given such statistical limitations, product uniformity can be difficult to attain. We can improve product uniformity statistically by reducing the molecular weight of the side chain but, as seen in Figure 1, this results in a greater degree of phase mixing and a lowered T_g. We can improve T_g by lowering the total PDMS content (Figure 2), but there clearly needs to be a minimum amount of PDMS to impart the desired RIE resistance.

The reactive-ion etch resistance of the graft copolymers is substantially improved over PMMA. Figure 3 shows a plot of film thickness remaining as a function of etching time for samples containing different amounts of PDMS. The graft MW was constant at 10K. Under the prevailing conditions, PMMA etched at a rate greater than 1800 Å/min., whereas the sample containing 25% PDMS of MW 20K etched at a rate of 50 Å/min. corresponding to an etch rate ratio of 36:1. These results are similar to those reported by Hartney *et al.* (12,13) for PCMS-PDMS block copolymers.

As seen in Figure 3, the initial etching rate is higher than the equilibrium rate which was attained after ~5 min. etching. We attribute this higher initial rate to the fact that a certain amount of time is needed to build up the protective oxide film. As expected, the etch rate decreased with increasing PDMS content (Figure 4), but only with a concomitant decrease in T_g (Figure 2). Such dichotomy again points to the fact that some compromise is required. If we define the acceptable etch rate ratio to be greater than 20:1, then the PDMS content must be $>25\%$, corresponding to $>9\%$ silicon in the polymer. This conclusion is in agreement with results of MacDonald *et al.* (18) who reported that a minimum of 10% silicon was needed to provide acceptable pattern transfer.

All samples yielded uniform, apparently homogeneous films when spun from chlorobenzene. SEM analyses show very smooth surfaces at 200K magnification indicating no features greater than 50 Å size. However, the surface develops a very distinct morphology during reactive-ion etching. Figure 5 shows evidence of spherical or worm-like features which are ~500 Å in diameter.

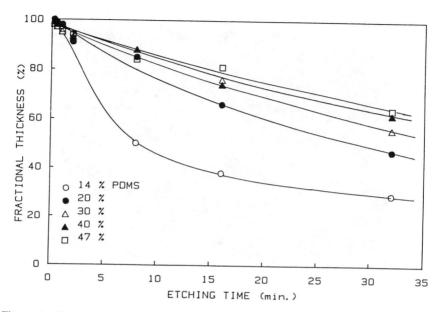

Figure 3.   Plots of thickness remaining vs. RIE time as a function of PDMS content at constant graft M.W. (10K).

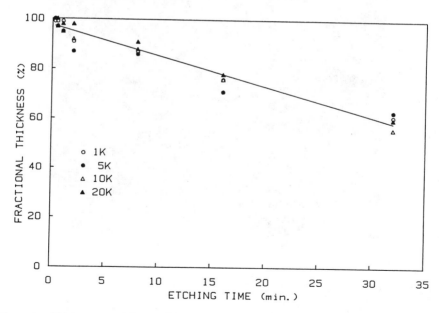

Figure 4.   Thickness remaining vs. RIE time as a function of PDMS graft M.W. for constant PDMS content (25-32%).

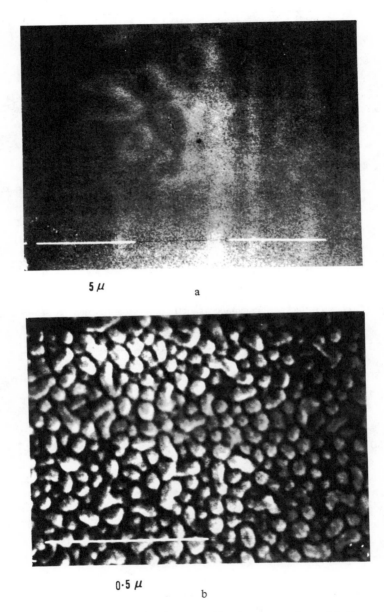

5 μ

a

0·5 μ

b

Figure 5.   SEM micrograph of the resist surface (a) after spin-coating and baking at 160°C;
            (b) after 5 min. O₂ RIE.

ESCA analysis of the surface prior to etching indicated that the surface composition is closely akin to that of PDMS. As seen in Table II, a sample which has been spun and dried under vacuum at room temperature (sample 26A) shows only 48.6% carbon when analyzed at a glancing angle of 10° increasing to 61.6% C at 90°. This leads us to conclude that the surface is PDMS rich, but deeper into the film, the composition approaches that of the block copolymer. Likewise, the silicon content decreases from 26.9% at 10° to 14.7 at 90°. Interestingly, post-baking at 160° for 1 hour appears to further concentrate PDMS towards the surface (sample 26B) in that the composition is now essentially invariant over all three glancing angles and is close to that expected for PDMS. These results are in agreement with the contact angle measurements of Smith and McGrath (17) and can be explained by the low surface energy of poly(dimethylsiloxane) which would cause it to locate at the surface during spin coating and subsequent drying.

Table II. ESCA Analysis of PMMA-*g*-PDMS Films

Sample	Angle	Carbon	Oxygen	Silicon
PMMA		71.4	28.6	
PDMS		50	25	
26A	10	48.6	24.5	26.9
	30	55.6	24.4	20.0
	90	61.6	23.7	14.7
26B	10	53.5	24.1	22.4
	30	52.3	25.0	22.9
	90	55.3	25.2	19.5
26D		25	50	25

After 5 min. RIE, the carbon content of the surface has sharply diminished with a concomitant increase in silicon content (sample 26D) which is consistent with the formation of an oxide layer. We note that the size of the surface features which develop (Figure 5) during RIE are considerable greater than the original domains which vary from 11 nm for graft MW of 5K (15% PDMS content) to 21 nm for 20K.

Further evidence of partial miscibility is seen from results on small angle X-ray scattering taken on samples of varying graft MW at constant PDMS content (15%) (see Table III). We see that the interdomain spacing varies uniformly with graft MW. However, the scattering intensity decreases with decreasing MW of the PDMS side chain at a much greater rate than can be accounted for simply by the shrinking domain size. This effect again leads us to conclude that there is partial miscibility below 25K MW PDMS side chains.

Table III. SAXS Analysis of PMMA-*g*-PDMS Films.

Sample	Interdomain spacing, nm	Scattering intensity, a.u.	Graft MW
7O	13.4	37	1K
7E	25.3	140	5K
7J	36.4	400	10K
11N	45.6	530	20K

The picture that emerges is of a system which does phase separate on a microscopic scale, but in which there is still a degree of partial miscibility, even at graft MWs of 20K. The surface is dominated by the PDMS because of the low surface energy of the siloxane chain which causes it to migrate to the surface. This surface agglomeration is aided by the prebaking process. RIE requirements dictate an optimum PDMS content of ~25% while the optimum graft MW appears to be around 20K. As we shall see below, this physical picture is substantiated by the radiation chemistry which is mirrored in the lithographic performance.

Figure 6 shows the lithographic response of sample 7Q to electron beam irradiation. Contrary to our initial expectation, the resist is negative acting with an interface gel dose (D_g) of 5 μC/cm^2. Figure 7 shows patterns generated in this resist and transferred to the 1.5 μm-thick planarizing layer by O_2 RIE. It was virtually impossible to obtain equal line and space features. As can be seen in Figure 6, the maximum thickness attainable was only on the order of 50-60% original film thickness.

The origin of these effects lies in the different radiation susceptibilities of the two copolymer components. PDMS is a negative-acting resist whose sensitivity has been reported to be ~5μC/cm^2 for a molecular weight of 71,800 (5). PMMA, on the other hand, is positive acting with a sensitivity of ~50 μC/cm^2. Thus, we would expect the grafted PDMS side chains to dominate the radiation chemistry at low doses. Although we had anticipated this, we had expected the lithography to be dominated by the radiation chemistry of PMMA, i.e., the ideal continuous phase, since crosslinking of the phase-separated rubber domains ought not to affect the dissolution of the continuous matrix. The fact that the lithography reflects the radiation chemistry of PDMS, supports the suggestion previously made that the PDMS does not exist exclusively in phase-separated domains, but is in fact partially miscible with PMMA and thus dominates the radiation chemistry of the matrix. At higher doses, crosslinking is offset by degradation of the PMMA which is reflected in the maximum in the contrast curve (Figure 6).

Since PDMS does not absorb in the deep UV, its partial miscibility with the PMMA continuous phase should be of no consequence as far as deep UV response is concerned. Indeed, the resist is positive acting in deep UV as seen in Figure 8. Broadly speaking, sensitivity is independent of the MW of the PDMS graft although there are marked differences in the developer strength required to develop the image. The various developers are summarized in Table IV. We again see that the sample with graft MW of 20K is most PMMA-like in that the developer strength is closest to that required for pure PMMA. At 1K, only very weak solvents can be used reflecting the greater degree of miscibility and solubility of the copolymer. Figure 9 shows patterns transferred into the thick planarizing layer. Excellent pattern quality is obtained.

Table IV. Developers for PMMA-g-PDMS Graft Copolymers

Sample	M_n of PDMS	Total PDMS wt.-%	Developer[a]
7Q	1K	32	IPA
7G	5K	31	IPA/MIBK 6:1
7L	10K	30	IPA/MIBK 3:1
11Q	20K	25	IPA/MIBK 3:2
PMMA			MIBK

[a] 45-60 s at room temperature; IPA = propanol-2;
MIBK = 3-methylbutan-2-one

CONCLUSIONS

The block copolymer principle has been extended to the design of positive resist systems. The degree of phase separation has been shown to be an important factor determining the lithographic response of the resist under conditions where the response is determined by the radiation chemistry of the continuous phase. When the discontinuous phase, i.e, the minor component, is more sensitive to the exposing radiation than the continuous phase (major component), and is partially miscible with the major component, its radiation chemistry will determine the lithographic response, particularly in cases where it crosslinks. Partial miscibility should not present a problem if the minor component does not absorb radiation at the exposing wavelength or if its radiation response is of similar tone.

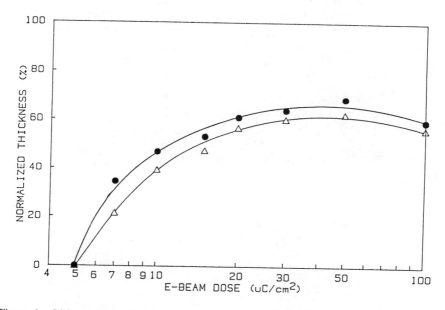

Figure 6. Lithographic response curve for sample 7Q (e-beam at 20 keV). Initial thickness: 3300 Å; ● developed 30 s in 1:1 MEK/MIBK; Δ 90 s in 1:3 MEK/MIBK.

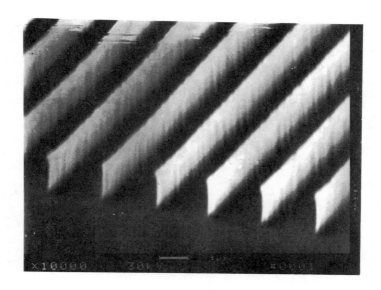

Figure 7. SEM micrographs showing features transferred via O_2 RIE into 1.5 μm planarizing layer. Continued on next page.

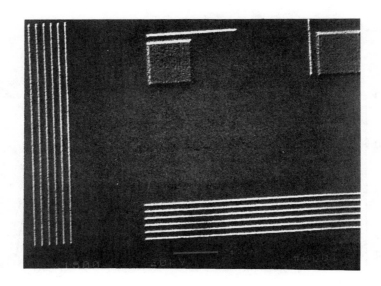

Figure 7.--<u>Continued</u>. SEM micrographs showing features transferred via O_2 RIE into 1.5 μm planarizing layer.

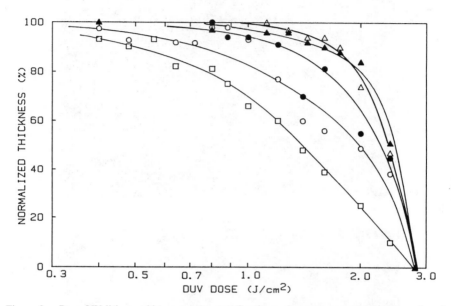

Figure 8.   Deep UV lithographic response curves for the graft copolymers as a function of graft
MW at constant composition (25-32%).   □ PMMA developed 45 s in MIBK.
○ 1K - IPA/45 s; ● 5K - IPA/MIBK 6:1/45 s;   △ 10K - IPA/MIBK 3:1/60 s;
▲ 20K - IPA/MIBK 3:2/60 s.

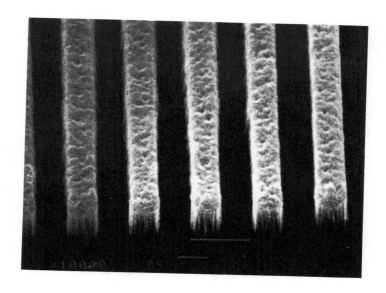

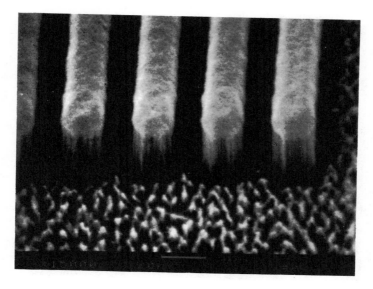

Figure 9.   SEM micrographs showing two-layer resist features transferred via O_2 RIE into a 1.5 μm-thick planarizing layer. Graft copolymer (top resist layer) patterned at 4 J/cm^2. <u>Continued on next page.</u>

Figure 9.--<u>Continued</u>. SEM micrographs showing two-layer
resist features transferred via O_2 RIE into a 1.5 μm-thick
planarizing layer. Graft copolymer (top resist layer)
patterned at 4 J/cm^2.

Literature Cited

1. Tarascon, R. G.; Shugard, A.; Reichmanis, E. *Proc. SPIE, Adv. Resist Technol. Process. III* **1986**, *631*, 40.
2. Taylor, G. N.; Wolf, T. M.; *Polym. Eng. Sci.* **1980**, *20*, 1087.
3. Taylor, G. N.; Wolf, T. M.; Moran, J. M.; *J. Vac. Sci. Technol.* **1981**, *19*, 872.
4. Gozdz, A. S.; Carnazza, C.; Bowden, M. J.; Ref. 1, p. 3.
5. Babich, E.; Paraszczak, J.; Hatzakis, M.; Shaw, J.; *Microcircuit Eng.* **1985**, *3*, 279.
6. Shaw, J; Hatzakis, M.; Paraszczak, J.; Babich, E.; *Microcircuit Eng.* **1985**, *3*, 293.
7. Hofer. D. C.; Miller, R. D.; Willson, C. G.; *Proc. SPIE, Adv. in Resist Technol.* **1984**, *469*, 16.
8. Miller, R.D.; Hofer, D. C.; McKean, D. R.; Willson, C. G.; West, R.; Trefonas, P.; In *Materials for Microlithography*; Thompson, L. F.; Willson, C. G.; Frechet, J. M. J, Eds.: ACS Symp. Ser. **1984**, *266*, 293.
9. Suzuki, M.; Saigo, H.; Gokan, H.; Ohnishi, Y.; *J. Electrochem. Soc.* **1983**, *130*, 1962.
10. Bowden. M. J.; Thompson, L. F.; Fahrenholtz, S. R.; Doerries, E. M.; *J. Electrochem. Soc.* **1981**, *128*, 1304.
11. Ito, H.; Pederson, L. A.; MacDonald, S. A.; Cheng, Y. Y.; Lyerla, J. R.; Willson, C. G.; *Proc. Reg. Conf. on Photopolymers: Principles, Processes and Materials*, Soc. Plast. Engineers, Ellenville, NY, Oct. 28-30, 1985, p. 127.
12. Hartney, M. A.; Novembre, A. N.; *Proc. SPIE, Adv. Resist Technol. Process. II* **1985**, *539*, 90.
13. Hartney, M. A.; Novembre, A. N.; Bates, F. S.; , see Ref. 11, p. 211.
14. Krause, S. J.; in *Polymer Blends*; Vol. 1, Paul, D. R., Newman, S., Eds.; Academic; New York, 1978.
15. Long, T. E.; Subramanian, R.; Ward, T. C.; Mc Grath. J. E.; *Polymer Prepr.* **1986**, *27(2)*, 258.
16. Kawakami, Y.; Murthy, R. A. N.; Yamashita, Y.; *Makromol. Chem.* **1984**, *185*, 9.
17. Smith, S. D.; McGrath, J. M.; *Polymer Prepr.* **1986**, *27(2)*, 31.
18. MacDonald, S. A.; Steinmann, F.; Ito, H.; Lee, W-Y.; Willson, C. G.; *Polym. Mat. Sci. and Eng.* **1983**, *49*, 104.

RECEIVED June 25, 1987

Chapter 12

Acid-Catalyzed Thermolytic Depolymerization of Polycarbonates: A New Approach to Dry-Developing Resist Materials

J. M. J. Fréchet[1], E. Eichler[1], M. Stanciulescu[1], T. Iizawa[1], F. Bouchard[1], F. M. Houlihan[1], and C. G. Willson[2]

[1]Department of Chemistry, University of Ottawa, Ottawa, Ontario K1N 9B4, Canada
[2]Almaden Research Center, IBM, San Jose, CA 95120-6099

The design of new condensation polymers which undergo acid-catalyzed thermolysis is explored with polycarbonates. The polymers are prepared by phase-transfer catalyzed polycondensation using active esters or carbonates and diols. Polycarbonates containing tertiary, allylic or benzylic diol units susceptible to elimination decompose thermally near $200°$ to volatile materials. Self developing resist materials can be designed by combining the active polycarbonates with photoactive triarylsulfonium salts or other similar compounds which generate strong acids upon irradiation. Exposure of the resist material creates a latent image which can be developed thermally with evolution of volatile carbon dioxide, alkenes, and alcohols.

The development of new materials for application as photoresists has seen numerous advances over the past few years as researchers strive to meet the industry's needs for materials of greatly improved properties through new conceptual designs. Of special interest to the synthetic organic chemist are improvements focused on increases in resolution and sensitivity as these may be prone to chemical solutions. In practice, enhanced resolution is required to achieve dimensional control while allowing for a reduction of the overall dimensions of active devices. Enhanced sensitivity is required to compensate for the loss of flux which generally results from the use of exposure instruments capable of providing higher resolution. Other critical resist properties such as etch resistance, adhesion, processing characteristics, etc., must also be considered though initial emphasis for the development of new designs may be placed on improvements in sensitivity and resolution [1,2].

The Chemical Amplification Approach

Significant advances in the field of materials which may be useful as E-Beam or X-Ray resists have been made based on the development of new chain growth polymers such as the polyalkenesulfones [3] and substituted polyacrylates [4,5].

Over the past few years we have been interested in the design of new types of resist materials which generally possess high sensitivities due to structural features which allow for the occurrence of radiation initiated repetitive processes. The three main approaches we have investigated to-date all maximize the use of available protons through "chemical amplification"; they are the following:

* Photoinduced changes in the physical properties of polymers [1,6].
* Photoinduced multiple molecular rearrangements [7].
* Photocatalyzed depolymerization or chain degradation reactions [8].

In all three of these designs, chemical amplification is the result of photoinitiated chain or catalytic reactions where irradiation is used only to initiate a chain reaction or to generate a catalyst within localised areas of a resist film.

The most relevant early work in the context of this study is the radiation induced depolymerization of poly(phtalaldehyde) [9]. In this case, depolymerization is due to a ceiling temperature phenomenon whereby radiation induced cleavage of the polymer causes it to revert fully to monomer. Poly(phtalaldehyde) is a material with a very low ceiling temperature which is only rendered stable at room temperature through the device of capping its chain-ends after low temperature polymerization, thereby preventing its spontaneous degradation when heated.

This very interesting approach leads to a potentially versatile resist which can be imaged with almost any source of energetic radiation. However, a significant drawback of the spontaneous depolymerization process may be the fact that monomer is evolved immediately upon irradiation, thus increasing the risk of contaminating the optics of the exposure tool. A better approach, which we have originally tested with the photoinitiated molecular rearrangement of certain polycarbonates containing protected glycidol moieties [7], would involve the photochemical generation of a latent image only, followed by a thermal self-development step outside the exposure tool (Figure 1).

Thermally Depolymerizable Polycarbonates

Our newly described family of thermally labile polycarbonates operate on a somewhat similar design [10]. The system is based on a two-component mixture consisting of a polymer with thermally labile bonds in its main-chain and a substance which can generate acid by exposure to radiation (e.g. triarylsulfonium salts [11]). Typical examples of the types of reactive polycarbonates we have prepared are polymers I, II, and III which are shown on page 141.

These polycarbonates undergo thermolysis at relatively low temperatures (150–250°C) while other less reactive polycarbonates are stable to much higher temperatures. For example, while polycarbonate I undergoes complete decomposition to volatile p-benzenedimethanol, isomeric C-8-dienes, and CO_2 near 200°C, the homopolycarbonate of 1,4–benzenedimethanol only undergoes a partial and complex decomposition at higher temperatures leaving an appreciable amount of charred residue behind.

A typical preparation of the polycarbonates is shown for polymer III in Scheme I. 1,4–Benzenedimethanol is activated by reaction with two equivalents of p-nitrophenyl chloroformate in pyridine and the resulting symmetrical dicarbonate is then used in a polycondensation with an equimolar amount of 2-cyclohexen-1,4-diol in a solid-liquid phase-transfer catalyzed reaction with 18-crown-6 as catalyst and solid anhydrous potassium carbonate as base. Alternately, the same polymer can be prepared by condensation of bis(4-nitrophenyl)-2-cyclohexen-1,4-ylene dicarbonate [12] with 1,4–benzenedimethanol or through a variety of similar polycondensations using diol bis-carbonylimidazolides [13,14].

All three types of polycarbonates I-III owe their thermolytic lability to their structural design which allows for low activation energy tertiary, benzylic, or allylic, transition states and which includes hydrogen atoms in positions β to the carbonate oxygen to enable elimination. In all cases the thermolytic decompositions are very clean reactions which usually proceed quantitatively. For example, in the case of polymer III, thermolysis affords only the three expected products as shown below in Scheme II.

Acid-Catalyzed Thermolysis of the Polycarbonates

Of particular relevance to this study is the knowledge that elimination of carbonates proceeds through a polar transition state [15]. Thus, the thermolysis of these activated carbonates should proceed through an acid-catalyzed process and polymers I-III would be expected to undergo thermolysis at temperatures well below 100°C in the presence of a catalytic amount of strong acid. In terms of resist imaging, these properties would be exploited as shown in Figure 2.

These expectations were confirmed fully in a model study involving the acidolysis of bis(4-nitrophenyl)-2-cyclohexen-1,4-ylene dicarbonate. This bis-allylic carbonate is expected to undergo a clean acidolytic cleavage to benzene, p-nitrophenol and carbon dioxide. Indeed when a sample of this model compound is treated with a catalytic amount of a strong non-nucleophilic acid such as trifluoromethanesulfonic acid, the decomposition proceeds as expected. This is most readily monitored by performing the acidolysis within an NMR tube on a solution of the dicarbonate as is shown quite graphically in Figure 3. Extension of this finding to a resist material is carried out readily as follows. A thin film of an active polycarbonate such as I containing a small amount of a radiation sensitive acid precursor such as a triphenylsulfonium salt is cast on an

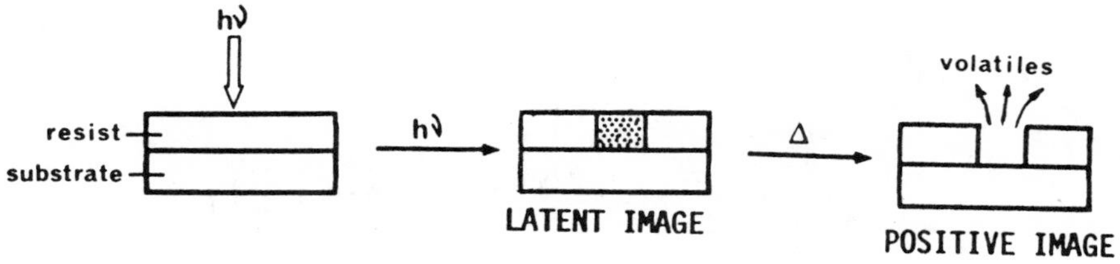

FIGURE 1. Thermal development of a latent resist image

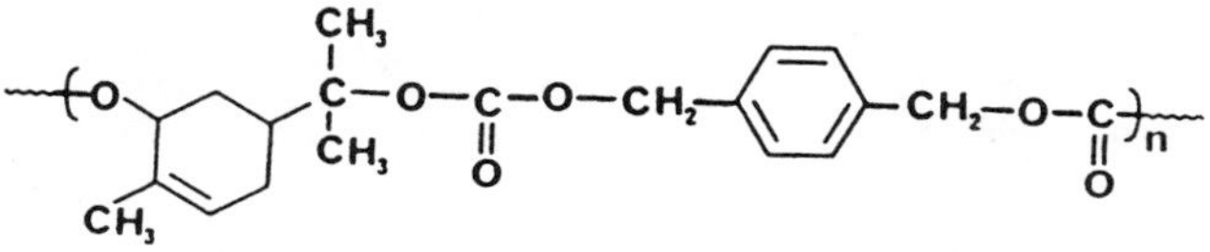

I

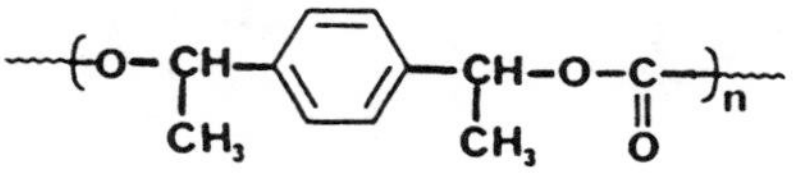

II

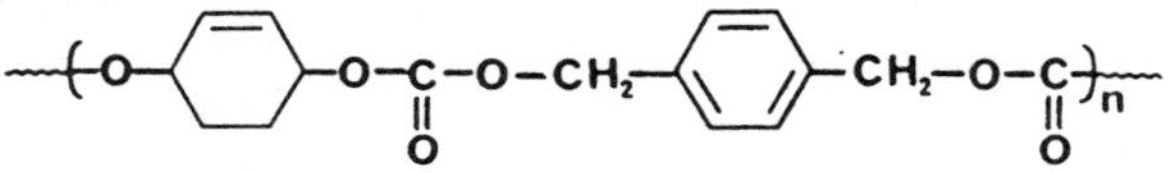

III

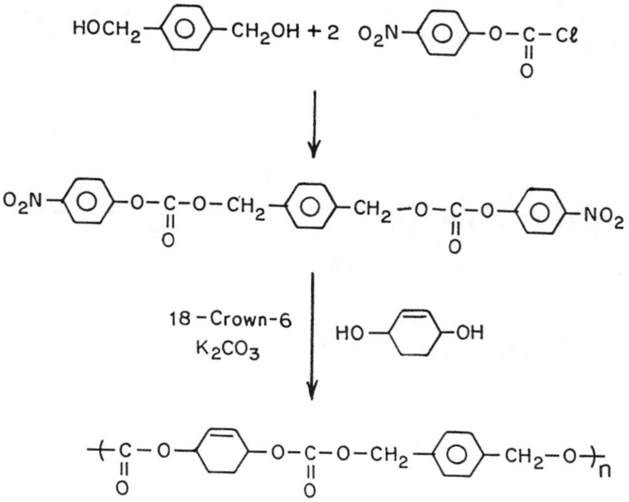

SCHEME I. Phase transfer catalysis in the preparation of polycarbonate III.

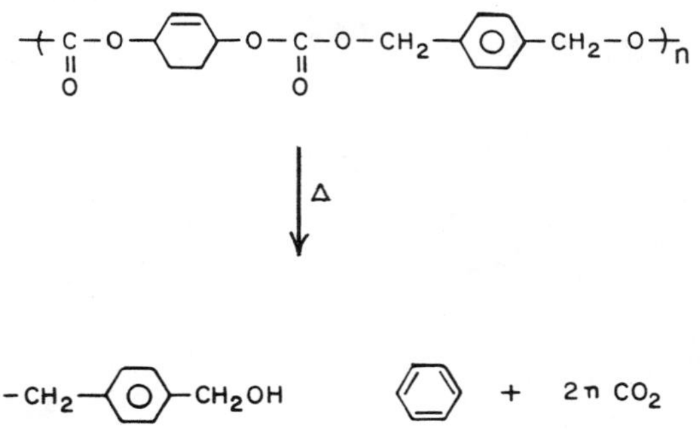

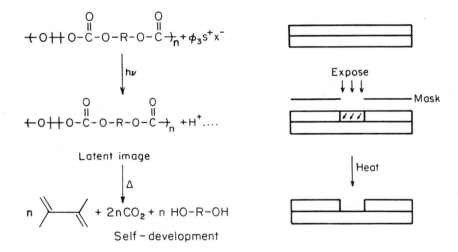

SCHEME II.   Thermolytic depolymerization of Polymer III.

FIGURE 2. Imaging of a two-component resist material containing a
          tertiary polycarbonate and an onium salt.

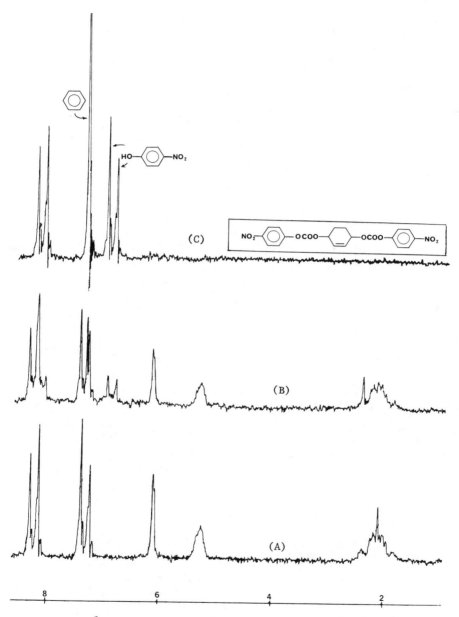

FIGURE 3. ^{1}H-NMR monitoring of the acidolysis of bis(4-nitrophenyl)2-cyclohexen-1,4-ylene dicarbonate. Spectrum A: starting dicarbonate in $CDCl_3$. Spectrum B: taken immediately after addition of a catalytic amount of CF_3SO_3H. Spectrum C: taken after 10 min. at RT and after D_2O exchange.

appropriate substrate. Selective exposure of portions of the film to an appropriate source of radiation causes the liberation of acid in the polycarbonate matrix with formation of a latent image. Examination of the exposed film of polycarbonate at this stage by Fourier-transform infrared spectroscopy suggests that no transformation other than that affecting the radiation sensitive catalyst has occurred. Full development of the image is then achieved by low temperature thermolysis, a process which does not affect the areas of the image where no acid has been produced. Though dry development should result, some of the products which are formed in the decompostion of the various polycarbonates may have a low volatility and thus it may be convenient to remove them under vacuo or through the use of a solvent which does not affect the unexposed polycarbonate.

Since a large number of polycarbonates possessing the desired structural features can be prepared from a variety of diols, it is useful to develop simple methods to predict their behavior as resist materials. To this effect, we have devised spectroscopic methods to follow the degradation of the polycarbonates under a variety of thermolysis or acidolysis conditions. For example, the thermolysis of the solid polymers can be followed conveniently by gas-chromatography-mass spectrometry. The thermolysis is a very clean reaction which proceeds as shown in Scheme III without side-product formation. Figure 4 shows the gas chromatographic trace obtained when polymer II is subjected to thermolysis near 250°; the products analyzed by the mass spectrometer have the expected structures as shown in Figure 4.

Though the polymers are meant to be used as thin film coatings, some useful information can be deduced from the monitoring of their acidolysis in solution within the confines of an NMR tube as was shown above for a model compound (Figure 3). For example, polycarbonate IV dissolved in deuterated chloroform undergoes acidolysis with loss of carbon dioxide upon addition of a catalytic amount of trifluoromethanesulfonic acid. The NMR spectrum of the polymer shows drastic changes with formation of a large amount of benzene and liberation of other phenolic moieties derived from bisphenol A. The thermolysis reaction can be accelerated considerably by slight heating after addition of the acid.

Resist Imaging Experiments

Imaging experiments can be done through irradiation either with a UV source such as a Perkin-Elmer 500 aligner at 254 nm or with an electron-beam. Typically, exposure is followed by a brief baking period at $50-70^\circ C$ depending on the exposure dose and the exact structure of the resist material. The potential of the system is exploited fully in the production of positive images though it may be also possible to produce negative images with certain structures. Figure 5 shows scanning electron micrographs of positive tone images obtained with a typical polycarbonate resist system. We are currently exploring more fully the properties of these systems as well as of novel radiation-sensitive compositions based on different families of polymers.

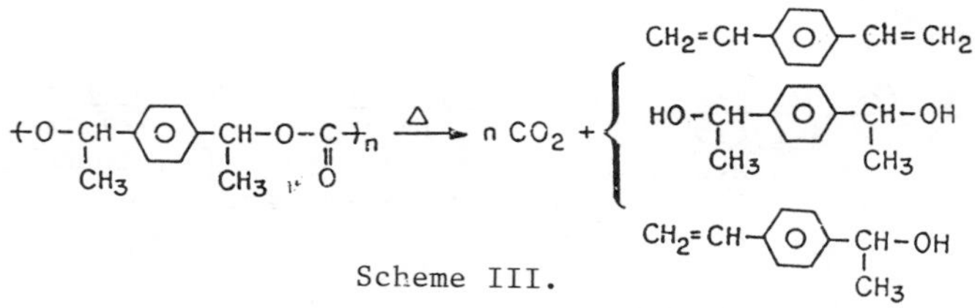

Scheme III.

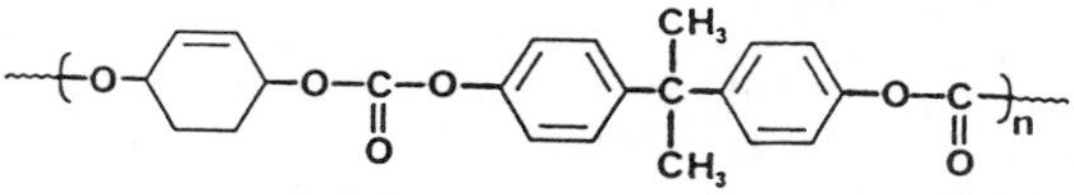

IV

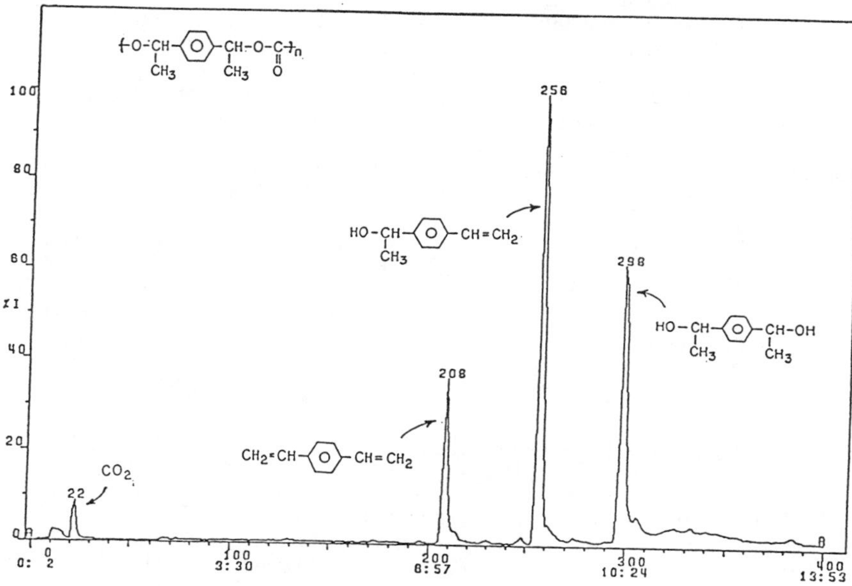

Figure 4. GC–MS analysis of the thermolysis of III.

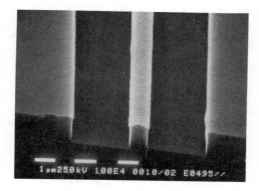

Figure 5. Positive image obtained from polycarbonate I.

Experimental

Synthesis of the bis-p-nitrophenylcarbonate of 1,4-benzene-dimethanol
A solution of 14.1 g (70 mmoles) of p-nitrophenylchloroformate in 40
mL of dry dichloromethane is added over a period of one hour to a
solution of 4.80 g (35 mmoles) of 1,4-benzenedimethanol in 5.53 g (70
mmoles) of dry pyridine and 60 mL of dry dichloromethane. After
stirring at room temperature overnight, enough dichloromethane is
added to the reaction mixture to bring the white compound formed into
solution. The organic phase is then washed successively with
distilled water, a 5% solution of hydrochloric acid and once more
with distilled water. After drying the organic phase over anhydrous
magnesium sulfate and removal of the solvent on rotatory evaporator,
a white solid is recovered. Recrystallization from dichloromethane
affords 9.11 g (56% yield) of pure product which is shown to be the
desired compound by spectroscopic and elemental analysis.
^{1}H-NMR: 5.32(s, 4H, CH_2); 7.53(s, 4H, CH, benzenedimethanol);
7.40(d,4H, CH nitrophenol); 8.30(d, 4H, CH nitrophenol).
^{13}C-NMR: 70.35(CH_2); 121.72(CH, nitrophenol, C #2); 125.32(CH,
nitrophenol); 128.98(CH, benzenedimethanol); 135.08(C,
benzenedimethanol); 145.43(C, nitrophenol); 155.40(C, nitrophenol);
152.40(C=O).
IR: 1767(C = O); 1617 and 1494(C = C); 1521 and 1345(NO_2 stretch);
1458(CH_2 scissoring); 1253,1215 and 1167(C-O-C); 863(= C-H out-of-
plane bending, aromatic).
MS: absent(M+); 286(VS, loss of CO_2 and nitrophenyl radical); 242(W,
loss of $2CO_2$ and nitrophenyl radical); 139(W, nitrophenol); 104(s,
loss of $2CO_2$ and 2 nitrophenyl radicals).
Analysis (C,H,N); calc: 56.42; 3.44; 5.98. Found: 56.29; 3.70, 5.84.

Reaction of the bis-p-nitrophenylcarbonate of 1,4-benzenedimethanol
with 2-cyclohexen-1,4-diol A mixture is prepared under argon
atmosphere, consisting of 1.1348 g (9.954 mmoles) of the diol, 4.6623
g (9.954 mmoles) of the bis-p-nitrophenyl carbonate 0.630 g of 18-
crown-6, 7.0 g of potassium carbonate and 15 mL of dry
dichloromethane. The reaction mixture is stirred and refluxed for 90
hours, after which time it is worked-up by the usual method. Three
precipitations into 1.5 liter of methanol are necessary to obtain
2.760 g (91% yield) of the pure white polymer. The cis:trans ratio
of the cyclohexenediol units in polymer III is shown to be 35:65 by
^{1}H-NMR reflecting the mixture used initially.
Analysis (C,H); calc.: 63.12, 5.30. Found: 63.36; 5.49.
^{1}H-NMR: 7.36 (s,4H,CH aromatic); 5.96 (m,2H,CH alkene); 5.14
(m,6H,CH-O and CH_2 benzenedimethanol); 2.15 (m,1.3H,CH_e-CH_e,trans
isomer), 1.92 (m,1.4H,CH_2,cis isomer), 1.76 (m,1.3H,CH_a-CH_a, trans
isomer)
^{13}C-NMR: 154.51 (C = O); 135.47 (s,C aromatic); 129.97 (d,CH alkene);
128.51 (d,CH aromatic); 71.09 (d,CH alicyclic); 69.14
(t,CH_2,benzenedimethanol); 25.25 (t,CH_2 alicyclic,trans isomer);
24.57 (t,CH_2 alicyclic, cis isomer)
IR: 1742 (C = O); 1452 (CH_2 scissoring); 1244, 1196 and 1079 (C-O-C);
792
(= C-H out-of-plane bending, cis 1,2-disubst.)
DSC: Tg = 65°C; 3 endotherms at 73,87 and 123°C

MW; Osmometry: Mn = 17,600
     GPC: Mn = 7,900, Mw = 15,000, D = 1.9

Acknowledgments

Financial support of this research by the Natural Sciences and Engineering Research Council of Canada and the IBM Corporation is gratefully acknowledged.

References

1. Willson, C.G.; Ito, H.; Fréchet, J.M.J.; Tessier,T.G.; Houlihan, F.M.;G; J. Electrochem. Soc. 1986, 133, 181.

2. Willson, C.G. in "Introduction to Microlithography" (L.F. Thompson, C.G. Willson, M.J. Bowden, Editors) ACS Symposium Series #219, Chapter 3, 1983.

3. Thompson, L.F.; Bowden, M.J., J. Electrochem. Soc., 1973, 120, 1722.
Bowden, M.J.; Thompson, L.F., Polym. Eng. and Sci., 1974, 14, 525.

4. Pittman, C.U.; Ueda, M.; Chen, C.Y.; Kwiatkowski, J.H.; Cook, C.F.; Helbert, J.N., J. Electrochem. Soc., 1981, 128, 1758.
Lai, J.H.; Helbert, J.N.; Cook, C.F.; Pittman, C.U., J. Vac. Sci. Technol., 1979, 16, 1992. Tada, T., J. Electrochem. Soc., 1979, 126, 1635.

5. Willson, C.G.; Ito, H.; Miller, D.C.; Tessier, T.G., Polymer Eng. and Sci., 1983, 23, 1000.

6. Fréchet, J.M.J.; Ito, H.; Willson, C.G.; Microcircuit Engineering 1982, 260. Ito, H.; Willson, C.G.; Fréchet, J.M.J.; US Pat. 4,491,628, 1985.

7. Houlihan, F.M.; Fréchet, J.M.J.; Willson, C.G.; Polym. Mat. Sci. Eng., 1985, 53, 268.

8. Fréchet, J.M.J.; Bouchard, F.; Houlihan, F.M.; Eichler, D.; Kryczka, B.; Willson, C.G.; Makromol. Chem. Rapid. Commun. 1986, 7, 121. Fréchet, J.M.J.; Bouchard, F.; Houlihan, F.M.; Kryczka, B.; Eichler, E.; Clecak, N.; Willson, C.G.; J. Imaging Sci. 1986, 30, 59.

9. Willson, C.G.; Ito, H.; Fréchet, J.M.J.; Houlihan, F.M., Proc. of IUPAC 28th Macromolecular Symposium, 1982, 448.
Ito, H.; Willson, C.G.; Polym. Eng. and Sci., 1983, 23, 1013.

10. Fréchet, J.M.J.; Bouchard, F.; Houlihan, F.M.; Kryczka, B.; Eichler, E.; Clecak, N.; Willson, C.G., J. Imaging Sci., 1986, 30, 59.

11. Crivello, J.V.; Lam, J.H.W., J. Polym. Sci., Polym. Chem. Ed.,
     1979, 17, 977

12. Fréchet, J.M.J.; Bouchard, F.; Houlihan, F.M.; Eichler, E.;
     Kryczka, B.; Willson, C.G., Makromol. Chem. Rapid Commun., 1986,
     7, 121.

13. Fréchet, J.M.J.; Houlihan, F.M., Bouchard, F.; Kryczka, B.;
     Willson, C.G., J. Chem. Soc. Chem. Commun., 1985, 1514.

14. Houlihan, F.M.; Bouchard, F.; Fréchet, J.M.J.; Willson, C.G.,
     Macromolecules, 1986, 19, 13.

15. Taylor, R., J. Chem. Soc. Perkin Trans. II, 1975, 1025.

RECEIVED April 8, 1987

Chapter 13

Sensitivity of Polymer Blends to Synchrotron Radiation

J. A. Jubinsky[1], R. J. Groele[1], F. Rodriguez[1], Y. M. N. Namasté[2], and S. K. Obendorf[2]

[1]School of Chemical Engineering, Olin Hall, Cornell University, Ithaca, NY 14853
[2]Department of Textiles and Apparel, Martha Van Rensselaer Hall, Cornell University, Ithaca, NY 14853

The sensitivity of poly(methyl methacrylate), PMMA, to x-rays is enhanced by the addition of poly(epichlorohydrin). The two polymers are miscible as shown by thermal analysis and by optical clarity of blends. Films of PMMA with 20 to 30% poly(epichlorohydrin) require only 1/2 to 1/4 the exposure to produce patterns compared to PMMA alone. Thinning-exposure curves also suggest a sensitivity increase of 3 or 4 times. Contrast suffers somewhat, but patterns are producible when a nitride-supported gold mask is used.

When the resolving power of photolithography is to be exceeded, the two technologies that are most often called upon are electron-beam and x-ray lithography. For purposes of discussion, the lower limit of features made by optical (deep UV) methods may be placed at about 0.5 um. Electron beams currently can be used to produce features from 1.0 to 0.25 um. Since backscattering is not a problem with x-rays, the minimum dimensions obtainable with x-rays in thick films may well be smaller than those obtained with electron beams. Moreover, x-ray lithography is efficient for large volume production of chips. Electron-beam technology has been widely used for some time in mask-making and for customized, low-volume chip production, so a considerable body of information has been acumulated on the response of polymers to 20- to 50-keV electrons. A similar body of information does not exist on the response to x-rays. There are currently three types of x-ray sources available to researchers: anodic, in which high-energy electrons act on a metal target; plasma, in which IR or UV laser radiation acts on a metal target; and synchrotron, in which a beamline is attached to a high-energy electron storage ring. Anodic sources are the least expensive type, but they are from 100 to 10,000 times less intense than the other sources.

0097–6156/87/0346–0149$06.00/0

As source strength is increased and exposure times
reduced, sensitivity is a less urgent matter. When the
response of a given resist to x-rays is being measured,
the wavelength of the radiation has to be taken into
account also. While synchrotron radiation is intense,
the energy distribution may not be concentrated in the
regions corresponding to efficient absorption by the
atoms making up the resist.

Several reports have been published on x-ray resists
in recent years. Haelbich et al. (1) compared the performance of PMMA using synchrotron radiation with that
using e-beam radiation. Yaakobi (4) et al. used a laser-
-ion x-ray source. The general topic of x-ray resists
has been reviewed by Lane (3) and Taylor (4). A very
optimistic view of the future for synchrotron-based
lithography was taken by Wilson (5). More guarded predictions of eventual application of x-rays have been made
by Broers (6) and Heuberger (7).

Poly(methylmethacrylate), (PMMA), is one resist
which is especially favored by researchers due to its
high resolution and contrast. Linewidths as small as 100
A have been produced with an extremely high dose of
x-rays (10 J/cm^2) using PMMA. However, even PMMA's
normal sensitivity of 600-1000 mJ/cm^2 is too slow for
commercial use. There have been many attempts to improve
the sensitivity of PMMA to electron beam irradiation
mainly by copolymerization (8). In the current paper, we
describe an alternative approach whereby a sensitive
polymer is physically blended with PMMA to increase its
sensitivity and yet maintain the good film qualities
associated with PMMA as a positive x-ray resist.

Poly(epichlorohydrin), CO rubber* (Hydrin), was
chosen for various reasons. The one reason was that CO
has been shown to be miscible with PMMA by Anderson based
upon differential scanning calorimetry (DSC) which showed
only one glass transition temperature (T_g) for the blend
(9). Since T_g is very sensitive to the disruption of the
local structure that results when two polymers are mixed,
the existence of a single glass transition temperature is
a good indicator of miscibility (10).

Recently, the miscibility of CO rubber with PMMA
over the entire range of concentrations and molecular
weights has been confirmed (11). There is a broadening
of the glass transition which reaches a maximum at about
30% PMMA. However, even at that concentration, the polymers are truly miscible. Studies of compatibility of CO
rubber and some copolymers have been reported with other
methacrylate polymers and also with acrylate polymers
over a wide variety of conditions (11, 12). Another
reason for blending with CO is that CO contains chlorine
which tends to enhance x-ray absorption, especially near
the absorption edge of 4.4 Å. It was hoped that the increase in absorption would produce more secondary

---

*ASTM abbreviation for this polymer.

electrons which would, in turn, lead to increased scissioning of PMMA.

Our exploratory work indicated that CO degrades rapidly on gamma radiation. Taylor et al. (13), using a very high molecular weight CO, has reported that CO crosslinks appreciably when exposed to $Pd_{L\alpha}$ x-rays. In the present work no decrease in solubility has been observed and only a very slight broadening of the molecular weight distribution has been seen even while the number average molecular weight decreased markedly.

Experimental Procedure

CO rubber (Hydrin 100) was obtained from the B.F. Goodrich Chemical Company. Based upon gel permeation chromatography (GPC) using tetrahydrofuran (THF) as a carrier solvent, the CO has a polystyrene equivalent number average molecular weight (M_n) of 303×10^3 and a weight average molecular weight (M_w) of 598×10^3. The glass transition temperature is $-20^\circ C$ (9). The PMMA used was Rohm and Haas A-100 ($M_n = 113 \times 10^3$, $M_w = 213 \times 10^3$). Some additional blends were made using PMMA from KTI ($M_n = 360 \times 10^3$, $M_w = 950 \times 10^3$).

Three mixtures of CO-PMMA were blended together for analysis of lithographic performance. Compositions of 20, 33, and 50% CO by weight were blended by dissolution in THF to form a 13% (total solids) solution. Cyclohexanone was added to the blend to form a 7% solution to impart proper viscosity and volatility for casting films. In the case of PMMA with $M_n = 950 \times 10^3$, CO was dissolved in hot chlorobenzene and then mixed with PMMA (also in chlorobenzene). Solvent choice was a matter of convenience.

The films were spun at 1250 RPM for 1 minute onto three-inch silicon wafers. The wafers were baked at $150^\circ C$ for 1 hour in a convection oven. Film thicknesses ranged from 0.8 to 1.3 μm. Pure PMMA films were also prepared using the same casting solvent mixture and baking conditions.

Gel permeation chromatography was used to determine polymer molecular weights. The model used was a Waters Associates 201 HPLC with 4 μStyragel columns (nominal pore sizes of 500, 10^3, 10^4, and 10^5 Å). THF was used as the carrier solvent. The molecular weights were reported as polystyrene equivalents with the exception of PMMA for which a PMMA calibration standard was used. Glass transition temperatures (T_g) were measured using a Perkin Elmer Differential Scanning Calorimeter (DSC) model DSC-2C.

The x-ray exposures were carried out using synchrotron radiation delivered by the National Synchrotron Light Source at Brookhaven National Laboratories. The beamline was built by IBM and has been described elsewhere (14, 15). In summary, the storage ring operates with electron energies at 750 MeV and a magnetic radius of 1.91 m. The current in the ring, which degrades with

time, ranges between 75 and 250 mA. The actual power
through the beryllium window, and incident on the wafer
ranged from 2.3 to 7.5 mW/cm^2 in the present work.
Samples were exposed to synchrotron radiation through a
gold-on-boron nitride mask and then developed using
various mixtures of isopropyl alcohol (IPA) and methyl
isobutyl ketone (MIBK). Development was conducted under
stirred conditions using a constant temperature bath at
23°C and was terminated by blow drying with Freon gas.
Film thicknesses were measured using a Tencor Alpha Step.
     Electron beam exposures were performed using a
modified RCA model EMV-3 transmission electron microscope. The aperture was opened to allow the beam to
spread over the entire 3-inch wafer giving a uniform
exposure. The accelerating voltage was 50 keV, and the
dose was measured using a faraday cup (9).
     Gamma irradiation was carried out at Cornell's Ward
Reactor Laboratory. Bulk polymer samples were irradiated
under nitrogen using ^{60}CO source at a dose rate of about
0.5 Mrad/hour.
     Laser interferometry was used to study the dissolu-
tion rates of selected films (16, 17). A 632.8 nm wavelength beam of unpolarized light from a 2 mW He-Ne laser
was directed on a submerged wafer that had been coated
with a polymer film. The angle of incidence was approximately 10°. Reflected light from the polymer-solvent and
polymer-substrate interfaces was directed toward a photocell which was coupled to a chart recorder. As the
polymer solvent interface began to recede due to
dissolution, constructive and destructive interference
from the interface took place, and a sinusoidal output
was recorded. By measuring the period of the sinusoidal
curve, the dissolution rate was determined (16, 17).

Results and Discussion

Lithographic performance of the CO-PMMA blends was evaluated by measuring the thickness of resist remaining for
various doses using arbitrarily chosen development times.
Contrast curves are shown for a 33% CO blend and for PMMA
in Figure 1. From the contrast curves, thinning curves
were constructed by plotting the unexposed normalized
thickness remaining versus the dose required for complete
development (Figure 2). For low x-ray doses the 33% CO
blend exhibited much less thinning than PMMA.
     There is a trend toward increasing sensitivity as
the percentage of CO is increased (Figure 2). For
example, for 20% thinning, the dose required to completely develop the 33 and 50% CO blends are 2 and 3
times less, respectively, than that needed for PMMA.
However, in the high dose regime where the allowable
percentage of thinning was reduced, the difference in
sensitivity narrowed. Unfortunately, the enhanced sensitivity of the CO blends occured at the expense of contrast which was reduced by 25 to 40% when blending. No
trend with respect to contrast among the CO blends was

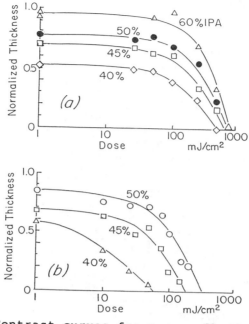

Figure 1.   Contrast curves for x-ray, flood-exposed films developed one minute in indicated mixtures of isoproply alcohol and methyl isobutyl ketone, (a) PMMA and (b) CO:PMMA, 1:2.

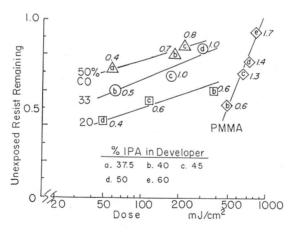

Figure 2.   Thinning curves for PMMA and various blends containing (a) 50%, (b) 33%, (c) 20%, and (d) 0% CO by weight.   The numbers in parentheses give the fraction of isopropyl alcohol in the solvent (balance of MIBK). The gamma value (contrast) also is given.

noticed. The 33% blend appeared to have the best contrast, but it was still significantly lower than that of
the PMMA.

The mechanism behind the increase in the sensitivity
of the blend appears to involve the scissioning and dissolution properties of CO. Adding CO to PMMA increases
the dissolution rate of the blend as compared to PMMA.
When the blend is exposed to radiation, the CO scissions
(faster than the PMMA) and its molecular weight is
decreased. The decrease in molecular weight of the CO is
apparently responsible for the increased dissolution rate
of the blends. This has been confirmed by comparing the
dissolution rates of unexposed CO-PMMA blends with two
different molecular weights of CO. A 33% (by weight) CO
blend containing CO with M_n = 86,000 dissolved 8 times
faster than a blend containing CO with M_n = 303,000.
Thus, the increase in the dissolution rate between the
exposed and unexposed regions of the blends appears to be
primarily due to the scissioning of the CO.

The original hypothesis that the CO enhances scissioning by supplying secondary electrons has not been
supported by this experiment. From the data, it appears
that the enhancement of dissolution rate by scissioning
of CO may account totally for the increased sensitivity
of the blends. This point could be clarified by a
further experiment in which the molecular weight of the
PMMA in the blend is independently monitored upon
exposure.

At high doses, the ability of the CO to influence
the dissolution rate was diminished since its absolute
molecular weight changed less significantly. It appears
that the main mechanism for increasing dissolution at
higher doses is not the degradation of CO, but the scissioning of PMMA. This accounts for the apparent similarity in behavior of PMMA and the blends at high doses.
Thus, the increased sensitivity of the blends is realized
only in the "forced developing" regimes.

The relative response of CO/PMMA blends to various
forms of radiation can be compared. In the first place,
gamma radiation of bulk samples of CO or PMMA results in
values of G(s) = 5.1 scissions/100 eV for CO compared to
0.8 for PMMA (Figure 3). It is interesting that CO
undergoes scissioning with little crosslinking (G(x) =
0.5 at most). It is presumably the ether linkage in
combination with the substituted chain carbon which leads
to scission. Poly(ethylene oxide), in contrast,
crosslinks readily, even in dilute solution as long as
oxygen is excluded (18). Taylor (11) found that the
disubstituted homolog, poly(cis-1,2-dichloromethyl-
oxirane), was unequivocally a scissioning compound, but
that CO appeared to crosslink somewhat. As mentioned
earlier, this may have been due to the higher molecular
weight CO which he employed.

The sensitivity of CO to electrons and x-rays was
characterized by comparing a blend with PMMA to PMMA

alone. In a beam of 50 keV electrons, a blend with a CO to PMMA ratio of 1:2 exhibits a slope about 2.9 times that of PMMA in a plot of $1/M_n$ versus dose (Figure 4). The polydispersity, M_w/M_n, of the blend (shown on the same plot) indicates that the CO degraded faster than the PMMA. The polydispersity of neat PMMA remains between 1.9 and 2.3 under the same experimental conditions.

Synchrotron radiation yielded a similar pattern to the e-beam result when the same blend and PMMA were compared (Figure 5). The slope for the blend is about 2.7 times the slope for PMMA.

While chlorine might be expected to enhance the sensitivity to x-rays more than either gamma radiation or electrons, the present work does not demonstrate this effect. The most probable reason is that the synchrotron beam was operating under conditions where the intensity was not high at the absorption edge for chlorine (4.4A). The dissolution behavior of CO-PMMA blends was examined for two molecular weights of PMMA (Table 1). In each case, the dissolution rate was increased by addition of CO. Two additional features can be noted. The amplitude of the oscillations in reflected light intensity appears to remain constant as the blends dissolve. This would indicate that the undissolved portion of the film is not undergoing swelling or extraction to any noticeable extent. The second feature, observable especially in the high molecular weight PMMA blend, is an increased offset between the maximum amplitude during dissolution and the reflection from the bare wafer after dissolution is complete (Figure 6). This offset has been interpreted as representing a transition layer on the surface of the dissolving film (19).

The increase in the offset means that the CO-PMMA blend has a thicker surface layer. On the other hand, the fact that the offset is only very slightly increased by high molecular weight CO with the lower molecular weights of PMMA means that the transition layer is probably not due to entangled CO molecules remaining on the surface.

Conclusions

The sensitivity of PMMA can be increased by adding CO in a blend. The sensitivity of the blend in comparison to the PMMA increased under forced developing conditions (i.e. longer developing times for development of lower doses). However, under severe conditions, the contrast of the blends suffered greatly which would make it difficult to form good lithographic patterns. At 15 to 20% thinning, the 33 and 50% CO by weight blends still show a 2- to 3-fold increase in sensitivity over that of PMMA. However, this occurred at the expense of a 25 to 40% reduction in contrast.

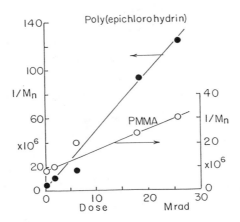

Figure 3.   Sensitivity to gamma radiation (Cobalt source).
● CO, G(s) = 5.1; ○ PMMA, G(S) = 0.8.

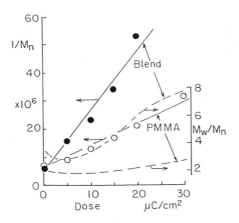

Figure 4.   Sensitivity to 50 keV electrons for (○)
PMMA and (●)CO:PMMA, 1:2.   The slope for the blend
is 2.9 times that for PMMA.

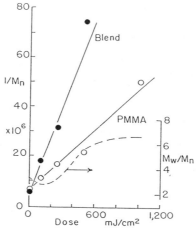

Figure 5.   Sensitivity to x-rays (flood exposure through berylium window) for same materials as in figure 4.   The slope for the blend is 2.7 times that for PMMA.   Polydispersity for PMMA ranges from 1.9 to 2.3 at all doses.

Table 1.   Dissolution in MIBK, 30°C

M_n of PMMA	PMMA alone		CO:PMMA, 1:4	
	Diss. Rate	f**	Diss. Rate	f
360×10^3	0.031	0.80	0.27	0.61
29×10^3	0.14	0.97	0.75	0.94

**Oscillation amplitude reduction factor (19), f = (a − b)/(a − b + 2s), where a and b are maximum and minimum reflected light intensities during dissolution and s is the offset (Figure 6).

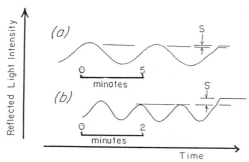

Figure 6.   Dissolution rates measured by interferometry using MIBK at 30°C for (a) high molecular weight PMMA and (b) CO:PMMA, 1:4.

Acknowledgments

This work was supported in part by the office of Naval Research. Some of the work was performed at the National Research and Resource Facility for Submicrometer Structures at Cornell which is partially supported by the NSF under grant ECS-8200312. Special thanks go to Dr. Jerry Silverman (IBM Corporation) for his help with the x-ray exposures at Brookhaven National Laboratories and Mr. Howard Aderhold (Ward Reactor Laboratory, Cornell) for the gamma radiation exposures.

Literature Cited

1. Haelbich, R. P., et al. J. Vac. Sci. Tech. 1983, B1(4), 1072.

2. Yaakobi, B., et al., Applied Physics Letters, 1983, 43, 686.

3. Lane, J. G., Proc. SPIE, 1984 448, 119.

4. Taylor, G. N., Solid State Tech. 1980, 23(5), 73.

5. Wilson, A. D., Solid State Tech. 1986, 29(5), 249.

6. Broers, A. N., Solid State Tech. 1985, 28(6), 119.

7. Moreau, W. M., Optical Engineering 1983, 22(2), 18.

8. Heuberger, A., Solid State Tech. 1986, 29(2), 93.

9. Anderson, C. C., Rodriguez, F., Proc. ACS Div. Polym. Matls,: Sci. & Eng. 1984, 51, 609.

10. Olabisi, O., in Encyclo. Chem. Tech. 3rd. ed., Wiley, New York, 1982, Vol. 18, p. 443.

11. Fernandes, A. C., Barlow, J. W., and Paul, D. R. J. Appl. Polym. Sci., 32, 5481 (1986).

12. Fernandes, A. C., Barlow, J. W., and Paul, D. R., J. Appl. Polym. Sci., 32, 6073 (1986).

13. Taylor, G. N., Coquin, G. A., Somekh, S., Polym. Eng. Sci., 1977, 17, 420.

14. Silverman, J. P., Haelbich, R. P., Grobman, W. D., Warlaumont, J. M., Proc. SPIE, 1983 393 99.

15. Haelbich, R. P., Silverman J. P., Warlaumont, J. M,, Nuc. Inst. Meth., 1984, 222, 291.

16. Rodriguez, F., Krasicky, P. D., Groele, R. J., Solid State Tech., 1985 28(5), 125.

17. Rodriguez, F., Groele, R. J., Krasicky, P. D., Proc. SPIE, 1985, 539, 14.

18. van Brederode, R. A., Rodriguez, F., Cocks, G. G., J Appl. Polym. Sci., 1968, 12, 2097.

19. Krasicky, P. D., Groele, R. J. Jubinsky, J. A., Rodriguez, F., Namaste, Y. M. N., Obendorf, S. K., Proc. 7th Int. Conf. on Photopolymers, SPE, 1985, p. 237.

RECEIVED April 8, 1987

RESIST MATERIALS AND PROCESSING
FOR OPTICAL LITHOGRAPHY

RESIST MATERIALS AND PROCESSING FOR OPTICAL LITHOGRAPHY

Optical lithography is the dominant production technology today for the fabrication of integrated circuits. As the size of individual features in microelectronic devices continues to shrink, the large infrastructure of optical technology will almost assuredly keep optical lithography effectively competitive with alternative lithographic technologies, even though we are approaching real physical barriers such as usable wavelength, numerical aperture of the lens, depth of field, etc.

Operating close to the resolution limit of optical lithography places stringent requirements on resist design and performance. The resist must provide the necessary linewidth control and resolution to enable the devices to be manufacturable. Some of the factors determining linewidth control, for example, that have impact on materials design can be broken down by imaging steps as follows:

Imaging Step	Controlling Factor	Materials Requirements
Incident image	Wavelength Laser intensity	Short wavelength sensitizers and resins. Materials capable of being exposed by high intensity sources.
Latent image	Resist absorption and contrast. Backscattered light from substrate	High absorption resists, contrast enhancement materials, and multilayer resist systems.
Developed image	Dissolution characteristics.	Solubility modifiers.
Pattern transfer	Resistance to transfer processes.	Thermal stability. Etch resistance.

The incident image is a direct function of the quality of the lens and the operating wavelength of the exposure tool. Given the current status of lens and source technology, advanced photolithographic hardware will be quite capable of operating around 250 nm as compared to 436 nm for current equipment. Furthermore, there is a high probability that these short-wavelength tools will use high-intensity laser sources which may also impact the design of resist materials.

The latent image or profile of exposed species in the resist will be modified from the incident or aerial image by the absorbing properties of the resist. These absorption effects have been used in "contrast-enhancement materials". The latent image can also be modified by light backscattered from the substrate. Multilayer resist techniques, in which an absorbing resist is placed between the imaging layer and the substrate, can eliminate this problem.

A developed image is produced when the developer removes either the exposed (positive tone) or unexposed (negative tone) areas of the resist. The variation in solubility rate with exposure dose has a large effect on the developed image. Hence, methods of modifying solubility characteristics are extremely important.

Finally, the developed image must be transferred into the semiconductor substrate, or in the case of multilayer structures, some underlying absorbing layer. Therefore, thermal stability and chemical etch resistance are also key materials requirements.

The chapters in this section describe new resist materials and processing considerations in which these various requirements are addressed. In some cases, several of the requirements are incorporated into a single resist design in a highly synergistic fashion.

All this creative work shows that there is sufficient effort being devoted to advanced optical materials today to eventually enable the manufacture of devices right down to the physical resolution limits of optical lithography.

Michael P. C. Watts
AZ Photoresist Products
American Hoechst Corporation
615 Palomar Avenue
Sunnyvale, CA 94086

Chapter 14

A New High-Sensitivity, Water-Developable Negative Photoresist

Anders Hult[1], Otto Skolling[1], Sven Göthe[2], and Ulla Mellström[2]

[1]Department of Polymer Technology, Royal Institute of Technology, S-100 44 Stockholm, Sweden
[2]AB Wilh. Becker, Box 2041, S-195 92 Märsta, Sweden

Polymers based on methylacrylamidoglycolate methylether (MAGME) have been synthesized and used as negative tone photoresists. MAGME containing polymers can undergo acid-catalyzed crosslinking by a selfcondensation reaction. P-Toluene sulfonic acid, a UV-deblockable sulfonic acid and a triphenylsulfonium salt have been used as catalysts. Acid-catalyzed crosslinking is another example of chemical amplification in photoresist sysems. These MAGME-polymers exhibit high sensivity but a limited linewidth resolution. They are soluble in harmless solvents like water and alcohols.

Much recent research has been focused on the development of new generation resist materials which possess improved sensitivity and enhanced resolution (1). An interesting approach to improve sensitivity involves the phenomenon of chemical amplification (2). This strategy has been demonstrated successfully for resist materials that undergo either acid catalyst hydrolysis (3) or polymerization (4). The key to these processes is photogeneration of strong acids. This can be achieved either by the use of onium salts (5) or latent UV-deblockable sulfonic acids (6). Both catalysts also generate a substantial amount of free radicals which may or may not interfere in the reaction. This paper will discuss another approach to chemical amplification, namely acid catalyzed self-condensation of acrylic polymers. These materials, based on methyl acrylamidoglycolate methylether, generate negative-tone images and consist of a water (or alcohol) soluble polymer which can undergo acid catalyzed crosslinking.

Experimental

Methyl acrylamidoglycolate methylether (MAGME) (American Cyanamid) was filtered while warm and recrystallized from xylene
(mp 70-73°C). All other monomers were freed from inhibitor on
an aluminium oxide column. p-Trimethylsilylstyrene was synthesized from p-chlorostyrene using a Grignard reaction and
chlorotrimethylsilane. Azobisisobutyronitrile (AIBN) was used
as free-radical initiator in all polymerizations and carbon
tetrabromide as chain-transfer agent. Polymerizations were
carried out at 60°C in a 50:50 mixture of toluene and butanol
under a nitrogen atmosphere. The reaction was carried out for
3,5 h and the formed polymer precipitated in cold diethylether.
It was then redissolved and repreciptated prior to further use.
once more IR spectra were recorded on a Perkin Elmer 1710
FTIR and NMR on a 200 MHz Bruker WP 200. FTIR was used to
determine the co-polymer composition.

p-Toluene sulfonic acid (I), a UV-deblockable sulfonic acid
(II) and triphenylsulfonium hexafluoroantimonate (III) were
used as acid catalysts in the crosslinking reactions (Figure
1). Films 20µm thick were casted on glass plates from methyl-
ethylketone solutions; thin films (SIµm) were spin-coated on
silicon wafers from a cyclohexanone solution of the polymer.
Irradiations were performed with an Oriel 82410 1000 W illum-
inator.

Results and Discussion

MAGME is a multifunctional acrylic monomer (Figure 2). It is
easily polymerized by a free-radical mechanism and can be copolymerized with several vinyl monomers (7). We have prepared a
variety of MAGME-containing polymers (Table 1) and studied
their ability to undergo acid-catalyzed crosslinking. Molecular
weight and molecular weight distribution was controlled by
addition of carbon tetrabromide which acts as a chain transfer
agent.

MAGME polymers can either be crosslinked in a selfcondensation
reaction or with a polyol (Scheme 1). The polyol can either be
blended or co-polymerized with MAGME. An example of such
monomer is 2-hydroxyethylmethacrylate. In this paper we will
only discuss selfcondensation of MAGME-polymers.

One of the most commonly used acid catalysts in organic reactions is p-toluensulfonic acid (PTSA). This acid was used to
evaluate the possibility of selfcondensation of MAGME-polymers.
Thick (20 µm) films were coated on glass plates and cured in an
oven at different temperatures and curing times. Data in Table
2 show the time required at lowest possible curing temperature.
In the case of PTSA it was difficult to achieve curing at temperatures below 120°C. This, and the fact that it took nearly

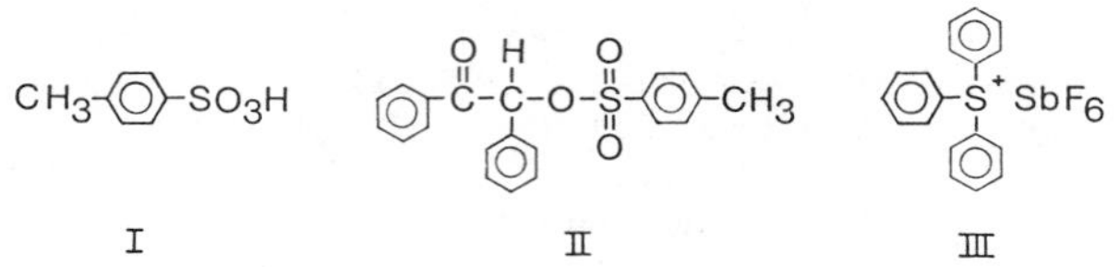

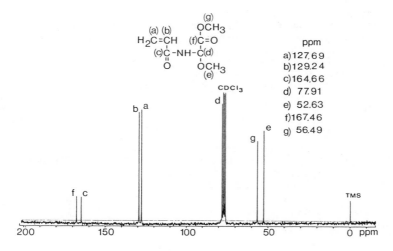

Figure 1.   I PTSA, II latent sulfonic acid, III triphenylsulfonium hexafluoroantimonate.

Figure 2. ^{13}C NMR of MAGME monomer.

TABLE 1

Polymer	% CBr$_4$ (w/w)	Conversion %	Mn	Mw/Mn	Tg C
MAGME	0.6	7	2.5 x 10^4	1.93	78
MAGME–MMA (26:74)	0.6	33	7.4 x 10^4	2.00	80
MAGME–MMA (48:52)	0.8	73	2.1 x 10^4	2.06	
MAGME–BA (39:61)	0.6	24	4.9 x 10^4	1.89	
MAGME–EHA (46:54)	0.6	34	3.3 x 10^4	1.86	
MAGME–STY	0.6	18	4.4 x 10^4	1.66	97
MAGME–TMSiSTY (50:50)	0.6	35	2.0 x 10^4	2.90	~100

All copolymerizations were carried out with 50:50 mixtures of the
two co-monomers. MMA-methylmethacrylate, BA - n-butylacrylate, EHA
- ethylhexylacrylate, STY - styrene, TMSiSTY - p-trimethylsilylstyrene.

TABLE 2

Crosslinking via self-condensation of poly-MAGME and its copolymers

Polymer	Film thickness (μ)	Catalyst	Conc. (w/w)	Irradiation time (sec)	Curing temp (C)	Curing time (min)
MAGME	1	III	10	3	100	2
MAGME–MMA (48:52)	20	I	0.3	–	120	20
MAGME–MMA (2:98)	20	III	5	4	100	5
MAGME–MMA (2:98)	20	II	5	1	100	5
MAGME–BA (39:61)	20	I	0.3	–	120	20
MAGME–TMSiSTY (50:50)	1	III	10	3	100	2

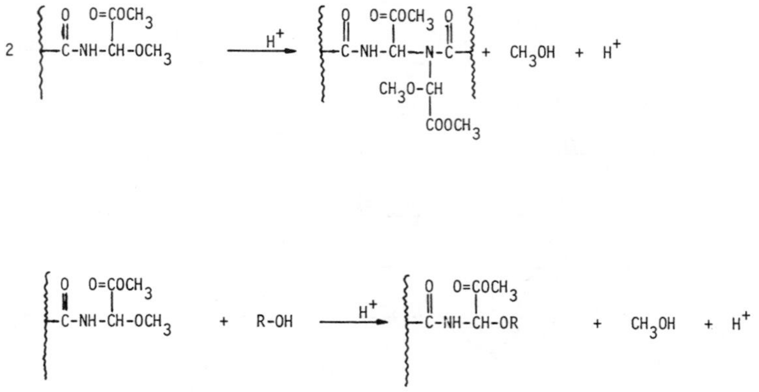

Scheme 1.   R is another Polymer chain.

20 minutes to cure the films indicated that the system was not
very acid sensitive. However, when the catalyst was changed to
the onium salt and UV-irradiated, both the curing temperature
and time decreased. This was also true at very low doses (~
10mJ/cm^2) of UV radiation.

This increased sensitivity is believed due to the fact that the
onium salt produces a much stronger acid, in this case HSbF$_6$.
Another contributing factor could be participation of free
radicals, formed during irradiation of the onium salt. To test
this hypothesis, experiments were performed with a latent UV-
deblockable sulfonic acid. This compound produces both PTSA and
free radicals when it is irradiated. Although the acid produced
was PTSA, the curing result was consistent with the result from
the onium salt experiment. These experiments indicates it is
the free radicals which are effective in crosslinking the
matrix. However, it may also just be a solubility effect, e.g.
catalysts II and III may be simply more soluble in the MAGME-
polymers than PTSA. Further experimentation is needed to determine whether it is a solubility effect or participation of free
radicals that explains the low sensitivity of PTSA. In the
experiments with pure PTSA, no increase in sensitivity was
observed when the PTSA concentration was increased above 0.3%
w/w.

In order to evaluate the possible use of MAGME-polymers in
resist applications, crosslinking studies were conducted on
thin films (1 μm), spinn-coated on silicon wafers. Poly-MAGME
is water soluble and most of its co-polymers are soluble in
alcohols (Table 3) making the materials relatively attractive
to work with from a production point of view.

TABLE 3

Solubility of poly-MAGME and its co-polymers

Solvent	MAGME	MAGME/MMA	MAGME/BA	MAGME/EHA	MAGME/STY
Water	x	-	-	-	-
Methanol	x	x	x	(x)	
Acetone	x	x	x	x	x
Ethylacetate	x	x	x	x	x
Chloroform	(x)	x	x	x	x
Toluene		x	x	x	x
N-heptane				x	

x = soluble, (x) = poor solubility, - = unsoluble, = soluble in isopropanol

One advantage with thin films is that the concentration of the photosensitive component can be increased. In contrast to the study with pure PTSA, sensitivity increased with increasing catalyst concentration. As a result, curing time could be decreased to 2 minutes at 100°C. The lithographic behaviour of the MAGME-polymers is similar to that of other photoresist systems based on a crosslinking mechanism. Due to the fact that the developer for the unexposed areas has a strong interaction with the crosslinked polymer, swelling becomes a problem. Figure 3 shows 2 μ lines and 4 μ spaces, which is probably close to the ultimate resolution of this system.

Co-polymers with p-trimethylsilylstyrene were also synthesized. This polymer showed a good resistance towards O_2-RIE. However, resolution in this resist is also controlled by swelling problems in the initial developing step. The MAGME-polymers can also be transformed to positive tone images. This is done by exposing the wafer to a base after it has been irradiated but before it has been thermally activated (Figure 4). The base (in our experiments we used ammonium hydroxide) consumes the acid and forms a latent catalyst image in the film, that can be activated by a subsequent flood exposure.

Although, the resolution of MAGME-polymers is limited to about 2 μm, they have several properties that make them attractive

Figure 3. Negative tone image made from poly-MAGME. Contact ptinted at 254 nm, using catalyst III. The SEM shows 2 μ lines and 4 μ spaces.

1. Coat

2. Expose

3. Treat with base

4. Flood-expose and heat

5. Develop

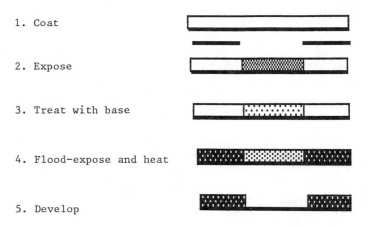

Figure 4. Positive tone process for MAGME-polymers.

for applications like circuit board fabrication, where high resolution is not required. The MAGME-polymers have high sensitivity (~ 10 mJ/cm^2) and they are soluble in harmless solvents which makes them attractive in a production environment.

Conclusion

Acid-catalyzed crosslinking is another example of chemical amplification in photoresist systems. It can be achieved via selfcondens-ation of MAGME-containing polymers. Crosslinking studies with PTSA, onium salts and a latent UV-deblockable sulfonic acid, indicate that free radicals participate in the crosslinking reaction also. These MAGME-polymers exhibit high sensitivity, but in common with most negative crosslinkable resist materials, a limited resolution. They are soluble in harmless solvents like water and alcohols.

Literature Cited

1. "Introduction to Microlithography", L.F. Thompson, C.G. Willson and M.J. Bowden, eds., ACS Symposium Series 219, (1983).

2. H. Ito and C.G. Willson, Polym. Eng. Sci., 1983, 23, 1012.

3. J.M.J. Fréchet, E. Eichler, H. Ito and C.G. Willson, Polymer, 1983, 24, 995.

4. A. Hult, S.A. MacDonald and C.G. Willson, Macromolecules, 1985, 18, 1804.

5. J.V. Crivello, "UV Curing: Science and Technology", S.P.Pappas ed., Technology Marketing Corporation, Stanford, Connecticut, 1978, p. 23.

6. G. Berner, R. Kirchmayr, G. Rist and W. Rutsch, SME Technical Paper, FC 85-446, 1985.

7. Technical Bulletin, American Cyanamid.

RECEIVED May 27, 1987

Chapter 15

Soluble Polysilanes in Photolithography

R. D. Miller[1], D. Hofer[1], J. Rabolt[1], R. Sooriyakumaran[1], C. G. Willson[1], G. N. Fickes[2], J. E. Guillet[3], and J. Moore[3]

[1]Almaden Research Center, IBM, San Jose, CA 95120-6099
[2]Department of Chemistry, University of Nevada, Reno, NV 89557
[3]Department of Chemistry, University of Toronto, Toronto, Ontario M5S 1A1, Canada

The drive toward improved resolution in lithographic processes has focused attention on dry processing techniques for the transfer of patterns created in thin imaging layers into thick planarizing layers. In this regard, silicon containing polymers have received considerable attention due to their ready accessibility and excellent O_2-RIE resistance. High molecular weight polysilane derivatives which contain only silicon in the polymer backbone have attracted interest recently because of their curious electronic properties. These materials are also radiation sensitive and polymer scissioning is the predominate process. As expected, the polysilanes are extremely resistant to O_2-RIE etching. In this paper we also discuss two multilayer applications for the production of high resolution patterns.

Future lithographic concerns are focused on the issues of improved process sensitivity and resolution. With regard to the latter, another feature has become apparent. As the industry moves toward smaller and smaller lateral geometries, the aspect ratios of the lithographic features necessarily increase due to minimum thickness requirements for the adequate coverage of topography. As this aspect ratio increases, it becomes increasingly difficult to maintain the desired geometries by classical wet development processes. For this reason, there has been increasing attention focused on dry, gas phase plasma processes which are intrinsically capable of highly anisotropic development (1). In this regard, O_2-RIE development seems ideally suited for multilayer schemes (2) such as the bilayer process illustrated in Fig. 1. In this example, a thick planarizing polymer layer is coated over the substrate with a thickness adequate to cover any surface topography. A thin layer of an appropriate imaging material is then coated over this material. This layer may be quite thin (it must be thick enough to be pinhole free) and hence can be imaged with high resolution. The imaged pattern is then processed by wet or dry development and transferred by anisotropic O_2-RIE into the planarizing polymer layer. By varying the plasma conditions, a variety of lithographic profiles can be generated.

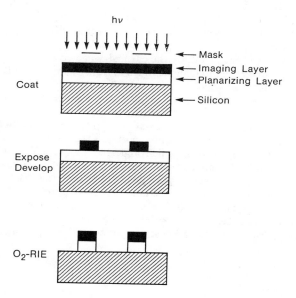

Figure 1. A bilayer process for the production of high resolution images utilizing O_2-RIE techniques for image transfer.

This type of process places unique demands on the imaging layer. Not only must it be sensitive to the exposing radiation, but it also must be resistant to the oxygen plasma used for the image transfer. Since most organic materials are readily attacked by the aggressive oxygen plasma (3), attention has focused on organometallic polymers containing elements such as silicon, boron, titanium, tin, *etc.*, which form nonvolatile, refractory oxides (4). The rapid formation of a thin metal oxide layer creates a barrier to subsequent oxygen etching. Since a large number of silicon containing polymers are synthetically available and SiO_2 forms a formidable oxygen etch barrier, these materials have been heavily investigated (5). In this regard, a relatively small amount of silicon (8-10% by weight) in the resist seems adequate for the formation of a suitable barrier layer (6).

In 1983 we began to study the chemistry and spectroscopy of substituted silane polymers and their potential applications to lithography (7-9). It was anticipated that materials of this type which contain only Si in the backbone would provide excellent etch barriers due to their relatively high silicon content.

Historically, the early attempts to prepare simple substituted polysilane derivatives produced only highly crystalline, intractable insoluble materials (10,11). We were intrigued by more recent reports on the preparation of soluble homo and copolymers (12-14) some of which appeared to undergo facile radiation induced crosslinking (14). In our initial studies, a number of unsymmetrically substituted, soluble high molecular weight polysilanes were produced by Wurtz coupling of the respective dichlorosilanes using sodium dispersion in toluene (7). Subsequently, we discovered that symmetrically substituted polysilanes containing long chain alkyl substituents could be also produced in the same manner and were quite soluble. Scheme I shows a representative sample of the soluble polysilane derivatives which have been prepared.

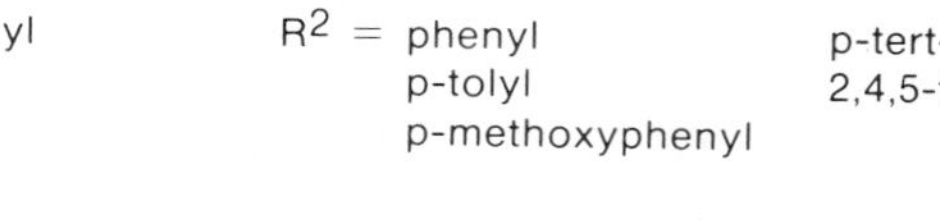

Aromatic

   R^1 = methyl R^2 = phenyl   p-tert-butylphenyl
                                   p-tolyl   2,4,5-trimethylphenyl
                                     p-methoxyphenyl

Aliphatic

   R^1 = methyl R^2 = n-propyl   n-dodecyl
                                     n-butyl   cyclohexyl
                                     n-hexyl

   R^1 = R^2 = n-butyl   n-octyl
                  n-pentyl   n-decyl
                  n-hexyl   n-tetradecyl
                  n-heptyl

Scheme I

The physical properties of these materials varied considerably with the substituents and have been described previously (7). The use of toluene as the polymerization solvent often resulted in low yields of high polymer (5-10%) particularly from more sterically hindered monomers (9). The effective polymerization of monomers of this type was critically dependent on the solvent. In this regard, initial studies on the polymerization of alkyl substituted dichlorosilanes showed that significant improvements in yield could often be realized by the use of solvent mixtures such as toluene-dimethoxyethane or toluene-diglyme (9). This dramatic improvement in yield suggested initially that the utilization of cation complexing solvents such as diglyme might be facilitating anionic propagation. More recent solvent studies on the polymerization of di-n-hexyldichlorosilane (Table I) do not seem to support this hypothesis. Examination of this data shows that as little as 5% diglyme by volume results in almost a six-fold increase in the crude polymer yield *versus* the use of toluene alone (entry 3). This was also accompanied by an increase in the high molecular weight portion of the bimodal polymer distribution. Attempts to confirm that the increase in yield was a result of cation complexation were unsuccessful, however, since the addition of the sodium cation complexing agent, (*i.e.*, 15 crown-5), resulted poorer yields of lower molecular weight material (entries 4 and 5). In addition, similar improvements in the yield and molecular weight distribution were obtained by the addition of a noncomplexing cosolvent such as n-heptane (entry 6). On the basis of this study, we feel that the results are best described as a bulk solvent effect rather than by specific cation solvation. Similar observations and conclusions have been recently reported by Zeigler and co-workers (15).

Much of the scientific interest in polysilane derivatives arises from their very curious electronic spectra (16). Even though the polymer backbone is comprised of only silicon-silicon sigma bonds, all high molecular weight polysilane derivatives absorb strongly in the UV. This transition has been described variously as a $\sigma\sigma^*$ or σ Si(3d) transition (16). Recent theoretical studies, however, have failed to support the significant involvement of silicon d orbitals in the excited state (17,18). The initial spectroscopic studies on a variety of unsymmetrically substituted, atatic polysilane derivatives both in solution and in the solid state resulted in the following conclusions (7): 1-both the λ_{max} and the ε_{Si-Si} are functions of the polymer molecular weight, increasing rapidly at first but approaching limiting values. This feature allows the direct comparison of high molecular weight samples with a variety of substituents; 2-simple alkyl substituted polysilanes absorb between 305-325 nm with red shifts occurring for the bulkier substituents; 3-aryl substituents which are directly attached to the silicon backbone cause a significant red shift to 335-350 nm which has been attributed to interaction between the backbone σ and σ^*orbitals with the substituent π and π^* orbitals (19,20). In this regard, the oxidation potentials of polymer films of the aryl substituted derivatives are also lower than those of alkyl substituted derivatives by ~0.5-1.0 V (21).

It now appears that these initial conclusions, while qualitatively correct, were somewhat oversimplified, and that the absorption spectra of polysilane derivatives can be effected by other factors. For example, it was noticed that while solutions of poly(di n-hexylsilane) (PDNHS) appeared quite normal (λ_{max} 316 nm), curious spectral changes occurred in the solid state (see Fig. 2) (22). The spectra of films of PDNHS which had been freshly baked to 100°C appeared quite similar to those observed in solution although the solid state spectra were understandably somewhat broader. However, upon standing at room temperature, the absorption at 316 nm

Table I. Polymerization of di n-hexyldichlorosilane: Solvent effects.

$$(C_6H_{13})_2\ SiCl_2 \xrightarrow[\text{Regular Addition}]{Na} -(\underset{\underset{C_6H_{13}}{|}}{\overset{\overset{C_6H_{13}}{|}}{Si}})_n-$$

Entry	Additive in Toluene	% Yield	$M_w \times 10^{-3}$	$M_n \times 10^{-3}$	$M_z \times 10^{-3}$	M_w/M_n	R	$\dfrac{\text{Area high } M_w \text{ polymer}}{\text{Area low } M_w \text{ polymer}}$
1	30% diglyme	37.3	1072.71 / 31.61	561.13 / 13.49	1756.21 / 58.42	1.91 / 2.34	1.42	
2	10% diglyme	36.0	1357.82 / 26.62	711.48 / 12.98	2084.48 / 46.93	1.91 / 2.05	2.61	
3	5% diglyme	33.7	1008.10 / 22.27	539.41 / 12.58	1524.26 / 34.17	1.87 / 1.77	3.41	
4	15-crown-5 0.5 mmol/mmol Na	17.5	6.39	4.52	10.20	1.42	—	
5	15-crown-5 1.1 mmol/mmol Na	12.9	6.38	4.61	10.02	1.38	—	
6	16% Heptane (Mole fraction 0.12)	26.6	1385.89 / 1.12	679.31 / 1.09	2179.36 / 1.14	2.04 / 1.03	9.61	
7	No Additive (Toluene only)	5.9	1982.00 / 1.20	1120.00 / 1.10	2873.00 / 1.40	1.77 / 1.13	3.12	

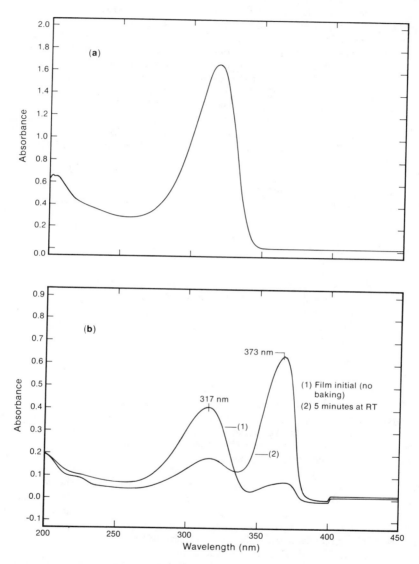

Figure 2. The electronic absorption spectrum of poly(di n-hexylsilane). (a) solution in cyclohexane, (b) a solid film (1) recorded after baking to 100°C 1 minute (2) after standing for 10 minutes at room temperature.

decreased markedly and was replaced by a strong, long wavelength absorption around 375 nm. This behavior was completely thermally reversible and the long wavelength absorption was destroyed upon heating. DSC analysis of the solid polymer showed a strong, reversible, endothermic transition at 41°C. This behavior was not limited to PDNHS, but was also observed in higher homologs such as the di n-heptyl, octyl, nonyl and decyl derivatives (23). IR, Raman and solid state NMR studies confirmed that the thermal transition at 41°C was due to crystallization of the hydrocarbon side chains (23). It was also suggested that this process caused the polymer backbone to be locked rigidly into a fixed conformation. IR, Raman and WAXD studies on stretch-oriented films showed conclusively that the backbone of PDNHS below 41°C was locked into a planar zig zag conformation and that this conformational change was associated with the spectral red shift (24). Above 41°C side chain melting occurred allowing the backbone to relax into a disordered, albeit still extended conformation with considerable backbone mobility. It now seems that the spectral values previously observed for unsymmetrical, atatic polysilanes are a result of conformational disorder in the polymer backbone rather than an intrinsic property of the chromophore.

$R_1 = C_2H_5-$, $n-C_4H_9-$, $iso\ -C_4H_9-$,
$(CH_3)_3C-$, $TMS-$, $n-C_6H_{13}-$

Since aryl substituted polysilanes absorb at longer wavelengths than their alkyl substituted counterparts, it was of interest to examine the spectral properties of symmetrically substituted diaryl polysilanes containing long chain alkyl substituents. The derivatives shown were synthesized in low yields (5-10%) by polymerization of the respective silyl dichlorides. To our surprise, the UV spectra of these materials were strongly red shifted to ~400 nm even in solution making them the most red shifted examples yet reported (see Fig. 3 for a representative example) (25). The electronic spectra of the substituted poly (diarylsilanes) were very similar both in solution and in the solid state and varied only slightly with temperature (−40 to 140°C). The long wavelength absorption was quite unexpected and could not be rationalized by simple electronic substituent effects. One possible, albeit tentative, explanation is that the bulk of these substituents is such that the polymer is forced into a planar zig zag conformation even in solution and this conformation is responsible for the spectral shifts. It is now clear that the absorption spectra of polysilane derivatives (see Table II) depend not only on the electronic nature of the substituents but also on the conformation of the backbone.

The dependence of both the λ_{max} and ϵ_{Si-Si} of the polymer on molecular weight suggests that any process which reduces the molecular weight will result in a bleaching of the initial absorption. This behavior is important for many lithographic applications (vide infra). In this regard, the irradiation of a 0.2 μm thick film of

Table II. UV spectroscopic data for substituted polysilane derivatives.

Polysilane	λ_{max} (nm)	
	Solution	Solid

$$\left(\begin{array}{c} R_1 \\ | \\ -Si- \\ | \\ Me \end{array} \right)_n$$

| | 305-325 | 305-325 |

R_1 = Alkyl (branched or normal)

$$\left(\begin{array}{c} R_1 \\ | \\ -Si- \\ | \\ R_1 \end{array} \right)_n$$

| | 313-320 | 313-320 |

R_1 = Alkyl (branched),
   n-C_4H_9-, n-C_5H_{11}-

$$\left(\begin{array}{c} R_1 \\ | \\ -Si- \\ | \\ Me \end{array} \right)_n$$

| | 335-350 | 335-350 |

R_1 = Aryl

$$\left(\begin{array}{c} R_1 \\ | \\ -Si- \\ | \\ R_1 \end{array} \right)_n$$

| | 315-320 | 370-380 |

R_1 = Alkyl, $C_6 \leq R_1 \leq C_{10}$ (normal)

| | 390-398 | 395-402 |

R_1 = Alkyl (branched or normal)

poly(cyclohexyl methylsilane) confirms that bleaching occurs upon irradiation (Fig. 4) (26). The reduction of molecular weight upon irradiation both in solution and in the solid state was confirmed by GPC analysis of the irradiated samples. An example of such an analysis performed on an irradiated film of poly(methyl dodecylsilane) is shown in Fig. 5. A similar analysis of an irradiated film of poly(methyl phenylsilane) also showed extensive molecular weight reduction, although a high molecular weight tail remained in the irradiated samples suggesting that concurrent crosslinking is occurring. All of the polysilanes investigated bleached significantly upon irradiation albeit the rate was somewhat dependent on structure. In this regard, polysilanes containing two large alkyl substituents seemed to bleach more rapidly than their less sterically hindered alkyl substituted counterparts.

The quantum yields for scissioning (Φ_s) and crosslinking (Φ_x) were determined for some representative polysilane derivatives both in solution and in the solid state (8). In all cases polymer scission is the predominant process and the Φ_s values ranged from ~0.2 to 1.0. For two cases, poly(methyl phenylsilane) and poly(cyclohexyl methylsilane) which were also examined in the solid state, the quantum yields were reduced by at least an order of magnitude from the solution values.

The photolysis of low molecular weight cyclic and acyclic silane derivatives has been shown *via* trapping experiments to produce substituted silylenes and silyl radicals (27,28). For this reason, the nature of the photochemical intermediates produced upon photolysis of a number of substituted silane high polymers was probed by irradiation in the presence of reagents such as triethylsilane, methanol and n-propanol (29). These materials have been previously shown to be effective traps for both silylenes and silyl radicals. In every case, disubstituted silylene adducts were isolated as the major products in 55-70% yield. Concurrently, evidence for the presence of silyl radicals was obtained by the isolation in lower yields of a variety of substituted disilanes when the trapping reagent was triethylsilane. The longer, linear silane derivatives were determined to be photolabile under the irradiation conditions (254 nm) and undergo subsequent photodegradation to yield ultimately the Si_2 species which inefficiently absorbs light at 254 nm. The evidence for the concurrent production of both silylenes and silyl radicals from the photolysis of high molecular weight polysilanes suggests that the mechanism of photodegradation (see Scheme II) may be similar to that postulated for the lower molecular weight, acyclic silanes. It is not known whether the silylenes are formed concurrent with or subsequent to radical formation.

Scheme II

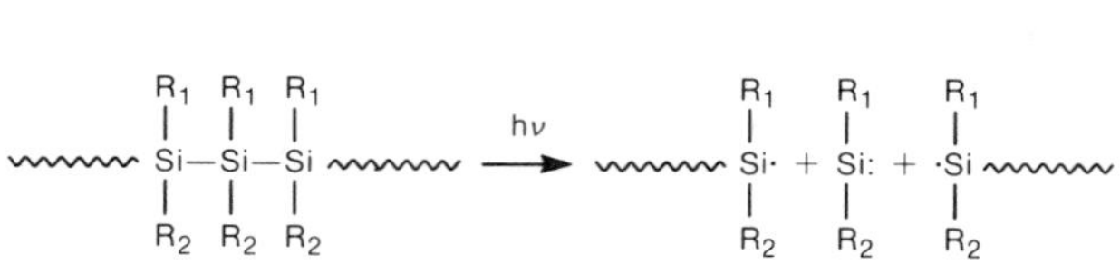

Lithography

Polysilanes by virtue of their high silicon content should provide an excellent etch barrier to O_2-RIE conditions. In this regard, the silicon content of the materials that we have examined ranged from 7-24%. We have studied in detail the etch resistance of poly(methyl phenylsilane) under O_2-RIE conditions and have compared it to a

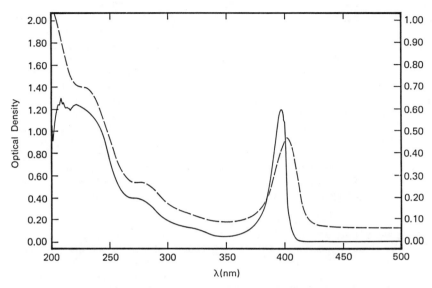

Figure 3. The electronic absorption spectra of poly(di-p-n-hexylphenylsilane): ____ solution in hexane, – – – solid film.

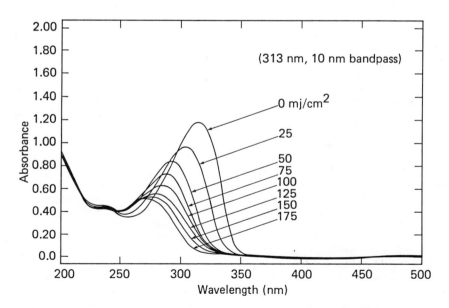

Figure 4. Photochemical bleaching of a film of poly(cyclohexyl methylsilane) upon irradiation at 313 nm. (Reproduced with permission from Ref. 26. Copyright 1984 The International Society for Optical Engineering.)

sample of hard baked AZ 1375J. The results of this experiment are shown in Fig. 6. During the etch period it was estimated that ~200Å of the polysilane layer were lost which corresponds to an etch ratio *versus* the AZ 1375J of ~50-100/1. Experiments using poly (cyclohexyl methylsilane) led to similar ratios. In the latter case, however, the surface of the polysilane became so soft during the early stages of etching, that the thickness could not be measured by mechanical means. This is presumably due to incomplete oxidation of the surface possibly leading to intermediate siloxane type polymers.   Upon continued etching, the surface rehardened and thickness measurements were again possible.   ESCA studies of the oxidized surface on another material, poly (di n-pentylsilane) showed a significant decrease in the C(1s) signal at 283.48 eV and a corresponding increase in the O(1s) signal. At the same time, there was an increase in the binding energy of the Si(2s) electrons of almost 3 eV which is consistent with the formation of an oxidized silicon species.   Comparison of the binding energy of the Si(2P) electrons in the oxidized material with data reported by Raider *et al.* (30) suggested the formation of SiO_x where x is estimated to be between 1.2-1.5.

Although we have utilized polysilanes in many lithographic configurations, at this time we will discuss only bilevel image transfer applications such as depicted in Fig. 1. For these applications, a thin layer (~2000Å) of a polysilane was coated over a thick planarizing polymer layer of a hard baked AZ type photoresist. The polysilane was imaged and developed prior to image transfer by O_2-RIE. We distinguish the processes by the nature of the initial development process which can either be dry or wet. In the former case, we have used the self developing characteristics of polysilanes exposed to intense light sources to produce the initial image.

In 1983, we reported that poly(methyl phenylsilane) and poly(methyl phenylsilane)-co-(dimethylsilane) produced self developed images upon irradiation with an excimer laser source at 308 nm (31). The self development of certain alkyl substituted polysilanes at 248 nm has also been reported by Zeigler and co-workers (32). Our initial experiments were complicated by the inability to remove all of the polysilane from the imaged area by irradiation. This residue subsequently formed an etch resistant barrier which prevented the efficient transfer of the images. The initial materials were known to have a significant crosslinking component upon irradiation (8), and it was felt that this feature might be responsible for the failure to completely remove the material at reasonable power levels. Since aryl substituted polysilanes absorb much more strongly in the deep UV region than their alkyl counterparts, they were highly desirable for DUV thin film imaging applications, if a material which could be completely removed could be identified.   In this regard, we felt that phenyl substituted polysilane derivatives which contained a bulky para substituent with no benzylic hydrogens might be less prone to crosslinking. For this reason, we prepared and tested poly(p-t-butylphenyl methylsilane) (PTBPMS). This material was a tough, brittle polymer with no observable glass transition below 150°C.   TGA analysis indicated that the polymer was stable to above 300°C in nitrogen. To assess its tendency toward crosslinking, PTBPMS was subjected to ^{60}Co γ radiolysis. Plots of $1/Mn$ and $1/Mw$ of the irradiated samples *versus* the dose (eV/g) yielded straight lines from which the G values for scission and crosslinking could be extracted (33). The G(s) and G(x) values for this polymer were determined to be 0.14 and 0.004, respectively. The high ratio of [G(s)/G(x) (~35)] confirmed that crosslinking was minimal under conditions of the irradiation.

The ablation rate of PTBPMS as a function of laser fluence was examined using a quartz crystal microbalance in a thermostated head which allowed the in situ

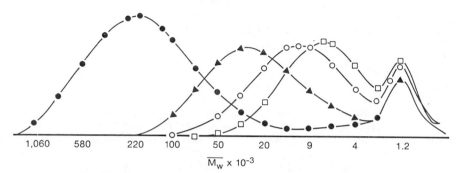

Figure 5. The change in molecular weight of a 0.006% solution of poly(dodecyl methylsilane) upon irradiation at 313 nm: ● 0 mJ/cm², ▲ 2 mj/cm², ○ 4 mJ/cm², □ 8 mj/cm²,

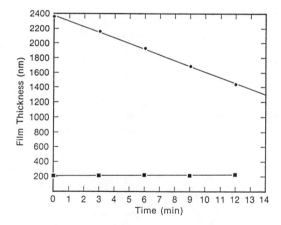

Figure 6. O₂-RIE etching of poly(methyl phenylsilane), O₂ 50 SCCM, −150V, 60 mTorr O₂ relative to hard baked AZ-1375 (240°C, 60m); ● AZ-1375 (hard baked), ■ poly(methyl phenylsilane).

monitoring of material loss as a function of dose (34). The optical density of this film was ~4-5 initially. Figure 7 shows that there is a threshold for ablation around 45 mJ/cm^2 − pulse. Below this critical fluence, there is actually a slight weight gain possibly due to the reaction of oxygen with reactive but nonvolatile fragments. Above the critical laser fluence, the rate of weight loss increased nonlinearly with the laser pulse energy as expected. Interestingly, a thick film of another material, poly (di n-pentylsilane), which had a much lower optical density at 248 nm (OD ~1.2) was not significantly ablated even at fluences of 75 mJ/cm^2 − pulse. However, when PDNPS was irradiated at 308 nm (which is near the absorption maximum at 313 nm), this material was readily removed at fluences above a 45 mJ/cm^2 − pulse threshold. This suggests that the energy deposited/unit volume is important for the removal of material which is consistent with a photothermal process. The rate removal of PTBMS at constant fluence is very similar in nitrogen or air and is slightly faster in pure O_2. It is, however, considerably faster in vacuum as would be expected for a process where small fragments are explosively ejected from the surface. PTBPMS can be used for the production of high resolution images as shown in Fig. 8 (33). In this case, the 2000Å thick polysilane layer was imaged with 550 mJ/cm^2 total exposure dose (248 nm). At this point, the pattern was clean and required no HF dip before transfer of the pattern using O_2-RIE. Using this all dry technique, submicron resolution was easily achieved.

Polysilanes can also be used in conventional bilayer processes where the imaging layer is wet developed prior to O_2-RIE image transfer (9,35). To demonstrate this principle, a thin film of an appropriate polysilane such as poly(cyclohexyl methylsilane) was spin coated over a thick planarizing layer of a hard baked AZ type resist. The resist was imaged in a PE-500 Microalign 1:1 projection printer using mid UV optics (100 mJ/cm ^{2}) and developed with isopropanol. The images were then transferred by O_2-RIE. Figure 9 shows submicron images with vertical wall profiles which were generated by this process. Etch parameters can be adjusted to vary the wall profiles as needed and Fig. 10 shows a sample which was deliberately overetched to produce an undercut profile suitable for metal liftoff processes. The use of other more sensitive polysilane derivatives such as poly (di n-pentylsilane) as the imaging layer allows the patterns to be developed at doses as low as 25-50 mJ/cm^2.

The use of dialkyl polysilanes derivatives such as poly(di n-hexylsilane) (PDNHS) which absorb around 375 nm in the solid state should permit imaging using a 365 nm source similar to that utilized in current I-line step and repeat tools. For this reason, a thin layer of PDNHS (2000Å) was coated over a planarizing layer of a hard baked AZ photoresist and exposed at 365 nm in a contact mode (150 mJ/cm^2). Subsequent development and O_2-RIE image transfer produced the images shown in Fig. 11. Prior to and during the exposure, it is important that the wafer be maintained below the transition temperature of PDNHS (~41°C) or the absorption at long wavelength will be destroyed as the backbone becomes disordered. This experiment demonstrates that polysilanes can also be utilized as effective photoresists for long wavelength exposures.

We have also demonstrated other lithographic uses for polysilanes as nonimageable O_2-RIE barrier layers (9), as short wavelength contrast enhancing materials (26) and more recently as sensitive, positive e-beam resists (36).

In summary, substituted silane high polymers are a new class of scientifically interesting radiation sensitive polymers with demonstrated potential as O_2-RIE resistant lithographic materials.

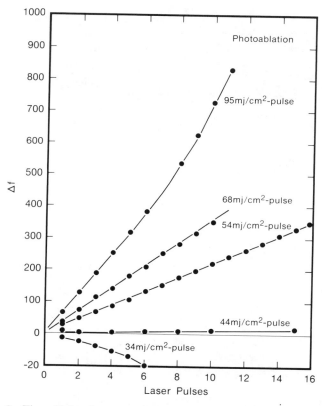

Figure 7. The self development of a thin film (~0.2 μm) of poly(p-t-butylphenyl methylsilane) upon irradiation at 248 nm (KrF excimer laser). Mass loss is proportional to the frequency change (Δf) of the quartz crystal microbalance.

Figure 8. Self developed submicron images produced in a film of poly(p-t-butylphenyl methylsilane) by irradiation at 248 nm (55 mJ/cm²-pulse, 550 mJ total dose). The images were transferred into 2.0 μm of a hard baked AZ photoresist by O_2-RIE.

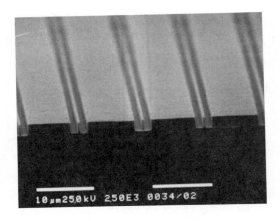

Figure 9. 0.75 μm features generated in a bilayer of 0.2 μm of poly(cyclohexyl methylsilane) coated over 2.0 μm of a hard baked AZ photoresist using mid UV projection lithography, 100 mJ/cm². Image transfer was accomplished using O_2-RIE. (Reproduced with permission from Ref. 35. Copyright 1984 The International Society for Optical Engineering.)

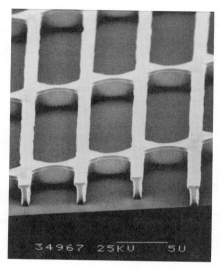

Figure 10. An overetched profile suitable for metal liftoff produced in a bilayer of 0.2 μm of poly(cyclohexyl methylsilane) imaged at 313 nm over 2.0 μm of a hard baked AZ photoresist. (Reproduced with permission from Ref. 35. Copyright 1984 The International Society for Optical Engineering.)

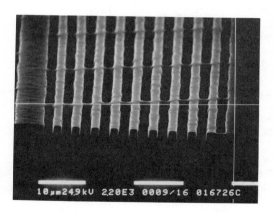

Figure 11. Images contacted printed in a bilayer of 0.2 μm PDNHS over 2.0 μm of a hard baked AZ photoresist, 150 mJ/cm^2, 365 nm; images were transferred by O$_2$-RIE.

Acknowledgments

The authors acknowledge the assistance of C. Cole and R. Siemens for the GPC and thermal analyses, respectively. R. D. Miller also gratefully acknowledges the partial financial support of the Office of Naval Research.

References

1. Introduction to Microlithography, Thompson, L. F.; Willson, C. G.; Bowden, M. J. eds., ACS Symposium Series, No. 219, American Chemical Society, Washington, D.C. (1983), Chap. 5.

2. Lin, B. J. ibid. (1983), Chap. 6.

3. Taylor, G. N.; Wolf, T. M. Polym. Eng. Sci. 1980, 20, 1087.

4. Taylor, G. N.; Wolf, T. M.; Stillwagen, L. E. Solid State Technol. 1984, 27, 145.

5. Reichmanis, E.; Smolinsky, G.; Wilkins, Jr., C. W. Solid State Technol. 1985, 28(8), 130, and references cited therein.

6. MacDonald, S. A.; Ito, H.; Willson, C. G. Microelectronic Engineering 1983, 1(4), 269.

7. Trefonas III, P. T.; Djurovich, P. I.; Zhang, X.-H.; West, R.; Miller, R. D.; Hofer, D. J. Polym. Sci., Polym. Lett. Ed. 1983, 21, 819.

8. Trefonas III, P. T.; West, R.; Miller, R. D.; Hofer, D.; J. Polym. Sci., Polym. Lett. Ed. 1983, 21, 823.

9. Miller, R. D.; McKean, D. R.; Hofer, D.; Willson, C. G.; West, R.; Trefonas III, P. T. in Materials for Microlithography, Thompson, L. F.; Willson, C. G.; Fréchet, J. M. J. eds., ACS Symposium Series No. 266, American Chemical Society, Washington, D.C., 1984, Chap. 3.

10. Kipping, F. S. J. Chem. Soc. 1924, 125, 2291.

11. Burkhard, C. S. J. Am. Chem. Soc. 1949, 71, 963.

12. Trujillo, R. E. J. Organomet. Chem. 1980, 198, C27.

13. Wesson, J. P.; Williams, T. C. J. Polym. Sci., Polym. Chem. Ed. 1980, 180, 959.

14. West, R.; David, L. D.; Djurovich, P. I.; Stearley, K. L.; Srinivasin, K. S. V.; Yu, H. J. J. Am. Chem. Soc. 1981, 103, 7352.

15. Zeigler, J. M. Polym. Preprints 1986, 27, 109.

16. Pitt, C. G.; in Homoatomic Rings, Chains and Macromolecules of the Main Group Elements Rheingold, A. L.; ed., Elsevier Publishers., New York, 1977, and references cited therein.

17. Bigelow, R. W.; McGrane, K. M. J. Polym. Sci. Part B: Polym. Phys. 1986, 24, 1233.

18. Bigelow, R. W. Chem. Phys. Lett. 1986, 126, 63.

19. Pitt, C. F.; Carey, R. N.; Loren Jr., E. C. J. Am. Chem. Soc. 1972, 94, 3806.

20. Pitt, C. G.; Bursey, M. M.; Rogerson, P. F. J. Am. Chem. Soc. 1970, 92, 519.

21. Diaz, A. F.; Miller, R. D. J. Electrochem. Soc. 1985, 132, 834.

22. Miller, R. D.; Hober, D.; Rabolt, J.; Fickes, G. N. J. Am. Chem. Soc. 1985, 107, 2172.

23. Rabolt, J. F.; Hofer, D.; Miller, R. D.; Fickes, G. N. Macromol. 1986, 19, 611.

24. Kuzmany, H.; Rabolt, J. F.; Farmer, B. L.; Miller, R. D. J. Chem. Phys. 1986, 85, 7413.

25. Miller, R. D.; Sooriyakumaran, R. J. Polym. Sci., Polym. Lett. Ed. (in press).

26. Hofer, D. C.; Miller, R. D.; Willson, C. G.; Neureuther, A. R. Proc. SPIE 1984, 469, 108.

27. Ishikawa, M.; Kumada, M. Adv. Organomet. Chem. 1981, 19, 51, and references cited therein.

28. Ishikawa, M.; Takaoka, T.; Kumada, M. <u>J. Organomet. Chem.</u> 1972, <u>42</u>, 333.
29. Trefonas III, P.; West, R.; Miller, R. D. <u>J. Am. Chem. Soc.</u> 1985, <u>107</u>, 2737.
30. Raider, S.; Flitsch, R. <u>J. Electrochem. Soc.</u> 1976, <u>123</u>, 1754.
31. Hofer, D. C.; Jain, K.; Miller, R. D. <u>IBM Tech. Disc. Bull.</u> 1984, <u>26</u>, 5683.
32. Zeigler, J. M.; Harrah, L. A.; Johnson, A. W. <u>Proc. SPIE</u> 1985, <u>539</u>, 166.
33. Schanabel, W.; Kuvi, J. <u>Aspects of Degradation and Stabilization of Polymers</u> Jellinek, H. H. G.; ed., Elsevier Publishers, New York, 1978, Chap. 4.
34. Miller, R. D.; Hofer, D.; Fickes, G. N.; Willson, C. G. Marinero, E.; Trefonas III, P.; West, R. <u>Polym. Eng. Sci.</u> 1986, <u>26</u>, 1129.
35. Hofer, D. C.; Miller, R. D.; Willson, C. G. <u>Proc. SPIE</u> 1984, <u>469</u>, 16.
36. R. D. Miller, presented at the Regional Technical Conference on Photopolymers: Principles - Processes and Materials, Ellenville, New York, 1985.

RECEIVED April 8, 1987

Chapter 16

Evaluation of Water-Soluble Diazonium Salts as Contrast-Enhancement Materials Using a g-Line Stepper

S.-I. Uchino[1], T. Ueno[1], T. Iwayanagi[1], H. Morishita[1], S. Nonogaki[1], S.-I. Shirai[2], and N. Moriuchi[2]

[1]Central Research Laboratory, Hitachi, Ltd., Kokubunji, Tokyo 185, Japan
[2]Hitachi Device Development Center, Ohme, Tokyo 198, Japan

Water soluble aromatic diazonium salts have been evaluated as photobleachable dyes for contrast enhancement lithography using a g-line stepper. The diazonium salts used in this experiment are zinc chloride salts of 4-diazo-2-ethoxy- N,N-dimethylaniline chloride 4-diazo-2,5-diisopropoxy- morpholino-benzene chloride and 4-diazo-2,5-dimethyl- N,N-dimethylaniline chloride.

An aqueous solution of diazonium salt and polyvinyl pyrrolidone is used as a contrast enhancement material. An improved resist profile is obtained with this CEL material when a g-line stepper is used. Resist contrast is discussed in terms of such optical characteristics as the quantum yield of bleaching and the molar absorption coefficient for the materials.

The thermal stability of diazonium salts in an aqueous solution is also examined.

1. Introduction

In recent years, rapid progress has been made in lithographic technology along with a concomitant reduction in the minimum feature size. The object of microphotolithography is to form high resolution resist patterns. Various techniques have been evaluated for improving the resolution of photoresist.

Griffing and West introduced the concept of contrast enhancement lithography (CEL) which improves the contrast of the resist process (1,2,3). The CEL process involves coating a CEL layer containing photobleachable dye on a conventional resist. A CEL layer is opaque before exposure but becomes transparent during exposure. When the areal image of a mask is incident on such a layer, the regions of the layer that are exposed to the highest intensity bleach through first, while those parts of the layer that receive the lowest intensity bleach through at a later time(1). Therefore, the degraded optical image caused by the lens system of the exposure apppratus can be improved by passing the exposure light through the CEL layer.

The contrast enhancement materials reported so far can be classified into three categories, namely nitrones(1,2,3), polysilanes(4), and a diazonium salt(5). However, the use of nitrone and polysilane in the CEL process presents a problem, because organic solvents are required in the film forming and

0097–6156/87/0346–0188$06.00/0

removal processes.   The use of such toxic solvents is not desirable
in the production environment.   Although the use of a diazonium salt
as a CEL dye has been reported by Halle(5),   the performance was not
satisfactory because of its   low   optical   density   at   exposure
wavelength.

This paper reports on water soluble diazonium salts which offer
good optical characteristics for the CEL process.

2. Experimental
2.1 CEL Materials

The diazonium   salts   used   in   the   experiment   are   listed   in
Table 1.   Two   of   them,   D2   and   D4,   were   synthesized   through
N,N-dimethylation,   nitrosation,   reduction,   and   diazotization   of
p-xylidine   and   o-phenetidine,   respectively.   The   details   of
synthesis for D4 are described here.

A solution   of   o-phenetidine (137g)   in water (250g) was mixed
with dimethyl sulfate (265g)   at   5°C.   After the solution was made
alkaline by addition of a 30% aqueous sodium hydroxide solution, the
reaction   mixture   was   extracted   with   ether.   The extract   was
distilled to yield N,N-dimethylamino compound, b.p.110°C/18mmHg.

N,N-dimethylamino   compound (133g)   in   acetic   acid (242g) and
water (300g)   was   mixed   with   a   30%   aqueous   sodium   nitrate
solution (280g) at   5°C.   The   precipitated   nitroso   compound   was
filtered out and washed with cold water.

To   a   stirred   mixture   of   the   nitroso   compound (40g),
methanol (200ml),   and   palladium   carbon (0.1g)   was   added   a   30%
aqueous sodium borohydride   solution under nitrogen bubbling.   The
reaction mixture was stired   untill it became colorless.   After the
solvent   and   palladium   were   removed,   the   reaction   mixture   was
extracted   with   ether   and   distilled   to   give   amino   compound,
b.p.105°C/1mmHg.

A mixture of amino compound (32g) and concentrated hydrochloric
acid   (58.5g) was cooled to below 0°C with crushed ice (40g) while a
30% aqueous sodium nitrate (45g)   was   slowly added.   A 50% aqueous
zinc   chloride   solution (54g)   was   added   to   the   solution and the
solution was kept below 10°C for   15   minutes.   The precipitated D4
was filtered off   and   the   solid   was   dissolved   in   a   1% aqueous
hydrochloric   acid   solution   and   mixed   with   a   50%   aqueous zinc
chloride solution (10g).   The recrystallized D4 was filtered off and
washed with ether and dried under a vacuum.   D2 was also synthesized
in the same manner.

2.2 Evaluation of CEL performance using diazonium salts

CEL   solutions   were   obtained   by   dissolving   poly(N-vinyl
pyrrolidone) and a diazonium salt   (D2, D3 or D4) in aqueous acetic
acid.   The solutions were spin-coated on a conventional photoresist
layer formed on a silicon wafer.   CEL   layers   on quartz substrate
were used for optical transmittance measurements.

Exposure was carried out with a Hitachi RA 501 g-line reduction
aligner.   UV   absorption   spectra   were   measured   on a Hitachi 340
spectrophotometer and infrared spectrophotometer.

The exposure   characteristic   curves   were obtained by plotting
normalized film thickness   versus   the   logarithm   of exposure time.
The improvement in resist contrast   was evaluated by comparing the
γ-value (slope   of   the   exposure   characteristic   curve)   for

conventional photoresist with that of the same system having a CEL layer on top of the photoresist.

After exposure, the CEL layer was removed using a water rinse. OFPR800 (Tokyo Ohka Co.) was used as the positive photoresist and developed with a 2.38% aqueous solution of tetramethylammonium hydroxide.

3. Results and discussion
3.1 Diazonium salts and binder polymers

Photobleachable dyes for water soluble CEL layers are required to have the following properties: (1) a strong absorption at the wavelength of the g-line stepper (436 nm), (2) a high solubility in water, (3) a high thermal stability in the aqueous solution and in the film.

Three diazonium salts, 4-diazo-2,5-dimethyl-N,N-dimethylaniline chloride zinc chloride (D2), 4-diazo-2,5-diisopropoxy- morpholino-benzene chloride zinc chloride (D3), and 4-diazo-2-ethoxy- N,N-dimethylaniline chloride zinc chloride (D4) which met the above requirements, were selected as the photobleachable materials.

The solubility of these compounds in water is dramatically improved by adding acetic acid. Since conventional novolak photoresist is insoluble in a mixture of acetic acid and water, the CEL layer can be coated without the need for a barrier layers. In addition, the CEL layer can be easily removed after exposure by rinsing it in water

It is neccessary that the binder polymers (1) be compatible with diazonium salts, (2) be uniform in film thickness when the layer is coated on the photoresist film, (3) show a high gas permeability to allow nitrogen photogenerated during the exposure of diazonaphtoquinone resist to diffuse. Many water soluble polymers were examined, and it has been found that poly(N-vinyl pyrrolidone) and its copolymers satisfy the above requirements.

The UV absorption spectra of the CEL layers which consist of these diazonium salts (D2, D3, and D4) and poly(N-vinyl pyrrolidone) are shown in Fig.1.

3.2 Bleaching reaction of diazonium salts

The UV transmittance of the D4-CEL layer increases with exposure time as shown in Fig.2. This increase in transmittance is caused by the decomposition of diazonium salt. The photochemical reaction of diazonium salts has been thoroughly investigated(6). The main reaction of diazonium salts in the solid phase can be described by the following fromula;

$$R-\langle\bigcirc\rangle-N_2Cl \xrightarrow{hv} R-\langle\bigcirc\rangle-Cl + N_2$$

Infrared spectra of the unbleached D4-CEL layer and the completely bleached layer are shown in Fig.3. The sharp absorption band at 2150 cm^{-1} which is assigned to N-N stretching vibration completely disappears after exposure (Fig.3). The broad bands at 3450 cm^{-1} and 3000 cm^{-1} are due to water and a dimethylamino group, respectively.

Similar infrared spectra were observed for both the D2- and D3-CEL layers.

Table 1 Diazonium salts evaluated as CEL dyes

SUB. NAME	R_1	R_2	R_3	λ_{max}(nm)	*ε (x10^4)
D 1	$-N(CH_3)_2$	$-H$	$-H$	380	3. 6
D 2	$-N(CH_3)_2$	$-CH_3$	$-CH_3$	394	3. 1
D 3	$-N\hspace{-2pt}\bigcirc\hspace{-2pt}O$	$-OCH(CH_3)_2$	$-OCH(CH_3)_2$	404	2. 7
D 4	$-N(CH_3)_2$	$-OCH_2CH_3$	$-H$	409	3. 1
GENERAL FORMULA	$R_1\bigcirc\!\!-N_2Cl\cdot n\ ZnCl_2$ with R_3 above and R_2 below				

* molar absorption coefficient in water at λ_{max}

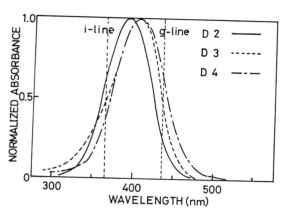

Fig.1 UV absorption spectra of the CEL layers containing diazonium salts.   Absorbance is normalized at the maximum absorbance of the CEL layer.

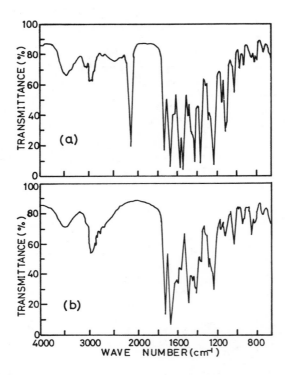

Fig.2 Infrared spectra of D4-CEL layer.   (a) unbleached layer (b)
bleached layer

3.3 Thermal stability of diazonium salt

It is well known that most of the diazonium compounds are thermally unstable and easily decompose in an aqueous solution. Therefore, the thermal stability of diazonium salts in an aqueous solution and in a film was evaluated by a kinetic analysis of the thermal decomposition and differential scanning calorimetry (DSC) analysis.

The thermal decomposition of diazonium salts in an aqueous solution is a first order reaction as shown in Fig.4. Arrhenius plots of diazonium salts in an aqueous solution are shown in Fig.5. The relationship between the decomposition temperature (Td) obtained from DSC curves and the decomposition rate constant (k) of the diazonium salts at $5^{\circ}C$ are shown in Fig.6. A good linear relationship is observed between Td and ln k for both the aqueous solution and the film from this figure. These linear relationships make it possible to predict the stability of diazonium salts in an aqueous solution and in film from the decomposition temperature Td in the solid state.

For example, Td for D2 is the highest in the three diazonium salts suitable for g-line exposure. The decomposition rate constant of an aqueous solution of D2 at $5^{\circ}C$ can be estimated to be 2.3×10^{-5} (hr^{-1}) from the linear relationship in Fig.6. This value means that 5% of the D2 in an aqueous solution will decompose after storage for 6 months at $5^{\circ}C$.

3.4 Bleaching characteristics

The bleaching curves of the CEL layers are shown in Fig.7. Two parameters which express the optical efficiency of the CEL layers can be derived from the bleaching curves. One is the T_{∞}/T_{0} ratio where T_{0} represents the initial transmittace of the CEL layer and T represents the transmittance of the completely bleached CEL layer. High contrast enhancement is provided by a high T_{∞}/T_{0} ratio(1). Since the three diazonium salts have a large T_{∞}/T_{0} ratio, a good improvement in resist contrast can be expected.

The other parameter relates to bleaching speed which is a function of molar absorption coefficient(ε) and bleaching quantum yield(ϕ). This parameter will be discussed in section 3.6.

3.5 Contrast and resist profiles

The performance of the CEL layer can be evaluated from the exposure characteristic curves for the conventional photoresist (OFPR800) and the CEL/OFPR800 layer systems shown in Fig.8. The CEL layer thicknesses used in these experiments were 0.8 um for D2, 1.0 um for D3, and 0.5 um for D4. A 1.0 um thick OFPR800 layer was used as the imaging layer.

The contast ratio is defined as γ_{CEL}/γ where γ represents the contrast of OFPR800 without a CEL layer and γ_{CEL} stands for the contrast of OFPR800 with the CEL layer. The contrast ratios for the three diazonium salt-CEL systems were 1.74 for D2, 2.62 for D3, and 2.52 for D4 (Fig.8). A good improvement in the resist image is expected from these contrast ratios.

Scanning electron microphotographs of a 1.0 um line/space

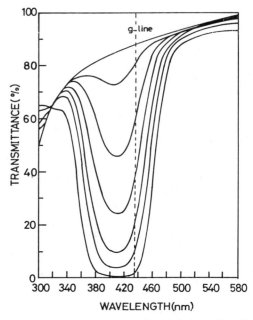

Fig.3 UV transmittance spectra of the D4-CEL layer.

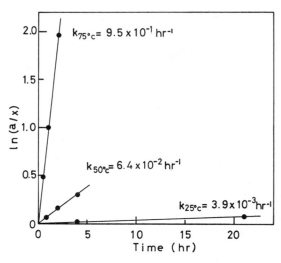

Fig.4 First order kinetics for the decomposition of the diazonium salts.

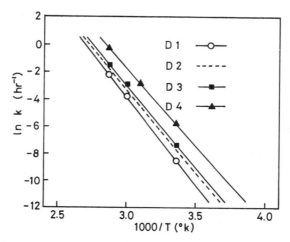

Fig.5 Arrhenius plot of the aqueous solution of the diazonium salts.

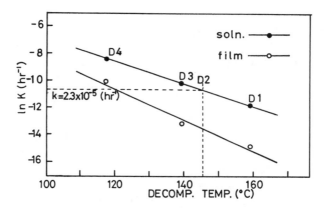

Fig.6 Relationship between decomposition rate and temperature in solid state for the diazonium salts.

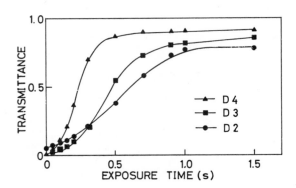

Fig.7 Bleaching curves of the CEL Layers.   CEL layer thickness is
0.8 um for D2, 1.0 um for D3, and 0.5 um for D4.

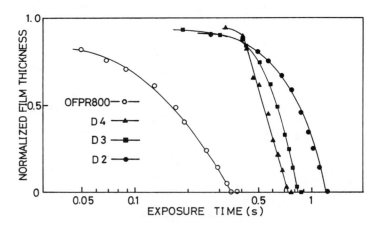

Fig8.   Exposure characteristic curves for the OFPR800 and
CEL/OFPR800 systems.   The film thickness of OFPR800 was 1.0um.
The thickness of each CEL layer was the same as in Fig.7.

resist pattern printed with and without a D4-CEL layer are shown in
Fig.9.   The photographs show that the residual film thickness and
wall profiles of the resist are improved with the diazonium salt-CEL
system.

3.6 CEL curve

According to Griffing and West, the main requirements for CEL
materials are as follows(1):   (1) the molar absorption coefficient
for the photobleachable dye must be as high as possible, (2) the
T_{∞}/T_0 ratio must be as large as possible, and (3) the quantum yield
of the bleaching reaction must be as high as possible.

These requirments were examined from the viewpoint of the
relationship between the resist contrast and such optical properties
of the photobleachable dyes as the quantum yield and the molar
absorption coefficient.

It was found that the contrast ratio (γ_{CEL}/γ) can be expressed
by using two parameters $\underline{P}$ and $\underline{a}$.   Parameter $\underline{P}$ is given by $\varepsilon \phi I_0 t_0$,
where ε is the molar absorption coefficient, ϕ is the quantum yield
of the bleachable material, and $I_0 t_0$ is the sensitivity of the
photoresist.   Parameter $\underline{a}$ is given by $T_{\infty}/T_0 - 1$, which represents the
transmittance ratio of the bleached layer to the unbleached one.
Similar parameters have been reported by Diamand and Sheats.(7)

The values of parameters $\underline{P}$ and $\underline{a}$ can be obtained from the
bleaching curves (Fig.7)   and   the   resist   sensitivity.   The
relationship between the contrast ratio and the two parameters ($\underline{P}$
and $\underline{a}$) is shown in Fig.10.   This curve is referred to as the CEL
curve.   It can be seen from this curve that the contrast increases
with increasing $\underline{a}$-values, and there is an optimum $\underline{P}$-value which
maximizes the contrast.

Contrast ratios can   be   calculated   from   the bleaching
curve (Fig.7) and the sensitivity of the photoresist, while the
observed contrast ratios can be directry obtained from the exposure
characteristic curves shown in Fig.8.   The calculated contrast
ratios were compared with the observed ratios in the CEL curve.   The
calculated values are in fair agreement with the observed values.
The CEL curve is thus thought to be a useful tool for designing CEL
dyes because it can predict the performance of the dyes used for CEL
materials.

A detailed discussion of this curve will be reported in a
subsequent paper.

4. Conclusion

Three kinds of water soluble diazonium salts were evaluated as
photobleachable dyes   for   the   enhancement   of   contrast   in
photolithography.   CEL layers formed from these deazonium salts and
polyvinyl pyrrolidone have good optical characteristics and they can
improve the ability of the conventional photoresist.

A more thermally stable diazonium salt or method of stabilizing
the aqueous solution is needed for the diazonium salt-CEL layers in
practical use.

A CEL curve was presented which can express the relationship
between resist contrast and optical properties such as molar
absorption coefficient and quantum yield.   This curve is thought to
be a useful tool for designing CEL materials.

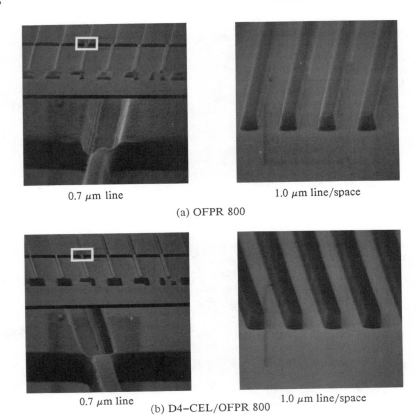

0.7 μm line   1.0 μm line/space

(a) OFPR 800

0.7 μm line   1.0 μm line/space

(b) D4-CEL/OFPR 800

Fig.9 Scanning microphotograph of photoresist patterns. (a) without D4-CEL layer, (b) with D4-CEL. Feature definition is enhanced in submicron structure.

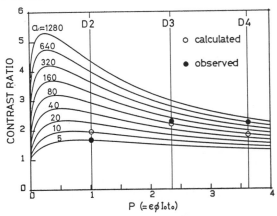

O CEL curve: relationship between resist contrast ratios ptical characteristic parameters of CEL materials(P and a).

Acknowledgments
    The authors would like to thank Toyozi Ohshima and Ryou-ichi Narushima of Hitachi Chemicals Co., Ltd. for synthesizing the diazonium salts, and to express our gratitude to Dr. Hiroshi Yanazawa of Hitachi Central Research Laboratory for making the optical exposure experiments used in this study.

References
1) B.F. Griffing, P.R. West, Poly. Eng. Sci., 23,
   947(1983)
2) B.F. Griffing, P.R. West, IEEE Electron Device Lett.,
   EDL-4, 14(1983)
3) P.R. West, B.F. Griffing, SPIE, 394, 33(1983)
4) D.C. Hofer,et al., SPIE,469, 108(1984)
5) L.F. Halle, J. Vac. Sci. Technol.,133, 323(1985)
6) Kosar: Light sensitive systems, Chapter 6,
   Jhon Wiley and sons, New York(1965)

7) J.J. Diamond, J.R. Sheasts, IEEE Electron Device Lett.,
   EDL-7, 383(1986)

RECEIVED May 6, 1987

Chapter 17

Thermally Stable, Deep-UV Resist Materials

S. Richard Turner[1], K. D. Ahn[2], and C. G. Willson[2]

[1]Corporate Research Laboratories, Eastman Kodak Company, Rochester, NY 14650
[2]Almaden Research Center, IBM, San Jose, CA 95120-6099

A new, thermally stable resist based on styrene copolymers with various photochemical acid generators is described. The title copolymers were prepared either by polymer modification or by copolymerization from monomers. Exposure of a resist formulated from the protected copolymers and a photoacid generator followed by baking results in efficient deprotection to produce the free phenolic copolymer in the exposed areas. The resulting latent image can be developed in either positive or negative tone upon proper choice of developer. The resulting relief images exhibit no detectable change in size or shape after extended heating at 200°C.

The increasing complexity and sophistication involved in integrated circuit manufacturing places rigorous performance demands on the polymeric resist materials used to delineate the circuit patterns. These demands have led to the development of new resist chemistry that allows improved resolution because of sensitivity to short wave length, deep UV light while at the same time providing greatly improved sensitivity. These new materials derive their sensitivity from a process that the authors term chemical amplification. One example of such a system is based on poly(p-t-butyloxycarbonyloxy)styrene (t-BOC styrene) (1).

The resist community has also continued attempts to improve more traditional systems. Resistance to deformation due to flow during high temperature processing is a property that is deficient in most currently available photoresists. The best of the widely used diazoquinone-novolac materials, for example undergo deformation at approximately 150°C (2-3-4). This deficiency in thermal stability is primarily due to the relatively low Tg of the novolac matrix resin. The Tg of these materials ranges between 70 and 120°C depending on structure and molecular weight (5). Image profiles that are stable to 200°C are desired for several processes used in semiconductor manufacturing. Several post exposure treatment schemes that improve the thermal stability of the novolac based resists have been proposed (6-7) but all of these schemes involve increased process complexity and higher cost.

Recently, a new family of high Tg, base soluble, phenolic copolymers based on N-(p-hydroxyphenyl)maleimide was reported (8). These copolymers were found to serve as replacements for novolac in diazoquinone based resists. Such formulations yield images that exhibit remarkable thermal resistance. The materials show no deformation upon exposure to 200°C after patterning. In this report, we describe a new, thermally stable, deep UV resist that combines the thermal stability of the new, high Tg, phenolic copolymers with the high sensitivity of the chemical amplification deprotection concept. This is accomplished by blocking the phenolic group of the high Tg, N-(p-hydroxyphenyl)maleimide polymers and copolymers with the t-butyloxycarbonyl (t-BOC) protecting group. When these materials are imagewise exposed in the presence of photoacid generating compounds such as sufonium salts, and then baked, the phenolic groups are deprotected via a thermolysis reaction that is acid catalyzed in a fashion analogous to the key process responsible for the function of t-BOC styrene (1-9). High resolution images have been obtained in either positive or negative tone depending on the developer. The resulting images exhibit no deformation upon heating to 200°C. A recent report has described a similar approach based on t-BOC protection of an unsubstituted maleimide copolymer (10).

Experimental

Instruments. Infrared spectra were obtained with an IBM IR/32 FTIR Spectrometer. NMR spectra were recorded on Varian EM390 and Brucker WP200 NMR Spectrometers in deuteriochloroform except as indicated. UV spectra were recorded on a Hewlett Packard Model 8450A UV/VIS Spectrometer using thin films cast on quartz plates. Mass spectra were obtained with a Hewlett-Packard 5995A GC/Mass Spectrometer. Gel permeation chromatogrames were obtained on a Waters Model 150°C GPC equippped with µ-styragel columns using THF and polystyrene calibration. Thermal analyses were carried out with a DuPont 951 and 1090 Thermal Analyzer at a heating rate of 10°C/min for DSC and 5°C/min for TGA measurement. Thermal analysis was carried out under inert atmosphere.

Preparation of N-(p-t-butyloxycarbonyloxyphenyl)maleimide I. The t-BOC monomer I was prepared by a reaction of p-t-butyloxycarbonyloxyaniline (11) with maleic anhydride according to the procedure of S. R. Turner (8). The detailed preparation method will be published elsewhere (12).

Copolymerization of I with styrene. A glass ampoule was charged with 1.45g (5m mol) of I, 0.53g (5m mol) of styrene, and 33 mg of AIBN (2 mol %) dissolved in 2.0 ml of dioxane. The ampoule was sealed under vacuum after a freeze-thaw cycle and the copolymerization was carried out at 58°C for 3 hours. The jelly-like polymer mixture was dissolved in NMP and the polymer was isolated by precipitation into methanol. After drying in vacuo, 1.76g (89%) of a white fibrous polymer were collected. The polystyrene equivalent molecular weight (Mw) is 1.3×10^6 by GPC.

t-Butyloxycarbonylation of poly(styrene-co-N-(p-hydroxyphenyl)maleimide). The synthesis of the starting copolymer has been previously described (8). The t-BOC protecting groups were introduced on the precursor polymers, two different molecular weights, using di-t-butyl dicarbonate (9-11-12). A solution of 5.87g of the copolymer (Mw = 16,100) in 70 ml of THF was cooled in ice water and 2.24g (20m mol) of potassium t-butoxide were added under nitrogen atmosphere. The pink solution was stirred for 20 min. at room temperature then 4.80g (22m mol) of di-t-butyldicarbonate were added and the cooled polymer mixture was stirred for 4 hours at room temperature. The t-BOC protected polymer was obtained by precipitating into water. Filtration and drying under vacuum afforded 6.35g (81% yield) of t-BOC-protected polymer. NMR, IR and UV were consistent with the proposed structure and identical to the copolymer prepared by direct copolymerization of the monomer.

Results and Discussion

The t-BOC protected copolymers were prepared both by copolymerization of the t-BOC protected hydroxyphenylmaleimide monomer with styrene and by modification of preformed phenolic copolymers of various molecular weights as shown in Scheme I. In both cases the copolymer compositions were found to be 1:1 based on NMR results and elemental analyses. The NMR and IR spectra obtained from copolymers from both routes were identical. The 13C and 1H NMR spectra of the modified polymer are shown in Figures 1 and 2. These data substantiate the completeness of the protection reaction of the preformed phenolic copolymer. The copolymers are presumed to be predominately alternating since these comonomers represent an example of the classic general alternating copolymerization case of an electron rich comonomer (styrene) and an electron poor comonomer (N-substituted maleimide) (13).

DSC analysis (Figure 3) of the copolymers shows a large, sharp endotherm at 152°C during the first heat. This endotherm and the concomitant mass loss are associated with the threshold like thermolysis of the t-BOC protecting group. A second heating of the sample shows a Tg at 235°C. The original substrate phenolic copolymers have been shown in a previous study to have high Tg's of similar values (8). The deprotected polymer undergoes thermal degradation beginning at about 300°C.

Thermal gravimetric analysis of the t-BOC protected copolymers shows a precipitous loss of 25% of the sample mass between 150 and 180°C then a plateau followed by slow decomposition above 300°C (Figure 4). These results mirror the DSC results. The first weight loss agrees well with that calculated for loss of CO_2 and isobutene (25.4%) and occurs coincident with loss of the 1755 cm^{-1} carbonate absorbance in the infrared and the appearance of the broad phenolic OH absorbance (Figure 5).

The thermolytic deprotection reaction is extremely clean. The infrared spectrum of the deprotected polymer is identical to that of the phenolic precursor as is the ultraviolet spectrum. The molecular weight of the phenolic copolymer precursor is unchanged by the t-BOC protection reaction/thermolysis cycle based on GPC data.

The t-BOC protected copolymers have good solubility in common organic solvents such as acetone, THF, chloroform and DMF but are insoluble in water, aqueous base and methanol. The deprotected, phenolic copolymers (or precursor phenolic copolymers) are very soluble in aqueous base, dioxane, THF, acetone and DMF and are insoluble in water and solvents of lower polarity such as chlorobenzene and toluene. Clearly the presence or absence of the t-BOC protecting group has a large effect on the solubility of the copolymer.

Imaging studies were done on copolymers prepared by the polymer modification route because of the availability of the precursor polymers of various molecular weights. The protected copolymers were compounded with triphenylsulfonium hexafluoroantimonate (13% w/w) in cyclohexanone. One micron thick films were spin coated on NaCl plates, baked at 140°C for 5 minutes to expel solvent and then subjected to infrared spectroscopic analysis before and after exposure. Exposure to 18 mJ/cm2 at 254 nm caused no change in the infrared spectrum. However, when the films were baked at 140°C for 120 sec. following exposure, deprotection was quantitative based on loss of the characteristic carbonate C = O absorption and

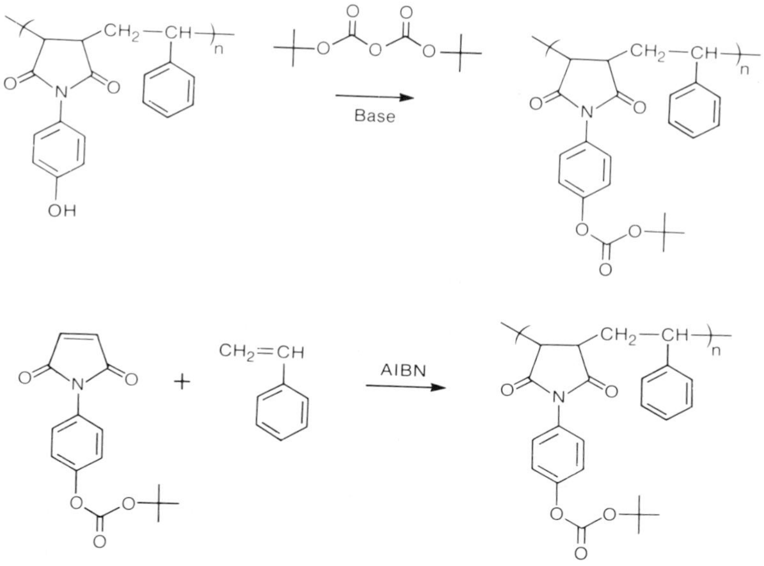

Scheme I

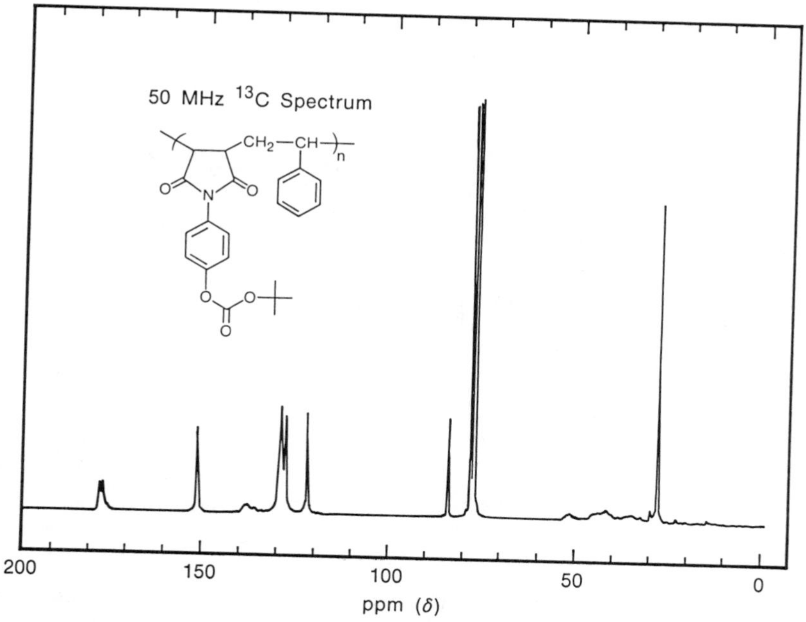

Figure 1. ^{13}C-NMR spectrum of poly (styrene-co-N-(p-t-butyloxy carbonyloxyphenyl) maleimide).

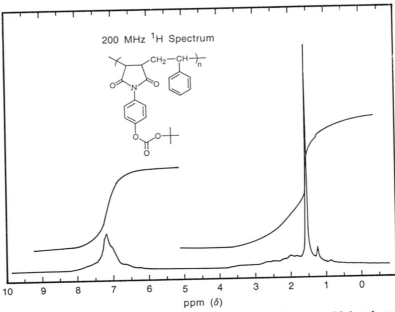

Figure 2. ¹H-NMR spectrum of poly (styrene-co-N-(p-t-butyloxy carbonyloxyphenyl) maleimide).

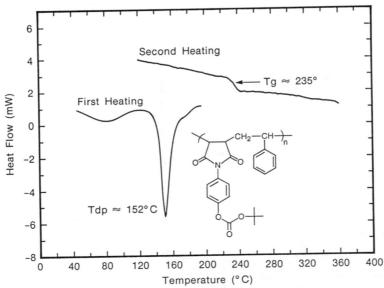

Figure 3. Differential scanning calorimetry analysis on the t-BOC protected copolymer. The endotherm at 152° on the first heating is associated with thermolysis of the t-BOC group. The second heating shows a Tg for the deprotected, phenolic copolymer at 235°.

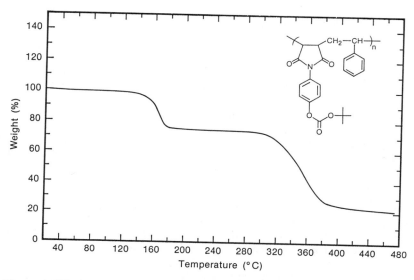

Figure 4. Thermal gravimetric analysis of the t-BOC protected copolymer. The mass loss at 150-160° is associated with thermolysis of the t-BOC group.

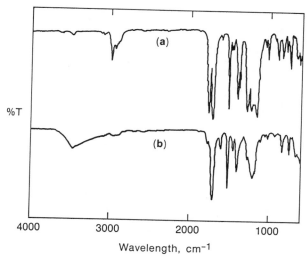

Figure 5. Infrared spectra of films of the t-BOC protected copolymer coated on NaCl plates before (a) and after (b) thermolysis of the protecting group. Note the loss of the carbonate carbonyl stretch at 1755 cm^{-1} and creation of the phenolic hydroxyl absorbance at 3400-3600 cm^{-1}.

appearance of the characteristic phenolic OH absorbance in the IR (Figure 6). The deprotected copolymer IR spectrum was identical to the spectrum of the unprotected precursor copolymer. Unexposed films baked under the same conditions showed no change in the IR spectrum.

The photoacidolysis was also documented by UV spectroscopy. The protected polymer has a weak absorbance at 280 nm. After exposure and baking, a more intense and red shifted band is produced (Figure 7). The photoacidolysis is very slow at room temperature but very fast above 100°C. The deprotection reaction and the resultant change in thermal activation for deprotection are shown in Figure 8.

Silicon wafers were coated with the formulation and baked at 100°C for 5 min. to give 0.9 micron thick films. These were exposed through a quartz resolution mask to narrow band 254 nm light. After exposure, the wafers were baked at 140°C for 2 minutes. Exposed and baked wafers were developed using organic solvent to give negative tone images and with aqueous base to give positive tone images. Resolution of 1.0 micron was achieved at 10 mJ/cm2. Optical micrographs of these images are shown in Figure 9.

The developed resist patterns were heated at 200°C for 1 hour in a convection oven. Scanning electron microscopy shows no thermal deformation in the negative tone images (Figure 10). The positive tone images lost thickness consistent with thermolysis of the t-BOC side chain but also do not show evidence of thermal flow deformation.

Thus, these new copolymers form the basis of a new, deep UV, thermally stable photoresist. This resist combines the high Tg and thermal stability of the styrene-N-(p-hydroxyphenyl)maleimide copolymer and the sensitivity of the chemical amplification design concept.

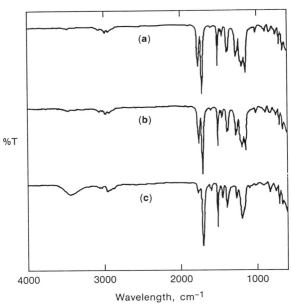

Figure 6. Infrared spectra of films of the resist formulation on NaCl plates a) after coating, b) after exposure to 18 mJ/cm2 of 254 nm radiation and c) after exposure and baking briefly at 140°C.

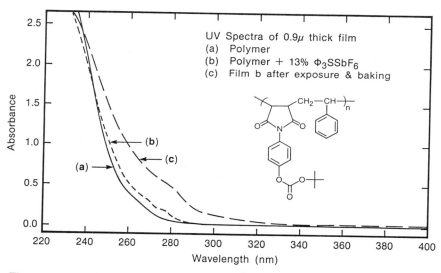

Figure 7. U.V. spectra of polymer films on quartz. Spectrum (a) is the t-BOC protected copolymer only, spectrum (b) is the resist formulation and spectrum (c) is the resist film after exposure and baking.

Chemical Amplification Concept

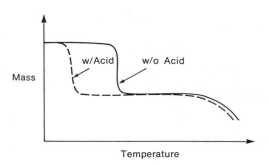

Soluble in Nonpolar Solvents
Insoluble in Polar Solvents

Soluble in Polar Solvents
Insoluble in nonpolar Solvents

Figure 8. Acid catalyzed thermolysis of the t-BOC protected copolymer is responsible for the change in solubility. The quantum efficiency for generation of the phenolic is the product of the efficiency of photoacid generation and the catalytic chain length.   Exposure generates a local concentration of acid. Subsequent heating to a temperature below that at which uncatalyzed thermolysis occurs allows local acidolysis of the t-BOC protecting group.

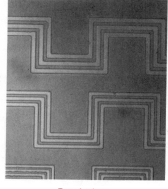

Negative   Positive

Figure 9. Optical micrographs of negative and positive tone resist images. The smallest features are 1 micron.

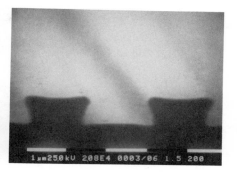

Start

200 °C – 30 min

Figure 10. Scanning electron micrographs of negative tone images before (left) and after (right) heating in air at 200°C. for 30 min.

Acknowledgment

The authors acknowledge R. Herbold for NMR spectra, R. L. Siemens for Thermal
Analysis and S. A. MacDonald for SEM work and for helpful technical discussions.

References

1. Willson, C. G.; Ito, H.; Fréchet, J. M. M.; Tessier, T. G.; Houlihan, F. M. J.
    Electrochem. Soc. 1986, 133, 181; Ito, H.; Willson, C. G. In Polymers in
    Electronics Davidson, T., Ed. ACS Symposium Series 242, Washington, DC,
    1984; p. 11.
2. Dill, F. H.; Shaw, J. M. IBM Journal Research and Development 1977, V21, 210.
3. Vazsonyi, E. B.; Vertesy, Z. Microcircuit Eng. 1983, 183, 338.
4. Johnson, D. W. SPIE, 469, Advances in Resist Technology, 1984; pg. 72.
5. Russell, G. Eastman Kodak Company unpublished results.
6. Hiraoka, H.; Pacansky, J. J. Vac. Sci. Tech. 1981, 19, 1132.
7. Allen, R.; Forster, M.; Yen, Y. T. J. Electrochem. Soc. 1982, 129, 1379.
8. Turner, S. R.; Arcus, R. A.; Houle, C. G.; Schleigh, W. R. Photopolymers:
    Principles, Processes, and Materials; SPE Reg. Tech. Conf. Proc. 35, Ellenville,
    New York, October, 1985; Polymer Engineering and Science, 1986, in press.
9. Fréchet, J. M. J.; Eichler, E.; Ito, H; Willson, C. G. Polymer 1983, 24, 995.
10. Osuch, C. E.; Brahim, K.; Hopf, F. R.; McFarland, M. J.; Mooring, A. M.;
    Wu, C. J.; Advances in Resist Technology and Processing III, SPIE 1986, 631.
11. Houlihan, F.; Bouchard, F.; Fréchet, J. M. J.; Willson, C. G. Canadian J. Chem.
    1985, 63, 153.
12. Ahn, K. D.; Willson, C. G. manuscript in preparation.
13. Mohamed, A. Aziz; Jebrael, F. H.; Elsabée, M. Z. Macromolecules 1986, 19, 32;
    Barrales-Rienda, J. M.; Gonzalez DeLa Campa, J. I.; Ramos, J. G.
    J. Macromol. Sci.-Chem. 1977, A11, 267.

RECEIVED June 1, 1987

Chapter 18

A Silicon-Containing Positive Photoresist Developable with Aqueous Alkaline Solution

N. Hayashi, T. Ueno, H. Shiraishi, T. Nishida, M. Toriumi, and S. Nonogaki

Central Research Laboratory, Hitachi, Ltd., Kokubunji, Tokyo 185, Japan

Mixtures of silicon compounds and a commercially available positive photoresist (OFPR-800, Tokyo Ohka Kogyo Co.) have been evaluated as a top imaging layer for a two-layer photolithographic system. We have found that a low-molecular-weight polyphenyl-silsesquioxane (PSQ) is compatible with OFPR-800 and that cis-(1,3,5,7-tetrahydroxy)- 1,3,5,7- tetraphenyl cyclotetrasiloxane (phenyl-$T_4(OH)_4$) is soluble in an aqueous alkaline solution. Taking advantage of these properties, we have been able to come up with a photoresist (ASTRO) formulated from PSQ, phenyl-$T_4(OH)_4$ and OFPR-800 which is developable with an alkaline solution and resistant to O_2-RIE. Good quality patterns of 1um line and space have been obtained as a result of a sequence starting with exposure (at 436 nm) followed by development of the top imaging layer with an aqueous alkaline solution and transferring the image by means of O_2-RIE to the bottom organic polymer layer.

Multi-level systems[1] allow the planarization of substrate topography formation of high aspect ratio patterns, and better resolution than single level resists. There is a drawback to such systems, however, and that is additional process complexity. To reduce this complexity, two-level resist systems have been reported[2]. The top imaging layer in two-level resist systems must exibit O_2-RIE resistance as well as desirable lithographic characteristics.

We previously reported that iodine containing compounds are resistant to oxygen plasma etching, and proposed two-level resist systems utilizing a mixture of iodine compounds and a commercially available positive photoresist[3]. However, this resist has low resistance to O_2-RIE.

Silicon compounds are well known as O_2-RIE resistant materials.

0097–6156/87/0346–0211$06.00/0

This property has been widely utilized, and many kinds of silicon-containing resists have been reported as a result. Most of the new resists developed use silicon containing novolac resins with a sensitizer(4)(5).

We have developed a silicon containing positive photoresist in the same manner that we developed an iodine containing resist previously. In other words, silicon compounds are blended into a conventional positive photoresist.

Experimental

Materials

As a positive working photoresist, novolac-naphthoquinonediazide type resist, OFPR-800(Tokyo Ohka Kogyo Co.) was used. Polyphenylsilsesquioxane (PSQ) was obtained under the name of Glass Resin GR-950 from Owens-Illinois Co. Cis-(1,3,5,7-tetrahydroxy)-1,3,5,7-tetraphenyl cyclotetrasiloxane (phenyl-$T_4(OH)_4$) was prepared by the following procedure reported by Brown(6): A cold mixture of phenyltrichlorosilane (196 g) and acetone (360 ml) was added slowly to an ice-water slurry (7100 g) with vigorous stirring. After the evolution of hydrogen chloride decreased, the solution was kept at $0°C$ for 2 days. A mixture of crystalline solid and resinous matter which precipitated in the solution was filtered and air-dried to give 95 g of solids. These solids were stirred with 200 ml of carbon disulfide in order to dissolve resinous materials, cooled to $0°C$, filtered, and washed with cold carbon disulfide to give 70 g of phenyl-$T_4(OH)_4$. Trimethylsililated phenols were prepared by lithiation of bromophenols with n-butyl lithium followed by reaction with trimethylchlorosilane. Trimethylsilyl-substituted polyvinylphenol was prepared by the same procedures as described above after bromination of polyvinylphenol. The other silicon compounds described in Table 1 were obtained from Shin-etsu Chemical Co. and Toray Silicone Co..

Contents of silicon compounds in new resists were presented by the weight ratio of additives as opposed to dry polymer weight calculated from the solid contents of OFPR-800.

Characterization of materials

Infrared spectra of polymer films were obtained on a Hitachi 260-10 Spectrometer. The films were formed from polymer solutions on NaCl discs. Molecular weight distributions were measured by gel permeation chromatography (GPC) using a Hitachi 635 Liquid Chromatograph equipped with Shodex A-804, A-802 and A-801 GPC columns. Tetrahydrofurane was used as the mobile phase.

Sensitivity

To measure the sensitivity, the resist solution was spin-coated onto a hard-baked (at $200°C$) OFPR-800 resist layer or a baked (at $200°C$) polyimide film (PIQ, Hitachi Chemical Co.). Irradiation was carried out using a high pressure mercury-xenon source (600W, Hanovia).

After exposure, the resist film was dipped in a NMD-3 developer (tetramethylammonium hydroxide solution, Tokyo Ohka Kogyo Co.), which was diluted with water to an appropriate concentration. Exposure characteristics were obtained by plotting normalized film thickness remaining vs exposure time under constant light intensity. Film thickness was measured by Alpha Step 200.

O_2-RIE

The O_2-RIE resistance of each material was determined using an ATM Inc. model AME 8181 or a modified NEVA FP-67A.

Results and discussion

The new silicon containing positive photoresists which consist of silicon compounds and a commercially available positive resist, OFPR-800, were evaluated as the top imaging layer in a two-level resist system. We sought silicon compounds specified by the following restrictions.
  1. Miscible with OFPR-800 without phase separation.
  2. Developable under the same conditions as those for OFPR-800.
  3. Well resistant to O_2-RIE.
  Many kinds of silicon compounds and silicone polymers were examined as materials to be used under these restrictions. The results are shown in Table 1. As seen in the table, low molecular weight silicon compounds have good compatibility with OFPR-800. With increasing the molecular weight, the weight ratio of compound blended without phase separation was decreased. However, low molecular weight compounds have poor resistance to O_2-RIE. We assume that this is a result of low molecular weight silicon compounds being vaporized by the heat generated in the etching process. Most of the silicone polymers in Table 1 had good resistance to O_2-RIE, but incurred phase separation after they blended to OFPR-800. We found that polysilsesquioxane oligomer GR-950 was the only available material for these purposes. We have further examined PSQ and OFPR-800 resists with respect to their compatibility, solubility in aqueous alkaline solution and resistance to O_2-RIE.

Compatibility

GR-950 is able to mix with OFPR-800 in an arbitrary ratio. It is a ladder type PSQ oligomer and its structure is shown in Fig.1. Another silicone resin, LP-103(Shin-Etsu Chemical Co.), has the same structure and higher molecular weight than GR-950 but is not miscible with OFPR-800. In order to certify the factors affecting compatibility, IR spectra were measured. Spectra of GR-950 and LP-103 are shown in Fig.2 and Fig.3 respectively. As can be seen in Fig.2, strong hydroxyl bands from silanol group(Si-OH) are observed at 3350 and 900 cm^{-1}. The IR spectrum of LP-103 in Fig.3, however, has no apparent hydroxyl band. LP-103 has higher molecular weight than GR-950, and therefore the hydroxyl groups it contains compared to other functional groups in the molecule is smaller than that of GR-950. The possibility of whether molecular weight affected

Table 1   Characteristics of Silicon Compounds. Compatibility in positive photoresist and resistance to O2-plasma etching and O2-RIE

Silicon Compd.	Structure	Compatibility (weight % per OFPR-800)	Resistance to O2-plasma etching	O2-RIE
o-hydroxyphenoxy-trimethylsilane		50	good	poor
3-(o-hydroxyphenoxy-propyl)trimethylsilane		50	good	poor
m-trimethylsilyl-phenol		50	good	poor
2,4-bis(trimethyl-silyl)phenol		50	good	poor
diphenylsilanediol		25	good	poor
hexamethylcyclo-trisiloxane		25		poor
octaphenylcyclo-tetrasiloxane		25 no compatibility in resist solution		

Name	Structure	Property	Rating
tetraethoxysilane	$Si(OCH_2CH_3)_4$	50	poor
tris-(2-methoxyethoxy) vinylsilane	$CH_2=CHSi(OCH_2CH_2OCH_3)_3$	50	poor
tetraacetoxysilane	$Si(OCOCH_3)_4$	50, gelation	
polydimethylsiloxane (Toray-silicone Co. SH-6018)	$-(\!-Si(CH_3)_2-O-\!)_n-$	48, phase separation	good
ethylsilicate-oligomer	$-(\!-Si(OEt)_2-O-\!)_n-$	8, phase separation	good
methyl-silsesquioxane (Glass resin GR-650)	$HO-(\!-Si(CH_3)(O)-O-\!)_n-H$	50, phase separation	good
phenylsil-sesquioxane (Glass resin GR-950)	$HO-(\!-Si(Ph)(O)-O-\!)_n-H$	200	good
(LP-103)		insoluble in resist solution	good
trimethylsililated-polyvinylphenol	$-(\!-CH_2-CH-\!)_n-$ with phenyl bearing OH and $Si(CH_3)_3$	insoluble in alkaline soln.	good

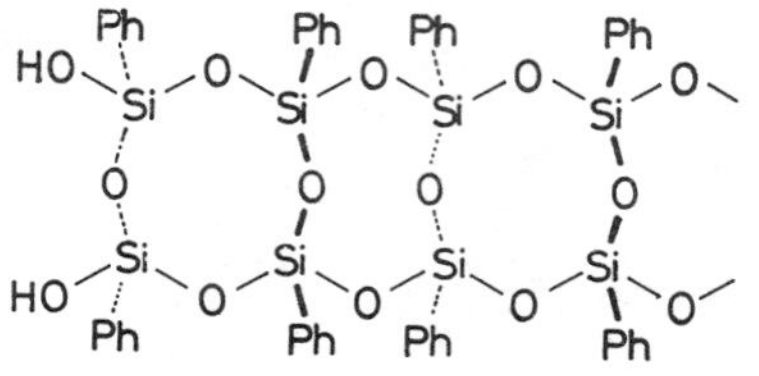

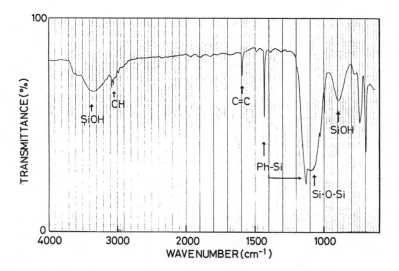

Fig.1 Structure of polyphenylsilsesquioxane(PSQ).

Fig.2 IR spectrum of Glass Resin GR-950.

compatibility with the photoresist was examined by eliminating silanol groups from GR-950. Terminal silanol groups were changed to trimethylsilyl ether groups through a reaction with trimethylchlorosilane and pyridine (Scheme 1). The IR spectrum of trimethylsilylated GR-950 is shown in Fig.4. As can be seen in Fig.4, hydroxyl bands at 3350 and 900 cm^{-1} disappear completely by trimethylsilylation. GPC analysis shows that molecular weight of this silylated GR-950 does not change from the original. Trimethylsilylated GR-950 was miscible with OFPR-800, but phase separation occurred when the resist film was formed on a wafer. From these results, it is clear that silanol groups of PSQ play an important role in ensuring that PSQ is miscible with a positive resist and makes a clear film without phase separation.

Solubility in aqueous alkaline solution

GR-950 is not soluble alone in the aqueous alkaline solution which is used as the developer of a positive photoresist. Further, we found a difference in the developing time of a resist film when using the same weight but a different lot of GR-950. Molecular weight distribution of these lots are shown in Fig.5. The IR spectrum of these samples were the same. Though lot A in Fig.5 has a higher average molecular weight, its developing time with alkaline solution is shorter than the case using lot B. Compared with the two molecular weight distribution curves in Fig.5, curve A shows a discernible shoulder at a molecular weight of about 500. It can be presumed that this shoulder is due to the existence of phenyl-$T_4(OH)_4$, the structure of which is shown in Fig.6. However, this can not be considered conclusive in determining the accuracy of GPC analysis for molecular weight determination. We therefore prepared phenyl-$T_4(OH)_4$ as a unit structure of PSQ and carried out further examinations. Molecular weight distributions of phenyl-$T_4(OH)_4$ and GR-950 were shown in Fig.7. As can be seen the figure, the shoulder of GR-950 distribution curve near molecular weight 500 corresponds to phenyl-$T_4(OH)_4$. Furthermore, we have found that phenyl-$T_4(OH)_4$ is soluble in an aqueous alkaline solution. The possibility of whether molecular weight affected the solubility in the aqueous alkaline solution was examined through preparing phenyl-$T_4(OH)_2(7)$. Two phenyl groups were substituted for the two hydroxyl groups of phenyl-$T_4(OH)_4$. Phenyl-$T_4(OH)_2$ was insoluble in an alkaline solution. It appears that phenyl-$T_4(OH)_4$ has a desirable proportion among silicon, phenyl group and hydroxyl group in terms of solubility in alkaline solution. From the molecular weight distribution of GR-950 in Fig.5, phenyl-$T_4(OH)_4$ is evaluated to comprise about 5 % GR-950. We assume that the different behaviors in development among some lots of GR-950 are caused by the differnt ratios of phenyl-$T_4(OH)_4$ they contain among them. The ratio of phenyl-$T_4(OH)_4$ contained is significant factor in the solubility in aqueous alkaline solution.

O_2-RIE behavior

The etching rates as a function of mixing ratio with GR-950 to OFPR-800 exposed to an oxygen RF discharge are examined. Curves obtained by plotting film thickness of relief image as a function of

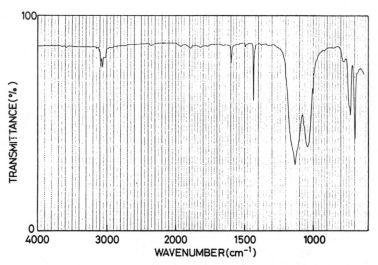

Fig.3 IR spectrum of Silicone Resin LP-103.

$$(-O-\underset{\underset{Ph}{|}}{\overset{\overset{|}{O}}{Si}}\!\!\!)_n OH \;+\; Cl\,SiMe_3 \;\xrightarrow{\;pyridine\;}\; (-O-\underset{\underset{Ph}{|}}{\overset{\overset{|}{O}}{Si}}\!\!\!)_n O\text{-}SiMe_3$$

Scheme I.

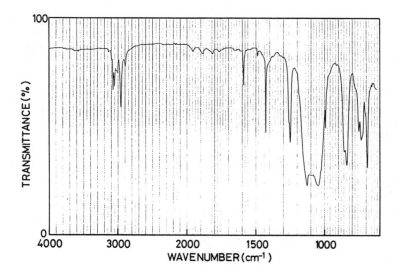

Fig.4 IR spectrum of trimethylsililated GR-950.

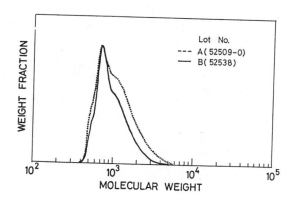

Fig.5 Gel permeation   chromatogram   of GR-950.   Difference of lot:
     A(#52509-0) and B(#52538)

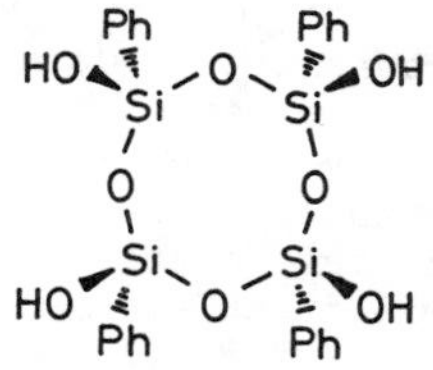

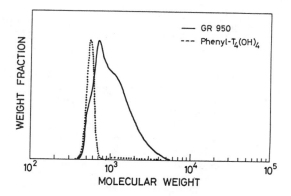

Fig.6 Structure of   cis-(1,3,5,7-tetrahydroxy)- 1,3,5,7-tetraphenyl
     cyclotetrasiloxane.   (Phenyl-$T_4(OH)_4$).

Fig.7 Gel permeation chromatograph of GR-950 and phenyl-$T_4(OH)_4$.

the etching time are shown in Fig.8. In this case, O_2-RIE was
performed by a modified NEVA FP-67A. RIE conditions were as
follows: Oxygen pressure was 2 m torr and RF supply was 100 W.
Curves obtained by plotting the decrease of film thickness after 40
min exposure to O_2-RIE as a function of weight ratio of GR-950
blended OFPR-800 are shown in Fig.9. The oxygen etching rates of
the resist film are markedly dependent upon the ratio of GR-950 it
contain. With increasing GR-950, film thickness loss rate
decreased. The addition of phenyl-$T_4(OH)_4$ makes no differance to
the case in which GR-950 is added. As seen in Fig.8, the resist
film in which OFPR-800 and GR-950 are mixed in a ratio of 2 to 1
(GR-950, 33 %) is 7 times as resistant to O_2-RIE as the parent
hard-baked OFPR-800 film under the same conditions. Considering the
process conditions for O_2-RIE, the ratio of OFPR-800 to PSQ for a
positive resist should be less than 2 to 1.

Sensitivity characteristics and O_2-RIE transfer of pattern

The photoresist which containes PSQ and a conventional positive
photoresist is named ASTRO (Alkaline developable Silicon containing
Top Resist with Oxygen RIE resistance). As described before, the
use of GR-950 alone could not be developed under the same conditions
for the parent positive photoresist. Using phenyl-$T_4(OH)_4$ in place
of GR-950, development of the exposed area could be done very
easily. Phenyl-$T_4(OH)_4$ is very soluble in an alkaline solution, but
the film thickness of the unexposed area becomes thinner after
development. We examined the possibility of using GR-950 and
phenyl-$T_4(OH)_4$ together. In these examinations, we found that the
ASTRO which consisted of OFPR-800, 45 % GR-950, and 30 %
phenyl-$T_4(OH)_4$ by weight, as opposed to the solid contents of
OFPR-800, had similar sensitivity characteristics to those of
OFPR-800. This composition is called ASTRO-7530. Sensitivity
characteristics of ASTRO-7530 and OFPR-800 are shown in Fig.10.
Development conditions for OFPR-800 were 24 sec dipping in a 2.38 %
solution of NMD-3 and 27 sec dipping in a 2.5 % solution of NMD-3
for ASTRO-7530. As shown in Fig.10, contrast and remaining film
thickness of the unexposed area of ASTRO-7530 were slightly inferior
to those of OFPR-800. In practice, however, only ASTRO-7530 can be
used the same way as a conventional positive photoresist.
        A preferred bi-layer process employing RIE transfer of a
pattern in ASTRO-7530 as a top layer used 200°C-baked PIQ as the
planarizing layer. Routine exposure and development of the top
layer, followed by O_2-RIE, transferred the pattern to the substrate.
Figure 11 shows the resulting nominal 1 μm lines and spaces
obtained. It is clear that ASTRO-7530 can act as an etching mask
and that fine patterns transfer to the bottom PIQ layer. In this
case, exposure was carried out by a reduction aligner Hitachi
RA-101L(at 436 nm). O_2-RIE was performed by a modified sputtering
machine NEVA FP-67A and RIE conditions were oxygen pressure of 2 m
torr, RF supply of 300 W and etching time of 12 min.

Conclusion

        Mixtures of ladder type silicon compounds PSQ and a commercially

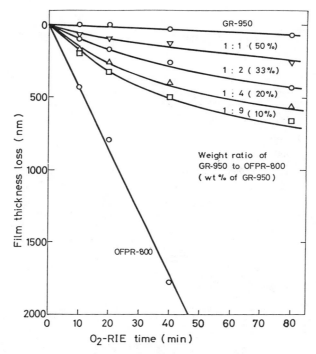

Fig.8 Plot of the film thickness loss vs O$_2$-RIE time.

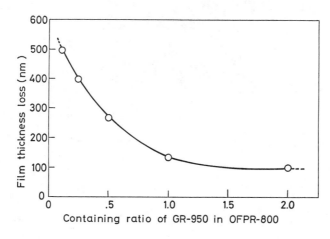

Fig.9 Plot of the film thickness loss after 40 min O$_2$-RIE vs containing ratio of GR-950 in OFPR-800(by weight).

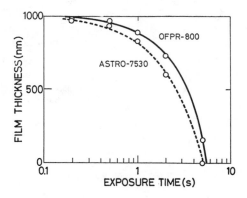

Fig. 10 Sensitivity characteristics of ASTRO-7530 and OFPR-800.

Fig. 11 SEM micrograph of 1 μm lines and spaces in ASTRO-7530 as a top imaging layer after O_2-RIE pattern transfer into the PIQ bottom layer.

available positive photoresist (OFPR-800) have been evaluated as the top imaging layer for a bi-layer photolithographic system. The results are as follows.

1 The silicon compound, polyphenylsilsesquioxane (PSQ) oligomer, is compatible with commercially available positive photoresist without phase separation. Terminal hydroxyl groups of PSQ are necessary to ensure that PSQ is miscible with positive photoresist and makes a clear film without phase separation.

2 Cis-(1,3,5,7-tetrahydroxy)-1,3,5,7-tetraphenyl cyclotetrasiloxane (phenyl-$T_4(OH)_4$), which can be looked upon as unit structure of PSQ, has been found to be soluble in an aqueous alkaline solution. The addition of phenyl-$T_4(OH)_4$ and PSQ oligomer to commercially available positive photoresists has resulted in improved developability with alkaline solution.

3 A mixture called ASTRO-7530, which consists of OFPR-800, 45 % PSQ oligomer (obtained as GR-950 from Owens-Illinois Co.) and 30 % phenyl-$T_4(OH)_4$, as opposed to the solid contents of OFPR-800, shows the same sensitivity characteristics as that of OFPR-800. The mixture is applied as the top imaging layer for a bi-layer photolithographic system. Good quality patterns of 1 μm lines and spaces have been obtained as a result of a sequence starting with exposure followed by development of the top imaging layer and transferring the image by means of O_2-RIE to the bottom organic polymer layer.

Acknowledgments

The authors would like to thank Dr. Kazuya Kadota for his helpful discussion. They also thank Dr. Hiroshi Yanazawa for his assistance in the exposure experiment using a reduction aligner.

References

1 J.M.Moran and D.Maydan, J.Vac.Sci.Tech., 1979, 16, 1620.
2 J.M.Shaw, M.Hatzakis, J.Paraszczak, J.Liutkus, and E.Babich, Polymer Eng.Sci., 1983, 23, 1054.
3 T.Ueno, H.Shiraishi, T.Iwayanagi, and S.Nonogaki, J.Electrochem.Soc., 1985, 132, 1168.
4 Y.Saotome, H.Gokan, K.Saigo, M.Suzuki, and Y.Ohnishi, J.Electrochem.Soc., 1985, 132, 909.
5 C.W.Wilkins,Jr., E.Reichmanis, T.M.Wolf, and B.C.Smith, J.Vac.Sci.Tech., 1985, B3, 306.
6 J.F.Brown,Jr., J.Amer.Chem.Soc., 1965, 87, 4317.
7 K.A.Andrianov, V.N.Emel'yanov, and A.M.Muzafarav, Dokl.Akad. Nauk.SSSR, 1976, 226, 827; Chem.Abstr., 1976, 85, 5744k.

RECEIVED May 5, 1987

Chapter 19

Intensity Dependence in Polymer Photochemistry

James R. Sheats and John S. Hargreaves

Hewlett-Packard Laboratories, Palo Alto, CA 94304

Intensity dependent bleaching of poly(methylmethacrylate) doped with acridine or diphenylbutadiene
and an acridine-containing methacrylate copolymer,
has been observed using excimer lasers (XeF, KrF).
Threshold-like irreversible bleaching was seen with
the copolymer; more gradual transient bleaching
with the mixtures. The implications for submicron
microlithography are discussed.

Despite frequent predictions of its imminent demise, optical lithography continues to be a prime contender in the current integrated
circuit fabrication arena; there are now several plausible candidates
for a viable half-micron optical lithography (1). There are two
fronts on which advances are being made. The capabilities of imaging
tools are being extended further into the ultraviolet, as a consequence of advances in lens design (2,3) and the introduction of
excimer lasers as illumination sources (2). Second, novel image
processing schemes have been introduced to enhance image quality
beyond what is available from the lens (4-7).
    One way to increase resolution is by use of a photobleachable
chemical system (applied as a thin polymer film on top of a conventional resist) whose response to light is strongly nonlinear (4-8).
As the image from the lens traverses this medium, higher doses (or
intensities) are attenuated to a lesser degree than lower doses (or
intensities), and so the sharpness ("contrast") of the image is increased before it is incident on the resist. The ultimate limit of
such an approach is a "threshold detector": a system resulting in
full exposure of the resist receiving incident doses (or intensities)
above some threshold value, and no exposure for incident doses (or
intensities) below that value. Although in this manner one should
obtain very high resolution (i.e., be able to print vertical-walled
resist features of some small width), linewidth control may be inadequate: the features may not have the width they were intended to
have (7). This is because the relative intensities at the nominal

feature edges for features of various sizes and with various neighbors are not the same, and so a specific threshold will not correspond to the line edge for all features (9).

One possible way to deal with this problem is to bias the mask: depending on the size of the feature and what is next to it, the mask dimension may be changed so that the printed feature will have the correct size. At first sight this would appear to imply a prohibitively complicated iterative calculation, similar to what has been discussed in connection with the electron beam proximity effect (10). However, the lateral diffraction spread that leads to the edge intensity variations in optical lithography is relatively short ranged, and corrections for features quite close to the diffraction limit may not require consideration of more than the adjacent feature, nor any iterative adjustment. Thus it may be possible to devise a mask biasing protocol that does not require an elaborate calculation, and this would make a threshold detector lithography system a potentially viable route to high resolution with adequate linewidth control.

Resolution may still be increased by use of a photobleachable dye that is not a threshold detector (4,6,7). While the ultimate resolution of such a process may be less than that of the threshold detector, considerable gains have been demonstrated (4,6,7). For moderately high resolution (e.g., 0.5 μm features using a lens of 0.42 numerical aperture at 365nm), computer simulations have shown that linewidth variations due to feature-dependent image variations can be reduced to the levels required in manufacturing ($\pm$10% of nominal)(7). This prediction is made for a photochemical system that undergoes "linear" photobleaching (the rate of reaction is linear in both intensity and concentration, though the transmittance is a nonlinear function of dose), but with the important property that the resist is unaffected by the dye exposure, and the dye is unchanged during the resist exposure (through the dye latent image). The same simulation model suggests that a process in which the dye and resist photochemistry occur simultaneously (4) will have considerably poorer linewidth control. Thus if a photobleaching system that is not a threshold detector is to be useful, it appears that it should provide separation of dye bleaching and resist exposure; the resist should see an unchanging image profile of the best contrast that the dye system can provide.

Up to now, most proposals for photobleaching image enhancement have relied on linear photochemistry, in which the transmittance is a function only of total dose, and not on the rate at which that dose is delivered. The kinetics of such linear photochemistry are well understood and have been described analytically (28). The exposure depends solely on a single parameter which is the product of extinction coefficient, quantum yield, intensity, and time. No increase in contrast can be obtained by changing extinction coefficient or quantum yield, since this merely scales the dose. Contrast can be increased only by increasing the initial optical density, which increases the dose requirement. Only with nonlinear (intensity dependent) photochemistry can one obtain steeper bleaching curves at a specified optical density.

Intensity dependent photochemistry could be useful in at least two distinct "modes". An optically bistable device (11) is an essentially perfect threshold detector for the incident light intensity; bistability with perfectly vertical transitions has been observed in a photochemical system (12). Thus it should be possible to construct an intensity dependent photochemical system that would serve as a threshold detector for use in optical lithography. Alternatively, intensity dependent mechanisms might be determined that would provide steeper bleaching curves than ordinary linear photobleaching for a given initial absorbance. The classification of bleaching behavior as "threshold detector" or "non-threshold detector" is of course somewhat arbitrary; we shall define a threshold detector as a system whose transmittance rises from near its initial value to a value at least ten times higher with an increase of intensity or dose of 10% or less.

An example of a potentially useful nonlinear system that is not a threshold detector is transient photobleaching, or absorption saturation (13). Regions of a film exhibiting transient photobleaching that are subjected to high incident intensities will quickly become bleached, and will then remain at a constant transmittance as long as the irradiating source is present. Regions that experience low incident intensities will remain at a constant low transmittance. Thus the underlying resist will be subject to a temporally unchanging image, unlike the case in which the bleachable film is irreversibly bleached with intensity independent photochemistry (4). The full contrast of the bleachable film is retained throughout the exposure of the resist, and there is no restriction that the exposure of the resist must require a smaller dose than that which bleaches the dye.

Photochemical (or photophysical) reactions may be either reversible or irreversible. If they are reversible, the system returns to its initial transmittance as soon as the light is turned off. An irreversible reaction causes a permanent change in the transmittance. In principle, there is no connection with whether a reaction is reversible or not and whether it is a threshold detector or not. In the experiments to be described, however, we have observed reversible (transient) bleaching with a smooth (but still steep) characteristic, and irreversible bleaching (in a different physical system) with threshold-like characteristics. The intensities for the two cases are quite different, but both are in the range such that pulsed lasers are needed; we have used excimer lasers operating at 350nm (XeF) and 248nm (KrF).

Experimental

Acridine and diphenylbutadiene were obtained from Aldrich; poly (methylmethacrylate)(PMMA) from KTI Chemicals (as a 9 wt.% solution in chlorobenzene). Solutions for spin-casting were prepared as follows: 1 ml of 1M acridine (in 2-ethoxy ethyl acetate) in 9 ml of 9 wt.% PMMA (thus 16 wt.% acridine in film); 0.3 g diphenylbutadiene in 7 ml of 9% PMMA solution (thus 30 wt.% diphenylbutadiene) (because this preparation was near its solubility limit, there is considerable uncertainty in the composition).

Films were prepared by spin-casting at 3000 rpm for 30 sec. on fused silica substrates (similar dimensions to conventional 3 inch diameter silicon wafers): the films were not baked. Films of poly [2-(9- acridyl)ethyl methacrylate-co-methyl methacrylate] (herein

referred to as "PAMA") were similarly prepared and baked for 30 min.
at 100°C. Thicknesses were estimated from absorption spectra.
     The synthetic route to acridine-containing polymers is out-
lined in Scheme I. 9-Methylacridine (II) was prepared via 9-ethyl-
malonate acridine by the method of Campbell et al (14), from
9-chloroacridine (I, Kodak), (mp. 117-8°C from ligroin). 9-(βhydro-
xyethyl) acridine (III) was synthesized via the condensation of
formaldehyde with II (15) (mp. 154-6°C, from ethanol-water). Metha-
cryloylchloride, freshly distilled, was reacted with III in dry THF
with pyridine as base to give IV, 2-(9-acridyl) ethylmethacrylate,
purified by chromatography with silica gel/toluene to give red gum
(IR 5.85 μm ester carbonyl stretch).
     The monomer was polymerized with methylmethacrylate using AIBN
in xylene. Typically, 5 g. of methylmethacrylate, 0.5 g. of acridine monomer, 0.05 g. of AIBN were dissolved in 5 ml of xylene in a
polymerization tube. After 3 freeze-pump-thaw cycles to remove
oxygen the tubes were sealed and placed in an oven at 60°C over
night. The pllymers were then precipitated into methanol several
times and thoroughly dried under vacuum. The mole fraction of
acridine present in the polymers was determined by UV spectroscopy
using acridine itself as a standard. The spectra of one of the
polymers in solution and in a film are shown in Figures 1 and 2;
the near-UV spectrum of the polymer in the film is very similar to
that in solution. Polymers with acridine mole ratios (relative to
methyl methacrylate monomer) of 0.023 (#1) and 0.070 (#2) were
obtained from feed ratios of 0.043 and 0.144 respectively.
     Excimer laser photolysis experiments were done with two
different setups. The first apparatus was identical to that used
in reference 5. A Lambda-Physik laser (nominal 10 nsec pulses) was
the source; the beam, after passing through a 12mm diameter aperture,
was split by a fused silica plate to provide a reference beam. The
main beam was focused by a 10cm focal length lens, and the sample
was placed after the focus. A 3.00mm diameter aperture was placed
immediately in front of the sample wafer; this aperture was at the
center of the beam. The intensity at the sample was varied over
about a factor of 50 by moving it with respect to the focus (from
6 to 45 cm); for still lower intensities the beam going into the
lens was obtained from the reflection from a fused silica plate.
     Transmittance was measured by UV-sensitive photodiodes (EG&G)
connected to a storage oscilloscope. Incident energy was measured
by a photodiode to which a portion of the incident beam was directed,
and this photodiode was calibrated by a Scientech calorimeter-type
power meter (which could be used reliably only in the average-power
mode due to thermal drift), which measured the energy transmitted
through the 3.00mm aperture for a given photodiode signal. To
measure transmittance, the two photodiode signals were measured both
with and without the sample in place for each positioning of the
3mm aperture with respect to the lens.
     The second setup used a Lumonics laser with an electrode set
that yielded nominal 35 nsec pulses. The beam transmitted by a 98%
reflecting dielectric mirror provided the reference; both incident
energy and transmittance were measured with a dual head Laser
Precision pyroelectric joulemeter. The sample position was moved
with respect to the focal plane of a 475mm focal length lens
(between lens and focus) for small variations in intensity, and

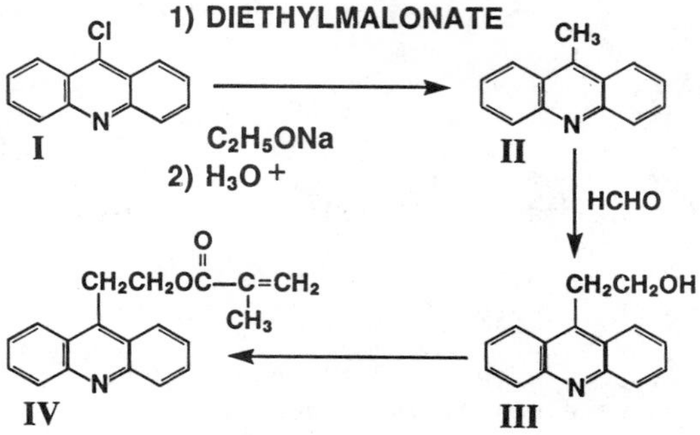

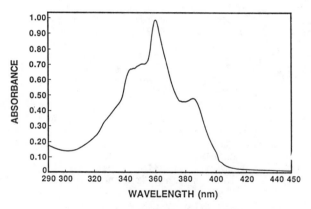

SCHEME I

Figure 1.   UV spectrum of PAMA #2, 0.185 g/liter in chlorobenzene (1 cm path). λ_{max}=360nm, absorbance 0.984. ε(monomer)=8.70x10^3 1/mol cm.

dielectric filters (CVI Laser Corp.) were used in the main beam (before the lens) for larger changes. The same 3.00mm aperture was placed directly in front of the sample as before, to obtain a region of uniform and known energy density.

Intensity is defined here as the energy density (fluence) divided by the nominal (i.e., manufacturer's literature) pulse length. It should be kept in mind that the temporal pulse shape is not gaussian, and may differ from one laser to another; the sample is actually subject to a range of intensities during the pulse, and the peak is greater than the value on the graph.

Results

Reversible (transient) Photobleaching. Reversible bleaching was observed only at 350nm (XeF laser). Figure 3 shows the transmittance of single 10 nsec pulses through a film of acridine/PMMA (~1.2μm). Each point is the average of several (usually about 5) pulses (but the same spot on the sample); the variation of the averaged data is not greater than the pulse-to-pulse variation from the laser (typically +10-15% or less), and there is never any trend from first to last pulse. Thus, the bleaching is transient only (though see below for much larger doses): by the time the next pulse is delivered (in a sequence at nominally constant incident intensity), the sample transmittance has returned to its original value. (This time was typically several seconds; a more detailed probing of the time dependence of the decay of bleaching was not possible.)

The data in Figure 3 were obtained by using a new spot for each new incident intensity. Using the same spot for more than one incident intensity gave the same results. However, after accumulating a large dose (always greater than 100 mJ/cm^2), irreversible changes were sometimes seen; the effect was mostly noticed for intensities of 500 kW/cm^2 or more. These irreversible effects were not quantitatively investigated. To be very safe, therefore, each spot used for Figure 3 was subject to no more than about 10 pulses, although it was demonstrated in one series of exposures that 75 pulses of 100-200 kW/cm^2 could be given to a single spot without observable damage.

Figure 4 shows similar data for diphenylbutadiene. Solubility limitations made it difficult to get a large optical density, and shot-to-shot fluctuations in the laser output were more severe for this experiment. The same bleaching effect is clearly evident, however, and it occurs at significantly lower intensity and with substantially less sharpness. Irreversible damage was again evident at intensities higher than those shown in Figure 4.

Irreversible Bleaching. Irreversible bleaching was observed at both 350nm and 248nm, but the most interesting effects from the lithographic viewpoint were obtained at 248nm. Figure 5 shows the transmittance of KrF laser pulses (35nsec) vs integrated dose, for various intensities incident on a sample of PAMA #2 whose initial transmittance (based on the ratio of extinction coefficients at 248 and 351nm measured in a much thinner film), was about 10^{-25} (~4.2 μm thick). There are actually two "thresholds" evident in this graph. For the higher intensities, the transmittance for a given

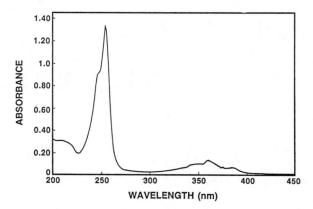

Figure 2. UV spectrum of PAMA #2 film on quartz. Assuming the
same extinction coefficient at 360nm as in chlorobenzene solution, and the density of PMMA (1.2 g/cm^3, Polymer Handbook), the
acridine monomer unit concentration is 0.73M, and the film
thickness 0.22 μm.

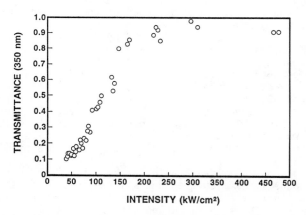

Figure 3. Transmittance of single XeF laser pulses (10nsec
nominal FWHM) vs. intensity (kW/cm^2) for acridine in PMMA (1.1M,
1.2μm thick) on quartz. Each point is from the average of
approximately 5 pulses, delivered to the same spot on the sample.

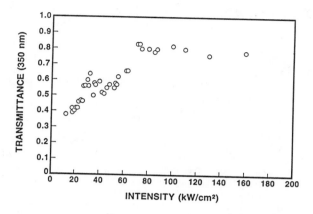

Figure 4.   Similar data to Fig. 3, for diphenylbutadiene.

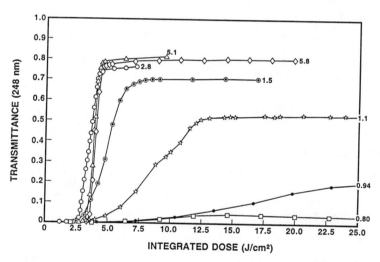

Figure 5.   Transmittance of KrF laser pulses (35nsec nominal FWHM) vs. integrated dose, at various incident intensities (in MW/cm^2), for a 4.2μm film of PAMA #2 on quartz. The higher intensity pulses (5.1 MW/cm^2 and above) were delivered singly; at lower intensities they were delivered in bunches of 10 or 100 at 25Hz. For such bunches, the transmittance recorded is the average over the bunch. In all cases, the data in the 'threshold" region came from single pulses.

incident intensity climbs very steeply as a function of integrated dose, and the curves for 2.8 MW/cm^2 and above meet our definition of threshold detectors, as a function only of dose. There is also a strong intensity dependent effect, however; at constant dose there is a very large increase in transmittance as the intensity rises. A vertical line drawn at 7.5 J/cm^2 shows that the definition of threshold detector is satisfied between 0.94 and 1.1 MW/cm^2. In a lithographic exposure, both effects are utilized: the portion of the image with a lower intensity also results in a lower dose, and vice versa.

Figure 6 shows several repetitions (different spots on the same wafer) for two intensities, showing the degree of repeatability of the data. All of the intensities shown in Figure 5 except 1.5 MW/cm^2 have this same repeatability. It is also interesting to note that for the highest intensities, the final transmittance is actually less than for the lower intensities; we have no explanation for this effect. Figure 7 shows a similar series of exposures for an intensity that is intermediate between the slowly varying behavior of <1.1 MW/cm^2 and the steep curves of >2.8 MW/cm^2. The spot-to-spot variation is much greater than at other intensities. This cannot be attributed to the laser (since all experiments in the series were done in the same run), and so probably reflects the effect of small film thickness variations in this region of rapidly changing intensity dependence.

The film remaining after extensive bleaching (where the bleaching curves of Figure 5 are flat) still contains some acridine as judged by the weak blue fluorescence, but the absorption spectrum shows only a small bump near 250nm on top of an otherwise feature-less curve.

Discussion

A large number of papers has appeared on the subject of excimer laser exposure of polymer films (16-21). Most of these have dealt with the phenomenon of photoablation. A few have observed intensity dependent photochemistry (22,23). The latter authors were concerned with the effect of exposure intensity on resist development char-acteristics. The utility of nonlinear photochemistry for image modification has not been explored except in our earlier commun-ication, in which strongly nonlinear irreversible bleaching was observed for KrF laser irradiation of acridine/PMMA films with 10nsec pulses (5).

We have indicated that intensity dependent phenomena may be useful in at least two distinct ways. One is to obtain something approaching a "threshold detector" resist response. To obtain a threshold development response in typical positive resists is difficult, since the development rate is in general a smoothly varying function of the photochemical reaction progress. The appl-ication of a layer of polymer with the bleaching characteristics shown in Figure 5 provides a way to obtain such threshold response with conventional resists, provided an excimer laser is used in the illumination system.

Photoblation is often spoken of as having a "threshold" (16). It is true that one typically observes a certain intensity (or narrow band of intensities) below which no ablation can be observed,

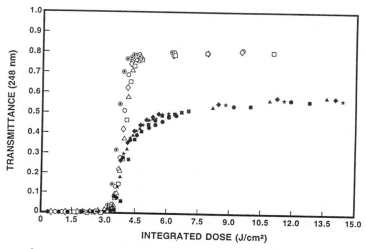

Figure 6. Similar data to Fig. 5, showing the data from 5 different sample spots for each incident intensity (2.8 and 9.1 MW/cm² for upper and lower curves).

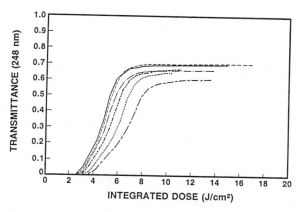

Figure 7. Similar data to Fig. 6, for 1.5 MW/cm² incident intensity.

and above which some is observed. However, the thickness removed
per pulse vs incident intensity curves for photoblation reported in
the literature do not closely approximate threshold detectors in the
sense defined in this paper, and are not as sharp as the curves in
Figure 7. Thus, the approach described here does provide an advantage not otherwise obtainable (such as, for example, by simply
ablating the resist directly).

The we have not yet explored the mechanism of the irreversible
bleaching in any detail. Some film removal is visually evident.
We cannot say what the relative mix of photochemical and thermal
effects is as yet. The literature suggests that significant photochemical reactions should occur due to 248nm irradiation of
acridine (25,26), but these are not the massive bond-breaking type
that characterize 193nm photoablation(16). The fluorescence yield
of acridine in PMMA is known to be about 0.2 (26) so considerable
heat is produced by the absorption of short pulses in the 100 mJ/cm^2
range; an estimate based on an approximate heat capacity formula
(27) is about 300°C. The excited state properties of acridine in
PMMA show a pronounced temperature dependence (26). It seems likely
that the bleaching arises from a combination of photochemical destruction of the acridine chromophore and polymer ablation.

The principal purpose of these experiments is to demonstrate
that extremely sharp thresholds in optical transmittance can be
obtained from laser photobleaching of a copolymerized absorber. It
is not clear that these results can be applied directly to wafer
processing conditions. Since the thermal characteristics of the
sample are probably quite important in our experiments, the results
obtained from a single polymer film on fused silica may not be the
same as those found with a multilayer film on silicon. Clearly
considerable additional experimentation is necessary before one can
consider practical applications.

The second application of intensity dependent processes is
in obtaining more modest resolution gains with good linewidth control
and without any mask biasing (using a reaction that is not a threshold detector). Theory does not lead one to expect extremely sharp
bleaching curves for the normal process of transient bleaching
(also called absorption saturation)(13); the nature of the curves
will be similar to those from ordinary photobleaching. The utility
of transient bleaching derives from the fact that a steady state is
rapidly reached, and the transmittance (for a given incident
intensity) does not change with incident dose. Thus, if an image
from a laser-illuminated camera is incident on a film with the characteristics shown in Figure 3, the image transmitted to the resist
underneath will be a "fixed" image; it will just be the product of
the incident image and the transmittance curve, regardless of dose.
In this way the main mechanistic advantages of the PIE process (7)
are retained while eliminating the necessity of a blanket exposure.
In addition, for reasons that we cannot explain, the curve of
Figure 3 is sharper than the theory of reference 13 would predict,
and is sharper than would be obtained from a linear photobleaching
system of the same initial optical density (28). Thus there appear
to be significant advantages in this type of nonlinear transient
bleaching, although more work is needed to understand it.

The fact that in the two systems investigated (Figures 3 and 4), irreversible effects appear at intensities not far above those required for transient bleaching suggests that cleanly obtaining the latter without the former may be difficult in many systems. Shorter laser pulses would be desirable for transient bleaching, in order to obtain the required high intensities for nonlinear effects without simultaneously involving the large amounts of energy that produce thermal and photochemical damage. There are many other 350nm absorbers that might be investigated. Solubility will be a major problem with many of them, which may be overcome by copolymerization as was done here with acridine.

Experimental evaluation of the lithographic utility of these systems requires imaging instruments using excimer lasers as light sources, which we do not have. However, such instruments have been reported in the literature (2,24)) and it is likely that they will become commercially available in the future. The XeF data (Figure 3) are particularly interesting in view of the fact that refractive optics can still be fabricated with conventional glasses at that wavelength, and the possibility of color center formation (a potential problem at 248nm) is eliminated. 248nm offers higher intrinsic resolution, and the bleaching curves of Figure 5 would, together with a 248nm stepper, provide the possibility of resolution around 0.2 - 0.25 μm (5). Many questions, such as the possible presence of defects due to ablation debris, must still be answered, however, before this goal is reached. Nevertheless, the remarkable performance characteristics shown in Figures 3 and 5 indicate that further exploration of intensity dependent photobleaching phenomena in polymers is well worth while.

Conclusions

Intensity dependent changes in transmittance of some thin polymer films, including a newly-synthesized acridine polymer, have been observed upon irradiation by excimer lasers at 350 and 248nm. There are two types of bleaching: permanent (irreversible) and transient (lasting no more than a second, and probably only as long as the laser pulse). Possible applications of these phenomena to microlithography are discussed; resolutions in the range of 0.2 - 0.5 μm appear possible. Further work on possible defect formation from debris and the requirement of excellent laser spatial and energy uniformity is required.

Literature Cited

1. SPIE Conferences on Advances in Resist Technology and Processing and Optical Microlithography. 1986, 631 and 633, 9-14 March, Santa Clara, CA.
2. Pol, V., Bennewitz, J.H., Escher, G.C., Feldman, M., Firtion, V.A., Jewell, T.E., Wilcomb, B.E. and Clemens, J.T. SPIE Proc. 1986, 633, 6-16.
3. Ushida, K., Anzai, S. and Kameyama, M. SPIE Proc. 1986, 633, 17-23.
4. West, P.R. and Griffing, B.F. SPIE Proc. 1983, 394, 33-38.
5. Sheats, J.R. Appl. Phys. Lett. 1984, 44, 1016-1018.

6. Vollenbroek, F.A., Nijssen, W.P.M., Kroon, H.J.J. and Yilmaz, B.
    Microcircuit Engineering 1985, 3, 145-151.
7. Sheats, J.R., O'Toole, M.M. and Hargreaves, J.S. SPIE Proc.
    1986, 631, 171-177.
8. Rothschild, M., Arnone, C. and Ehrlich, D.J. SPIE Proc. 1986,
    633, 51-57.
9. Neureuther, A.R., Hofer, D.C. and Willson, C.G. Microcircuit
    Engineering, 1985, 53-60.
10. Parikh, Mihir J. Appl. Phys. 1979, 50, 4371.
11. Neyer, A. and Voges, E. IEEE J. Quantum Electronics 1982,
    QE-18, 2009-2015.
12. Zimmermann, E.C. and Ross, John J. Chem. Phys. 1984, 80, 720-
    729.
13. Hercher, M. Appl. Optics 1967, 6, 947-954.
14. Campbell, A., Franklin, C.S., Morgan, E.N. and Tivey, D. J.
    J. Chem. Soc. 1958, 1145.
15. Homberger, A.W. and Jensen, H. J. Amer. Chem. Soc. 1926, 48,
    800. Eisleb, A. Chemical Abstracts 1937, 31, 5802.
16. Srinivasan, R. J. Vac. Sci. Technol. 1983, B1, 923-926.
17. Srinivasan, R. and Leigh, W.J. J. Amer. Chem. Soc. 1982, 104,
    6784-6785.
18. Geis, M.S., Randall, J.N., Deutsch, T.F., Efremow, N.N.,
    Donnelly, J.P. and Woodhouse, J.D. J. Vac. Sci Technol. 1983,
    B1, 1178-1181.
19. Andrew, J.E., Dyer, P.E., Forster, D. and Key, P.H. Appl. Phys.
    Lett. 1983, 43, 717-719.
20. Koren, G. and Yeh, J.T.C. Appl. Phys. Lett. 1984, 44, 1112-
    1114.
21. Latta, M., Moore, R., Rice, S. and Jain, K. J. Appl. Phys.
    1984, 56, 586-588.
22. Kawamura, Y., Toyoda, K. and Namba, S. Appl. Phys. Lett. 1982,
    40, 374-375.
23. Rice, S. and Jain, K. IEEE Trans. Electron Devices 1984, ED-31
    1-3.
24. Jain, K. and Kerth, R.T. Appl. Optics 1984, 23, 648-650.
25. Albert, A. The Acridines St. Martin's Press, New York, 1966,
    p. 242.
26. Kasama, K., Kikuchi, K., Uji-Ie, K., Yamamoto, S.A. and Kokubun,
    H. J. Phys. Chem. 1982, 86, 4733-4737.
27. Berry, R.S., Rice, S.A. and Ross, J. Physical Chemistry,
    Wiley, New York, 1980, p. 807.
28. Diamond, J.J. and Sheats, J.R. Electron Dev. Lett. 1986,
    EDL-7, 383-386.

RECEIVED April 8, 1987

Chapter 20

Effects of Additives on Positive Photoresist Development

R. C. Daly, T. DoMinh, R. A. Arcus, and M. J. Hanrahan

Eastman Kodak Company, Rochester, NY 14650

The addition of specialized small molecules to a polymer coating is the functional basis for most photoresists. Conventional positive-working photoresists function owing to the difference in solubility caused by the imagewise exposure of a small molecule naphthalene diazoquinone sulfonate ester (NDS). The presence of this small molecule dramatically inhibits the dissolution of the novolac binder while its photodecomposition accelerates the binder dissolution in aqueous base. An attempt to formulate a poly(4-hydroxy styrene)-based resist was less than completely successful because the difference in the rates of dissolution were too small to be used to give high contrast images. Other small molecules were added to the NDS/novolac resist and these were also found to have a profound effect upon the performance of the resist, particularly the development properties. When it was necessary to obtain higher dissolution rates, several triazoles and sulfonamides were found to improve the rate of development in the exposed areas without causing unacceptable thickness losses in the unexposed areas. Dyes incorporated to minimize problems of reflection and scattered light were also found to alter the dissolution behavior of the resist coating.

Making a polymer relief image commonly requires two processes. First, there is a photochemical process which alters the solubility of the exposed areas relative to the unexposed areas. This is followed by the actual dissolution of the most soluble areas during development. Historically, studies of photoresists have emphasized the photochemical aspects of image formation rather than the dissolution process. The central theme of this paper is the very large effects on dissolution rate that can result from adding small molecules to the matrix.

Three distinctly different types of compounds have been added to a polymeric binder and studied for photoresist applications.

0097–6156/87/0346–0237$06.00/0

These are naphthalene diazoquinones as photoactive compounds, benzotriazoles as development enhancement agents, and dyes to reduce
exposure from reflected light. The latter compounds have been added
to resist formulations which already contain a photoactive compound
in addition to the polymer binder.

The resist system that we have explored is based upon the photochemistry of naphthalene diazoquinone in a base-soluble polymer
matrix. This type of system is of interest because it is the basis
for all commercial near-UV photoresists. The photochemistry is old
but recently has been studied in detail ($\underline{1}$). Reaction scheme 1 shows
the series of reactions and rearrangements that occur after the
photolysis of a diazoquinone in the presence of water.

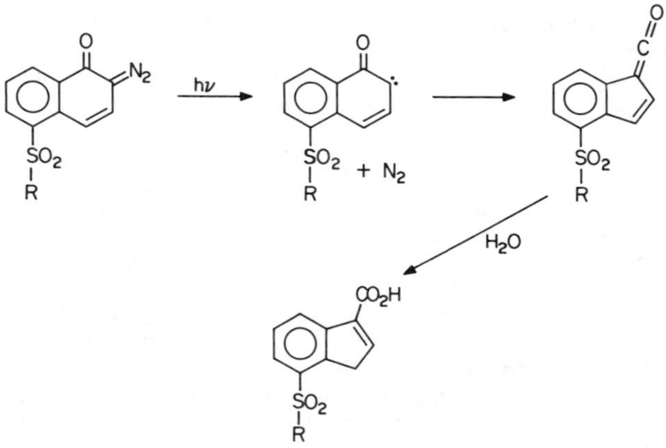

Scheme 1

There are three major events in the reaction sequence. First
the naphthalene diazoquinone is destroyed. Second, nitrogen gas is
released. Third, an acid is produced.

However, the photochemistry itself does not make a relief image.
Rather it is used to modify the solubility of the polymeric binder.
The diazoquinone compounds used in resists are referred to as dissolution inhibitors or photoactive components (PAC's). The addition
of a diazoquinone molecule dramatically inhibits the dissolution
rate of a thin film of a novolac resin. Upon exposure, the dissolution rate of the novolac based resist is considerably faster than
the rate for the novolac alone. The accelerated dissolution rate
may be caused by formation of acid and its subsequent ionization
during development or by enhanced diffusion of the developer into
the coating because of changes caused by the formation and fate of
the nitrogen ($\underline{2}$).

When working with this type of resist to make VLSI devices there
are several problems that come up, particularly the need for improved
thermal stability from the resist images and a method to control
light that is reflected or scattered in a nonimagewise manner. The
function of a development enhancement agent is to increase the dissolution rate of the photoresist so that polymers with better physical properties but slow dissolution rates may be used. The dyes

mentioned above are being incorporated in resists in hopes that they can absorb the unwanted light and help to produce higher resolution images.

Results and Discussion

Table I contains dissolution rate data for an m-cresol novolac, 1, and for the novolac containing 15 wt% of the diazoquinone 2. Using

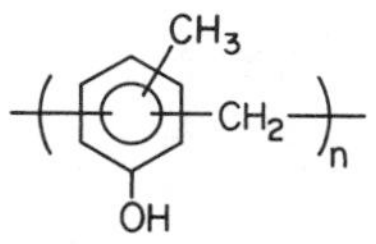

1

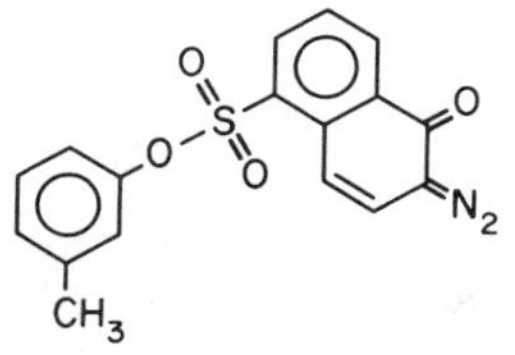

2

Table I. Development Rates

	Novolac, 1 μm/min	pHOSt, 3 μm/min
Polymeric binder	0.15	6.15
Polymeric binder + 15 wt% 2 unexposed	0.02	2.50
Polymeric binder + 15 wt% 2 exposed	1.70	10.10
Discrimination $\frac{DR\ exposed}{DR\ unexposed}$	85	4

Note: The developer was Kodak micro positive developer 934 diluted to a 1.5 wt% level of $(CH_3)_4NOH$ and used at 20°C.

a 1.5% solution of tetramethyl ammonium hydroxide, Me_4NOH, developer the novolac dissolves at a rate of 0.15 μm/min. The addition of the photoactive model compound decreases this rate to 0.02 μm/min. The mechanism of this dissolution inhibition is not clearly understood. However, upon exposure the rate of dissolution is increased to 1.70 μm/min, a rate substantially faster than the original novolac. The developer discrimination between exposed and unexposed novolac coatings containing 2 is above 80. It is this high discrimination that has allowed the naphthalene diazoquinone chemistry to dominate positive resist chemistry for almost 25 years.

The effect of the PAC's is not limited to novolac binders. Poly(4-hydroxy styrene), pHOSt, 3, has been suggested as a binder resin for resist formulations. The primary advantage of a resist based upon poly(4-hydroxy styrene) is the high glass transition temperature of the polymer (Tg = 187°C). A high Tg polymer has the potential to give a resist that is better able to withstand the harsh environment of ion implantation and plasma etching. The disadvantage of a high Tg is that all of the solvent cannot be removed without decomposing the PAC. The addition of a PAC to pHOSt causes

dissolution inhibition, and after exposure the dissolution is accelerated. However, the discrimination/dissolution rate ratio is only
4, a factor of 20 lower than for the novolac resist.

$$\left(CH_2-CH\right)_n$$

OH

3

   Significant differences are observed in the performance of
novolac and poly(4-hydroxy styrene) as resist binders. Though the
two polymers have equivalent activation energies of about 10 kcal/
mole for dissolution at workable rates, the pHOSt is slightly more
soluble in strong base (requires pH>12 using Me$_4$NOH) than is novolac
(requires pH>12.5 using Me$_4$NOH). pHOSt behaves as a stronger acid
than does novolac. The pHOSt dissolves much faster in aqueous base
than does the novolac at approximately the same Mw of 10,000.
Another major difference is that the discrimination between exposed
and unexposed coatings is considerably less for the pHOSt resist.
The rate of development is always faster for the pHOSt but the
discrimination between the exposed and the unexposed resists is
smaller. Comparison of the dissolution rate curves for novolac and
pHOSt in Figure 1 gives a good view of the difference in response of
the two resist systems. It is difficult to formulate the pHOSt into
a simple resist because there is too much thickness lost in the
unexposed areas when a developer is used that is appropriate to
develop the exposed areas in a reasonable time.
   However, there are compelling reasons for wanting to have a
higher Tg resist that would be better able to withstand harsh treatment during the fabrication of the device. One of the reasons that
the Tg's of the novolac resins are generally low is that they are
quite low in molecular weight (Mw between 2000 and 6000). This low
molecular weight allows for easier synthesis and much faster development times but at a cost in resist performance. Careful attention
to the process of making the novolac can overcome the first problem
but the resulting resists made from high molecular weight resins
(Mw's greater than 12,000) dissolve too slowly in the dilute developers needed to prepare high resolution images.
   One way to speed the dissolution is to add other components to
the resist. In particular the addition of benzotriazoles and sulfonamides to a resist formulation produces a resist that appears to be
more sensitive (3-5). Since sensitivity is a combination of photospeed and dissolution, an increase in sensitivity can imply either a
lower exposure or a faster development from the original exposure.
The data in Table II give a simple picture of the effect of several
benzotriazole additives, 4, on the dissolution rate of a conventional

4

Table II.   Enhanced Development

Additive	Time to clear (sec)	Thickness loss (Å)
None	38.0	400
4: R_1,R_2,R_3=H	18.0	100
4a: R_1=Cl;R_2,R_3=H	16.5	100
4b: R_1,R_2,R_3=Cl	7.5	100

Note: The developer was Kodak micro positive developer 934 diluted 1:1 with deionized water. Puddle development took place at 20°C for these 1.2-μm coatings.

novolac/NDS resist formulation. Time-to-clear, the figure of merit in Table II, is the development time needed to dissolve all of the resist in an exposed area. In each case the additive was used at 12 wt% of the solids. Putting a benzotriazole into the coating significantly increases development speed. Adding the electron-withdrawing chlorine substituents to the benzotriazole further hastens development. It is assumed that it is the increased acidity of the N-H which causes this effect. The benzotriazoles seem to only effect the dissolution rate of the exposed areas; there is no increase in thickness loss in the unexposed areas. Less PAC needs to be destroyed to give a developable resist image since the film is more soluble to begin with.

Mechanistically, there are three ways that these molecules could effect the speed of the resist: they could 1) make the photochemistry more effective, 2) chemically assist in the production of acid or 3) simply alter the dissolution of the thin film.

It is relatively easy to follow the course of the photochemistry since the initial photochemical event results in the destruction of the quinone diazide chromophore. For similar resist formulations with and without additives there was no change noted in the rate of bleaching of the UV absorption of the quinone diazide. The position, shape and extinction coefficient of the absorption were not altered by the additives. With these experiments in mind, it is very unlikely that the additives are involved in the photochemistry.

If the development enhancement agents react chemically with the photoproducts of the NDS, it should be possible to postulate the reaction products and a mechanism for the enhanced response. Faster development might result if the benzotriazole competes effectively

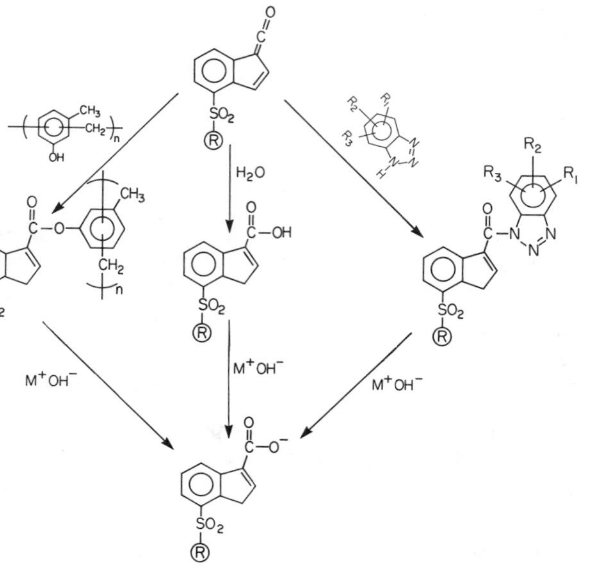

Scheme 2

with any parasitic side reactions of the indene ketene intermediate.
Scheme 2 shows the desired reaction of the ketene along with one of
several undesired reactions and a competing reaction of benzotriazole.
The undesired reaction is the formation of an ester with the novolac
resin. This type of reaction is particularly bad since the ester
saponifies too slowly in the developer to yield acid during development. Since the photoactive component normally has several reactive
groups the novolac resin becomes crosslinked and less soluble due to
this reaction. If the benzotriazole reacts quickly enough to reduce
the amount of ester formed and forms an amide product that is very
readily hydrolized in the developer, then there will be a net gain
in the effectiveness of the photoactive component.

     It should be possible to see the chemical products in the
infrared spectra of the thin film. In order to look at the small
quantities of material involved, it was necessary to do in situ
exposures of the resist coatings on silicon wafers in a Fourier
transform infrared spectrophotometer. The technique was capable of
following the loss of quinone and the formation of ketene with considerable success. By purging the wafer in the chamber for some
time in the presence of dry nitrogen, it was possible to observe a
stable ketene signal even hours after the exposure. While these
experiments were not quantitative, they did give two pieces of

information. First, the reaction of the ketene with novolac must be quite slow in comparison with its reaction with water. Second, the stability of the ketene signal was not effected by the additives. There was no indication of new species in the FTIR spectra. The combination of these two phenomena makes it highly unlikely that chemistry occurs between the reactive intermediates and the additives. Experiments in the solid state but without the novolac resin indicated that the proposed amide intermediate could be observed for benzotriazole in a glassy quinone diazide matrix.

The increased rate of development for exposed coatings with additives is, however, quite pronounced. This can be seen most easily in the traces from a Perkin Elmer dissolution rate monitor. The data in Figure 2 show that at all exposure levels the resist containing a chlorobenzotriazole development enhancement agent requires a shorter development time than does the unaltered resist. The converse of this process is that the resist with the DEA requires shorter exposure times when the same development time and process are used. For example, only 25 mJ of exposure are required to give an image that clears at 30 sec for the resist with DEA while a 70 mJ exposure is needed for the model resist itself. The contrast of the resist image is also improved by the addition of a benzotriazole DEA. Using the relatively long exposure times needed for quick development in track equipment, the contrast is always higher for the resist with DEA, Figure 3. Since the thickness loss in the unexposed areas is little effected by the DEA, the improved contrast is probably an artifact of increased developer discrimination between exposed and unexposed resist.

Another class of small molecules that may be added to positive photoresists are dyes. Control of the scattered and reflected light in the resist coating is the common reason for adding a dye to a resist formulation. Initially, the NDS photoactive component absorbs less than one half of the incoming light at 436 nm. This means that there will be significant amounts of light reflected from the substrate even as the exposure begins. During the exposure, the NDS absorbance is bleached, allowing more light be be scattered and reflected. Since the different microelectronic substrates have different reflectivities, the resist will receive differing exposures depending upon the substrate under the area being exposed. Several other factors, including variations in thickness, only serve to make the problem of obtaining uniformly exposed resist lines more difficult.

Several approaches to making images over reflective topography use dyes that absorb at the exposure wavelength. Two of these processes involve putting a dye in a separate layer under the photoresist. The use of a thin dyed layer (an antireflection coating) under the resist keeps the reflected light from reentering the resist (6). However, this layer must be coated and removed, possibly requiring several extra steps, and care must be taken to see that the resist and the antireflection layer do not mix. Placing the dye in a thick layer under the resist offers the advantage of addressing the topography of the device wafer surface (7). A thick planarizing layer is coated to reduce the discontinuities of the substrate and to present a uniform surface for the resist coating. Both of these dyed layers suffer from the problem of how to transfer the image from the resist on top through the under layer. While these concepts

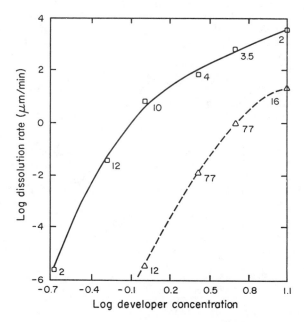

Figure 1.   Development occurred at 20°C with $(CH_3)_4NOH$ developer:
(---△-) novolac; (——□-) pHOSt. The numbers at the data points
represent discrimination between the dissolution rate for exposed
and unexposed resist formulated from the bender with 15 wt% 2.

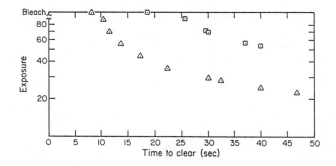

Figure 2.   The model resist, 13 wt% PAC in novolac, was developed
at 21°C with 1.5% $(CH_3)_4NOH$ developer: □ = no DEA, △ = 13 wt% of
4b.

are workable they present formidable engineering difficulties. Also, since the dyes absorb reflected light which might otherwise enter the resist, extra exposure is required to achieve the desired image when exposing resist on top of a dyed layer. However, significant improvements in the image uniformity over reflective topography can be seen with either of these dyed processes.

Putting a dye into the resist is the most direct way to reduce the amount of reflected light (8). Here the dye in the resist coating competes directly with the photoactive component for the incoming light, thereby causing an even larger speed decrease than seen in the previous two cases. However, significant performance improvements have been shown for dyed resists, particularly in the area of reduced notching over Al.

However, the dyes also alter the dissolution properties of the resist (9). Figure 4 shows the absorption spectra of a cyanine dye, 5, which has been used in a positive resist to reduce the image

5

6

distortion caused by reflections. Important features of this dye are that it has a very high extinction coefficient near 436 nm and that it has very little absorbance in the mid- and deep-UV. When the dye is added to Kodak micro positive resist 820 (820 resist), there is a significant improvement in the linewidth control and also a large decrease in the dissolution rate of the resist. Figure 5 shows the variation of the dissolution rate as dye is added. At a dye level of 0.25% the dissolution rate is less than 60% of the undyed resist. Other dyes have been found that accelerate the dissolution of the dyed coatings. The dyes that aid dissolution are generally those that have functional groups that dissolve readily in aqueous base, such as sulfonamides, phenols and acids. Figure 6 is a plot of the dissolution behavior of a dissolution-accelerating chalcone sulfonamide dye, 6. The acidic sulfonamide proton should ionize in the basic developer and hasten the dissolution of the exposed and already dissolving films. Those dyes that inhibit dissolution seem to be molecules that can react or complex with the acid product of the photochemistry. For example, the ionic ammonium function of the cyanine could interact with one or two of the indene carboxylic acids to form complexes that would be much less soluble in the aqueous base developer.

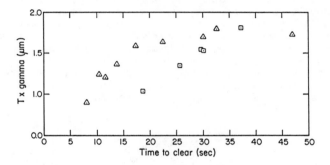

Figure 3.   The model resist, 13 wt% PAC in novolac, was developed
at 21°C with 1.5% $(CH_3)_4NOH$ developer: ▫ = no DEA, ▲ = 13 wt% of
4b.

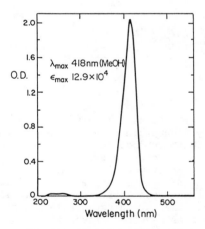

Figure 4.   UV absorption spectra of cyanine dye 5.

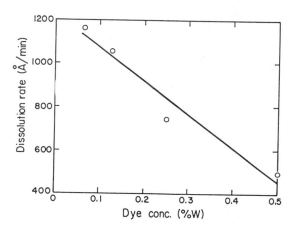

Figure 5. Dissolution rate for 820 resist containing dye 5 in 934 developer.

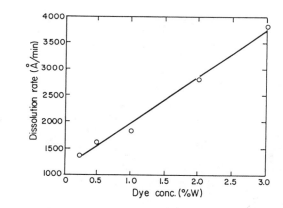

Figure 6. Dissolution rate for 820 resist containing dye 6 in 934 developer.

The best development discrimination occurs with the dissolution inhibiting dyes, while the higher speed occurs with the development accelerating dyes, Table III.

Table III. Performance of Dyed 820 Resist

	Control	Cyanine, 5	Chalcone, 6
Dye conc. %W	None	0.25%	1.0%
Dissolution rate Å/min	1290	810	1910
Relative rate	1.00	0.63	1.48
Stepper exp. msec (436)	140	180	160
Thickness loss	3%	1%	5%
Selectivity	28	31	11

Summary

The purpose of this work was to show several of the ways in which small molecules effect the dissolution of a polymer film. This phenomena is significant for microresists because it represents the basis for micro imaging and because several new ways of improving or altering the resist performance depend upon adding new chemicals to the resist formulation. The intent of these materials may not be to alter the resist dissolution but the potential for change is ever present.

Experimental

The novolac used in this study was prepared by the base catalyzed solution condensation of m-cresol with formaldehyde ($Mn = 1800$, $Mw = 9300$, $Mw/Mn = 5.07$). The poly (p-hydroxy styrene) was PHP-6817-24, obtained from Marusen Oil Co. ($Mn = 3900$, $Mw = 10200$, $Mw/Mn = 2.62$). Kodak micro positive developer 934 (934 developer) (predominantly tetramethyl ammonium hydroxide in water) was used at various dilutions with deionized water. M-cresyl naphthalene diazoquinone sulfonate was prepared by a pyridine catalyzed condensation of m-cresol with naphthalene diazoquinone sulfonyl chloride.

The simple dissolution rate results were obtained using a laser interferometer with a 15 mw/cm^2 He-Ne laser at normal incidence to the wafer surface in the agitated developer bath. The reflected beam was directed by a beam splitter onto a photocell. The photocell output was fed through a Keithly series 500 interface into an IBM-PC. The more complex dissolution data were collected on a Perkin-Elmer dissolution rate monitor using 934 developer at a 1:1 dilution with deionized water at a temperature of 21°C. The stepped exposures were obtained using a calibrated multidensity chrome stepwedge.

Acknowledgments

The authors wish to thank Sandra Finn for FTIR studies and both Kathlene Hollis and Caroline Little for running many of the dissolution rate experiments.

Literature Cited

1. Pacansky, J.; Lyerla, J. R. Polym. Eng. Sci. 1985, 20, 1049.
2. Hinsberg, W. D.; Willson, C. G.; Kanazawa, K. K. SPIE Proc. 1985, 6, 593.
3. Western Electric G.B. Patent 1 317 796, 1972.
4. IBM French Patent 2 325 076, 1976.
5. Eastman Kodak U.S. Patent 4 365 019, 1982.
6. Brewer, T.; Colson, R.; Arnold, J. J. Appl. Photogr. Eng. 1981, 7, 184.
7. O'Toole, M. M.; Liu, E. D.; Chang, M. S. SPIE Proc., 1981, 128, 275.
8. Brown A. V.; Arnold, W. H. SPIE Proc., 1985, 259, 539.
9. Bolsen, M.; Buhr, G.; Merren H. J.; van Werden, K. Solid State Technol. 1986, Mar, 83.

RECEIVED June 15, 1987

Chapter 21

Importance of the Interface Condition upon Photoresist Image Adhesion in Microelectronic Device Fabrication

J. N. Helbert[1] and N. C. Saha[2]

[1]Motorola Bipolar Technology Center, Mesa, AZ 85202
[2]Semiconductor Research and Development Laboratory, Motorola SPS, Phoenix, AZ 85008

Microelectronic device fabrication currently relies primarily upon photoresist processing for integrated circuit pattern delineation. Adhesion of polymeric photoresist patterns, especially those of micron and submicron dimensions, to the required fabrication substrates is of paramount importance. Photoresist image adhesion problems encountered in device fabrication have been solved by chemical interfacial treatments. Current new trends in microelectronic adhesion technology will be described and discussed with emphasis upon the chemical nature of the interface involved as determined by ESCA.

Integrated circuit pattern image adhesion is one of the most important polymeric resist performance parameters(1,2). Resist lithographic design and performance are important, but if the resist is not process compatible in an adhesion sense, it is of academic interest only. Fortunately, this parameter can usually be satisifed by substrate interfacial chemical treatments without changing the chemical nature of the photoresists themselves. Since positive photoresists, mixtures of novolak resins and diazoquinone photoactive dissolution inhibitors, are the most plagued by problems with image adhesion (see Figure 1), we will focus all discussion upon treatments which solve adhesion problems for this generic class of resist materials.

Surface chemical changes created by these surface chemical reactions will be monitored by Electron Spectroscopy for Chemical Analysis (ESCA)(3), a very powerful spectroscopic technique for investigating surface compositions extending from 1-20 monolayers in depth from the surface. When the spectrometer is equipped with angle resolution capability, it also offers a means of non-destructive depth profiling of the upper 50A of substrate layers.

EXPERIMENTAL
Wafer Testing

The resists used in this work are all conventional positive
photoresists. They are all proprietary formulations, but generically
they are composed of a mixture of (1) novolak resins, (2) photoactive
components of the diazoquinone type, (3) leveling agents and/or
surfactants, and (4) glycol-based spinning solvents. The specific
resists tested were Polychrome 129, Hunt 204, KTI-II, Dynachem OFPR
800, and AZ 1350. They were all applied by conventional spinning
technique by either an Silicon Valley Group (SVG) track, or by
Headway or Solitec manual spinner systems.
    The commercially available adhesion promoters (Petrarch Systems
Inc.) tested were applied as dilute solutions 0.3 - 7% by weight in
acetone or xylene, or as for hexamethyldisilazane (HMDS), either as a
liquid or vapor (Imtec Star 2000).

ESCA Analysis

ESCA measurements were carried out in a PHI Model 5300 ESCA
spectrometer using magnesium K-alpha X-rays. The base pressure in the
analyzing chamber was 1×10^{-9} torr or better during analysis. In the
PHI Model 5300 system, the variable take-off angle measurements are
performed by rotating the sample on an axis through the sample
surface. The measurements can be made from grazing angles of 3° to
10° past normal.

RESULTS AND DISCUSSION
Single Crystal Silicon

The reason for studying this substrate is one of ease of ESCA
analysis and detection sensitivity considerations. A substrate was
needed, where detection of carbon and silicon-containing surface
species would not be over-shadowed by signals from the bulk species,
even at low angles. Of course, this substrate contains a native oxide
layer less than 20A thick, but this surface is similar to that of
other silicon dielectric substrates differing only in growth
temperature. Although the integrated circuit (IC) is built upon a
single crystal silicon wafer, this substrate and interface are
usually not seen by the patterning photoresist in new MOS fabrication
processes; however, this may not be the case for older product
process flows.
    Before determining surface condition treatment effects, the
surface sensitivity enhancement of angle resolved ESCA surface
analysis is demonstrated in Fig. 2. The figure clearly shows the O,
Si, and C surface concentrations to be dramatically different than
the respective bulk concentrations. The silicon concentration in the
top surface layer measured at 5° ESCA take-off angle, for example, is
approximately 2.5X less than that obtained at a 75° take-off angle,
which represents an approximate depth of 50A from the surface. Most
of the surface data presented here is at the low take-off angle of 5°
to concentrate primarily upon the wafer surface condition in the
first 2-3 atomic layers, and how it affects polymeric photoresist
adhesion.

ADHESION

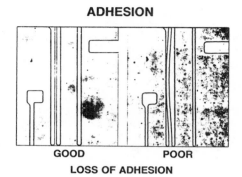

GOOD   **POOR**

LOSS OF ADHESION

Figure 1. Optical micrograph of "lifted" photoresist test
           image.

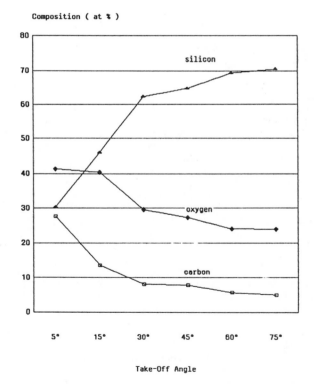

Figure 2. ESCA profile data for Y58 blank wafer.

The ESCA results do not indicate a dramatic change in C/Si ratios following HMDS treatment, liquid or vapor phase. O/Si ratios of HMDS treated samples, particularly after vapor phase treatment, do indicate a detectable lowering of the relative oxygen surface concentration compared to blank wafers, consistent with dehydration which is known to also occur with HMDS treatment ($\underline{1}$).

Figure 3a shows a typical high resolution C 1s ESCA spectrum recorded for a blank silicon wafer (Y58) at 5° take-off angle. Deconvolution after background subtraction, gives three well defined peaks at 285.5+/- 0.1 eV, 287.0 and 289.7 eV. The peak at 285.5 eV is assigned to a $-CH_x$ (hydrocarbon) species, the 287.1 eV peak to $-CH_2O$ (e.g. alcohol or hydroperoxide) species, and the 289.7 eV peak to a $-CO_2-$ (carboxylic acid or ester) adsorbed species. These carbon compounds are found on nearly all substrates, and are thought to be adsorbed from the processing ambient.

Figs. 3 (b & c) show C 1s ESCA spectra from HMDS (SVG)/Y58 and HMDS (*2000) treated Y58 silicon wafers. Interestingly, the C 1s spectra for the HMDS (*2000)/Y58 sample consist mainly of a single peak with only a small shoulder at higher binding energy (BE). The BE of the main C 1s peak is 284.5 eV, which is 1.0 eV lower than the adsorbed hydrocarbon peak in the blank wafers. This peak is attributed to the $-CH_3$ surface groups resulting from the HMDS treatment, covalently anchored to the surface. We favor covalent bond formation because overnight storage of the sample in the high vacuum (2×10^{-10} torr) of the ESCA analyzer chamber did $\underline{not}$ reduce the C 1s peak intensity. This would not have occurred $\overline{if}$ the signal was due to just adsorbed HMDS or the dimer $(CH_3)_6Si_2O$, which usually forms due to surface hydrolysis; the signal would not have been as stable, the observation for some control substrates with no surface treatments.

The C 1s spectral changes (Figs. 3a and 3c) strongly suggest that the vapor phase HMDS treatment cleans the wafer very efficiently by removing all three types of adsorbed carbon compounds, followed by the spreading of the methyl blanket from the surface HMDS reaction across the substrate. Since the samples were handled in air and spectra recorded at least 24 hours after vapor priming, this process must be considered to be a very good surface stabilizing treatment. Furthermore, the liquid phase reaction of HMDS from the wafer track could not remove the adsorbed species as completely, nor was it capable of rendering the substrate resistant to readsorption of carbonaceous contamination.

Figures 4(a-c) show three ESCA Si 2p spectra resulting from the (a) Y58 blank, (b) HMDS (SVG track) and (c) HMDS (*2000)/Y58 wafer, respectively. All spectra were recorded at 5° take-off angle. These spectra clearly indicate the evolution of a new peak between the elemental silicon and the SiO_2 peaks. The growth is very pronounced in the case of the HMDS (*2000)/Y58 treated wafer. The new Si 2p peak, arising from HMDS treatment of the wafer, is centered at 101.8 eV (see Figures 4b and 4c). The other five peaks present in the spectrum of the blank wafer are assigned to Si^o, SiO, Si_2O_3 and SiO_2 species ($\underline{4}$).   The new peak at 101.8 eV is assigned to $(CH_3)_3$ Si-O- type of Si species formed on the surface due to the HMDS reaction. The presence of this new Si 2p peak is consistent with the earlier interpretation of the C 1s ESCA data.

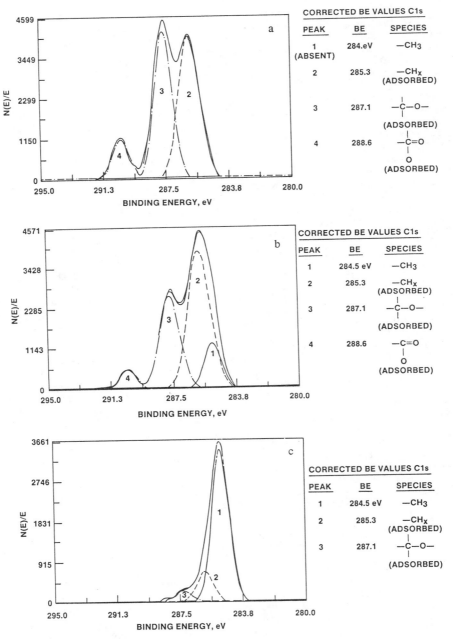

Figure 3.   (a) C 1s ESCA spectra for blank silicon wafer (Y58).
(b) C 1s ESCA spectra for SVG conventional liquid HMDS-treated
silicon wafer.   (c) C 1s spectra for Star 2000 vapor HMDS-treated
silicon wafer.

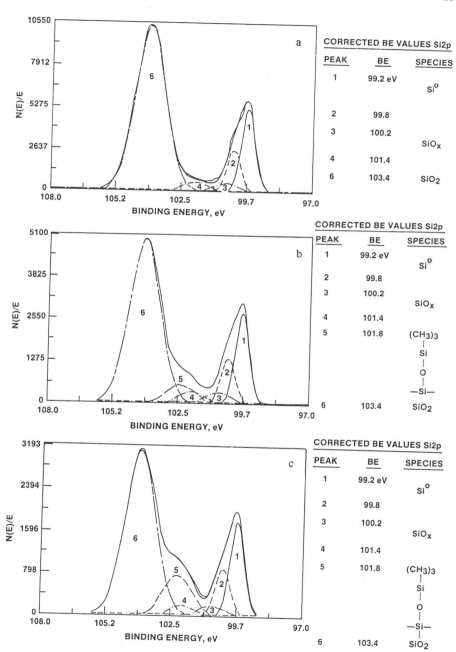

Figure 4. (a) Si 2p ESCA spectra from the untreated blank Y58 wafer. (b) Si 2p ESCA spectra from HMDS (liquid)-treated Y58 wafer on a SVG track. (c) Si 2p ESCA spectra from Star 2000 vapor primed Y58 wafer.

SiO_2 Substrates

Since SiO_2 substrates appear frequently during IC fabrication, the
adhesion test results for this substrate are important. Four types of
oxides have been extensively tested. They are (1) thermal oxide grown
at $1100^\circ C$, (2) softer oxides processed by conventional spin-on-glass
technology, (3) phosphorus doped liquid phase chemical vapor
deposition (LPCVD) oxide, and (4) low temperature ($200^\circ C$) plasma
deposited oxide (PEO).

Adhesion has been achieved on these oxides through a variety of
processes. Conventional liquid phase application of HMDS, however,
was not adequate for the latter three substrates listed above.
However, it did provide adequate photoresist adhesion for thermal
oxides. For the last three substrates, a double adhesion promoter
process was needed and developed. This process has been incorporated
into actual device fabrication processing.

The double promoter process involves the successive application
of liquid promoter solutions of vinyltrichlorosilane (VTS) and
3-chloropropyltrimethoxysilane, followed by successive cure cycles in
dry N_2 at $90^\circ C$ before photoresist application.

Later, the successful but somewhat complex double promoter
process was replaced by the "vapor phase" HMDS process in the Star
2000. Then superior resist image adhesion was obtained on all four
oxide substrates with all the photoresists tested.

What are the mechanisms of these two successful promoter
processes? To answer this question, ESCA analysis was employed.

ESCA survey scans recorded for thermal oxide surfaces indicate
the presence of Si, C and O. Again, the simple elemental compositions
do not show any dramatic change due to the surface chemical
modifications by the HMDS treatment, except for the decrease in O/Si
ratio. But, the ESCA data do show that the chemical nature of these
elements changes significantly, i.e., C/O containing molecular
impurities are being replaced by a covalently bonded Si/O/C
containing stabilization layer. As seen for Y58 wafers, ESCA spectra
resulting from Si 2p transitions for HMDS treated oxide wafers also
depict the evolution of a new peak at 101.8 eV. Again, this peak is
most prominent in the Star 2000 HMDS treated oxide sample, and is
assigned to the Si 2p peak present on the oxide surface due to $(CH_3)_3$
Si-O- surface species.

When a conventionally applied VTS solution is used, adhesion is
increased to SiO_2 substrates,(5) and no trace of Cl remains at the
surface. ESCA results for VTS treated PSG oxides show reproducible
increases in carbon concentration as expected from $-Si-O-Si-(CH=CH_2)$
surface reaction products. Consistent with this hypothesis,
broadening of the ESCA Si 2p peak at lower BE is also observed. The
broadened Si 2p spectrum can be simulated by two Gaussian curves with
1.8 eV full width at half maximum (FWHM) centered at 103.0 and 103.7
BE. The 103.7 eV peak would be the same as that observed for the
blank with the second peak at 103.0 eV attributed to the Si product
of the VTS surface reaction.

ESCA results for double promoted oxides are less definitive than
those previously described, but definitely show differences in the
Si 2p spectra, hence, indicating that again surface changes are
occurring due to the irreversible surface chemical reactions. At the

same time, these oxide surfaces are being at least partially cleaned of C/O containing atmospheric contaminants.

HMDS Treated Substrate Comparison

Several enlightening comparisons can be made from the data of Table I. First an estimate of the vapor phase HMDS surface reaction efficiency vs that of the liquid phase can be obtained. The features of the vapor phase and liquid phase treatments are found in Tables II and III. From Table I, the values for Y58 substrates are 0.22 and 0.11, respectively, thus indicating an approximate 100% greater efficiency towards silanol conversion to trimethyl silyl labelled reaction product for the vapor treatment. An approximate 30% gain is obtained for the oxide substrate comparison.

Table I. Comparison of HMDS Coverage on Y-58 and Oxide Surfaces

METHOD	SURFACES	AREA RATIO [(CH3)C-SI-O-]/SIO2	C/SI RATIO [Cls PEAK #1/Si2p PEAK #6]	TOTAL C/SI RATIO
VAPOR PHASE HMDS (*2000)	Y-58	0.22 ± 0.02	3.2 ± 0.2	0.8 ± 0.1
	OXIDE	0.13	~ 3-4	0.8
LIQUID PHASE HMDS (SVG)	Y-58	0.11	2.9	1.0
	OXIDE	0.10	~ 5	1.1
BLANK	Y-58	0	NA	1.1
	OXIDE	0	NA	1.1

Next, it is interesting to look at the results in the fourth column of the table. Here, the theoretical C ls/Si 2p ratio should be 3, the stoichiometry of the labelling trimethyl silyl group. The values found for Y58 substrates are 3.2 and 2.9, while larger values between 3-5 were observed for the oxide substrates. The latter large

Table II.   Star 2000 Vapor Phase HMDS Features and Results

FEATURES:

    1. LOW PRESSURE BAKE PREPARATION

    2. 150 C DRY NITROGEN CYCLED IN-SITU DEHYDRATION BAKE

    3. VAPOR PHASE HMDS TREATMENT AT 150 C

    4. 250 WAFERS/40 MINUTES

    5. LOW HMDS USE- COST REDUCTION

ADHESION TESTS POSITIVE WITH:

    1. THERMAL OXIDE(1100 C)

    2. SILICON NITRIDE

    3. SPIN ON GLASS

    4. PECVD OXIDE(200 C)

    5. PSG OXIDE

    6. POLYSILICON

Table III.   Conventional Processing Steps

1. DEHYDRATE BAKE: NITROGEN-PURGED OVEN OR HOT PLATE AT 1 ATMOSPHERE

2. AIR TRANSFER

3. LIQUID HMDS TREATMENT: NEAT OR 30% SOLUTION

4. CURE IN NITROGEN OVEN OR HOT PLATE AT 90 C,

OR NO CURE JUST RESIST PREBAKE(SEE #6)

5. AIR TRANSFER

6. PHOTORESIST COAT/PREBAKE AT 90-105 C

values reflect a greater peak deconvolution uncertainty for obtaining this ratio for higher temperature oxide surfaces, which have less well resolved ESCA spectra and/or may contain additional chemisorbed HMDS species-- thus leading to higher ratios.

Finally, it is interesting to follow the total C/Si ratio of the last column of Table I. The overall result of vapor treatment creates a surface with a 27% reduced overall carbon surface concentration. It is also felt the trimethyl silyl blanket created by the in-situ gas phase HMDS reaction has a much more stable carbon surface population, as well as, producing a reduced carbon surface layer concentration. This effect combined with the dehydration of the substrate (1), which is well known to occur, creates a surface condition of the wafer conducive to low day to day variation in wafer surface condition. That is, changes in humidity and ambient carbon impurity contaminant concentrations will have little influence upon the wafer surface leading to improved photoresist adhesion reproducibility.

Photoresist and Processing Effects

Although the photoresists employed were generically similar in formulation composition, they exhibited significantly different adhesion image "lifting" results. The majority of the results are for standard track applied liquid promoter processing. "Lifting" was observed for first generation resists, like PC 129 and KTI II, on SiO_2 substrates. Hunt 204 images have lifted on oxide surfaces, and in one case, not even double promoter processing could solve the problem. The representative second generation resist tested, OFPR-800, did not suffer from "lifting" even on blank wafers, although image edge "lifting" or poor image edge acuity did occur for images on control wafers.

When vapor priming was used on the Star 2000, no "lifting" occurred for any of the resists on the "lifting-susceptible" substrates tested. This dramatic result must be attributed to the process of that system. The in-situ dehydration bake of that system is far superior to that of older processing. There, the wafer was (1) dehydration baked in dry-N_2 convection ovens, (2) cooled in air, and (3) track adhesion promoted and resist coated in fab area ambient. Obviously, the wafers could be rehydrated or surface contaminated in the older processing scheme. The Star 2000 treated wafers, could be stored in ambient for 1-3 days without image "lifting" occurring, therefore, these processes stabilized the wafer surfaces.

Conclusions

Superior resist adhesion or resistance to resist image "lifting" has been achieved by (1) conventional liquid phase double chemical treatments, (2) single chemical liquid phase treatments, and (3) vapor phase HMDS treatments. These successful but different adhesion processes have one point of commonality. They all dehydrate, clean, and chemically stabilize or passivate surfaces that require photoresist adhesion. Furthermore, the observed improved adhesion must be the result of stronger forces than simple short range van der Waals forces. Electrostatic forces extending farther than the first monolayer at the interface must be hypothesized to be occurring for these systems as well. The wafers containing micron sized resist images with properly prepared surfaces resist lifting during extensive high pressure (>30 psi) nitrogen gas dry and water rinse cycles. Furthermore, if the only forces of attraction were van der Waals, lifting would not be observed for over promoted wafers, vapor or liquid phase treated, as is experimentally observed for wafers with excessive multiple promoter treatments or vapor treatment times. Of course, acid–base forces of attraction cannot be ruled out either (6).

References

1. K.L. Mittal, Solid State Technology, 89 May (1979).
2. J.N. Helbert, R.Y. Robb, B.R. Svechovsky, and N.C. Saha, Proceedings of Symposium on Surface and Colloid Science in Computer Technology, 5th International Conference on Surface and Colloid Science, in press (1985).
3. T.A. Carlson, "Photoelectron and Auger Spectroscopy," Plenum Press, New York, 1975.
4. F.J. Grunther, P.J. Grunther, R.P. Vasquez, B.F. Lewis, and J. Maserjian, J. Vac. Sci. Technol., 16 (5), 1443 (1979).
5. J.N. Helbert and H.G. Hughes, in "Adhesion Aspects of Polymeric Coatings," K.L. Mittal, editor, Plenum Press, New York, 499 (1983).
6. H. Yanazawa, Colloid Surfaces, 9, 133 (1984).

RECEIVED April 8, 1987

Chapter 22

Polymer Processing to Thin Films for Microelectronic Applications

Samson A. Jenekhe

Physical Sciences Center, Honeywell, Inc., Bloomington, MN 55420

Thin solid films of polymeric materials used in
various microelectronic applications are usually
commercially produced by the spin coating deposition
(SCD) process. This paper reports on a comprehensive
theoretical study of the fundamental physical
mechanisms of polymer thin film formation onto
substrates by the SCD process. A mathematical model
was used to predict the film thickness and film
thickness uniformity as well as the effects of
rheological properties, solvent evaporation, substrate
surface topography and planarization phenomena. A
theoretical expression is shown to provide a universal
dimensionless correlation of dry film thickness data
in terms of initial viscosity, angular speed, initial
volume dispensed, time and two solvent evaporation
parameters.

Synthetic polymers have long been used as insulating dielectric
materials in electronic components (1-4). Recently, however,
diverse applications of polymeric materials in microelectronics
and solid state devices, components and systems are emerging and
growing (5-8): lithographic resists for integrated circuit (IC)
fabrication; intermetal dielectric layers for IC and microelectronic
interconnect and packaging; protective coatings; planarization
layers; dopant diffusion layers; implant masks; sensing materials
in microsensors; thin film wave guides; optical data storage and
magnetic recording media; etc. In most of these current and future
applications of polymeric materials in microelectronics the polymers
must be used in the form of thin solid films deposited onto
substrates. Traditional techniques (9) of polymer processing to
free standing or supported films are incompatible with the fragile
substrates (e.g. silicon or gallium arsenide wafers) and planar
processing technology of the microelectronics industry.
Spin coating deposition (SCD) is the primary commercial process
for forming thin films of the various polymeric materials used in
the electronics industry. Yet, very little is known about the
fundamental physical processes of polymer thin film formation on

substrates by SCD which includes hydrodynamics, rheology of polymer solutions, solvent mass transfer, surface and interfacial phenomena, heat transfer and the interplay of these processes. The theoretical and experimental studies presented in part here were motivated by the need to understand the underlying basic mechanisms of polymer thin film formation from solutions using the SCD process (10-13). It is also hoped that results of such studies would be of practical interest to both the specialty polymer manufacturers in the chemical industry on the one hand and the polymeric materials users in the electronics industry on the other.

Theoretical Modeling of SCD

The deposition of thin solid polymer films by the SCD process is illustrated in Figure 1. A fixed volume of a viscous solution of the polymer to be deposited is placed on a flat substrate, such as a silicon wafer, and then rapidly rotated. Depending on the solvent volatility from the solution and such parameters as initial volume of solution, initial concentration of solids, the substrate temperature relative to the boiling point of the solvent, rotational speed, and spinning duration, a wet or dry polymer film results. In most cases the consistency of the polymer film at the end of spinning ranges from a highly viscous liquid to a wet solid. The as-deposited film is therefore usually treated by post-spinning bake at a higher temperature or in vacuum to remove any remaining solvent in order to produce a dry solid film. Two film thicknesses are therefore to be distinguished. The as-deposited wet film thickness h or H($H = h/h_0$, where h_0 is some initial thickness, which can be taken as the volume of solution dispensed divided by wafer area) obtained at the end of spinning and the dry film thickness h^* or H^* which results from further post-deposition drying. The two thicknesses can be related by

$$h^* = \psi h \tag{1a}$$
$$H^* = \psi H \tag{1b}$$

where ψ is a wet film contraction factor which is a function of initial volume fraction of solids in solution ϕ_0, solvent evaporation rate during spinning Λ_s, and other variables. We note that if there is no solvent evaporation during spinning, i.e. $\Lambda_s = 0$, $\psi = \phi_0$; if all the solvent evaporated during spinning, $\psi = 1$. Thus

$$\phi_0 \leq \psi \leq 1 \tag{2}$$

If the volume fraction of solids in solution at the end of spinning is ϕ_1, then $\psi = \phi_1$ exactly.

The dry film thickness h^* or H^* is usually the experimentally and readily measured one in industrial practice although techniques are available for measuring the wet film thickness. On the other hand, h or H is the theoretically predicted one. This distinction between wet and dry film thicknesses enables the delineation of effects of solvent evaporation: those effects during spinning which are due to a finite Λ_s; and those effects due to post-SCD drying which can be measured by ψ.

The formal SCD modeling problem is to theoretically predict h or H and its radial uniformity, in terms of the significant SCD process parameters, including rotational speed ω, duration of spinning t, initial kinematic viscosity of solution ν_0, initial volume of solution dispensed, evaporation rate Λ_s, initial volume fraction of solids in solution ϕ_0, substrate surface topography, etc. Mathematically, the problem is one of coating flow of a polymer solution from which the solvent is evaporating on a rotating disk. The complexity of the mathematical problem, even for a numerical solution, has forced the necessity of treating theoretical models, with various simplifying assumptions, which nevertheless yield useful solutions and provide understanding of the SCD process.

Effects of Rheological Properties

In the earliest theoretical study of the SCD process Emslie et al (14) obtained an analytical solution given by,

$$H = \left[1 + 4\ \frac{\omega^2 h_0^2 t}{3\nu_0} \right]^{-1/2} \tag{3}$$

for a Newtonian fluid, without taking solvent evaporation into account, and concluded that a radially uniform film can always be obtained. However, a similar analysis of the case of a power-law fluid by Acrivos et al (15) lead to the conclusion that uniform films cannot be obtained for non-Newtonian liquids. Neither analytical solution nor explicit film thickness profiles was obtained by Acrivos et al. Polymer solutions are generally non-Newtonian liquids which cannot be described adequately by either the Newtonian or the power-law model except at respectively low and high shear rates. The non-uniform shear field inherent in the flow problem, with shear rates varying from zero at the center to a maximum at the edge of the disk, means that the power-law model and the resultant conclusions may not be valid.

We have assessed the effects of the rheological properties of polymer solutions on the film thickness and its radial uniformity and the dynamics of the SCD process by obtaining the explicit film thickness profiles for the Carreau model (12), viscoplastic liquids (13), Newtonian liquids (12) and power-law fluids (12). Flack et al (16) have also theoretically investigated the SCD process for electron beam resist (PMMA) solutions in chlorobenzene using essentially a Carreau-type non-Newtonian viscosity equation. Their results also demonstrate the significant effects of rheological properties on film thickness and film thickness uniformity. All the polymer solutions of interest in commercial microelectronics processing, such as polyimide coating and resist solutions, can be described by the Carreau non-Newtonian viscosity equation (10):

$$\eta\ (\dot{\gamma}) = \eta_0 \left[1 + (\lambda \dot{\gamma})^2 \right]^{(n-1)/2} \tag{4}$$

where η is the non-Newtonian viscosity, η_0 is the zero-shear rate viscosity, λ is a characteristic time constant, $\dot{\gamma}$ is the shear rate and n is a dimensionless power law index. Note that the viscosity

of Equation 4 exhibits shear-thinning at high shear rates and a
constant viscosity at low shear rates. Typical non-Newtonian
viscosity data for commercial polyamic acid solutions, the soluble
precursors of polyimides, have been published (10).

Effects of Solvent Evaporation

Attempts to use the analytical result of Equation 3 to correlate
experimental data have consistently failed (17). Consequently,
empirical and semi-empirical models which include various factors
to account for evaporation and non-Newtonian behavior have been
proposed (17) but these too have not been able to satisfactorily
fit the available data. We have considered the coating flow problem
with simultaneous solvent evaporation (11). In the regime of interface mass transfer controlled evaporation, i.e. at high solvent
concentration, the fluid mechanics problem can be decoupled from the
mass transfer problem via an experimental parameter α which measures
the changing time-dependent kinematic viscosity due to solvent
evaporation. An analytical expression for the film thickness has
been obtained (11):

$$H* = \psi \left[1 + \frac{(2 + \alpha)}{2} \tau \right]^{-1/(2+\alpha)} \tag{5a}$$

$$\tau = \frac{4\omega^2 h_o^2 t}{3\nu_o} \tag{5b}$$

Planarization and Effects of Substrate Surface Topography

Many of the substrates onto which polymer thin films are deposited
by the SCD process usually have a non-planar surface topography due
to microelectronic structures already etched on them. It is
generally desired that the polymer film provide planarization of
such underlying microelectronic structures. However, perfect
planarization cannot be obtained. The degree of planarization
achieved depends on both the initial polymer solution properties and
the geometry of the microelectronic features (18). Using the
variables defined in Figure 2 and Rothman's (18) definition of
degree of planarization ε

$$\varepsilon = 1 - h_2/h_1 \tag{6}$$

we have theoretically predicted the degree of planarization ε and
related it to important SCD variables, initial polymer solution
properties, post-SCD drying and processing, the solid film
properties, and the underlying surface topography. For example, it
can be theoretically shown that (19)

$$\varepsilon = \phi_1 = \psi \geq \phi_o \tag{7}$$

$$\varepsilon = H* \left[1 + \frac{(2+\alpha)}{2} \tau \right]^{1/(2+\alpha)} \tag{8}$$

where notations in Equations 7 and 8 are as defined previously.

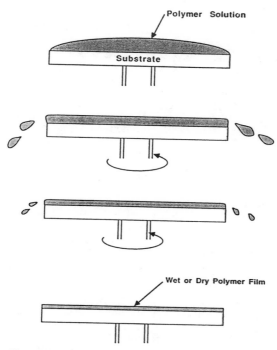

Figure 1.   The deposition of polymer thin films by the SCD process.

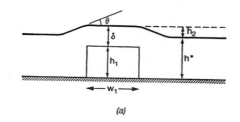

(a)

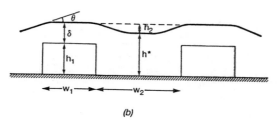

(b)

Figure 2.   Planarization of microelectronic structures by SCD polymer thin films: (a) Isolated line feature; (b) proximate line features.

Results and Discussion

The radial film thickness profiles at different durations of
spinning and for the four different rheological models (Newtonian,
Power-law, Carreau and Viscoplastic) were obtained by numerical
methods (12-13). Figure 3 shows representative film thickness
profiles for Newtonian (n = 1) and non-Newtonian (n < 1) liquids.
It was found that both Newtonian and Carreau liquids always gave
radially uniform films at sufficiently long spinning times even for
initially non-uniform film thickness profiles. Power-law and
viscoplastic materials gave highly non-uniform films even for
initially uniform film thickness profiles. Since all the polymer
solutions of interest in microelectronic applications can be
described by non-Newtonian Carreau viscosity equation it can be
concluded that uniform polymer thin films can be produced by the
SCD process.
    The theoretical result of Equation 5 provides a universal
dimensionless expression for the correlation of experimental dry
film thickness data in terms of the four variables: initial
viscosity, angular speed, initial volume dispensed or film thickness
and time and two solvent evaporation parameters. Note that Equation
5 reduces to Equation 3 when there is no solvent evaporation,
i.e. α = o. Also, note the forms of the predicted H^{*} dependencies
on the four variables, at large dimensionless number τ. The
parameter ψ can be determined from Equation 1, estimated from
Equation 2 or obtained by a fit of data. The rheological and
evaporation parameter α can be obtained from a fit of $H^{*}(\omega)$, $H^{*}(\nu_o)$,
$H^{*}(t)$ or $H^{*}(h_o)$ data or from concentration dependent viscosity
data (11). Figure 4 shows a good fit of Equation 5 to literature
data (17). All the available data (17,20) including our own (11)
have been successfully correlated with this expression.
    This universal film thickness equation immediately unifies all
prior apparently conflicting film thickness data correlations. For
example, one of the recurring sources of controversy and confusion
in the literature is the observed variability of the exponent p
in empirical and semi-empirical correlations of the form $h^{*} \sim 1/\omega^{p}$.
Some have generally assumed that p = 1/2 while others argue that
p = 2/3 (11,17). On the contrary, the observed values of the
exponent p is in the range of 0.40-0.82 for various photoresists,
electron beam resists, polyimide coatings, etc.
    Note that the predicted value of p = 2/(2+α), at sufficiently
large values of the dimensionless number τ (>1). This means that
p can have continuous values, between 0 and 1, which depend on the
degree of solvent evaporation and change in the viscosity of the
solution during the SCD process. However, solvent evaporation is
not the only source of the observed variability in the reported
values of p. Figure 5 shows the plot of $H^{*}(\omega)$ of Equation 5 for
fixed values of the two evaporation parameters. A unique (single)
curve cannot be obtained because of the variability of B_1, a
parameter which is a combination of the remaining three variables
when H^{*} versus ω experiments are done. Similar illustrations of

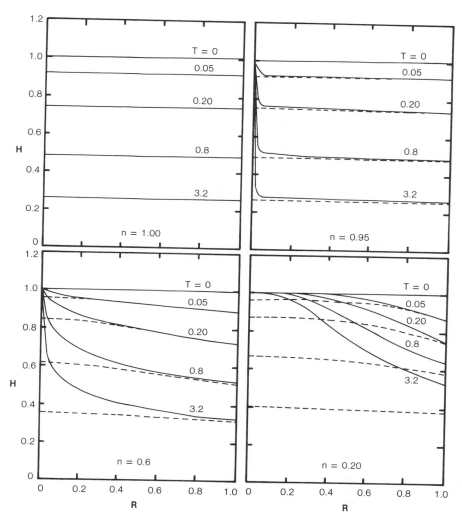

Figure 3.   Calculated dimensionless wet film thickness profiles for Newtonian and non-Newtonian liquids at selected dimensionless spinning time T defined in ref. (12). Power-law (——) and Carreau (----) models. The rheological parameter n is the same as the power-law index defined in Equation 4:   n = 1 indicates a Newtonian liquid and n < 1 indicates a non-Newtonian liquid.   R is a dimensionless radius along the radii of the wafer; it is given by r/r_o, where r is any radius and r_o is the radius of the water.

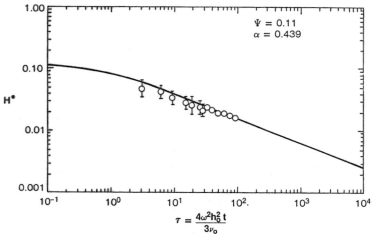

Figure 4.   Experimental and calculated dry film thickness of
            polyimide thin films deposited on silicon wafers.
            The line is Equation 5 and the data is from Daughton
            and Givens [Ref. (17); Figure 11].

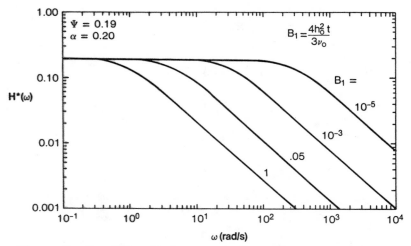

Figure 5.   Dry film thickness versus angular speed correlation
            [Equation 5] at fixed values of solvent evaporation
            parameters.   Different curves result from different
            values of B_1 and this shows the non-uniqueness of
            thickness versus speed correlations.

$H*(\nu_0)$, $H*(t)$ and $H*(h_0)$ can be given to show the universality of
the $H*(\tau)$ correlation and plot of Figure 4.   Further correlations
of literature data similar to Figure 4 and additional results and
details will be presented elsewhere (21).

The theoretical results of Equations 7 and 8 show the predicted dependence of degree of planarization on initial solution properties and on SCD variables and parameters. Note that Equation 7 gives a lower bound on the achievable ε as identical to the initial volume fraction of polymer in solution prior to SCD. Solvent evaporation during SCD enhances the degree of planarization above this minimum. Strategies for achieving high degrees of planarization are provided from the theoretical results. The theoretical details of the modeling of planarization of microelectronic structures by polymer films produced by SCD will be presented elsewhere (19).

Literature Cited

1. A.R. Von Hippel, Dielectric and Waves, Wiley, New York, 1954.
2. P.E. Bruins, Ed., Plastics for Electrical Insulation, Interscience Publishers, New York, 1968.
3. J.J. Licari, Plastic Coatings for Electronics, McGraw-Hill, New York, 1970.
4. C.A. Harper, Electronic Packaging With Resins, McGraw-Hill, New York, 1961.
5. J.H. Lai, S.A. Jenekhe, R.J. Jensen and M. Royer, Solid State Technology 27 (1), 165-171 (1984); ibid, 27 (12), 149-154 (1984).
6. S.A. Jenekhe and J.W. Lin, Thin Solid Films 105, 331-342 (1983).
7. T. Davidson, Ed., Polymers in Electronics, ACS Symp. Series No. 242, Am. Chem. Soc., Washington, D.C., 1984.
8. E.D. Feit and C.W. Wilkins, Jr., Eds., Polymer Materials for Electronic Applications, ACS Symp. Series No. 184, Am. Chem. Soc., Washington, D.C., 1980.
9. D.J. Sweeting, Ed., The Science and Technology of Polymer Films, Wiley, New York, Vol. 1, 1968; Vol. 2, 1971.
10. a. S.A. Jenekhe, Polym. Eng. Sci. 23, 830-834 (1983).
    b. S.A. Jenekhe, Polym. Eng. Sci. 23, 713-718 (1983).
11. S.A. Jenekhe, Ind. Eng. Chem. Fundam. 23, 425-432 (1984).
12. S.A. Jenekhe and S.B. Schuldt, Ind. Eng. Chem. Fundam., 23, 425-432 (1984).
13. S.A. Jenekhe and S.B. Schuldt, Chem. Eng. Commun., 33, 135-146 (1985).
14. A.G. Emslie, F.T. Bonner and L.G. Peck, J. Appl. Phys. 29, 858-862 (1958).
15. A. Acrivos, M.J. Shaw and E.E. Petersen, J. Appl. Phys., 31, 963-968 (1960).
16. W.W. Flack, D.S. Soong, A.T. Bell and D.W. Hess, J. Appl. Phys. 56, 1199-1206 (1984).
17. W.J. Daughton and F.L. Givens, J. Electrochem. Soc. 129, 173-179; 2881-2883 (1982).
18. L.B. Rothman, J. Electrochem. Soc. 127, 2116-2220 (1980).
19. S.A. Jenekhe, to be submitted to J. Electrochem. Soc.
20. R.J. Jensen, J.P. Cummings and H. Vora, IEEE Trans. Vol. CHMT-7, 384-393 (1984).
21. S.A. Jenekhe, to be submitted to J. Electrochem. Soc.

RECEIVED April 8, 1987

Chapter 23

Stress-Dependent Solvent Removal in Poly(amic acid) Coatings

C. L. Bauer and R. J. Farris

Polymer Science and Engineering, University of Massachusetts, Amherst, MA 01003

The development of residual stresses in poly(amic acid) coatings (precursors to polyimide films) was investigated using force-temperature experiments. Results indicate that the residual solvent content is dependent on the stress level. Experiments were performed using poly(N,N'-bis(phenoxyphenyl)pyromellitamic acid) at temperatures between 60° and 80° C to insure no cycloimidization. By altering the balance between internal stress and the driving force for solvent evaporation further solvent removal may occur.

Because of their extremely good thermal and electrical properties, thin polyimide films are widely used in the electronics industry. One commonly used polyimide is poly(N,N'-bis(phenoxyphenyl)-pyromellitimide). This may be prepared from the reaction of pyromellitic dianhydride (PMDA) and oxydianiline (ODA) in a two step process (Figure 1). The first step involves a solution reaction forming the poly(amic acid) (PAA). After solvent removal this material can be thermally cyclized to the polyimide (PI). To improve properties, it is often annealed at temperatures up to 400° C.

In its final state, this polyimide cannot be processed; thus processing occurs before the imidization step. PAA solutions may be spin-coated onto the appropriate substrates and then thermally treated. This sequence establishes the molecular order in the material (1,2) and in-plane orientation (3). Further development of residual stresses is also associated with this coating process.

Stresses in solvent based coatings arise from the differential shrinkage between the thin film coatings and the corresponding substrates. These stresses are due to volume changes associated with solvent evaporation, chemical reaction (i.e. cyclization in polyimide formation) and differences in thermal expansion coefficients of the coating and substrate (4,5). The level of residual stress depends on the material properties such as modulus, residual solvent content and crosslinking (5) and its thermal-mechanical history.

0097–6156/87/0346–0270$06.00/0
© 1987 American Chemical Society

Initially after coating a flat, planar substrate, as solvent is removed, stresses do not develop in the film since the liquid freely shrinks. When sufficient solvent has evaporated, the material gels and eventually solidifies. Assuming complete adhesion to the substrate, the material is then constrained in the plane of the film but can freely contract normal to this plane. As further solvent evaporates, the material shrinks and develops stresses in the plane due to the constraints. There are no shear stresses at the coating/substrate interface, except at the edges. These shear stresses at the edges result from an equilibrium force balance and decay to zero away from the edge. Because there exists a strong interplay between stress and swelling (solvent content) it may be impossible to remove all the solvent during the evaporation step. A steady state stress level and residual solvent content may be attained (5,6) and the stress in some coatings has been shown to be related to the volume fraction of solvent (7). Additional solvent may be removed by exposing the material to higher temperatures which contributes to further shrinkage.

The solvent removal process in coatings has been addressed as a transport problem (8,9) and from a mechanics viewpoint. Several methods have been developed to measure residual stresses in coatings. For organic coatings many of these methods utilize plate or beam deflection (6, 10-13). Only recently have they been applied to PI coatings.

In the formation of PI films, the material undergoes a solvent removal step plus several thermal treatments to cyclize the polymer and anneal it. Each of these processes contributes to the development of residual stresses. Goldsmith, et. al. (14) have shown that the resulting stresses from curing PI films are independent of film thickness and the maximum room temperature stress developed for a fully cured film is 70 MPa.

With several stages of processing, it may be possible to optimize the final stress state. Therefore it is important to understand the material behavior through all the processing stages. In this paper, the importance of the coating process in PI films and the interrelationship of stress level and swelling in PAA films was investigated by a direct measurement of the stress.

Experimental

Materials. The polyimide precursor used was DuPont PI5878 poly(amic acid) and is based on PMDA-ODA. N-methylpyrrolidinone (Aldrich #M7960-3) (NMP) was used as the solvent.

Force/Displacement-Temperature Experiments. Force as a function of temperature was obtained by placing a sample in a vertical glass oven and gripping both ends, with one end attached to a load cell and the other to an adjustable mount. A slow nitrogen flow was introduced through the bottom and an RTD probe was placed near the middle of the sample. Output from the probe and load cell was recorded on a Bascom-Turner Instrument model 4000. With a sample in place several heating and cooling cycles were performed at various loads. Typically, the temperature was rapidly increased to 80° C, maintained at this level for a given time frame and then cooled.

Samples consisted of a strip of rubber (6 cm x 1.5 cm x 0.25 mm)
with the ends glued between aluminum foil. This was then dipped into
a 15% PAA/85% NMP solution and placed into the force-temperature
(F-T) apparatus, gripping only the aluminum tabs.

To obtain displacement (strain) as a function of temperature,
the load cell in the above apparatus was replaced with a linear
variable differential transducer.

To determine if any cycloimidization occurred with these
heating/cooling cycles, infra-red spectra were obtained and inspected
for imide bands and the absence of N-H stretch.

Linear Mass Experiments. Using long strips of natural rubber (60 cm
x 2 mm x 0.5 mm) coated with a 15% PAA solution in NMP, the mass per
unit length of the sample was obtained as a function of temperature.
The technique involves measuring the tension on the sample and the
time of flight of a traveling wave on the sample (M. Chipalkatti,
J. Hutchinson and R.J. Farris, *Rev. Sci. Instr.*, in press.). These
quantities are then related to the mass/length using the wave
equations for a flexible string which yields:

$$mass/length = tension/(velocity)^2$$

Results and Discussion

Results of force-temperature experiments for a PAA solution on rubber
are shown in Figure 2. For the first cycle, the initial load is that
of the rubber since the coating is still in the liquid state and
cannot support a load. At this load the rubber is just below its
thermoelastic inversion point and its contribution to the force
change is negligible.

As the sample is heated, it experiences a positive thermal
expansion. Shrinkage does not occur until a critical amount of
solvent is removed and the coating solidifies, at which point the
force increases as solvent is removed. When cooled to room
temperature, the force (at 25° C) has increased by 250g due to
solvent loss in the PAA coating. Considering only the coating on the
two planar faces of the rubber, the change in load for one cycle is
125g/coat.

By adjusting one of the mounts in the F-T apparatus, the load
was reduced to 100g, approximately the same starting force as in the
first cycle. On heating and cooling the sample for the second cycle,
the path traversed is similar to that of the first cycle. If instead
the load was left unchanged after the first cycle (that is at 375g)
the heating/cooling curve reversibly follows the upper path (cooling)
of cycle-1 (between 375g at 25° C and 165g at 85° C).

For comparison with other samples the final cross sectional area
of the coating (~1.4 cm x ~10 μm) was used to compute stresses in the
films. For the above experiment, the stress change due to shrinkage
with a one dimensional constraint was found to be 8 MPa/coat.
Assuming planar symmetry, for a two dimensionally constrained film
the shrinkage stress would approach 16 MPa in equal biaxial tension.

Results in Figure 2 suggest that the evaporation process is
dependent on the stress level. Figure 3 illustrates this further.
Force-temperature cycles were performed as before, except after the

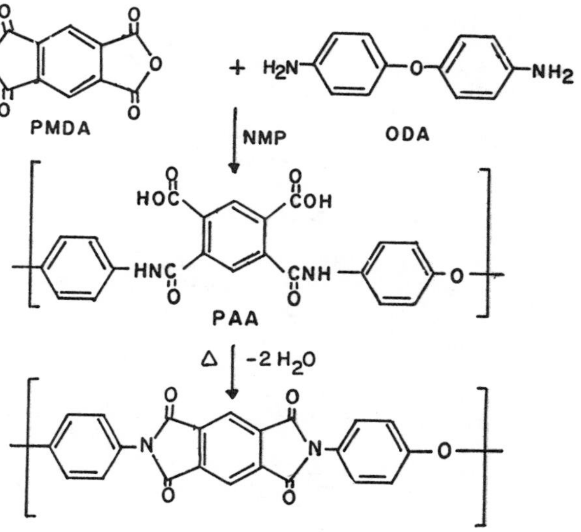

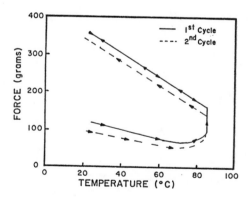

Figure 1.   Reaction scheme for formation of polyimide.

Figure 2.   Force versus Temperature for Poly(amic acid) solution on rubber, sample 1.

first cycle the applied load was manually increased above any
previous level. On heating to 80° C, little shrinkage occurred after
20 min. For the next cycle (C) the load was adjusted back to the
original level (50g). On reheating the sample, solvent evaporation
occurs causing shrinkage forces which approach that of the first
cycle. These results indicate that a coupling exists between the
internal stress in the film and the amount of solvent the material
can retain. If the internal stress balances the driving force for
solvent removal at a given temperature, further solvent can be
removed by reducing the stress.

Temperature should also play a similar role, where a higher
temperature should be needed to remove additional solvent. This is
confirmed in an experiment like that described above, except the
temperature is increased after the second cycle instead of reducing
the load. In this case when the temperature was increased to 110° C
additional shrinkage was observed. After this point, if the
temperature is increased to 150° C, the load does not increase since
the thermal expansion dominates the behavior. However, at
temperatures above 150° C the load increases due to shrinkage caused
by the cyclization.

To determine if this balance between stress level and swelling
is a non-equilibrium effect, long-term experiments were performed.
Figure 4 shows the determination of steady state shrinkage stress.
A sample was heated under nitrogen to 62° C and kept at this
temperature (10 hr.) until the stress reached a constant value
(8.3 MPa). (On cooling to room temperature, the stress reached
10 MPa) Probably over an infinite period of time, slightly more
solvent loss would occur; however, in a reasonable time frame, a
"quasi"-steady state is obtained as illustrated.

The temperature was then increased to 84° C. The initial
decrease in stress with this temperature change is due to thermal
expansion of the material. If the sample is kept at this
temperature, the stress again gradually increases. This increase is
due to further solvent evaporation and subsequent material shrinkage.
This additional drying is influenced by two factors. First, since
the change in stress due to temperature is equivalent to a change due
to mechanical deformation, then as noted above, a reduction in stress
allows further solvent to be removed. Secondly, the higher
temperature increases the driving force for solvent evaporation.

In another set of F-T experiments the stress was varied and the
temperature held constant (Figure 5). The sample was heated to 80° C
and reached a constant stress of 9.5 MPa; after which the sample was
cooled and the stress manually reduced from 10 MPa to 1 MPa. After
heating for 4 hrs at 80° C the stress increased to 5.4 MPa but had
not attained a level value. In part, this can be due to viscoelastic
recovery, but since there is a measurable change in sample mass
between the two runs the stress change is probably due to shrinkage.
In comparing Figures 4 and 5 there is a noticeable change in the rate
of shrinkage for the different temperatures. However, the steady
state stress level is not appreciably different for the two
temperatures.

To further understand the solvent removal process, strain as a
function of time (and temperature as a function of time) was
obtained. Results are illustrated in Figure 6. A constant load of

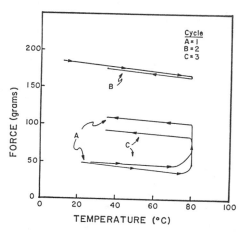

Figure 3.   Force versus Temperature for Poly(amic acid) solution on rubber, sample 2.

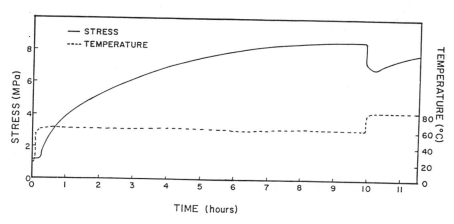

Figure 4.   Stress and Temperature versus Time for Poly(amic acid) solution on rubber, sample 3.

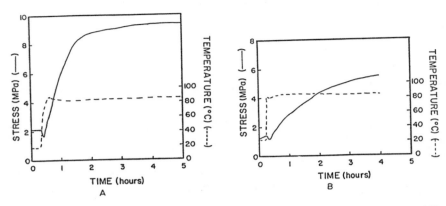

Figure 5. Stress and Temperature versus Time for Poly(amic acid)
solution on rubber, sample 4. A - first run, B - second run,
stress reduced manually to 1 MPa.

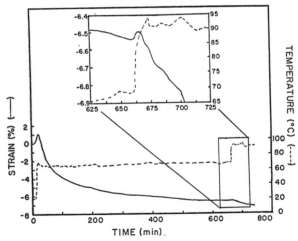

Figure 6. Strain and Temperature versus Time for Poly(amic acid)
solution on rubber, sample 5.

115g was applied to the sample. On heating at 65° C, the strain decreased with time as expected. After 10 hrs the strain leveled at 6.5%. When the temperature was increased to 80° C the strain increased by 0.04% due to thermal expansion. This corresponds to a thermal expansion coefficient of 3 E(-5) cm/(cm·°C); whereas Kapton (a polyimide based on PMDA-ODA from DuPont) has an expansion coefficient of 2 E(-5) cm/(cm·°C) (15). Further heating at 80° C results in a decrease in strain which must be due to shrinkage and not creep.

After the equilibrium strain has been reached, if the applied load is increased there is no further change of strain with time; but if it is decreased, the strain again decreases with time. As with force-temperature experiments there is a detectable change in mass between runs to indicate solvent loss.

To correlate stress changes with solvent concentration changes, linear mass experiments were performed. Preliminary results support the conclusions above and further experimentation is in progress to quantify these changes. Using this technique plus direct sample weighing, for each successive heating/cooling/load reduction cycle less solvent is removed to attain the steady state stress level. Because the solvent acts as a plasticizer, the material modulus is reduced. Then as solvent evaporates, the modulus of the film increases (with a final modulus of 1.5 ± 0.4 GPa). Therefore, less solvent needs to be evaporated to generate the same force as in earlier cycles.

Conclusions

These results indicate that the solvent removal process in poly(amic acid) film formation is dependent on the residual stress level. Solvent evaporation creates volume changes which in turn generate shrinkage stresses (~ 8 MPa at room temperature). This will continue until there is a balance between the internal stress and the driving forces for solvent removal. Further solvent may be removed if this balance is altered by a change in stress or temperature.

Literature Cited

1. N. Takahashi, D.Y. Yoon and W. Parrish, *Macromolecules*, 17, 2583 (1984).
2. T.P. Russell, *J. Polym. Sci., Polym. Phys. Ed.*, 22, 1105 (1984).
3. T.P. Russell, H. Gugger and J.D. Swalen, *J. Polym. Sci., Polym. Phys. Ed.*, 21, 1745 (1983).
4. A.G. Evans, G.B. Crumley and R.E. Demaray, *Oxid. Metals*, 20 (5-6), 193 (1983).
5. K. Sato, *Prog. Org. Coat.*, 8, 143 (1980).
6. D.Y. Perera and D.V. Eynde, *J. Coat. Technol.*, 56 (718), 69 (1984).
7. S.G. Croll, *J. Appl. Polym. Sci.*, 23, 847 (1979).
8. R.F. Eaton and F.G. Willeboordse, *J. Coat. Technol.*, 52 (660), 63 (1980).
9. J. Holten-Anderson and C.M. Hansen, *Prog. Org. Coat.*, 11, 219 (1983).

10. E.M. Corcoran, *J. Paint Technol.*, <u>41</u> (538), 635 (1969).
11. S.G. Croll, *J. Coat. Technol.*, <u>50</u> (638), 33 (1978).
12. T.S. Chow, C.A. Liu and R.C. Penwell, *J. Polym. Sci., Polym. Phys. Ed.*, <u>14</u>, 1311 (1976).
13. R.N. O'Brien and W. Michalik, *J. Coat. Technol.*, <u>58</u> (735), 25 (1986).
14. C. Goldsmith, P. Geldermans, F. Bedetti and G.A. Walker, *J. Vac. Sci. Technol.*, <u>A1</u> (2), 407 (1983).
15. "Kapton Polyimide Film" Summary of Properties, DuPont Technical Bulletin, E-50533 (1982).

RECEIVED April 8, 1987

Chapter 24

Adhesion and Yield of Polyacrylate-Based Photoresist Lamination in Printed-Circuit Fabrication
Influence of Substrate Thickness and Preheat Treatment

Eric S. W. Kong

Hewlett–Packard Laboratories, Palo Alto, CA 94304

During fuser-roll resist lamination processes, the copper-clad FR-4 substrate surface temperature was found to be inversely proportional to the substrate thickness. This temperature fluctuation has resulted in changes of adhesive forces in the copper/resist interface which in turn can affect the yield in the printed circuit manufacturing processes. By using infrared preheat treatment on the substrate prior to fuser-roll lamination, the adhesion was found to be improved in the copper/resist interface. This adhesion improvement was found to be reflected in yield increase in fine line printed circuit fabrication.

As a standard procedure during image transfer processes in printed circuit fabrication, dry film photoresist is normally applied to a cleaned copper substrate using heat and pressure from a fuser roll.[1-3] Resist lamination using a filled silicone-rubber coated fuser roll is a common practice[1,3]. Schematics showing the dry film lamination process are shown in Figures 1 and 2. The resist lamination must be carried out in a yellow safe light environment. Control of airborne contamination (Class 1000 or better clean rooms) as well as control of the temperature (ca. 25 °C or lower) and humidity (50% or lower RH) are important factors for fine line image transfer at high yields.

Ripsom and Wopschall suggested a linear relationship between logarithmic Riston 3600 resist viscosity and temperature[4]. At 25 °C, the resist viscosity was reported to be 10^9 poises. At 100 °C, the viscosity would decrease to $10^{5.5}$ poises. Hence, there is over 3 orders of magnitude drop in resin viscosity as the temperature is raised from room temperature to 100 °C, which is 5 degrees below the typical fuser roll lamination

0097–6156/87/0346–0279$06.00/0

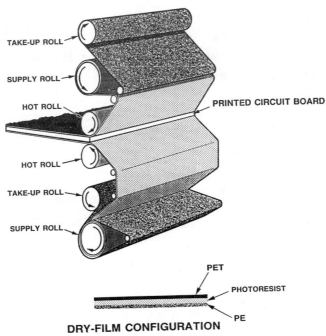

DRY-FILM CONFIGURATION

Figure 1.   Three-dimensional schematic of the hot roll lamination
process: PET is poly(ethylene terephthalate); PE is polyethylene.

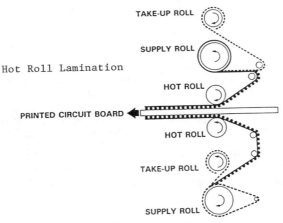

Figure 2.   Two-dimensional schematic of the hot roll lamination
process: PET is poly(ethylene terephthalate); PE is polyethylene.

temperature. Ripsom and Wopschall[4] have also reported the glass transition temperature (T_g) to be between 46 and 64 °C. Hence, heating the resist to temperatures near 100 °C while applying a pressure of 40 PSI$_g$ can effectively force the softened resist to bond firmly onto the copper surface of the FR-4 substrate. It has been reported, however, that "interfacial voids" between the copper/resist interface are rather difficult to avoid after the resist lamination.[4]

A survey of the literature[1-5] suggested that preheating the FR-4 substrates prior to resist lamination could improve the adhesion substantially. Specifically, a positive correlation was reported to exist between good resist adhesion and copper surface temperature[5] at lamination. Secondly, moisture can often cause resist lifting and resist breakdown during plating,[1,4] hence it is recommended that moisture are to be removed by preheating the panels just prior to lamination.

One obvious motivation behind industrial research is to improve the yield or connectivity of the printed circuitry. In this investigation, two main variables were studied concerning their influence on the yield of fine line circuitry: substrate thickness and preheat conditions. Comparative data from the literature as well as from private communications will also be critically reviewed.

Experimental

Materials.

The photoresist was acrylate-based Riston 3600 series supplied by E.I. du Pont de Nemour and Company, Wilmington, Delaware. This is a resist which can be developed in aqueous medium. This resist is negative working in the sense that it would predominantly undergo crosslinking polymerization upon ultraviolet light irradiation.

The copper clad Fiberglass-reinforced epoxy FR-4 laminates were supplied by Nelco Products, Fullerton, California. The copper clad thickness was 35 microns (1.4 mils = 0.0014 inch) which is equivalent to 1.0 oz. of copper per square inch. Laminates of the thicknesses 10, 21, 31, 59, and 93 mils were utilized in this investigation.

The Adhesion Tests

The pull tests were performed in order to characterize the adhesive forces between the copper/photoresist interface. The tensile tests were measured using a screw-driven UNITE O MATIC tester manufactured by United Calibration Corporation, Garden Grove, California. The tensile loading forces were measured using a 10 lb. load cell manufactured by Interface Company, Scottsdale, Arizona. Calibrations were done using standard weights of 5.0 and 10.0 lbs. supplied by the National Bureau of Standards. Tensile deformation was detected by a linear variable differential transformer (LVDT). The pull-test specimens were prepared adhering studs of area 0.07 square inch (0.45 sq. cm) directly onto the photoresist (0.0015 inch or 0.038 mm thick) by cyanoacrylate adhesives. The cyanoacrylate adhesive was supplied by 3M Company, Minneapolis, Minnesota, with the Trade Name CA-8 Scotch-Weld. Prior to applying the adhesive onto the stud, the latter was degreased and surface roughened by the following procedure: grinding by sandpaper; degreasing by isopropanol; treatment by Hydrohone process; and a final degreasing by trichloroethylene.

Surface Treatment

A pumice scrubber was used prior to lamination to remove cupric oxide from the copper surface of FR-4 substrates. The scrubber was manufactured by Resco Equipment North America, Montreal, Canada. Abrasive jets at 26 PSI_g of pumice and water cleansed the malleable copper surface. (Pumice is 73% silicone dioxide and 12% aluminum oxide, plus small amount of other minerals). The mechanical action of the pumice also created a roughened, mushroom-like morphology on the copper (evidence from scanning electron micrographs) which enhanced the adhesion with the resist. The copper "surface roughness" right after the scrubbing had a root-mean-square value of 3.0 ± 1.0 microns and the contact angle was 70 ± 10 degrees while interacting with deionized water. The pumice treated copper surface held a continuous film of distilled water for 30 seconds. As a FR-4 substrate passed through the scrubber, it was subjected to 8% sulfuric acid cleansing (to rid of cupric oxide); soft water rinse; pumice jet scrubbing; air knife treatment; deionized water rinse; and finally a drying process at 45°C.

Scanning Electron Microscopy: All pumice treated copper surface morphology were examined under a scanning electron microscope equipped with an x-ray analyzer. The electron microscope was Model JSM 840 supplied by JEOL Limited, Tokyo, Japan and the x-ray analyzer was Model TN-5500 manufactured by Tracor Northern, Middleton, Wisconsin. This technique was used extensively at many stages of the printed circuit fabrication, especially at the point right after resist lamination and development.

Profilometer and Contract Angle Measurements: The copper surface after Pumice treatment was also examined by profilometry and characterized by contact angle measurement. The profilometer was supplied by Sheffield Measurement Division of Dayton, Ohio. The contact angle measurements were made using a microscope device supplied by Gilmont Instruments, Great Neck, New York.

Dry Film Lamination: The resist lamination was performed using a Model 712 Hot Roll Laminator supplied by MacDermid, Inc., Waterbury, Connecticut. The heart of the lamination system is an aluminum-cored fuser roll which is about 2.5 inches in diameter and is coated by 0.1 inch thick additives-filled silicon rubber. Figure 3 shows a schematic of the fuser roll. All dry film lamination experiments were performed using a conveyor speed of 20 inches per minute; an air-assisted pressure of 40 PSI_g and a fuser roll temperature of 105°C. The hold time between surface treatment and resist lamination was kept constant at 60 minutes.

In the thermal profile measurement experiment, a fine-wire thermocouple is soldered onto the center of an 8 inch by 11 inch copper clad FR-4 printed circuit board (The solder was an alloy of 63% tin and 37% lead which melts at 183°C). The thermocouple used was an American Wire Gage 30 chromel-alumel wire which had a maximum use temperature of

482°C.   The supplier of the thermocouples was Omega Engineering, Stamford, Connecticut.

Computer Simulation: The empirical thermal profiles were simulated using a Model 9236 computer supplied by Hewlett Packard, Palo Alto, California.

Ultraviolet Curing/Printing: The laminated boards were subjected to ultraviolet light irradiation using an exposure unit (Model OB-1600 FPD) supplied by Optical Radiation Corporation, Azusa, California.   The exposure unit provides a collimated light source. All of its specifications are summarized in Table 1. Figure 4 shows a schematic of the optics in this system. In order to measure the amount of radiant energy per unit area (milliJoules per square centimeter) impinging onto the photoresist, a radiometer (Model IL740A) was utilized.   The photoresist research radiometer was supplied by International Light, Newburyport, Massachusetts.   All resists were subjected to 75 mJ/cm^2 of radiant energy.   A hold-time of 10 minutes was allowed to elapse before the developing process began. The artwork used was IPC pattern A-38.[6] The pattern has conductor widths between 2.5 and 4.5 mils.

The developer: The exposed dry film was developed using 1% sodium carbonate solution in a vertical processing developer supplied by Circuit Services, Minneapolis. The pH of the developing solution was kept at 11.0 ± 0.2 at all times during the development process as monitored by a in-line pH meter/controller (Model 31171-00) manufactured by Hach Company, Loveland, Colorado.   The conveyor speed was 5.0 feet per minute.   The developing sump temperature was kept at 85 ± 2°F (ca. 29°C); while the rinse temperature was kept at 80 ± 2°F (ca. 27°C). The developing chamber spraying pressure was 29 PSI_g and the two stages of rinse chamber deionized water pressure was 12 PSI_g and 18 PSI_g respectively.

Optical Microscopy: The resist pattern was examined under an optical microscope normally at the stage shortly after the development.   The Nikon Optiphot microscope was supplied by Nikon Instrument Division of Garden City, New York.

Etching: Etching of the copper was performed in a Chemcut etcher at a temperature of 130 ± 5°C (ca. 54°C).   The etcher was Model 547-20, supplied by Chemcut Corporation, State College, Pennsylvania.   The chemistry of the etching solution was that of cupric chloride/hydrochloric acid and hydrogen peroxide. The conveyor speed was operated at 4.0 feet per minute.

Stripping: The resist was stripped or removed using a Chemcut stripper placed in tandem with the Chemcut etcher. The Model number was also 547-20. Potassium hydroxide solution was used as the stripping medium at a temperature of 150°F (ca. 66°C). The conveyor speed was 7.0 feet per minute.

The electrical tests: In order to test the yield of the print and etch process, electrical tests were performed on the printed circuits looking for both open and short connections. Dedicated fixtures for the electrical tests were supplied by Everett/Charles Test Equipment, Santa Clara, California.

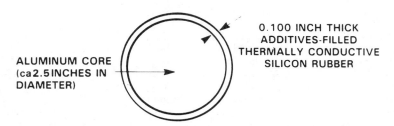

ALUMINUM CORE
(ca 2.5 INCHES IN
DIAMETER)

0.100 INCH THICK
ADDITIVES-FILLED
THERMALLY CONDUCTIVE
SILICON RUBBER

"FILLED" $(SiR_2O)_n$ RUBBER HAS A THERMAL CONDUCTIVITY
OF ca. 3.0 TO 4.0 BTU-INCH/HOUR Ft^2 °F AND A DUROMETER
HARDNESS VALUE OF 50 TO 60.

Figure 3.   Configuration of a "hot roll".

Table I.   Specifications of the ultraviolet light exposure system.

• **LIGHT SOURCE:**	1600 WATT PRESSURIZED SHORT-ARC MERCURY/XENON LAMP
• **LIGHT INTENSITY:**	> 20 mW/cm² AT EMISSION OF 330 to 440 nm AT LEVEL OF PHOTOTOOL
• **LIGHT COLLIMATION:**	< 2° HALF ANGLE
• **LIGHT DECLINATION:**	< 1.5°
• **LIGHT UNIFORMITY:**	± 10% FROM POINT TO POINT OVER AREA OF 12 X 12 in² (304.8 X 304.8 mm²)

Results and Discussion.

All FR-4 substrates were examined under the electron microscope after the pumice surface treatment to analyze the morphology as well as determine whether there were any residual pumice particles left on the substrate. A "mushroom-like", surface-roughened morphological relief was observed on the copper surface. Scanning electron micrographs of a pumice treated copper surface indicated striations that reflect the conveyorized nature of pumice treatment. No trace of the elements silicon or aluminum were observed by the x-ray analyzer, hence indicating a total removal of pumice from the substrate. The copper surface roughness was observed to have a root-mean-square value of 3.0 ± 1.0 microns and a contact angle of about 70 degrees when that surface interacted with distilled water. The fact that this copper surface held a continuous water film for at least 30 seconds indicated that very little organic contaminants were left on the copper right after the pumice/sulfuric acid/water rinse treatment.

Figure 5 shows the time-temperature profiles for substrates of various thicknesses interacting with the fuser/hot roll. In this specific experiment, no resist was laminated: only the direct thermal interaction between the bare board and the fuser roll was studied. The data clearly indicate an inverse relationship between board thickness and surface temperature. For example, a 10 mil thick (1 mil is ca. 25.4 microns) substrate shows a peak temperature of 93.3 ± 1.5°C (From this point on the error band of plus and minus 1.5 degrees will not be reiterated). As a thicker substrate, for example, 31 mil thick panel was fed through the laminator, a much lower peak temperature of 74.0°C was registered. This result is undoubtedly due to the much higher thermal mass of the 31 mil board compared to that of the 10 mil substrate: both have the same amount of copper cladding materials (35 microns thick) but the former have about three times the amount of Fiberglas/epoxy composite. Further increasing the board thickness to 93 mils resulted in a further decrease in surface temperature to 68.7°C. Gravimetrically, a panel size of 11 inches by 8 inches would weigh 51.9g for a 10-mil board; 84.8g for a 21-mil board; 114.7g for a 31-mil board; and 278.9g for a 93-mil board. As the data in the following discussion will demonstrate, this surface temperature fluctuations have a direct influence on the adhesion properties with the resist and even the fabrication yield of the printed circuits.

Figure 6 shows the thermal profiles during Riston 3615 photoresist lamination. Since the resist served as a thermal energy absorber/insulator, the surface temperature of the substrate became lower compared to the first series of experiments shown in Figure 5. In general, the surface temperature with the resist lamination is about 5 to 6 degrees Celcius below that of the situation in which the board is in direct contact with the fuser roll. Again, it was found that the resist/copper interfacial temperature decreased as the board thickness increased. For the thinnest board of 10 mils, the interfacial peak temperature was 87.3°C. For board thickness of 22 mils, that temperature decreased to 74.5°C. For a board thickness of 31 mils, the interfacial peak temperature further dropped to 69.9°C. For a board thickness of 59 mils, that peak temperature further dropped to 62.5°C. From this point however, the temperature decrease became more and more asymptotic.

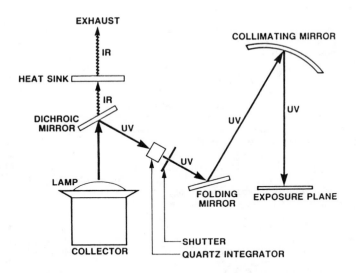

Figure 4.   Optics of an ultraviolet exposure unit.

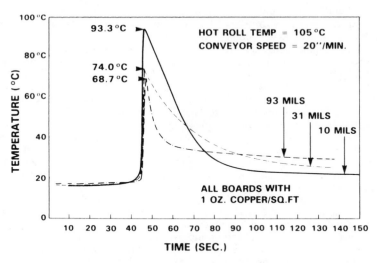

Figure 5.   Thermal profiles for FR-4 substrates passing through
the hot roll laminator.

That is, further increasing the board thickness will only reflect in smaller and smaller decrease in surface temperature. We found, e.g., further increasing the board thickness to 93 mils only resulted in a peak temperature of 61.5 °C, not that big of a difference from the value obtained by the 59 mils board. We will see in our discussion later that thermal modeling confirms this observation.

Changing the resist thickness was expected to play a role in varying the interfacial lamination temperature. It was expected that a thicker resist/insulator would absorb more thermal energy and hence resulted in a lower interfacial temperature. This may indeed be the case but the magnitude of difference may be too small to be significantly reflected in the empirical data. In Figure 7, we show the interfacial temperature to stay roughly at 69.9 °C, regardless whether the resist was 1.0, 1.5, or 2.0 mils. The FR-4 substrate used in this series of experiments was 31 mils thick.

By means of infrared preheating the substrate during it conveyorized voyage to encounter the fuser roll, the surface temperature of the substrate can be raised (See Figure 8), depending on the temperature of the heating elements. For instance, a infrared chamber setting of 150 °C would result in heating up a 31 mils thick (8 inch by 11 inch) panel from room temperature to 50 °C, at a conveyor speed setting of 20 inches per minute. This surface temperature of 50 °C is critical: this temperature is above the Tg of the acrylate-based photoresist, which is reported to be between 46 and 64 °C. At this elevated temperature, it is expected that the adhesion promoter molecules in the resist formulation would be thermally activated to improve the bonding of the copper/resist interface. On direct contact with the fuser roll the peak temperature was further promoted to 84 °C. Further increasing the infrared chamber setting to 200 or even 250 °C would further elevate the surface temperature of the board closer and closer to that of the hot roll temperature of 105 °C. These higher surface temperatures would further activate the flow or rheological behavior of the resist to conform onto the copper surface, hence a better adhering interface.

It is well known in polymer science literature[9,10] that increased temperature will result in a lower viscosity in the resin. This lower viscosity is often translated into higher chain mobility or segmental motion in the macromolecules. We expect therefore a preheat treatment could result in a stronger adhesion between the polymer photoresist and the copper substrate. This expectation has actually been borne out by experiments. In our adhesion pull tests, adhesion force between resist/copper interface was determined to be ca. 5.3 lb (error band about plus or minus 10%). Preheating the surface temperature to 50 °C (150 °C setting) would result in improving the adhesive force to 6.8 lb (stud size kept constant at 0.07 square inch). Preheating the surface tmeperature to 64 °C (200 °C chamber setting) further improved the adhesion force to 7.2 lb. Finally, preheating the panel surface temperature to 80 °C further improved the adhesion force of the interface to 7.4 lb. It is noticed, however, as long as the board is preheated from room temperature to a temperature above the Tg of the photoresist, there was a quantum jump in adhesion improvement: from 5.3 to 6.8 lb or 28.0% increase. The further improvement from surface temperature of 50 to 80 °C was only gradual: from 6.8 to 7.4 lb or 8.8%. These data suggested that as long as the resin

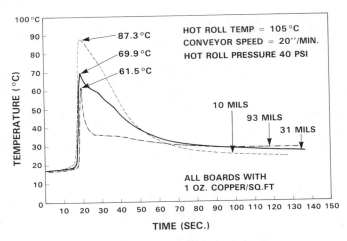

Figure 6. Thermal profiles for FR-4 substrates going through resist lamination.

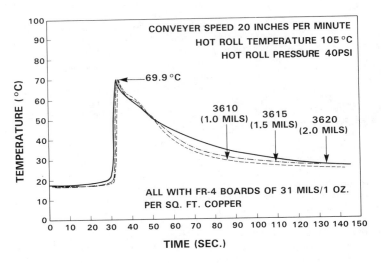

Figure 7. Influence of photoresist thickness on the thermal profiles.

reached its Tg or above, the rheological behavior would be improved to result in a significantly better bonded interface with copper.

Besides adhesion, preheat was also found to be beneficial in improving the photoresist integrity and side-wall profiles. These parameters tend to be difficult to characterize quantitatively. However, the optical micrograph and observations in scanning electron microgrphs lend support to the statement above.

The surface temperature/adhesion properties were found to be related to the yield during the fabrication and manufacturing of the printed circuits. Yield analysis[6-7] from electrical tests showed that thicker panels have lower yield compared to thinner panels and that more defects (mostly open circuits) were detected in substrates that have not gone through the preheat treatment raising the board surface to 80 °C. For example, thicker FR-4 substrates (59 mils) had an average of 10 defects (open circuits) per board whereas thinner substrates (10 mils) only had an average of 2 defects per board. This observation is undoubtedly due to the surface temperature fluctuations. In order to remedy this shortcoming, we prescribe a preheat treatment. This has resulted in improvement in adhesion as well as in yield for boards going through the print and etch process: for conductor width of 2.5 mils the yield was only 20% (averaged of 10 experiments) without the preheat treatment. With preheat, this yield data was improved from 20% to 25%. Going from without preheat to with preheat, yield improved from 70% to 90% for 3.5 mils wide conductors. Similarly, yield improved from 90% (without preheat) to 98% (with preheat) for 4.5 mils wide conductors. All these yield improvement are manifested due to the ability to control the surface temperature to one very close to the fuser roll temperature.

To substantiate the confidence level on our empirical data, data simulation using thermal modeling was exercised. The classical diffusion equation[11] and the following parameters were utilized: The thermal conductivity of copper is 3.88 watts/°C-cm whereas that of the Fiberglass/epoxy is 0.0028 watts/°C-cm. The heat capacity (Cp) of copper is 0.09215 cal/°C-g whereas that of the Fiberglass/epoxy composite is 0.22 (30 °C Cp) and 0.45 cal/°C-g (120 °C Cp).

We found a near perfect fit of experimental data to the theoretical curve as indicated by dotted line in Figure 9. Of great interest is the observation that theory predicts an asymptotic decrease of temperature after the panel thickness reaches 50 mils or above. This prediction was again borne out by experiments.

Summary

Due to surface temperature fluctuations during lamination and the resulted changes in photoresist/copper interface adhesion, printed circuit yield is inversely proportional to FR-4 substrate thickness. Preheating the substrates, however, would improve the adhesion and result in a higher yield for the printed circuits.

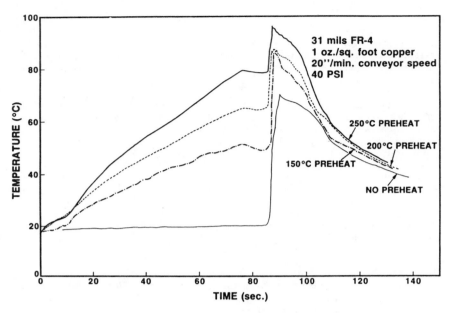

Figure 8.   Thermal profiles of Riston 3615 lamination as influenced
by infrared preheat.

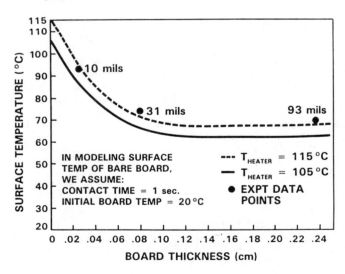

Figure 9.   One-dimensional thermal modeling of heat diffusion.

Acknowledgment

The author would like to acknowledge the support and technical help from Dr. Peter Robrish in performing computer modeling of the thermal profile data.   The discussion and useful information supplied by Dr. Robert Wopschall is deeply appreciated.   The technical assistance from Mr. Marshall Goins and Mr. Philip Avery is gratefully acknowledged.

References

1   Clark, R.H., Handbook of Printed Circuit Manufacturing, Van Nostrand Reinhold Company, New York, Pages 174-177 (1985).

2   Einarson, N.S., Printed Circuit Technology, Printed Circuit Technology, Burlington, Massachusetts, Pages 83-84 (1977).

3   Armstrong, E. and Duffek, E.F., 'Image Transfer' in Printed Circuit Handbook edited by C.F. Coombs, second edition, McGraw-Hill, New York, Pages VI-17, (1979).

4   Ripsom, J.R. and Wopschall, R.H., 'Optimizing Dry Film Photoresist Application for Fine Line Innerlayers'. Proceedings of ther AES Merrimack Valley Printed Circuit Workshop, March (1985).

5   Schantz, L.E. and Pusch, J.H., 'Infrared Baking of positive Photoresist Application for Hybrid Microcircuit Fabrication'. Technical Report BDX-613-1256, Bendix Corporation, Kansas City, Missouri (1976).

6   IPC Technical Report: 'Leading Edge Manufacturing Technology Report'. Report no. IPC-TR-578, Evanston, Illinois, September (1984).

7   Bartlett, C.J. and Schoenberg, L.N., 'A Quantitative Method for Evaluation of Fine Line Circuit Board Processes'. IPC Technical Review, Page 11, August/September(1980).

8   Turley, A.P. and Herman, D.S., 'LSI Yield Projections Based Upon Test Pattern Results: An Application to Multilevel metal Structures'. IEEE Trans. on Parts, Hybrids, and Packaging, Volume PHP-10, no. 4, pp. 230-234, December (1974).

9   Wallig, L.R., 'Dry Film Photoresists' in The Multilayer Printed Circuit Board Handbook, edited by J.A. Scarlett, pp. 111-153, Electrochemical Publications, Ayr, Scotland, United Kingdom (1985).

10   DeForest, W.S., 'Photoresist: Materials and Processes'. Pages 163-212, McGraw-Hill, New York (1975).

11   Shewmon, P.G., "Diffusion in Solids'. McGraw-Hill, New York (1963).

RECEIVED April 8, 1987

Chapter 25

Simulation of Resist Profiles for 0.5-μm Photolithography at 248 nm

R. K. Watts, T. M. Wolf, L. E. Stillwagon, and M. Y. Hellman

AT&T Bell Laboratories, Murray Hill, NJ 07974

One of the methods under development at AT&T Bell Laboratories for submicron lithography is deep ultraviolet projection photolithography. (1) Fine line definition is obtained by use of 248 nm light and a lens of large numerical aperture. Because of the large chromatic aberration of the quartz lens a spectrally line-narrowed krypton fluoride excimer laser is used as a light source.

Microposit 2400 resist, manufactured by Shipley Co., has been shown to be sensitive at this short wavelength (2) and is being employed with the deep UV stepper. We report here the results of resist profile modeling for submicron photolithography at 248 nm. Various model parameters needed as input data were measured to characterize exposure and development of the resist.

Determination of Resist Parameters.

Most resists are designed for exposure at wavelengths longer than the 248.4 nm radiation provided by a KrF laser source. Wolf and coworkers (3) have found that the choice of a positive resist for use at this wavelength is limited. They evaluated a number of positive resists. Only Microposit 2415, and its newer analog, Microposit 2400-17, were compatible with the anticipated exposure time of 0.5 to 1.0 seconds for resist sensitivity of 100 to 200 mJ/cm^2 needed with the new exposure tool developed by Pol and coworkers. (1) The resist consists of three components: a resin, a photoactive compound or PAC (which acts as a dissolution inhibitor), and a solvent. Upon exposure, the PAC is destroyed, and this allows the resist film to dissolve in the aqueous basic developer.
    The SAMPLE program (4) is used to simulate exposure and development of features in the MP2400-17 positive resist. The three resist parameters A, B and C defined by Dill and coworkers (5) and the solubility rates of the resist films as a function of exposure dose (or PAC content) must be determined to perform the simulation.

Optical Absorbance Measurements.

The resist parameters A, B and C are normally determined from
optical absorbance measurements of exposed and unexposed resist
films. (5)   The B parameter is obtained from the absorbance of a
film that is given a sufficient exposure dose to destroy all the
photoactive compound.   The quantity A+B is obtained from the
absorbance of an unexposed resist film and C is determined from the
initial slope of an absorbance versus exposure time plot.
     Figure 1 shows the ultraviolet (UV) spectra of an unexposed
MP2400-17 film and a film exposed to 470 mJ/cm^2. The films were
spin coated onto quartz substrates and baked at 90°C for 30 minutes
in a forced-air oven. The film thickness after baking was 0.465 μm.
Exposure was accomplished by direct irradiation of the film to the
narrowed output of a Math Sciences EXL-100 KrF excimer laser.
Thirty pulses were delivered at an average dose of 15.7 mJ/cm^2 per
pulse.   The spectrum of the quartz substrate was automatically
subtracted from the spectra of the films on the substrates to
obtain Figure 1.   The absorbance of the two films is the same at
wavelengths below 310 nm and the solid curve in Figure 1 represents
the absorbance of both films.   The absorbance at 248 nm is 0.49 for
both films.   Destruction of the PAC by the 248 nm radiation, as
evidenced by the disappearance of the longer wavelength absorption
in the irradiated film, presumably occurs after nonradiative energy
transfer.
     A value of 2.4 μm^{-1} is obtained for B using values of 0.49 for
the absorbance and 0.465 μm for the film thickness (see ref. 5 for
details of calculation).   Since the film does not bleach at 248 nm,
the A parameter is 0 and the C parameter cannot be determined in
the normal way (5) from the absorbance data.   We estimated the C
parameter using a chromatographic technique that will be described
later.

Preparation and Exposure of Films.

This section describes the preparation and exposure of the films
used in the chromatographic and dissolution rate measurements.
MP2400-17 films, 0.465 μm thick, on top of a 1.5 μm thick hard-
baked (HB) HPR206 film were irradiated with a KrF excimer laser
that will be described below.   The HB206 film was prepared by spin
coating a HPR206 film onto a 3 inch diameter silicon wafer and
baking the film at 210°C for 1 hour.   The MP2400-17 film was spin
coated on top of the HB206 film and the bilevel film was baked at
90°C for 30 minutes.   The baking was done in a forced-air oven.
The bottom HB206 film totally absorbed any light that passed
through the MP2400-17 film, eliminating reflections from the
substrate.
     Figure 2 shows a schematic of the system that was used to
flood expose the resist films.   A Quanta Ray EXC-1 KrF excimer
laser was used for all exposures.   The laser output at 30 keV was
30-55 mJ/pulse/cm^2 with a 0.33% fluorine, 5% krypton, and 94.67%
neon gas mixture at 45 psig.   The beam, at 100 pulses/sec., was
reduced to a 3 mm diameter spot using aperatures.   A 5 mm diameter
light pipe was used to scramble the beam and improve uniformity.

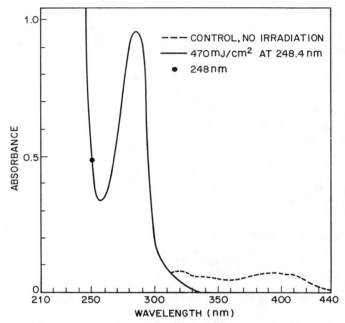

Figure 1.   UV spectra of 0.465 μm-thick Microposit 2400-17 photo-resist on quartz.   The curves correspond to doses of 0 and 470 mJ/cm^2.

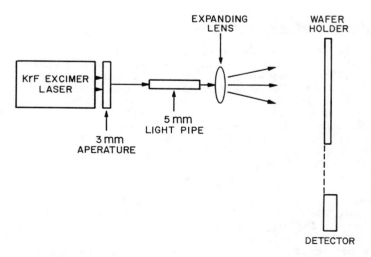

Figure 2.   KrF excimer laser exposure apparatus.

The beam was then made divergent with a lens. The wafer holder was positioned to ensure a uniform exposure of the resist. In line with the wafer holder was a Scientech 365 Power and Energy Meter equipped with a 2.5 cm diameter thermopile detector. Power/pulse/cm^2 was obtained while the laser was in a free running mode. The number of pulses needed for a particular dose was calculated. The laser was then triggered and controlled with a Hewlett Packard Model 3314A Pulse Generator.

Gel Permeation Chromatography.

The parameter C is defined as the fractional PAC decomposition rate per unit intensity (5) and can be determined from the slope of a plot of ln M versus dose where M is the normalized PAC concentration. Gel permeation chromatography (GPC) was used to determine average values of M for exposed and unexposed Microposit 2400-17 films.

The coated wafer was placed in a petri dish and enough tetrahydrofuran was added to completely cover the wafer. The tetrahydrofuran dissolved the MP2400 resist film and the resulting solution was collected in a 5 cc volumetric flask and brought to the mark with additional solvent. Prior to the solution addition, a known constant volume of benzene solution was pipetted into the volumetric flask. The benzene serves two purposes; first it acts as a marker to correct for flow rate variations and secondly its calculated area serves to normalize injection variations.

After filtration, a constant volume of solution was injected into a Waters Model 244 Gel Permeation Chromatograph containing five Microstyragel columns with porosities of 10^6,10^5,10^4,10^3 and 500Å, and two columns with porosities of 100Å. The columns separated the resin from the PAC and the amounts of each species eluting from the columns were monitored simultaneously by both a differential refractive index (DRI) and UV (254 nm) detector. A Minc 23 computer (Digital Equipment Corp.) controlled the experiment and analyzed the data to obtain molecular weights and normalized areas of all species. Of most importance in this study was the calculation of the normalized area of the PAC peak from the UV detector.

Figure 3 shows the UV detector output for the unexposed MP2400-17 resist film (curve A) and for a film that was exposed to 418 mJ/cm^2 of 248 nm irradiation (curve B). The two major peaks in the unexposed film chromatogram are due to the PAC and the benzene internal standard. The chromatogram of the exposed film shows a small peak at the elution volume expected for the PAC. The area of this peak is not altered in chromatograms of films exposed to doses larger than 418 mJ/cm^2 leading us to believe that this peak may be due to the novalac resin and should be treated as a background absorbance. The indene carboxylic acid photoproduct apparently did not elute during the GPC analysis. The origin of the smaller peak that elutes between the peaks due to the PAC and benzene is unknown.

Figure 4 shows the differential refractive index traces obtained for an unexposed film (trace A) and a film exposed to 300 mJ/cm^2 (trace B). The peak labeled PAC is attributed to the unphotolyzed photoactive compound. This peak disappears after

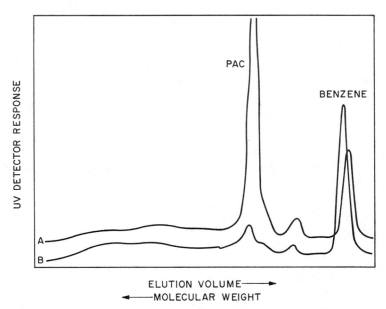

Figure 3.   Gel permeation chromatograms of (A) unexposed Microposit 2400-17 and (B) Microposit 2400-17 exposed to 418 mJ/cm^2.
The UV detector output at 250 nm is shown.

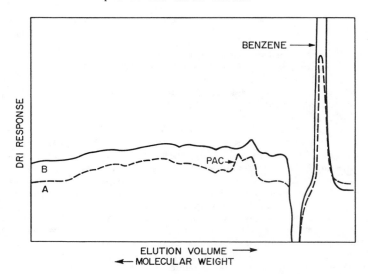

Figure 4.   The DRI detector output for (A) unexposed Microposit
2400-17 and (B) Microposit 2400-17 exposed to 300 mJ/cm^2.

300 mJ/cm^2 irradiation and no major new peaks appear in the exposed film chromatogram. This further supports our hypothesis that the small peak in chromatogram B of Figure 3 is due to the novolac resin and not to the PAC.

Values for M were calculated using the PAC peak areas determined for several irradiated films as shown in Table I. The area of the PAC peak in each chromatogram was normalized to the area of the internal standard benzene peak and corrected by subtracting the normalized area of the background peak that was determined from the chromatogram of the film exposed to 418 mJ/cm^2. The PAC concentration M is proportional to the peak area and was normalized by setting the area of the peak for the unexposed film to 1.0. Figure 5 shows a plot of log M versus dose. The solid line is a least-squares fit to the data and C was determined from the slope of this line to be 0.013 (mJ/cm^2)$^{-1}$.

Table I.   Calculation of M from the GPC peak area data

Dose (mJ/cm^2)	Peak Area*	Peak Area – Background**	M
0	3.19	2.65	1.00
11.7	2.86	2.32	0.88
14.0	2.80	2.26	0.85
23.3	2.23	1.69	0.64
30.0	2.17	1.63	0.62
46.5	1.98	1.44	0.54
52.2	1.77	1.23	0.46
60.0	1.75	1.21	0.46
69.8	1.49	0.95	0.36
78.3	1.25	0.71	0.27
90.0	1.32	0.78	0.29
120.0	1.11	0.57	0.22
417.6	0.54	–	–

* Areas are normalized to benzene internal standard peak area.
** Background area equal to 0.54.

Measurements of Dissolution Rate.

The dissolution rates, R, of the resist films in the developer were measured using a laser end-point detection system. (6) Figure 6 shows the output of the device for an ideal case. The change in film thickness between maxima (or minima) Δt is given by

$$\Delta t = \lambda/2n \qquad (1)$$

where λ is the wavelength of light and n is the refractive index of the film. R can be measured by calculating Δt and measuring the time span between maxima. For this study the developer was a 3.5 : 1 mixture of water and AZ 400K concentrate at 27.1°C.

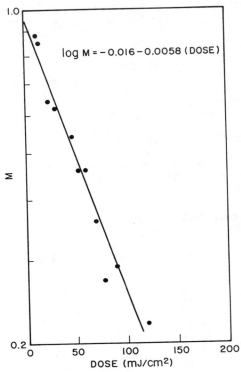

Figure 5. Log of the normalized PAC concentration (M) as a function of exposure dose.

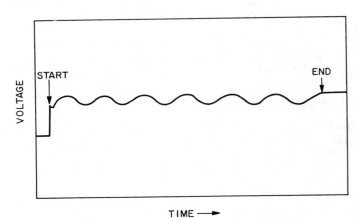

Figure 6. Response of the film thickness monitor for a 1.4 μm-thick film of poly(methylmethacrylate) dissolving in methyl ethyl ketone at 21.8 °C.

Figure 7 shows the typical output of the end-point detection device for the dissolution, at 27.1°C, of a photolyzed MP2400-17 film on a 1.5 μm thick HB206 film. The initial MP2400-17 film thickness was 0.465 μm. Values of 1.55 and 6328Å were used for n and λ (He-Ne laser), respectively. The value of n was determined using a similar end-point detection device during oxygen plasma etching of MP2400-17 films, and measuring Δt directly with a profilometer. The value for n was calculated using Equation 1. The output for the dissolution of the bilevel film appeared to be ideal, but the HB206 film apparently slowly dissolved or swelled in the developer and the output continued to oscillate after the top Microposit resist layer had dissolved, as shown in Figure 7. The dissolution rate was measured as a function of depth into the film for several photolyzed MP2400-17 films on a HB206 film. The dose was estimated as a function of depth into the resist according to $I(z) = I(0)\exp(-Bz)$. Thus, several pieces of data (dissolution rate as a function of dose) were obtained from each dissolution rate curve. The value of the PAC content corresponding to each dose was obtained from Figure 5 and the resulting data (dissolution rate R vs. M) were plotted in Figure 8. The solid line in Figure 8 is a fit of the data to Equation 3.

Modeling.

For positive photoresists the model of Dill (5) is used to simulate exposure and development. In this model the illumination intensity I at 248 nm and the normalized concentration of photoactive compound are given by the two coupled equations,

$$\frac{\partial I}{\partial z} = -I(x,z,t)[AM(x,z,t) + B]$$

$$\frac{\partial M}{\partial t} = -I(x,z,t)M(x,z,t)C \qquad (2)$$

in terms of the exposure parameters A, B, and C. The values of A, B and C used in the simulation were 0, 2.4 μm^{-1} and 0.013 $(mJ/cm^2)^{-1}$, respectively.
        The rate of dissolution in the developer solution was modeled by the expression

$$R = \exp(E_1 + E_2 M + E_3 M^2) \qquad (3)$$

The three parameters E_1, E_2, and E_3 were found by fitting Equation 3 to the measured values of R as a function of PAC concentration, as shown in Figure 8. The values obtained are $E_1 = 5.25$, $E_2 = 0.45$, $E_3 = -4.5$.
        The SAMPLE program (4) was used to obtain resist profiles. Exposure at 248 nm with a lens of numerical aperture NA = 0.38 and partial coherence σ = 0.7 was assumed. The substrate was 1200 Å of SiO_2 on thick HB206 photoresist on silicon. That is, the sensitive resist is part of a trilevel structure. Exposure dose was set to 120 mJ/cm^2, and development times were about 2 min.

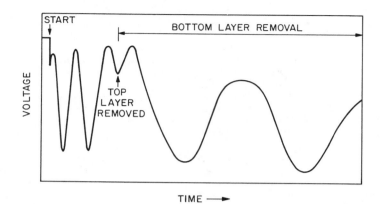

Figure 7. Response of the film thickness monitor during dissolution of a 0.465 μm-thick film of Microposit 2400-17 on a 1.5 μm-
thick HB206 film.   The solvent is a 1:3.5 mixture of AZ400K
developer and water at 27.1 °C.

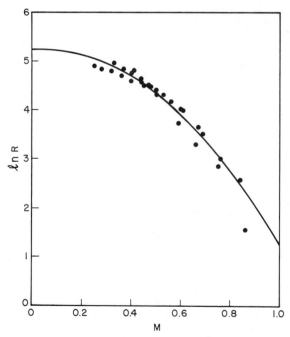

Figure 8.   Plot of ln R (in Å/sec) vs. normalized PAC concentration, M.   The solid line is a fit of the data to Equation 3.

Figures 9, 10, 11 and 12 show the results for 1.0, 0.5, 0.4 and 0.35 µm equal line/space patterns, respectively. Curves are shown for perfect focus and for defocus of 1 µm and 2 µm. These profiles illustrate the best results that can be obtained with the system, since a lens free from aberrations is assumed in the calculation. The figures show that we cannot expect to achieve vertical resist profiles at these resolutions. The sloped profiles underscore the importance of optical density and photobleaching with exposures at 248 nm. A 0.465 µm thick film of Microposit 2400-17 has an optical density of about 0.5. As a consequence the upper regions of the resist receive a larger dose than the bottom. This broadens the top of the exposed regions. An improvement in resist profile should be obtained from a reduction of optical density which would allow a more uniform exposure throughout the resist film. Photobleaching at the exposing wavelength would also improve resist contours because of a contrast enhancement effect. Depth of focus is somewhat less for 0.5 µm lines and spaces than for 1 µm lines and spaces, although in practice a depth of focus of ±1 µm could be assumed for both patterns. The presence of the shallow standing wave pattern is due to reflections from the SiO_2 interlayer. Figure 13 shows the experimental results corresponding to the simulations. There is an asymmetry in the line profile probably because of a tilt of the lens. The patterns are slightly overexposed, causing the spaces between lines to be wider than the lines. These figures show the non-vertical profiles predicted by the simulation.

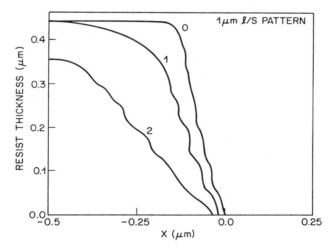

Figure 9. Simulation of a line profile for a 1.0 µm line/space pattern. The edge of the slit is at x = 0. Profiles are shown for defocus values of 0, 1, and 2 µm.

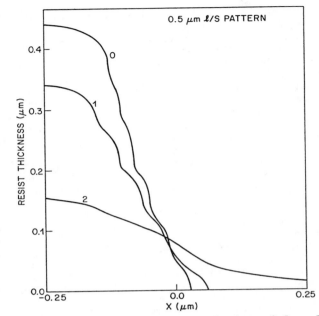

Figure 10.   Simulation of a line profile for a 0.5 μm line/space pattern.   The edge of the slit is at x = 0.   Profiles are shown for defocus values of 0, 1, and 2 μm.

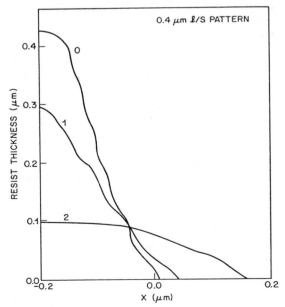

Figure 11.   Simulation of a line profile for a 0.4 μm line/space pattern.   The edge of the slit is at x = 0.   Profiles are shown for defocus values of 0, 1, and 2 μm.

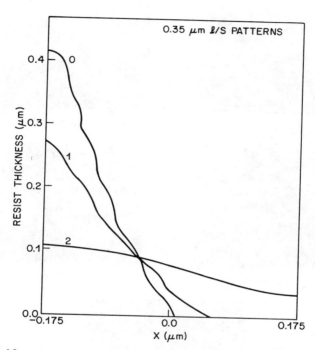

Figure 12. Simulation of a line profile for a 0.35 μm line/space pattern. The edge of the slit is at x = 0. Profiles are shown for defocus values of 0, 1, and 2 μm.

Experimental MP2400–17 Resist Profiles

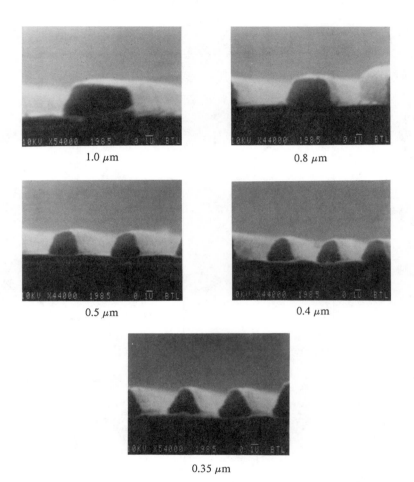

1.0 μm

0.8 μm

0.5 μm

0.4 μm

0.35 μm

Figure 13. SEM photographs of 1.0, 0.8, 0.5, 0.4, and 0.35 μm line/space patterns.

Acknowledgments

The authors thank J. Bennewitz for the deep UV stepper exposures. We also thank Andrea Pastuck for the SEM photographs and Tanya Jewell for providing the value for the refractive index of the MP2400-17 resist film.

References

1. Pol, V.; Bennewitz, J. H.; Escher, G. C.; Feldman, M.; Firtion, V. A.; Jewell, T. E.; Wilcomb, B. E.; Clemens, J. T.; "Excimer Laser-Based Lithography: A Deep Ultraviolet Wafer Stepper" SPIE Conf. on Microlithography, March 13, 1986.
2. Lin, B. J.; Proc. Electrochem. Soc., Electron and Ion Beam Science and Technology, International Conf., 1979, 78-5, 320. ;
3. Wolf, T. M.; Hartless, R. L.; Shugard, A.; Taylor, G. N.; "Evaluation of Resists for Lithography at 248 nm. I. Positive Tone Resists", Int. Symp. on Electron, Ion, and Photon Beams, May, 1986.
4. Oldham, W. G.; Nandgaonkar, S. N.; Neurether, A. R.; O'Toole, M IEEE Trans. Electron Dev. 1979, ED26, 717.
5. Dill, F. H.; Hornberger, W. P.; Hauge, P. S.; Shaw, J. M.; IEEE Trans. Electron Dev. 1975, ED22, 445.
6. Introduction to Microlithography, Thompson, L. F.; Wilson, C.B. Bowden, M. J. eds; Ch. 3. ACS Symposium Series No. 219, 1983.

RECEIVED May 13, 1987

ETCH RESISTANCE OF POLYMERS IN PLASMA ENVIRONMENTS

ETCH RESISTANCE OF POLYMERS IN PLASMA ENVIRONMENTS

The requirements on lithography and resist processing for microcircuit fabrication are becoming increasingly stringent as the dimensions of circuit features in microelectronic devices shrink below one micron. It is important to note however, that despite the continued shrinking of the lateral dimensions, the vertical dimension has remained virtually unchanged. Therefore, the ability to generate images with high resolution is not enough. To permit fabrication of devices on wafers using stepped topography, these images must also have high aspect ratios.

This requirement has created problems for single-layer resist processing schemes because the variations in resist thickness, which occur as a consequence of having to cover steps on the wafer surface, result in linewidth variations in the developed image. These difficulties prompted the development of a variety of multi-layer resist processing schemes which obviate the problem by separating the imaging and planarizing functions of the resist. The common feature of all such processes is the initial deposition of a thick resist layer which planarizes the wafer surface. This permits imaging to be done in a thin resist layer which is deposited on top of the planarizing layer. In the tri-layer approach, these two resist layers are separated by a thin intermediate film of a material such as SiO_2 which acts as an etch mask for oxygen reactive-ion etching of the thick underlying layer. The degree of process complexity associated with the tri-layer approach can be reduced by eliminating the intermediate layer, but this requires the imaging resist to be highly resistant to the dry-etching environments used to transfer the image into the planarizing layer.

The past several years have seen the development of several resist systems capable of meeting such a demand. Inorganic resists originally held promise of utility in such applications although processing difficulties appear to have resulted in a general disaffection with these materials. The first paper in this section discusses the current status of inorganic resists. Considerable work has been directed toward organometallic polymers containing elements such as silicon or tin which form a refractory oxide in an oxygen plasma. In resist application, the oxide forms at the surface of the resist and protects the underlying material against further erosion. Organosilicon polymers in particular have been widely investigated, primarily because of the rich variety of materials and processing schemes afforded by organosilicon chemistry. In order to optimize processing of these materials, it is important to understand the chemistry of surface passivation and how organosilicon chemistry may be employed to develop better processing methods. This theme is amplified in the papers that follow.

The potential of multi-level resist processing has been clearly demonstrated. The time is now ripe to implement it into industrial practice.

Y. Ohnishi
Fundamental Research Laboratories
NEC Corporation
Miyazaki, Miyamae-ku
Kawasaki 213, Japan

Chapter 26

Inorganic Resist for Bilayer Applications

Akira Yoshikawa and Yasushi Utsugi

NTT Electrical Communications Laboratories, 3-1, Morinosato, Wakamiya, Atsugi-shi, Kanagawa 243-01, Japan

The present status of research on Ag/Se-Ge inorganic resist technology is reviewed. Emphasis is placed on inorganic/organic bilayer resist technology and its application to LSI fabrication. Bilayer resist processing technologies are briefly discussed and a dry-development scheme using reactive-ion etching is shown to eliminate the undercutting usually associated with wet development. Patterning characteristics including linewidth accuracy, depth of focus, and alignment accuracy are examined. Defocus tolerance is shown to be superior to that of polymeric resists being larger by 2~3 μm. The inorganic resist is sensitive to synchrotron radiation where the short photoelectron range in the resist results in very high resolution.

An inorganic resist utilizing the known Ag-photodoping effect in amorphous chalcogenide films was first developed in 1976 as a product of extensive research on the physics and applications of amorphous chalcogenide materials (1). Continued research efforts have been made to exploit the potential of this inorganic resist which has many advantageous features over conventional organic polymeric resists. These include high-resolution, dry-processing capability, wide applicability to a variety of exposure tools, and attractive material properties. These include no deformation due to swelling and high optical absorption (2). The inorganic resist is currently of considerable interest in LSI microlithography, mainly due to the fact that it exhibits the highest resolution of all existing resist materials (3) and shows excellent performance as the top layer of a bilayer resist system (4-6).

This paper describes the present status of research in Se-Ge inorganic resist technology with emphasis focused on bilayer applications. First, the fundamental characteristics of the inorganic resist are briefly summarized. Then the bilayer process technology is described including various dry-processing techniques, e.g., dry development. Patterning characteristics, including linewidth control, defocus tolerance and alignment accuracy, are discussed and satisfactory results of some applications to LSI fabrication processes are shown. Finally, some recent findings on the sensitivity of inorganic resists to synchrotron radiation are presented.

Fundamental Characteristics of Inorganic Resists

The inorganic resist is a bi-layer system composed of a thin Ag or Ag-compound film such as Ag-Se, on top of an amorphous Se-Ge film. The film thickness of the Ag and Se-Ge layers is typically 7 nm and 0.2 μm, respectively. A composition at about $Se_{80} Ge_{20}$ was found to be most suitable for resist applications.

0097-6156/87/0346-0309$06.00/0
© 1987 American Chemical Society

Photoirradiation of this stacked film causes the Ag to diffuse into the Se-Ge film in a process referred to as Ag-photodoping. This diffusion, which is characteristic of amorphous chalcogenide materials, causes the Se-Ge film to become insoluble in alkaline solutions and the resist thus functions as a negative-type photoresist. The photodoped regions also are resistant to fluorine-based plasmas enabling dry development of the resist (7). Dry development has also been achieved using a reactive-ion etching (RIE) technique and is discussed later.

The sensitivity of the resist to the mercury g- and i-lines varies between 80 and 200 mJ/cm^2, depending on the developing conditions, film thicknesses and Se-Ge composition. The γ-value is about 4.5. The spectral photosensitivity of the inorganic resist is shown in Figure 1 as the photon energy dependence of the normalized remaining thickness after sufficient exposure ($\sim$5 J/cm^2) and development. The optical absorption coefficient of the Se$_{80}$ Ge$_{20}$ glass is also shown. E_{gopt} refers to the optical band gap energy. Figure 1 clearly shows that the Se-Ge inorganic resist is sensitive to light of shorter wavelength than that corresponding to the optical absorption edge of the Se-Ge material. Sensitivity to deep UV light from KrF (249 nm) and ArF (193 nm) excimer lasers has also been confirmed.

The e-beam sensitivity is not particularly high being about 300 μC/cm^2 for an accelerating voltage of 20 to 30 kV which is comparable to PMMA. However, the contrast is extremely high with a γ-value equal to 8 (8). The resist has been reported to be sensitive to ion beams (9) and to soft x-ray irradiation as well. Thus, the inorganic resists are applicable to all non-optical lithographies.

BILAYER RESIST PROCESSING

This section deals with process technologies associated with the bilayer process. Particular emphasis will be given to the development of an all-dry processing scheme including dry development and dry deposition of the Ag-compound sensitizing layer.

Bottom Layer Formation

Figure 2 shows the outline of the bilayer resist process in which a conventional positive photoresist (Tokyo Ohka, OFPR-800) was used as the bottom layer. To improve accuracy in detecting alignment marks, the bottom layer resist covering the alignment marks was removed by exposure and development. The optical absorption coefficient of the inorganic resist, shown in Figure 1 for light in the UV and deep UV region, is about two orders of magnitude larger than that of organic polymeric resists. This high optical absorbance of Se-Ge films can eliminate the linewidth variations due to standing-wave effects and reflections from the topographic features on the underlying substrate. Therefore, it is not necessary to increase the absorbancy of the bottom layer by hard baking for long times at high temperature. In the present study, the bottom layer was baked at only 150°C. An advantage of low-temperature baking is that subsequent removal of the bottom layer later in the processing sequence is relatively simple.

Inorganic Resist Deposition

Deposition of the Se-Ge film onto the bottom polymeric layer was performed using a magnetron-type sputtering method in an Ar atmosphere with throughput of 25 wafers/hour. The temperature of the polymer surface during sputtering did not increase significantly. There was no evidence of deformation or crack generation in either the polymer layer or the deposited Se-Ge film. It is important to monitor the Se-Ge film composition since it significantly influences resist sensitivity as well as pattern quality (1,2). Our studies using fluorescence x-ray analysis have shown that the difference in composition between the target and the film can be kept to less than 1% and good reproducibility and uniformity across the wafer can be obtained. Adherence to the polymer layer was satisfactory and there were no problems with

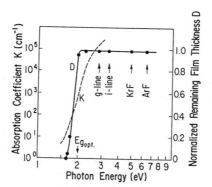

Figure 1.   Spectral photosensitivity of the $Se_{80}Ge_{20}$ inorganic resist. The optical absorption coefficient is also shown (2).

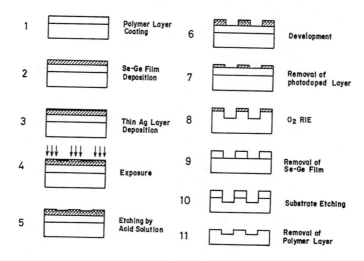

Figure 2.   Inorganic/organic bilayer resist processes.

peeling. Ag-sensitization of the Se-Ge layer was performed by immersing the film into an aqueous solution containing $Ag(CN)_2$ for about 1 min. The composition of the Ag-sensitized layer thus formed was about $Ag_{1.5} Se_1$.

Development

After exposure, the remaining Ag-sensitized layer on the unexposed region was removed by dissolution in diluted aqua-regia. This treatment significantly dominated pattern accuracy and quality. The unexposed or unphotodoped regions were then etched away using aqueous alkaline solutions or gaseous fluorine-based plasmas, such as CF_4 or C_2F_6.

Pattern Transfer

The resist patterns were transferred into the bottom organic polymer by reactive-ion etching using oxygen as an etching gas (O_2 RIE). The etch rate ratio of inorganic/organic layer was greater than 100 to 1. The substrate temperature had to be cooled in order to achieve accurate pattern transfer with no undercutting in O_2 RIE (10). In LSI fabrication, the top inorganic resist must be removed from the polymer layer prior to etching the substrate to avoid possible process contamination by Ag.

DRY PROCESSING WITH INORGANIC RESIST

The inorganic resist comes very close to the goal, long sought by resist designers of achieving an all-dry process. The initial resist coating is made by "dry" rf-sputtering, not by "wet" spin-coating, and the resist can be developed by a dry-development process which was first demonstrated in these materials. The following section deals with recent advances in dry-processing technology related to inorganic resists.

RIE Development

The Ag-photodoped depth in the Ag-Se/Se-Ge inorganic resist is fairly shallow under ordinary exposure conditions being typically 20 to 30 nm. As a consequence, undercutting is more or less unavoidable during development [using isotropic wet or plasma etching techniques] (See Figure 3). Such undercutting limits the dimensional accuracy of delineated patterns. To overcome this problem, an anisotropic RIE technique using fluorocarbon gases and their mixtures with oxygen has been developed (11).

It was found that while undoped Se-Ge films etched slightly in pure C_3F_8 gas, they etched rapidly in both CF_4 and CF_4+O_2. However this was accompanied by substantial undercutting. Lowering the gas pressure to reduce undercutting resulted in deterioration of the resist surface due to ion bombardment. Satisfactory results were obtained using a mixture of $C_2F_6+O_2$, where the O_2 was added to increase the etch rate. An RIE-developed inorganic resist pattern is shown in Figure 4, together with that of a wet-chemical-developed pattern, for comparison. The bending at the edges of the wet-developed pattern is probably due to stress relief following development. Sensitivity and contrast are almost equal for both RIE and wet development. These results clearly indicate that the RIE technique provides a very beneficial means of development.

Dry Ag-Sensitization

In an effort to develop a completely dry lithographic process, we replaced the solution-deposited $Ag(CN)_2$ film with an evaporated or sputtered film of Ag_2 Se. Both approaches gave comparable results for Ag sensitization on subsequent exposure. The only wet-processing step

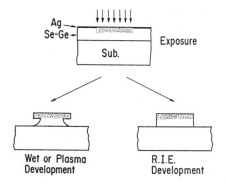

Figure 3.   Schematics of developed resist pattern profiles.[11]

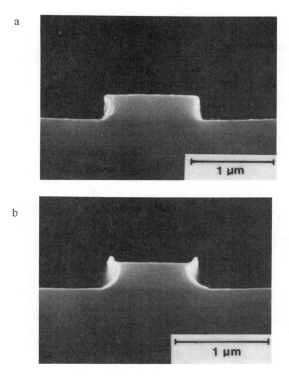

Figure 4.   SEM micrographs showing inorganic resist pattern profiles developed by (a) RIE and (b) wet chemical method.[11]

currently remaining is the aqua-regia treatment to remove undoped Ag or Ag-doped Se-Ge layer and studies are continuing to develop an equivalent dry process.

Patterning Characteristics

The resist exhibits very high resolution and is well within the lithographic resolution capability of conventional exposure tools such as the g-line or i-line steppers in use today.

Linewidth Control

Linewidth measurements as a function of exposure time using a g-line stepper with a 0.35NA lens indicate that ±10% linewidth control is attainable within ± 13% and ± 6% exposure time latitude for 0.8 μm and 0.6 μm line and space patterns, respectively (6). These values are considerably larger that those for organic polymeric resists. Linewidth versus exposure time curves show no plateau with prolonged exposure (5) indicating that the edge-sharpening effect does not extend the exposure latitude in a practical process.

The linewidth differences between the mask and the corresponding resist pattern were precisely evaluated for isolated line and isolated space patterns. The deviation in linewidth versus linewidth shows a maximum or minimum with inorganic resists (13). This behavior is in contrast to that for organic polymeric resists, in which the linewidth deviations gradually increase as the mask feature size decreases. This difference may be explained in terms of self-compensation of the optical proximity effect due to lateral Ag diffusion.

Defocus Tolerance

In order to examine the defocus tolerance of inorganic resists, the finest line and space pattern was evaluated by defocusing in the region of 0-5 μm (6). Figure 5 shows the defocus tolerance as a function of the resolution limit thus determined, together with simulated results using modulation transfer function (MTF) values as parameters. Circles and triangles show sufficiently and insufficiently resolved line and space pattern width. From the separate experiments for determining the resolution limits, the required contrast for pattern formation for inorganic and organic resists can be represented by 40-50% MTF and 80-90% MTF, respectively. Simulated curves obtained for 40-60% MTF agree with the experimental results. This fact indicates that the defocus tolerance of the inorganic resists is superior to that of polymeric resist by 2-3 μm which constitutes an important advantage of the former.

Alignment Accuracy

Precise alignment between reticle and wafer patterns is an extremely important requirement for LSI lithography. In the case of inorganic/organic bilayer resists, the thick bottom polymer coating on groove marks frequently makes alignment signals asymmetric. To achieve high overlay accuracy, it is better to remove the bottom layer on the alignment marks. Deposition of the inorganic resist on the mark reduces the noise in alignment signals and an alignment accuracy of 0.1 um (2σ) was attained, which is the same as that using uncoated marks.

APPLICATION TO LSI FABRICATION PROCESS

The inorganic/organic bilayer resist technology has been successfully applied to the Al interconnection process in bipolar LSI fabrication, including the 1 kbit RAM and several repeater ICs for optical communication systems (6). The minimum feature size of the patterns was 1.5 μm for lines and 1.0 μm for spaces. In order to further evaluate the lithographic quality of the inorganic system, the resistance of Al lines with an overall length of 20 mm and thickness of 0.8 μm was measured for 69 patterns on a wafer. From the histogram of the measured resistance values, a fairly low σ-value of 0.08 μm, was obtained as the standard deviation in linewidth.

SYNCHROTRON RADIATION EXPOSURE CHARACTERISTICS

Recent studies have shown that the inorganic resist is sensitive to synchrotron radiation(SR). (Saito, K.; Utsugi, Y.; Yoshikawa, A. submitted to Appl. Phys. Lett.) Experiments were performed using SR emitted from a 2.5 GeV electron storage ring of the Photon Factory at the National Laboratory for High Energy Physics. The spectrum of the SR flux ranged from 0.2 to 1.3 nm in wavelength. The sensitivity was roughly the same as that for PMMA. An example of replicated patterns is shown in Figure 6. Fine, 50 nm-wide patterns were also formed using a fine diffraction pattern at the mask edge. Such high resolution is mainly due to the fact that the photo-electron ranges in the resist are considerably shorter than those in polymeric resists.

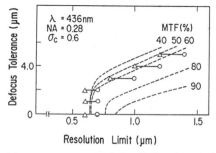

Figure 5.   Defocus tolerance versus resolution-limit characteristics. Dotted curves are simulated results (6).

Figure 6.   Inorganic resist pattern replicated by synchrotron radiation exposure. Minimum width is 0.5 μm.

SUMMARY

Inorganic resists have many advantages over conventional polymeric resists, especially in regard to resolution, linewidth control latitude, focus depth tolerance, and dry processability. These features make them very suitable for application to submicron VLSI fabrication. Efforts to improve processing technologies will open the way for practical use in LSI fabrication.

Literature Cited

1. Yoshikawa, A.; Ochi, O.; Nagai, H.; Mizushima, Y. Appl. Phys. Lett. 1976, 29, 677.
2. See for example, Mizushima, Y.; Yoshikawa, A. In Japan Annual Reviews in Electronics, Computers & Telecommunications, Amorphous Semiconductors Technologies & Devices; Hamakawa, Y., Ed.; OHM*North-Holland: Tokyo, Amsterdam, 1982; p 277.
3. Tai, K. L.; Vadimsky, R. G.; Ong, E. Proc. SPIE 1982, 333, 32.
4. Tai, K. L.; Sinclair, W. R.; Vadimsky, R. G.; Moran, J. M.; J. Vac. Sci. Technol. 1979, 16, 1977.
5. Tai, K. L.; Vadimsky, R. G.; Kemmerer, C. T.; Wagner, J. S.; Lamberti, V. E.; Timko, A. G. J. Vac. Sci. Technol. 1980, 17, 1169.
6. Utsugi, Y.; Yoshikawa, A.; Kitayama, T. Microelectronic Eng. 1984, 2, 281.
7. Yoshikawa, A.; Ochi, O.; Mizushima, Y. Appl. Phys. Lett. 1980, 36, 107.
8. Yoshikawa, A.; Ochi, O.; Nagai, H.; Mizushima, Y. Appl. Phys. Lett. 1977, 31, 161.
9. Wagner, A.; Barr, D,; Venkatesan, T.; Crane, W. S.; Lamberti, V. E.; Tai, K. L.; Vadimsky, R. G. J. Vac. Sci. Technol. 1981, 19, 1363.
10. Namatsu, H.; Ozaki, Y.; Hirata, K. J. Vac. Sci. Technol. 1982, 21, 672.
11. Yoshikawa A.; Utsugi, Y. In Japan Annual Reviews in Electronics, Computers & Telecommunications, Vol. 16, Amorphous Semiconductors Technologies & Devices; Hamakawa, Y. Ed.; CHM*North-Holland: Tokyo, Amsterdam, 1984; p 168.
12. Yoshikawa, A.; Hirota, S.; Ochi, O.; Takeda, A.; Mizushima, Y. Jpn. J. Appl. Phys. 1981, 20, L81.
13. Nakase, M.; Utsugi, Y.; Yoshikawa, A. J. Vac. Sci. Technol. 1985, A3, 1849.

RECEIVED July 6, 1987

Chapter 27

Enhancement of Dry-Etch Resistance of Poly(butene-1 sulfone)

William M. Mansfield

AT&T Bell Laboratories, Murray Hill, NJ 07974

The etch resistance of poly(butene-1 sulfone) in fluorocarbon-based plasmas can be enhanced by prior treatment of the surface in an oxygen plasma. This pretreatment inhibits or retards the depolymerization reaction that characterizes normal etching in fluorocarbon plasmas, thereby permitting formation of a surface-modified layer which exhibits a substantially reduced etch rate. Pretreating PBS in an oxygen plasma enables it to be used subsequently in selective reactive-ion etch processes involving fluorocarbon plasmas to delineate submicron, anisotropically etched patterns.

Poly(butene-1 sulfone (PBS) is a highly sensitive, high-resolution electron-beam resist (1-2) which is used primarily as a wet-etch mask in the fabrication of chrome photomasks. PBS has found little use as a dry-etch mask because of its lack of etch resistance in plasma environments (3-8). This primarily stems from the fact that PBS depolymerizes in such an environment which greatly enhances the rate of material loss from the film. Moreover, depolymerization is an activated process which causes the etching rate to be extremely temperature dependent. Previous work (3,7) has shown that the etch rate of PBS in fluorocarbon-based plasmas varies by orders of magnitude for temperature differentials of less than 30°C.

It has been found, however, that the etch rate of PBS can be reasonably controlled in both oxygen and CF$_4$/O$_2$ plasmas if the substrate temperature is kept below room temperature (9). This fact has been utilized to reduce the defect density in the manufacture of chrome photomasks by exposing the developed PBS pattern to a low-temperature oxygen plasma (descum) prior to wet-etching the chrome. We have now found that the plasma-etch resistance of PBS in a CF$_4$/O$_2$ plasma can be markedly enhanced at room temperature simply by exposing the resist to a short oxygen plasma pretreatment prior to exposure to the fluorinated plasma. This effect can be used in a variety of pattern transfer processes to controllably generate submicron features on wafers and masks. This paper examines the parameters associated with this effect, proposes a mechanism to account for the results and delineates some possible pattern transfer processes.

EXPERIMENTAL

Sample Preparation

All experiments were carried out on three-inch silicon wafers. Details of the various steps involved in the processing of PBS which include spin-coating, exposure, development, and baking are proprietary, and are available under license from AT&T. Etching rates of films spun on silicon wafers were determined by measuring film thickness as a function of time, using a Nanometrics/Nanospec AFT Micro Area thickness gauge. Selected measurements were verified with a stylus profilometer. Auger and scanning electron microscope analyses (SEM) were made on silicon wafers which had been coated with sputter deposited tantalum/gold/tantalum metallization prior to resist application.

0097–6156/87/0346–0317$06.00/0

Plasma Etching

All plasma exposures were carried out in an IPC (International Plasma Corporation) 2005 capacitance-coupled barrel reactor at 13.56MHz. The reactor was equipped with an aluminum etch tunnel and a temperature controlled sample stage. Pressure was monitored with an MKS capacitance manometer; RF power was monitored with a Bird R.F. power meter and substrate temperature was measured with a Fluoroptic thermometer utilizing a fiber optic probe which was immune to R.F. noise.

Auger Analysis and Scanning Electron Microscopy

The surface composition of the uncoated, patterned wafer was analyzed by Auger spectroscopy. Line quality and surface detail at each key process step were examined by scanning electron microscopy using samples overcoated with 100Å of a platinum/gold alloy.

Gaseous Etchants

Two fluorocarbon mixtures were used in plasma etching studies. The first consisted of 96%CF_4 and 4%O_2 while the second contained 50% helium, 49%CF_4 and 1%O_2.

RESULTS AND DISCUSSION

PBS/Substrate Etching in Fluorocarbon Plasmas

The results obtained on etching PBS in fluorine-based plasmas vary quite considerably depending on the temperature of the substrate. Below 15°C, both the resist and substrate are attacked (etched), by the etchant species while at an intermediate temperature (15-23°C), the phenomenon of reversal etching is observed (8) in which the exposed substrate is not etched at all, while the resist (and subsequently the substrate underneath the resist) is readily etched by the plasma thus giving rise to a negative-tone pattern. At higher temperatures, (>25°C), both the resist and substrate are etched by the plasma as in the low-temperature case, except that etching completely stops after several seconds.

These apparent contradictions can be rationalized in terms of a model which incorporates plasma-induced polymerization along with depolymerization. PBS has long been known to exhibit a marked temperature-dependent etch rate in a variety of plasmas. This is clearly seen in the previously published Arrhenius plots (3,7) for two different plasma conditions (Figure 1). This dependence is characteristic of an etch rate that is dominated by an activated material loss as would occur with polymer depolymerization. The latter also greatly accelerates the rate of material loss from the film. Bowmer et al. (10-13) have shown in fact that poly(butene-1 sulfone) is thermally unstable and degrades by a depolymerization pathway. A similar mechanism had been proposed by Bowden and Thompson (1) to explain dry-development (also called vapor-development) under electron-beam irradiation.

The reactive species present in a plasma possess sufficient energy to break the weak main chain carbon-sulfur bonds in PBS forming propagating radicals which, at the high temperatures encountered in the plasma environment, rapidly initiate depolymerization. The etchant species can also abstract hydrogen atoms from the hydrocarbon moiety (butene) in the polymer chain facilitating formation of fluorinated species at the surface of the resist. Depolymerization in this case thus results in sequential elimination of SO_2 and fluorinated butene derivatives. Since the gaseous products are continuously removed in the system vacuum, the reaction should proceed in the direction of complete depolymerization resulting in total material loss. These decomposition reactions occur at a sufficiently high enough rate under ordinary conditions to render PBS ineffective as a plasma-etch mask, i.e., the resist is removed long before etching of the substrate is complete.

This relatively straightforward process becomes more complicated at elevated temperatures where the rate of depolymerization is correspondingly higher. We speculate that the evolved

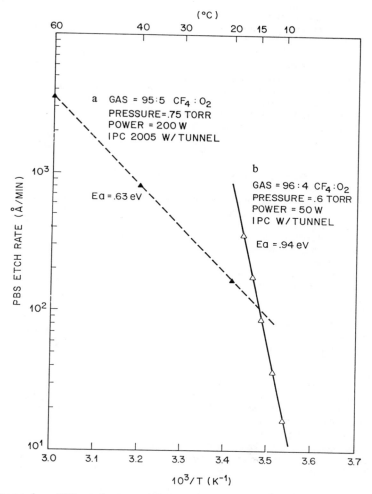

Figure 1.   PBS etch rate versus inverse temperature in fluorocarbon plasma.   Curve a, Ref. 3; curve b, Ref. 7.

species can participate in a plasma-induced polymerization reaction which competes with the etching reaction. We further suggest that this competing reaction is initiated only when the concentration of reactive species (which includes species derived from the fluorinated etchant as well as fluorinated degradation products) reaches some threshold level. We would expect such a threshold to depend on both the system configuration (which determines the residence time of the reactants in the plasma), and the rate at which the fluorinated species are generated by depolymerization. In our experimental setup, the pumping parameters and flow rate were fixed as were the pressure and power. This left substrate temperature as the only variable. Thus we would expect the rate of depolymerization to increase with increasing temperature, eventually reaching a point where the rate of gaseous evolution becomes sufficient to permit plasma polymerization to occur. Accordingly we would expect etching to stop because of the build-up of a deposited fluorinated layer on the surface of the resist and the substrate.

In order to check the validity of this model, we analyzed the surface of both the resist and substrate by Auger spectroscopy. Figures 2A and 3A show Auger spectra of the resist and substrate, respectively, prior to plasma exposure. Surprisingly, only two of the elements common to the resist composition (C,S) are evident in Figure 2A. The expected oxygen peak associated with the SO_2 structure is absent. In fact, all spectra taken in the course of this work show the absence or reduction of the expected oxygen peak at the resist surface (see Figures 2B, 6A, 6B, 6C). Work is currently ongoing to explain this phenomenon as recent ESCA results indicate that the oxygen associated with the SO_2 structure is indeed present at the surface. We speculate that the absence of the peak in these spectra results from some complex interaction of the Auger beam with the resist surface. For this reason, extreme care must be taken in using these spectra for quantitative analysis, and it is pointed out that these spectra will only be used to indicate the existence of the carbon and sulfur components at the PBS surface in this work. Inversely, the spectrum of the substrate surface (Figure 3A) reflects the presence of oxygen, presumably due to the formation of the native oxide (Ta_2O_5) at the surface. Samples which were ion-milled and analyzed while in the Auger chamber indicate that the oxide film is approximately 100Å thick. The presence of the oxygen peak even in the etched samples (Figures 7B and 7C) is attributed to the fact that these samples were exposed to ambient air for more than an adequate amount of time between etching and Auger analysis to allow the oxide to regrow.

Samples were subjected to a fluorocarbon plasma for 30 seconds and 90 seconds, respectively, at temperatures spanning the three regions referred to above. Figures 2B and 2C show the changes in the Auger spectrum of the resist surface for plasma exposure (100W, 0.5 Torr) in the intermediate temperature range, i.e., at 16°C. The corresponding substrate spectra are shown in Figures 3B and 3C. These spectral changes are characteristic of all three temperature regimes, except that the changes occurred at different rates depending on the particular temperature studied. Thus, whereas it took 90sec for the sulfur peak in the spectrum of the resist to disappear at low and intermediate temperatures (leaving only the fluorinated hydrocarbon (see Figure 2C)), it disappeared after only a few seconds plasma exposure at high temperature (> 25°C).

As noted earlier, the resist is etched during exposure at low and intermediate temperatures and is eventually removed entirely. On the contrary, etching completely stops after a few seconds at high temperatures. We attribute this to the fact that in the former case, depolymerization does not produce "monomer" at a sufficiently high enough rate to sustain polymerization at a rate greater than the plasma-induced etching rate of the resist. Thus although we see a build-up of fluorinated material at intermediate temperatures and below, its thickness and/or density is presumably not sufficient to prevent etching of the underlying resist. At high temperatures, the fluorinated surface layer builds up at a considerably greater rate because of the additional contribution from high rate plasma-induced polymerization. The similarity of the spectra after prolonged etching at low temperature with that at high temperature is understandable if we assume that SO_2 does not take part in the plasma polymerization reaction. This would seem to be a reasonable assumption in view of the results of Brown and O'Donnell on the gas-phase copolymerization of butene-1 and SO_2 in which they showed that the temperature for formation of PBS from the gas phase is around 0°C. Hence it

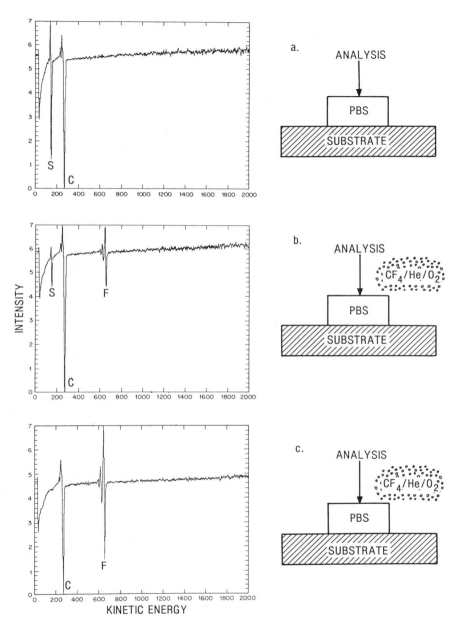

Figure 2. Auger spectra of PBS surface post (a) no plasma exposure (b) 30 second and (c) 90 second fluorocarbon plasma exposure.

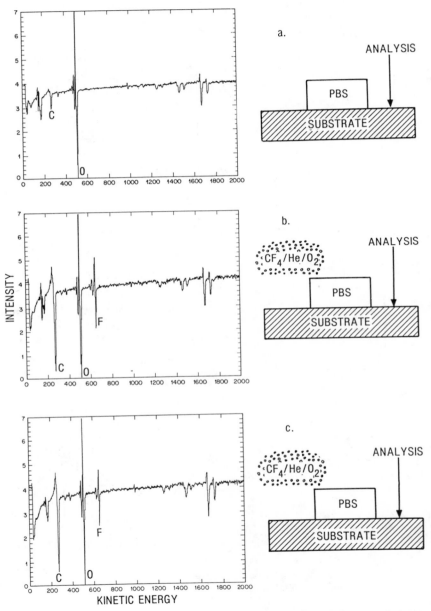

Figure 3. Auger spectra of tantalum substrate surface post (a) no plasma exposure (b) 30 second and (c) 90 second fluoro-carbon plasma exposure.

would not be possible for SO_2 to participate in a plasma-initiated copolymerization reaction at the temperatures used in our studies. We therefore suggest that the actual spectrum in Figure 2C which was observed after 90sec at a temperature of 16°C, reflects the elemental composition of the *resist* surface, whereas the similar spectrum observed at higher temperature (25°C) reflects the composition of the polymerized, *deposited* material.

Figures 4A and 4B are micrographs of the resist patterns after 30 and 90 seconds plasma treatment at 16°C corresponding to the spectra in Figs 2 and 3. It will be recalled that this is the temperature at which the phenomenon of reversal etching is observed. The micrographs clearly show evidence of attack at the resist surface together with surface deposition. Moreover, since we did not observe etching of the exposed substrate, the deposited material must effectively cover and protect it from further attack. Auger analysis of this surface revealed only C, O and F (Fig 3B) so it is not surprising that such a material would protect the substrate from further etching. We speculate that in this complex plasma environment deposition onto the substrate is favored over deposition onto the resist surface and that this difference in deposition rates accounts for the reversal etching mode. Thus, while the substrate does not etch, the resist continues to etch and will eventually result in a tone-inverted etched image.

Given the complicated kinetics associated with the process, we were not able to obtain reproducible etch rates under our experimental conditions. Depolymerization introduces a degree of process complexity that precludes useful application of polymers such as PBS in dry-etching environments.

PBS/SUBSTRATE ETCHING IN FLUOROCARBON PLASMAS WITH PRIOR OXYGEN EXPOSURE

PBS Etching in Oxygen Plasma

Figure 5 shows a plot of etching rate of PBS in an oxygen plasma versus $1000/T(°K)$. The apparent activation energy obtained from a least squares plot of the data was 0.2 eV, which is a factor of two or three lower than that observed for etching in the fluorocarbon plasma discussed in the previous section. Analysis of the exposed substrate shows no evidence of film deposition (Fig. 7A and 7B). The low activation energy energy for etching in oxygen would suggest that depolymerization is inhibited or retarded in the oxygen plasma.

Subsequent Fluorocarbon Plasma Treatment

Figure 8 shows a plot of thickness of PBS removed versus time in both the CF_4/O_2 and $CF_4/He/O_2$ plasmas for samples priorly exposed to an oxygen plasma (100W, 0.5 Torr, 3 minutes 16°C). The etching curves in the fluorocarbon plasma are characterized by two distinct regions. Initially, the etch rate of PBS is quite high being comparable to that of samples not subjected to pretreatment in O_2 plasma (cf. Figure 1). The etch rate then quickly diminishes to a low constant value of 12 ± 2Å/min (for $CF_4/He/O_2$ and 29 ± 5Å/min in CF_4/O_2. When the linear removal rate, obtained from a least-squares plot of the thickness removed versus plasma exposure time, is plotted as an Arrhenius expression at different temperatures (Figure 9), an activation energy of zero is obtained.

It would therefore appear that pretreating PBS in an oxygen plasma not only substantially lowers the etching rate in the fluorocarbon-based plasmas, but also eliminates the temperature sensitivity of the process. In order to follow the changes in surface composition, PBS patterns on tantalum/gold/tantalum-coated wafers were subjected to a three-minute oxygen plasma exposure (100W, 0.5 Torr, 16°C) followed by a 90 sec exposure to the fluorocarbon plasma under conditions identical to those used previously (100W, 0.5 Torr, 16°C). The Auger spectra of the resist and substrate surfaces are shown in Figures 6 and 7, respectively. Figure 6C shows that even after prolonged exposure to the fluorocarbon plasma, a significant sulfur peak still remains which is quite contrary to the case discussed earlier involving samples not pretreated in the O_2 plasma, which showed the complete extinction of the sulfur peak at the surface (see

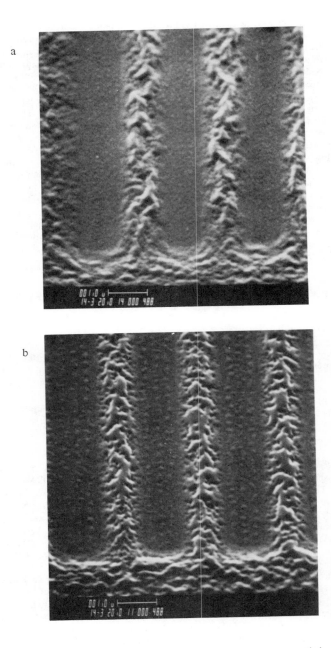

Figure 4. PBS features on tantalum substrate post (a) 30 second
and (b) 90 second fluorocarbon plasma exposure.

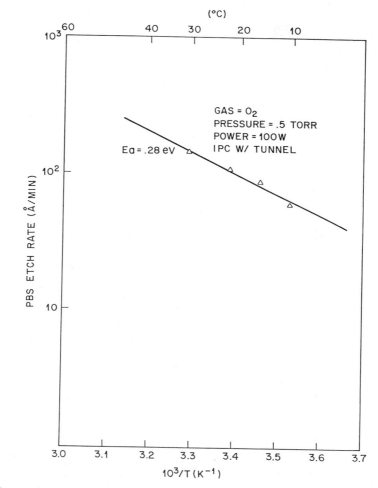

Figure 5.   PBS etch rate vs inverse temperature in an oxygen
plasma.

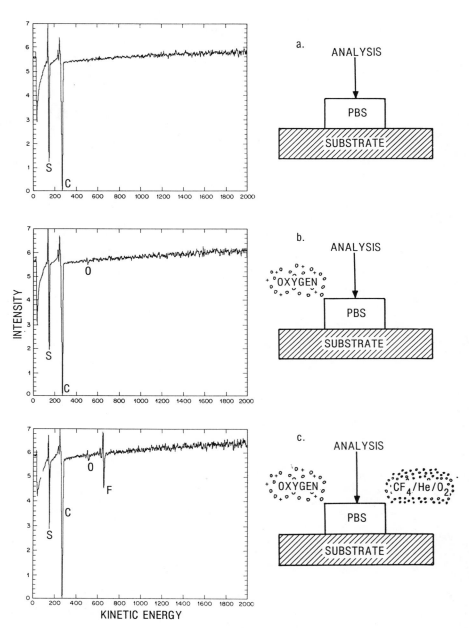

Figure 6.   Auger spectra of PBS surface post (a) no plasma exposure (b) 3 minute oxygen plasma exposure and (c) 3 minute oxygen plasma and 90 second fluorocarbon plasma exposure.

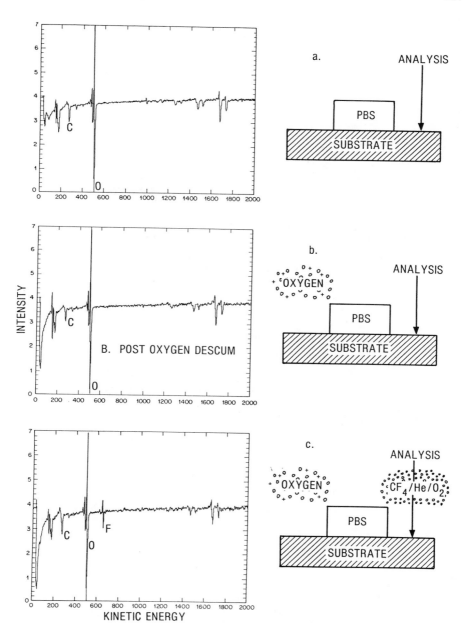

Figure 7. Auger spectra of tantalum substrate surface post
(a) no plasma exposure (b) 3 minute oxygen plasma exposure and
(c) 3 minute oxygen plasma and 90 second fluorocarbon plasma
exposure.

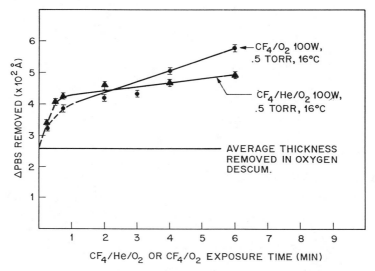

Figure 8. PBS thickness removed vs time in fluorocarbon
plasma post 3 minute oxygen plasma exposure.

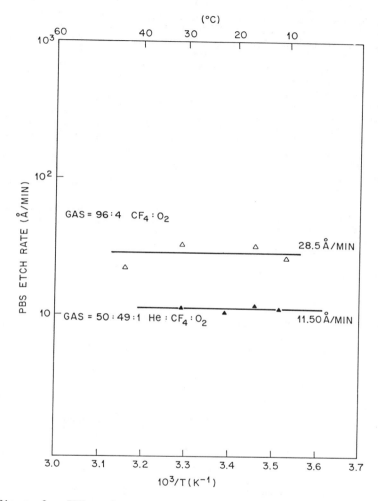

Figure 9. PBS etch rate vs inverse temperature in fluoro-carbon plasma post 3 minute oxygen plasma exposure.

Figure 2C). Likewise, Figure 7C shows the spectrum of the partially etched tantalum surface whose composition appears similar to the virgin surface (Figure 7A). There is evidence of some fluorine in the surface resulting from the probable ethant species. Figure 10 is a micrograph of the resist features at the conclusion of etching. It shows no surface irregularities, and the features are little changed from their unprocessed state. Under these conditions, the substrate can be completely etched with minimal loss of PBS.

We speculate that pre-exposure to an oxygen plasma creates oxygenated species, perhaps peroxides, at or near the resist surface which terminate depolymerization during subsequent exposure in fluorocarbon plasma. Post-oxygen plasma exposure dissolution studies indicate that active species diffuse to a considerable depth in the resist film (200-800Å). It is possible that these species create termination sites at these depths, thus explaining the initial high etch rates in fluorocarbon plasma. The top surface layers are removed by the fluorocarbon plasma before the termination sites are reached. Once termination is reached, a permanent layer of fluorinated species builds up at the surface which results in permanent stabilization which cannot be destroyed by baking. Evidence of this altered surface layer is seen in post-enhancement treatment dissolution and thermal degradation studies. For films subjected to the two-step enhancement process, a measurable residual film, approximately 200Å thick, always remains after either dissolving the film in a good solvent or subjecting it overnight to a temperature greater than its thermal degradation temperature ($>140°C$). No residual film is observed in the case of unprocessed samples nor for samples subjected to either oxygen or fluorocarbon plasma alone.

PATTERN TRANSFER PROCESSES

Single Level Process

Table 1 summarizes the plasma etching conditions which enable PBS to be used as an etch mask in a fluorocarbon plasma environment. It should be noted that the conditions cited result in isotropic etching of the substrate. In order to maintain good resolution ($<1.0\mu m$) without excessive linewidth loss, the substrate must be relatively thin.

Table 1

Single Level Enhancement/Etch Process

Step	Gas	Conditions
1. DESCUM	O_2	High press/low bias/ moderate temperature*
2. Enhancement/Etch	CF_4/O_2 $CF_4/He/O_2$	High press/low bias

* Typically the temperature is $-15°C$ so that controllable and reproducible etch rates can be achieved.

Trilevel Process

Figure 11 shows schematically the processing sequence using PBS as the imaging layer in a trilevel scheme (14). The etch resistance of the imaged layer is enhanced using the two-step process outlined in the previous section, and the pattern is then transferred by RIE into the underlying thin oxide layer and subsequently into the planarizing layer. The resulting etched patterns in the planarized layer for coded 0.75μm lines spaced 0.25 μm apart are shown in Figure 12.

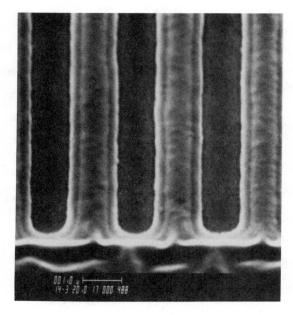

Figure 10. PBS features on tantalum substrate post 3 minute oxygen plasma and 90 second fluorocarbon plasma exposure.

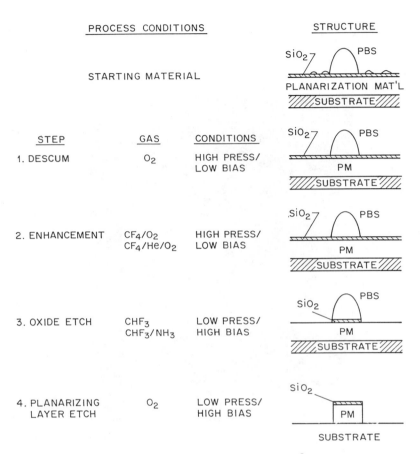

Figure 11.   PBS tri-level pattern transfer process.

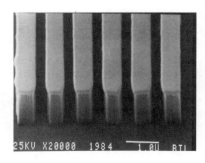

Figure 12.   Submicron tri-level features generated with PBS
enhancement process (0.75 um lines/0.25 um spaces).

CONCLUSIONS

Several conclusions may be drawn from this work:

1. Under normal conditions, etching of PBS in fluorocarbon plasmas is dominated by depolymerization which results from scission of the main polymer chain caused by the activated species in the plasma.

2. Depolymerization of PBS in fluorocarbon plasmas results in system-dependent etching rates which are difficult to reproduce. This is due to plasma-initiated polymerization of the gaseous degradation products which competes with the degradation reaction.

3. Pretreatment of PBS films in an oxygen plasma allows the creation of a chemically altered surface during subsequent fluorocarbon plasma treatment which provides enhanced etching resistance.

4. Surface-enhanced PBS films can be used in selective anisotropic reactive-ion etch processes to reproducibly delineate high aspect ratio features with submicron resolution.

LITERATURE CITED

1. Bowden, M. J.; Thompson, L. F. *Applied Polymer Science* 1973, *17*, 3211-3221.

2. Thompson, L. F.; Bowden, M. J. *Electrochem. Soc.* 1973, *120*, 1722.

3. Harada, Katsuhiro *J. of Applied Polymer Science* 1981, *26*, 3395-3408.

4. Curtis, B. J.; Brunner, H. R.; Ebnoether, M. *J. Electrochem Soc.* 1983, *130, No. 11,* 2242-2249.

5. Taylor, G. N.; Wolf, T. M. *J. of Polymer Eng. and Science* 1980, *20*, 1086.

6. Schnabel, Wolframe; Sotobayashi, Hideto *Prog. Polym. Sci.* 1983, *Vol. 9*, 297-365.

7. Bowden, M. J. *Solid State Tech.* 1981, *V 24 (6)*, 73.

8. Yamoyaki, T.; Wtakabe, Y.; Suzuki, Y.; Nakata, H. *J. Electrochem Soc.* 1980, *127(8)*, 1859.

9. Bowden, M. J.; Pease, R. F. W.; Yau, L. D.; Frackoviak, J.; Thompson, L. F.; Skinner, J. G. In *Microcircuit Engineering*; Asmed H.; Nixon, W. C., Eds.: Cambridge University Press, Cambridge, 1980.

10. Brown, J. R.; O'Donnell, J. H. *Macromolecules* 1970, *3*, 265.

11. Brown, J. R.; O'Donnell, J. H. *Macromolecules* 1972, *Vol. 5, No. 2*

12. Bowmer, T. N.; O'Donnell, J. H. *Polymer* January 1981, *Vol. 22*, 71-74.

13. Bowmer, T. N.; O'Donnell, J. H.; Wells, P. R. *Polymer Bulletin 2* 1980, 103-110.

14. Moran, J. M.; Maydan, D. M. *J. Vac. Sci. Technol.* 1979, *16(6)*, 1620.

RECEIVED May 27, 1987

Chapter 28

Degradation and Passivation of Poly(alkenylsilane sulfone)s in Oxygen Plasmas

A. S. Gozdz, D. Dijkkamp, R. Schubert, X. D. Wu[1], C. Klausner, and Murrae J. Bowden

Bell Communications Research, Red Bank, NJ 07701–7020

The passivation and degradation of poly(3-butenyltrimethylsilane sulfone) (PBTMSS), a sensitive, positive electron-beam resist has been studied in a parallel plate and barrel type plasma reactor. The resist is passivated in an oxygen plasma only when the oxygen pressure and/or the cathode self-bias exceed certain critical values during the initial exposure of the polymer to the plasma environment. Several oxygen plasma pretreatment procedures have been developed leading to the formation of a high quality silicon dioxide surface passivated layer. The oxide layer was characterized by IR, Auger electron spectroscopy, and Rutherford backscattering spectroscopy. The passivation efficiency is determined by the relative rates of surface oxidation and polymer oxidation which are dependent upon the concentration of active species (ions and neutrals) in the plasma.

The reactive environments employed in the plasma-enhanced deposition and etching processes which are widely used at various stages of integrated circuit manufacturing (1), place stringent requirements on the resist materials used in these processes. For example, oxygen reactive-ion etching (RIE) is often used in two-layer microlithography to transfer the image from a thin imaging resist layer into an underlying thick planarizing layer. This requires the top resist to be highly resistant to the oxygen plasma environment with an etch rate ratio between the bottom and top layer greater than 10/1. Such requirements have led to the development of organosilicon polymers - long known for their thermal and chemical resistance - as resists for this application (2-4). These materials are oxidized upon exposure to an oxygen plasma resulting in the formation of a thin surface layer of silicon oxide which passivates the surface (5-7) and greatly retards the rate of resist removal. This surface oxide layer is slowly sputtered during RIE, but the overall resist etch rate is much lower than that of the planarizing layer. Etch rate ratios as high as 100:1 can be obtained.

The majority of organosilicon polymers crosslink efficiently upon e-beam exposure (8) and thus behave as *negative*-tone resists. Poly(alkenylsilane sulfone)s, on the other hand, have been found to degrade easily upon e-beam exposure, which places them in a small class of *positive*

[1]Current address: Department of Physics, Rutgers University, New Brunswick, NJ 08903

0097–6156/87/0346–0334$06.00/0

organosilicon electron-beam resists (*10-13*). In view of their high e-beam sensitivity (1 to 2 μC/cm^2 at 20 keV), and superior resolution, (i.e., compared to the negative silicon-containing resists which are generally characterized by inferior resolution and linewidth control because of the extensive swelling and thinning during solvent development), these materials presage potential application in the fabrication of microelectronic devices with submicron geometries via direct e-beam lithography (*9*).

Initial processing experiments showed however, that under typical O$_2$ RIE conditions (power = 0.1 to 0.2 W/cm^2, self-bias = -250 to -350 V, pressure = 5 to 20 mTorr O$_2$), these resists are not very resistant, particularly under prolonged etching. Effective pattern transfer may require etch times of 15 to 30 min (*11,12*). which are sufficient to cause extensive degradation of the resist. Such behavior is reminiscent of poly(olefin sulfone)s such as the well-known PBS e-beam resist, which is etched 5-7 times faster than polystyrene in an oxygen plasma (*5*).

Further studies have shown, however, that under appropriately chosen plasma etching conditions, poly(alkenylsilane sulfone)s can be passivated in an oxygen plasma like other organosilicon polymers (*12,13*), but the passivation process depends strongly on the plasma processing parameters. In this paper, we report the results of our studies on the degradation and passivation processes of a typical poly(alkenylsilane sulfone), viz., poly(3-butenyltrimethylsilane sulfone) (PBTMSS) (*13*) in oxygen plasmas.

Experimental

Materials. Poly(3-butenyltrimethylsilane sulfone) (PBTMSS) was synthesized by free-radical copolymerization of 3-butenyltrimethylsilane with liquid sulfur dioxide (molar ratio 1:9) initiated with azobisisobutyronitrile (AIBN) at 35°C in a sealed glass ampoule. The detailed preparation procedure and properties of this copolymer have been reported elsewhere.(*13*)

Sample preparation. Thin films of PBTMSS for Rutherford backscattering spectroscopy (RBS) and general plasma etching studies were spun on polished silicon wafers from a 3.5% solution in chlorobenzene using a photoresist spinner. The films were baked for 10 to 20 min. at 105-120°C in air. PBTMSS films for Auger electron spectroscopy (AES) studies were spin-coated on silicon wafers previously coated with 2000 Å of gold. Films for IR studies were spin-coated onto NaCl plates.

Oxygen Plasma Etching. PBTMSS films were etched in a Cooke Vacuum Products parallel plate RIE system operating at 13.56 MHz. The stainless steel reactor was 30 cm in diameter and 35 cm high. Both electrodes, 12.5 cm in diameter and spaced 9 cm apart, were covered with a Si wafer and both were water-cooled. The RF power could be coupled either to the lower electrode (RIE mode) or to the upper electrode (sputter mode, SME). The required oxygen pressure and its flow rate were maintained by the mass flow and exhaust valve controllers. Composition of the gas mixture during RIE was monitored using a quadrupole mass spectrometer (Quadrex 200, Inficon Leybold-Heraeus, Inc.). The pressure was reduced to the 10^{-6} Torr range using a flow-restricting orifice and a turbo-molecular pump. Other parameters of the etching process are reported in the text.

Resist films were also etched in a barrel-type plasma reactor (Plasmod, March Instruments, Inc.) operating at 13.56 MHz at a power density of approx. 0.007 W/cm^3 and with an oxygen pressure of 0.85 Torr.

Surface Analysis. The resist etching process was studied by measuring changes in the resist thickness versus etching time using a mechanical stylus surface profiler (Alpha-Step 200, Tencor Instruments, Inc.).

Infrared spectra were recorded on a Perkin-Elmer Model 983G double-beam spectrophotometer in the transmission mode using 3500 Å thick PBTMSS films spin-coated and processed on polished NaCl plates. Spectral subtraction and absorbance correction to account for the decreased film thickness were used to isolate the silicon oxide absorption band at about

1050 cm^{-1}. The equivalent thickness of silicon oxide surface layers was calculated by comparing the absorbance at 1050 cm^{-1} with that of thermally grown oxide films of known thickness. The change in absorbance per unit thickness obtained from the calibration curve was $1.5 \cdot 10^{-4}$ a.u./Å.

Glancing angle ($\alpha = 80°$) Rutherford backscattering spectrometry (RBS) with a 2.0 MeV He^+ beam was used to obtain the depth-resolved elemental composition of the films. In order to obtain sufficient counting statistics, and at the same time minimize beam-induced damage, each spectrum was accumulated from 20 different spots on the sample. The He^+ ion beam dose was kept below 1 $\mu C/cm^2$ per spot. Spectral simulation was performed using the computer code RUMP (14).

Auger electron spectroscopy (AES) data were obtained on a Perkin-Elmer Scanning Auger Multiprobe PHI 600 using a 10 keV, 5 nA electron beam being rastered at 400X, and a 3 keV, rastered 1.1 nA argon ion beam. These electron and ion beams were of sufficiently low power so as not to thermally decompose the polymer when the beams were used in the raster mode of operation. The area sputtered by the argon ion beam was about 2 by 2 mm. The e-beam survey area had a rectangular shape with sides about 0.4 by 0.5 mm; it was located in the center of the ion-sputtered crater. Sputtering rate for the initial PBTMSS film under experimental conditions used was 140 ± 20 Å/min. Exposure to the electron beam alone did not cause any detectable change in surface composition.

Results

Oxygen Plasma Etching. The change in thickness versus time for a 3600 Å-thick PBTMSS film during O_2 RIE is shown in Figure 1 (curve A). The RIE parameters were: power: − 0.12 W/cm^2 , bias: −400 V, O_2 pressure - 20 mTorr, flow rate - 10 sccm. These parameters correspond to what we shall refer to as the "standard" RIE process parameters throughout the remainder of the text. As seen in Figure 1, the thickness decreased by 30 to 70%. Accompanying this reduction in thickness was the appearance of numerous defects in the form of pinholes and bubbles in the film, particularly after prolonged etching. This behavior was similar to that of the previously reported terpolymer of allyltrimethylsilane with 1-butene and sulfur dioxide (11,12).

Reducing the oxygen pressure and/or the bias increased both the rate of resist removal and the number of defects in the residual film. On the other hand, increasing the self bias resulted in a significant decrease in the etching rate (Figure 1, curves B & C). The number of defects was also significantly reduced. In fact, for a bias greater than −600 V, an entirely defect free surface was obtained provided the etching time was less than 30 sec. With prolonged etching at high bias, numerous defects again appeared in the etched film as shown by dashed curves in Figure 1. If, however, the bias was lowered back to the standard value of −400 V following 10 to 30 sec at high bias (−600 to −900 V), etching could be continued without any subsequent defect formation. In fact, films subjected to this short high-bias pretreatment were able to withstand longer than 30 min. standard RIE processing without the formation of surface defects, thus facilitating excellent RIE pattern transfer.

In another set of experiments, PBTMSS samples were etched in the reactor which has been configured for sputtering, i.e., with a relatively high RF power (0.8 W/cm^2) coupled to the upper target electrode. The wafer potential in this mode is low (80 to 130 V). The resist thickness was reduced by about 800 to 1000 Å after 1 min. of such treatment (Figure 1, curve D). However, the surface of the polymer was smooth and defect-free. It was also found that a PBTMSS film pretreated by this method would withstand subsequent standard O_2 RIE pattern transfer in a manner similar to the sample pretreated by the high-bias process, i.e., without the formation of surface defects. For convenience this pretreatment process will be referred to as "sputter-mode etching" (SME).

We found that there is an RF power density threshold below which this SME passivation pretreatment does not occur. In our RIE/sputtering system at p=20 mTorr O_2, this threshold was about 0.25 W/cm^2 (30 W) (Figure 2). At lower power densities, the resist was again seriously damaged and mostly volatilized. At higher RF power densities but with longer etch

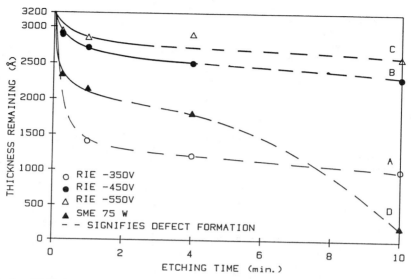

Fig. 1. PBTMSS film thickness vs. time during RIE and SME at 20 mTorr O_2. Bias: A: −350 V; B: −450 V; C: −550 V; D: sputter mode etch (SME) at 75 W.

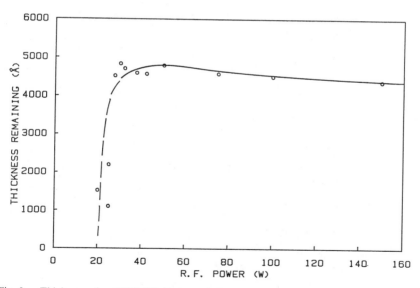

Fig. 2. Thickness of a PBTMSS film remaining after 1 min. of SME etching (20 mTorr O_2, 10 sccm) versus RF power.

times (greater than 2-4 min), the resist began to decompose with the formation of numerous bubbles occurs during the prolonged high-bias RIE etching. We found that a passivation time of 15 to 60 s in the SME mode at a power of 50 to 75 W was sufficient to permit subsequent pattern transfer by "standard" RIE.

The SME passivation pretreatment has two major drawbacks: 1) it introduces additional process complexity in cases where the RIE system cannot be conveniently converted to a sputtering mode, and 2) it causes a relatively large decrease in the resist thickness which may lead to loss of linewidth control.

We also studied the etching of PBTMSS in a barrel-type oxygen plasma reactor at a much higher oxygen pressure (850 mTorr) than that used in the RIE process (10 to 20 mTorr). The etching curve of normalized resist thickness remaining as a function of time is shown in Figure 3, curve A. Contrary to the SME process, the loss of thickness during plasma barrel etching was always less than ca. 300 Å even after prolonged exposure to the oxygen plasma. The process again created a surface oxide layer which was durable enough to withstand subsequent "standard" O_2 RIE (Figure 3, curve B). No surface defects were present either after etching in the barrel reactor or after subsequent RIE, provided a barrel reactor pretreatment of ~ 2 min. was employed.

This result suggested that poly(alkenylsilane sulfone)s might undergo passivation under standard RIE conditions of bias and power but at significantly higher oxygen pressures than the standard 20 mTorr used in the earlier studies. The results from a series of experiments designed to determine the range of process parameters (oxygen pressure and self-bias) which cause rapid and defect-free passivation of the polymer are summarized in Figure 4. All results refer to etching time of 1 min. A combination of low oxygen pressure and low bias resulted in inefficient passivation of the resist, leading instead to rapid conversion to gaseous products and/or the formation of numerous pinholes. On the other hand, prolonged etching at high bias caused excessive heating of the surface and bubble formation. However, the combination of processing variables encompassed by the shaded area in Figure 4 favored formation of a defect-free surface, provided etching times were less than 1 minute. For example, a defect-free surface could be obtained under standard conditions of bias (-400 V) and power (0.12 W/cm^2) provided the oxygen pressure was greater than 100-140 mTorr. It is important to note, however, that extended etching under these conditions will lead to undercutting of the planarizing layer which is why it is necessary to revert to the "standard" low-pressure conditions to obtain anisotropic etched profiles. All samples pretreated under conditions encompassed by the shaded area withstood subsequent "standard" RIE pattern transfer etching with negligible thinning and without the formation of defects in the etched film. A comparative summary of the four passivation pretreatment methods for PBTMSS is given in Table I.

Mass Spectrometry. The etching and passivation process was studied by measuring the concentration of volatile gaseous products in the plasma by quadrupole mass spectrometry. Sulfur dioxide gives rise to a characteristic peak in the mass spectrum of the degradation products at m/z=64 corresponding to SO_2^+ resulting from the oxidation of these poly(olefin sulfone) polymers. Since there are no other interfering mass peaks, the change in the peak intensity at m/z 64 vs. time during etching provides a reliable measure of the progress of resist degradation. Other mass peaks, such as m/z 18 (H_2O^+), 28 (CO^+, but also N_2^+), 32 (O_2^+), 44 (CO_2^+), and 48 (SO^+) were more prone to interference from other fragmentation products or were less sensitive than the SO_2^+ peak.

Changes in the concentration of SO_2 in the plasma during RIE at various cathode bias values as well as during SME at the same pressure are shown in Figure 5. The change in intensity of the m/z 64 peak with time depended markedly on the bias conditions and electrode configuration and mirrored the changes in the resist thickness shown in Figure 1. SO_2 was rapidly eliminated during the initial phase of low-bias RIE (curves A-C), although the extent of

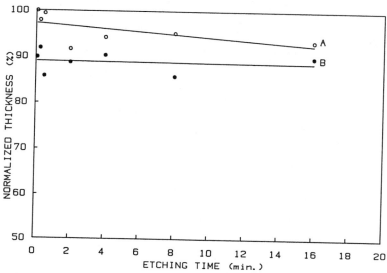

Fig. 3.   Thickness of PBTMSS after etching in a barrel plasma reactor (A) (p_{O_2} = 850 mTorr) versus time. The samples were then etched for 10 min. under O_2 RIE conditions (B) (p_{O_2} = 20 mTorr, U = −400 V).

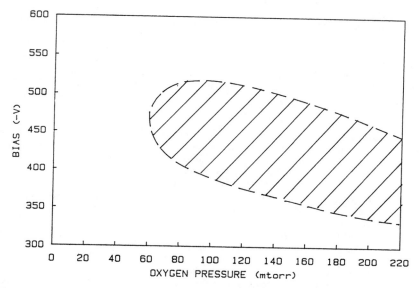

Fig. 4.   Effect of pressure and bias on surface quality following 1 min. O_2 RIE. Shaded area shows combination of pressure and bias that result in a defect-free surface.

Table I. Comparison of PBTMSS Passivation Methods

Parameter	Passivation Method			
	Sputter Mode	Barrel Reactor	Hi-Pressure RIE	Hi-Bias RIE
O_2 press., mTorr	20	800-1000	100-200	20
bias, -V	80-130	10-30	400	600-900
power, W	50-100	low	50	50
time, s	60	120	30-60	10-30
loss of thickness, Å	800-1000	150-200	200	200-300
thickness loss after 10 min. RIE	800-1000	300-400	~300	~300

degradation (as judged from the peak intensity) decreased with increasing bias. Above −500 V, very little SO_2 was detected in the plasma (curves D and E) indicating a low extent of degradation. This is in agreement with the low thickness loss observed under high-bias conditions.

Curves F and G in Figure 5 represent changes in the SO_2 concentration during SME at a RF power of 50 and 100 W, respectively. The intensity of the m/z 64 peak increased sharply at the very beginning of the process during which about 800 to 1000 Å of the resist were volatilized. Thereafter, it decreased asymptotically to the background level. The amount of SO_2 liberated initially depended on the RF power being somewhat higher at 100 W (curve G) than at 50 W (curve F).

Infrared spectroscopy. The IR transmission spectra of PBTMSS films (original thickness: 3500 Å) in the range 900 to 1200 cm^{-1} after passivation by various methods are shown in Figure 6. The difference spectra between the initial and etched (for 1 and 10 min.) films corrected for changes in film thickness are also plotted in Figure 6.

Curves 1a, 1b, and 1c show the IR spectra after 0 (initial), 1, and 10 min. etching under standard RIE processing conditions (−400 V, 20 mTorr O_2 10 sccm) respectively. The corresponding difference spectra are represented by curves 1d and 1e. The major difference was the appearance of a peak at 1070 cm^{-1} which corresponded to the Si−O stretching frequency. We attribute this to SiO_x formed at the surface of the resist by oxidation in the plasma environment. The intensity of this peak increased with etching time corresponding to increased conversion of Si to SiO_x.

The remaining spectra in Figure 6 show evidence of similar formation of SiO_x for the different passivation treatments. The extent of oxide formation as judged from the intensity of the peak at 1070 cm^{-1} depended on the particular passivation treatment as did stability to an additional 10 min. "standard" RIE. For example, SME resulted in a much greater extent of oxide formation (curve 2d) than high-bias (curves 3d and 4d) or barrel etching (curve 5d).

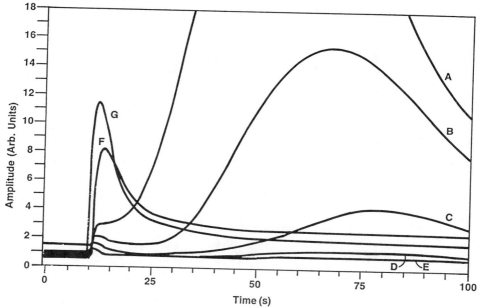

Fig. 5.   Concentration of sulfur dioxide in an oxygen plasma measured by mass
spectrometry vs. time during PBTMSS etching under various conditions. RIE (20
mTorr O_2 10 sccm): A: -320 V, B: -350 V, C: -380 V, D: -450 V, E: -550 V; SME
(20 mTorr, 10 sccm O_2); F: 50 W, G: 100 W

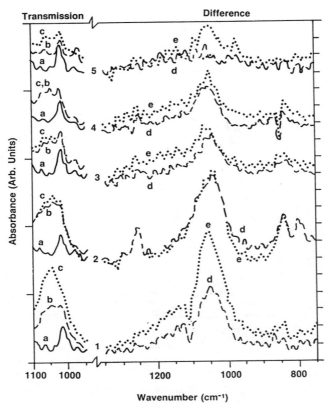

Fig. 6.   IR transmission (a-c) and difference (d,e) spectra of O_2 plasma etched PBTMSS
films listed in Table II. Samples: 1 - RIE −400 V; 2 - SME; 3 - high-pressure RIE;
4 - high-bias RIE; 5 - barrel reactor etched. Curves a are for the initial film, b for
the film treated as given in Table I, c after additional RIE at 20 mTorr and −400 V
for 10 min. Curves d=b−a, e=c−a.

Contrary to the increase in oxidation observed during standard RIE, films subjected to one of the passivation pretreatments (with the exception of the barrel reactor process) showed little further increase in oxide conversion beyond the initial pretreatment conversion. Only the films pretreated by high pressure oxygen plasma in the barrel reactor showed a further increase in oxide conversion on subsequent "standard" RIE (curve 5e).

It is recognized that transmission IR only presents an integrated picture of the silicon conversion throughout the whole thickness of the resist. However, the bulk of the oxidation occurs at the surface, and the equivalent oxide thickness obtained from the calibration curve for thermally grown oxide films can be taken to represent the approximate thickness of the plasma-treated films. The thickness of the oxide film formed by the various pretreatments is summarized in Table II.

Table II. Thickness of Surface SiO_x and PBTMSS After

Plasma Passivation process (see Tab.I)	Passivation			After RIE		
	absorbance at $1070cm^{-1}$ a.u.	thickness of		absorbance at $1070cm^{-1}$ a.u.	thickness of	
		SiO_x, Å	resist %		SiO_x Å	resist %
RIE −400 V	0.0108	72	30-60	0.0192	128	⁻10
SME 75 W	0.0138	92	76	0.0156	104	76
barrel	0.0042	28	96	0.0072	48	88
hi-bias RIE	0.0072	48	92	0.0084	56	92
hi-press RIE	0.0072	48	95	0.0084	56	92

Auger Electron Spectroscopy (AES). AES coupled with argon ion-beam sputtering was used to obtain atomic concentration depth profiles of etched PBTMSS films. Poly(olefin sulfone)s present considerable experimental difficulties for surface analytical techniques involving high energy particle beams because of their facile decomposition when irradiated with such high energy radiation. However, with the precautions outlined in the experimental section, spectra could be obtained which corresponded to the actual composition of the surface.

An Auger electron survey spectrum of an initial PBTMSS film showed the expected composition of the surface layer, namely C, O, Si, and S as shown in Figure 7, curve A. After the exposure to an oxygen plasma the surface composition of the film was vastly different as shown in the AES survey (Figure 7, curve B). Essentially, only Si and O were present in the surface layer. Moreover, the shift in the AES Si peaks from 92 and 1619 eV to 76 and 1606 eV indicated that the Si was present as SiO_x as opposed to elemental or organosilicon. Monitoring both Si peaks (labeled as Si_e (92 eV) and Si_o (76 eV) enabled us to follow the growth of the oxide layer as a function of time for the different passivation treatments.

The atomic composition of the unetched PBTMSS surface shown in Figure 8 closely corresponds to the molecular formula of the polymer. The slight surface enrichment in Si and O and corresponding depletion in C may be due to a small degree of oxidation during prebaking. Figure 9 shows the concentration profiles for the different elements as a function of depth for a film pretreated for 1 min. in an oxygen barrel reactor. The results show that the composition of the surface layer is variable to a depth of about 300 Å, thereafter conforming to that expected of the virgin polymer. The surface layer is composed mostly of inorganic silicon and oxygen with small amounts of organic silicon and carbon.

The sputtering time required for the atomic composition of the surface layer to reach that of the unetched, i.e., virgin PBTMSS depended on the passivation pretreatment process. The time increased in the order SME > high bias > barrel reactor reflecting the differences in thickness of the oxide layer. These results are also in qualitative agreement with the IR results presented in Figure 6 and Table II.

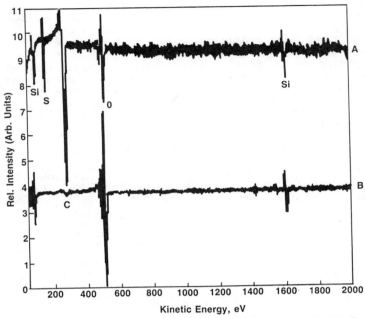

Fig. 7.   Auger electron survey of PBTMSS surface before (curve A) and after O_2 RIE (1 min, −600 V, 20 mTorr O_2) (curve B)

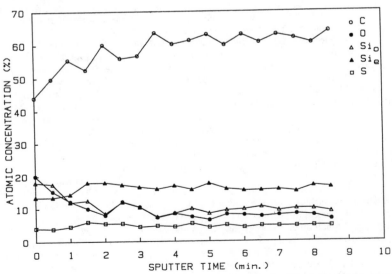

Fig. 8.   AES atomic concentration depth profile for an untreated PBTMSS film on Au/Si (1 min. sputtering $\simeq$ 140 Å).

Rutherford Backscattering Spectroscopy. The high energy of the He^+ ions used in RBS gives rise to electronic energy loss (bond breaking) through the entire thickness of the polymer film. Therefore, the precautions described in the Experimental section were taken to minimize loss of PBTMSS film during the measurement. Nevertheless, the spectrum of the untreated film (Figure 10) shows a sloped Si substrate edge below 0.8 MeV which is caused by thinning of the film during the measurement. However, the simulation based upon the elemental composition of PBTMSS ($C_7H_{16}O_2SSi$) shows excellent agreement above 0.8 MeV.

Figure 11 shows an RBS spectrum of a PBTMSS film after O_2 RIE (1 min, −800 V, 20 mTorr). The substrate signal in this and other spectra is markedly lower and it is similar to the simulated spectrum which indicates that oxygen plasma treatment makes the film more radiation resistant. In addition, the surface layer of plasma treated films does not contain measurable amounts of sulfur. The thickness of this sulfur-free layer, estimated by comparison with simulated spectra, is about 50 Å after RIE and SME and 25 Å after the barrel etching. The data also show that the same layer is enriched in oxygen and silicon which was confirmed by simulation. These results compare favorably with those obtained from AES spectra.

Discussion

Passivation and Etching of PBTMSS. Poly(olefin sulfone)s traditionally have shown very poor resistance to plasma environments. This primarily stems from the tendency of these polymers to decompose with extensive depropagation of the polymer chain in a high-energy radiation environment, particularly at elevated temperature, leading to enhanced rates of material loss. The onset of thermal depolymerization of PBTMSS in air determined by thermogravimetric analysis is approx. 170°C *(13)*, but radiation-induced depolymerization occurs at much lower temperatures depending on the ceiling temperature (T_c) of the polymer. By analogy with the 1-olefins, T_c for formation of PBTMSS is probably in the neighborhood of +60°C which is comparable to the temperature at the wafer surface during RIE. Thus we would expect PBTMSS to be degraded in an RIE environment in a manner similar to the other poly(olefin sulfone)s.

Our results indicate that PBTMSS does indeed rapidly degrade (volatilize) in the O_2 RIE environment, particularly at low bias and low oxygen pressure. Increases in the mass spectrum peak intensities at m/z equal to 1, 2, 17, 18, 28, 44, 48, and 64 are entirely consistent with the production of SO_2 and 3-butenyltrimethylsilane, the latter being oxidized in the oxygen plasma to CO (m/z=28), CO_2 (44), H_2O (18) and SiO_2. A uniform passivating SiO_x layer does not form under these conditions reflecting a much greater rate of polymer degradation compared to surface oxidation (passivation).

At higher oxygen pressures or at higher bias (>−600 V) however, the situation appears to be reversed, i.e., the rate of surface oxidation is considerably greater than the rate of polymer degradation resulting in efficient surface passivation. At high bias, for example, very little SO_2 is eliminated during the initial stages of etching (curve E, Figure 5), i.e., for etching times <2min, while at the same time, a surface layer of SiO_x approximately 50 Å thick is formed. These conditions clearly favor surface oxidation of the silicon which protects the underlying resist against further erosion. At longer etching times, significant degradation does occur which we attribute to thermal degradation of the underlying resist caused by the high temperature attained during high-bias RIE. If however, the power dissipated in the plasma is reduced by lowering the bias potential thus lowering the maximum temperature reached by the substrate during etching, then the temperature of the substrate remains below the decomposition temperature of the polymer, and the oxide film formed during the high-bias pretreatment facilitates defect-free pattern transfer.

It appears then that the relative rates of the reactions taking place at the surface of the resist under high-bias conditions are quite different from those occurring under low bias. In part this is related to the overall energy discharged in the plasma. Since a fully matched network was used in these studies, the RF power determines the electrode self-bias at constant

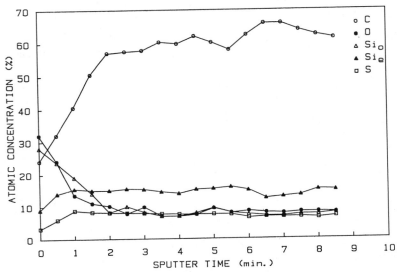

Fig. 9.   AES atomic concentration depth profile for a **PBTMSS** film on Au/Si. Film etched for 1 min. in a barrel reactor at 850 mTorr O_2.

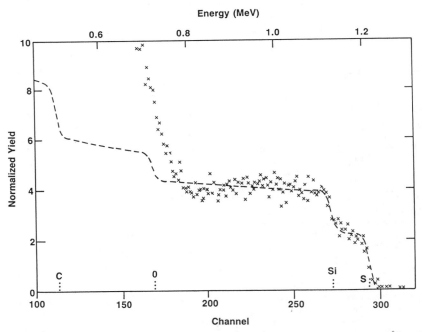

Fig. 10. RBS spectrum of untreated PBTMSS film on Si (initial thickness 5500 Å). The dashed line represents a simulated spectrum for a 5500 Å thick film of $C_7H_{16}O_2SiS$

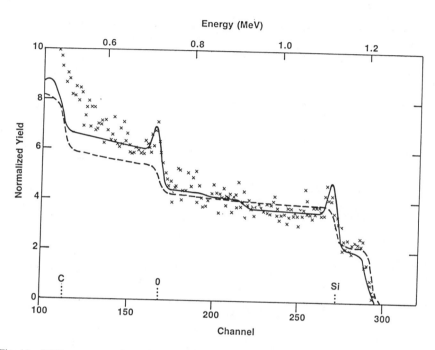

Fig. 11. RBS spectrum of a PBTMSS film after RIE (1 min., −800 V, 20 mTorr). The lines represent simulated spectra of untreated PBTMSS (dashed line) and a three-layer system composed of 50 Å of Si_3CO_8, 1000 Å of $C_7H_{16}O_3S_{0.9}Si$ and 4000 Å of PBTMSS (solid line).

gas pressure. These three interdependent macroscopic parameters determine the microscopic properties of the plasma such as the concentration and energy distribution of active species, viz., excited neutrals, ions and electrons. Steinbruchel (15) in fact has shown that at constant O_2 pressure, the concentration and energy of both the ions and neutral species increases with increasing RF power. At the same time the electron temperature decreases.

It is generally believed that the active neutrals are responsible for oxidation. It would therefore appear that the surface passivation at high bias is a direct result of the higher concentration of neutral active oxygen species. We speculate that at low pressure and low bias, the concentration of neutral active species is too low relative to the concentration of ionic species. The latter possess sufficient energy to break the C—S bonds in the main chain with concomitant depolymerization leading to rapid material loss. Thus degradation prevails over passivation. One might expect similar results at high bias given that the concentration of both ions and excited neutrals increases with increasing bias. The fact that passivation is favored suggests that either the increase in power results in different proportions of the neutral and ionic species, or alternatively, the kinetic order of passivation is greater than that of degradation.

If this model is correct, we should be able to effect passivation at low bias simply by increasing the oxygen pressure. As seen in Figure 4, increasing the oxygen pressure to 100-200 mTorr at a bias of −400 V indeed results in efficient surface passivation. Likewise, efficient passivation occurs in a barrel reactor where the energy of any ionic species is very low and the oxygen pressure is relatively high.

The RF power during SME is higher than during RIE which should result in a higher concentration of active oxygen in the plasma. Further, the energy of ions bombarding the resist is much lower since a large negative bias develops at the counter-electrode, and the bias at the sample is of the order of several tens of volts as indicated by a very thin plasma sheath and voltage measurements. According to our previous arguments, such conditions should favor rapid oxidation with little sputtering. Surface studies do indicate that the top layer of the polymer is strongly oxidized although there is considerable thickness loss during SME pretreatment. This high initial rate of film removal is probably caused by the large flux of high-energy electrons to the substrate (16) under SME conditions which causes self-development of the resist before passivation occurs.

The etching of polymers at high oxygen pressures in barrel-type reactors is thought to be a purely chemical process (1,5,7,16). A fully oxidized 200 Å-thick layer of PBTMSS which corresponds to the thickness lost during barrel etching, would afford a 35 Å-thick layer of SiO_2, which is in excellent agreement with the IR, RBS and thickness loss measurements. Since there is no sputtering component and almost no surface heating in the barrel reactor, the resist is oxidized under very mild conditions. However, the oxide layer is very thin and not very compact, thus it is further compressed and thickened by the subsequent RIE process.

Contrary to the observations reported by Bagley et al. (7) on the etching of an organosilsesquioxane in a barrel reactor, the PBTMSS etching process does not proceed beyond the surface layer, probably because much higher elasticity of the PBTMSS network causes collapse of the micropores through which the active species diffuse into the film.

Conclusions

Poly(alkenylsilane sulfone)s undergo efficient passivation in oxygen plasmas containing a high concentration of active oxygen species. These conditions can be attained by using either high oxygen pressure or high RF power in the plasma discharge. Several different oxygen plasma passivation procedures have been developed which lead to the formation of a resist surface which is defect-free and covered with a 30 to 90 Å-thick layer of mixed silicon oxide with a minimal thickness loss of the resist. Spectroscopic plasma diagnostics methods should provide additional information on the effect of plasma conditions on the passivation and degradation mechanisms of this sensitive organosilicon polymer.

Acknowledgments

We wish to acknowledge helpful discussions with W. E. Quinn on the mechanism of oxygen plasma etching.

Literature Cited

1. Einspruch, N. G.; Brown, D. A., Eds., "*VLSI Electronics Microstructure Science*", vol. 8, "*Plasma Processing for VLSI*", Academic: New York, 1984.
2. Reichmanis, E.; Smolinsky, G.; Wilkins, Jr., C. W. *Solid State Technol.* Aug. 1985, **28**(8), 130.
3. Hatzakis, M. *Solid State Technol.* Aug. 1981, **24**(8), 74.
4. Lin, B. In *Introduction to Microlithography*, Thompson, L. F.; Willson, M. J.; Bowden, M. J., Eds., ACS Symp. Ser. **219**, American Chemical Society: Washington, DC, 1983; Chapter 6, p 288.
5. Taylor, G. N.; Wolf, T. M. *Polym. Eng. Sci.* 1980, **20**, 1087.
6. Chou, N. J.; Tang, J.; Paraszczak, J.; Babich, E. *Appl. Phys. Let.* 1985, **46**, 31.
7. Bagley, B. G.; Quinn, W. E.; Mogab, C. J.; Vasile, M. J. *Materials Lett.* 1986, **4**, 154.
8. Babich, E.; Hatzakis, M.; Paraszczak, J; Shaw, J. paper presented at the XX Organosilicon Symposium, April 18-19, 1986, Tarrytown, NY,
9. Bowden, M. J.; O'Donnell, J. H. In *Developments in Polymer Degradation - 6*, N. Grassie, Ed., Elsevier, London 1985, Chapter 2, p 21.
10. Kilichowski, K. B.; Pampalone, T. R. U. S. Pat. 4,357,396 (1982).
11. Gozdz, A. S.; Craighead, H. G.; Bowden, M. J. *J. Electrochem. Soc.* 1985, **132**, 2809.
12. Gozdz, A. S.; Craighead, H. G.; Bowden, M. J. *Proc. SPE Reg. Tech. Conf. Photopolymers* Oct. 28-30, 1985, Ellenville, NY, p 157.
13. Gozdz, A. S.; Carnazza, C.; Bowden, M. J. Proc. SPIE, 1986, **631**, p 2.
14. Doolittle, L. R. *Nucl. Instr. Methods Phys. Res.* 1985, **B9**, 344.
15. Steinbruchel, Ch.; Curtis, B. J.; Lehmann, H. W.; Widmer, R. *IEEE Trans. Plasma Sci.* 1986, **PS-14**, 137.
16. Chapman, B, "*Glow Discharge Processes*" Wiley-Interscience, New York, NY, 1980, p. 322.

RECEIVED April 30, 1987

Chapter 29

A Single-Layer, Multilevel Resist: Limited-Penetration Electron-Beam Lithography

S. A. MacDonald, L. A. Pederson, A. M. Patlach, and C. G. Willson

Almaden Research Center, IBM, San Jose, CA 95120-6099

A novel process is described which uses a single coating step but achieves the function of more complex multilayer oxygen etch transfer imaging systems. The key features of the system involve confining the radiation chemistry to the upper surface of the resist film through use of retarding potential, e-beam lithography. The radiation chemistry is exploited to allow selective surface silylation of the resist film. Dry development of the image by anisotropic oxygen reactive ion etching generates the high resolution high aspect ratio images typical of multilayer resist processing.

In a typical direct write e-beam lithographic system, the resolution of a dense line-space array is often limited by the effect of electrons backscattered from the substrate. In these arrays, the backscattered electrons from one exposed line increase the net exposure density in an adjacent line. While this problem of non-uniform exposure can be corrected by varying the exposure dose within the pattern, this form of proximity correction requires sophisticated algorithms and extensive computer facilities.

One technique for reducing this proximity effect is to replace the single layer of resist with a multilevel resist (MLR) system. In this case, a thin radiation sensitive resist is separated from the substrate by a thick layer of organic material (1). The utility of this approach has been demonstrated experimentally in several laboratories and also investigated by various computer modeling studies.

Another method for minimizing the proximity effect involves changing the electron beam acceleration potential from the 10 to 20 KV range that is typically used in direct write applications. L. D. Jackel and co-workers have proposed that increasing the acceleration potential to 100 KV will minimize the need for proximity corrections as the backscattered electrons will be averaged over a larger area (2). While this approach may reduce the effect of backscattered electrons, it will also significantly reduce the resist sensitivity since a smaller number of electrons will be deposited in the resist at 100 KV.

0097–6156/87/0346–0350$06.00/0

On the other hand, R. F. W. Pease and others have argued that decreasing the beam energy to less than 5 KV should reduce backscattering effects because all of the incident electrons will be confined to a thin volume at the resist-air interface (3). This design predicts minimal backscattering simply because the electrons never hit the substrate. One disadvantage to confining the e-beam exposure to the top surface (100 nm) of a typical resist is that one is limited to working with extremely thin films.

As it is difficult to design a high resolution electron optical system which operates at very low accelerating voltage, these low beam energies are obtained by using a retarding potential in a standard 20 KV system. In this approach, the electrons are accelerated to a high potential in a conventional column. They then are retarded to the desired landing potential at the substrate by an electrostatic field located in front of the target.

The work discussed in this paper combines the limited penetration e-beam lithography described above, with our previously reported oxygen plasma developable resist system (4). This combination results in the formation of a "2-layer MLR" type of structure within a single layer of resist. In this work, a 1.0μ thick layer of resist is exposed to low landing energy electrons such that the radiation chemistry is confined to the resist-air interface. In a subsequent step, the exposed resist is treated with a silylating agent to introduce silicon into the surface of the exposed regions. This generates a thin, patterned layer of an organo silicon species within the top surface of the resist. When the exposed and silylated resist film is placed into an anisotropic RIE chamber, the regions containing silicon are not etched. Correspondingly, the unexposed and not silylated areas are etched to substrate. This dry develop process, which generates a negative tone relief image, is outlined in Figure 1.

These experiments confirm the potential advantage of combining low energy e-beam exposure with a dry develop resist scheme proposed in 1984 by G. N. Taylor, L. E. Stillwagon, and T. Venkatesan (5). At that time lack of a high resolution, low voltage e-beam exposure system prevented these workers from experimentally verifying the concept.

Results and Discussion

Modifications to Electron Beam System. The retarding potential field was introduced into our vector scan system by attaching a 0-to-20 KV, 2 mAmp external power supply to the wafer holder. This is shown diagrammatically in Figure 2. Whereas the wafer is at ground potential in the standard system, in our modified system the net potential at the wafer can vary from 0 to 20 KV. Thus, the electrons are accelerated down the column at a standard operating potential, exit the final lens and are retarded by the variable electrostatic field near the wafer plane.

As shown in Figure 1, the variable power supply is located outside of the e-beam chamber and simply wired to the wafer holder. Since the chamber is metallic and operates under vacuum, additional insulators are required. The first is a feed-through to bring the high voltage lead through the chamber wall and into the evacuated region. We used an automobile spark plug for this task. A second insulator is required to isolate the wafer holder from the metallic stage. This was accomplished by incorporating a 2 inch thick ceramic disk into the wafer holder.

Although this method of generating a retarding field at the wafer plane is straightforward and requires minimal modifications, it does introduce a few

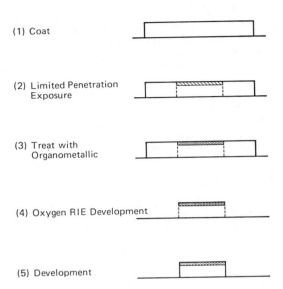

(1) Coat

(2) Limited Penetration
    Exposure

(3) Treat with
    Organometallic

(4) Oxygen RIE Development

(5) Development

Figure 1. Overall scheme that uses limited penetration e-beam lithography and dry development to produce a single layer MLR system.

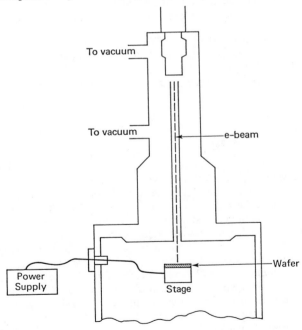

To vacuum

To vacuum   e-beam

Power
Supply   Stage   Wafer

Figure 2. Overview of modified electron beam system.

limitations. For example, with the wafer holder directly wired to the power supply, the wafer holder can not be transported by the standard vacuum handling mechanism. In our case, we brought the e-beam chamber to atmospheric pressure, placed a wafer directly into the chamber and then evacuated the system. Since it takes several hours to pump the system down to its operating pressure of 10^{-5} Torr, only two or three wafers can be processed in a day.

In this approach, the excellent stability of the external power supply minimizes aberrations within the field near the wafer plane. However, the final field size at the wafer will vary with the net acceleration potential. The specific field size was determined by measuring the distance between four gold marks (using the laser controlled stage) at each landing energy. This value was then compared to the distance observed when the bias on the wafer holder was zero.

The rational for modifying the e-beam system in this fashion is that low beam energy should result in confining the exposing radiation to the top surface of the resist. Figure 3 shows the calculated energy distribution within the film, as a function of accelerating potential. For example, Figure 3 predicts that the energy distribution at 10 KV should be fairly uniform through a 1.5μ thick film, while at 3 KV all of the energy should be deposited in the top 0.25μ of the resist. Given the general trend in Figure 3, one would anticipate that lower landing energies would result in confining the exposure to a thinner region the resist-air interface.

In this work we have looked at 3 KV for several reasons. Figure 3 predicts that proximity effects within a line-space array should not be observed if a 1.0μ thick film is exposed with 3 KV electrons. This net landing energy can be readily obtained by operating the column at 15 KV and applying a 12 KV bias to the wafer holder. Finally, at 3 KV the e-beam detection system, which is required to focus the beam, is still operational.

Resist Lithography. As discussed in the introduction, e-beam exposure of the top 100 nm of a 1.0μ thick resist does not usually result in a relief image that is developed to the substrate. However, this approach has been used successfully in ion beam lithography. Venkatesan and co-workers used a focused indium ion beam to write a pattern onto the surface of an organic polymer (6). When the resist was subjected to oxygen RIE, the surface of the ion-implanted regions was oxidized to indium oxide, which functions as an etch barrier. The resist areas that were not implanted with indium ions were etched to the substrate, forming a negative tone image.

In this work we have used a dry develop resist system that has been previously described (4). This system is based upon the radiation-induced generation of reactive functionality within the resist film which reacts in a subsequent step with an organosilicon species. As a result, the organosilicon material is selectively and covalently incorporated into the exposed regions. When the treated film is placed into an oxygen RIE chamber, the regions that do not contain silicon are etched to the substrate while those areas containing the organosilicon species are not attacked, generating a negative tone image. The polymers shown in Figure 4 are examples of materials that undergo a significant change in chemical reactivity upon exposure.

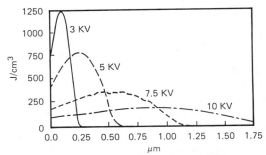

Figure 3. Calculation of the energy distribution within a resist film, at several accelerating potentials. The resist-air interface is located at 0.0 μm.

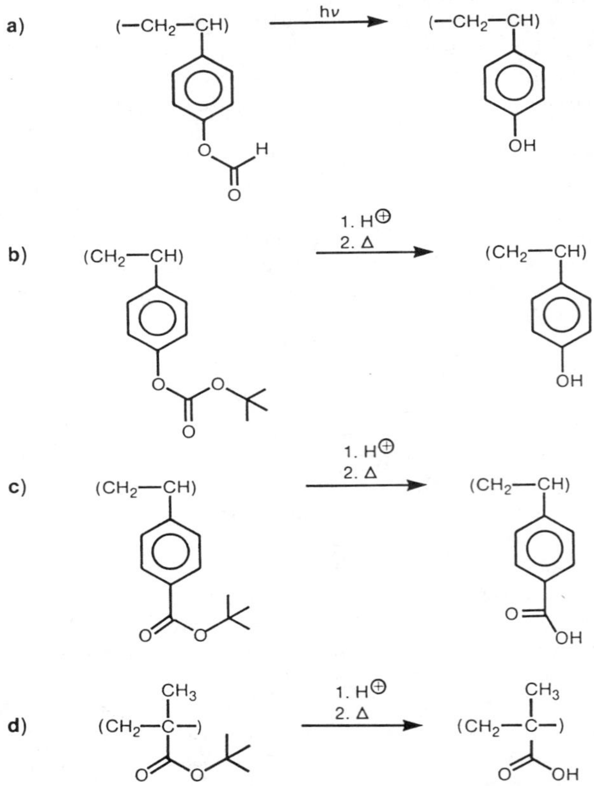

a) (—CH₂—CH) $\xrightarrow{h\nu}$ (—CH₂—CH)

b) (CH₂—CH) $\xrightarrow{\begin{array}{c}1.\ H^\oplus\\2.\ \Delta\end{array}}$ (CH₂—CH)

c) (CH₂—CH) $\xrightarrow{\begin{array}{c}1.\ H^\oplus\\2.\ \Delta\end{array}}$ (CH₂—CH)

d) (CH₂—C—) $\xrightarrow{\begin{array}{c}1.\ H^\oplus\\2.\ \Delta\end{array}}$ (CH₂—C—)

Figure 4. Several resists that undergo a change in chemical reactivity upon radiolysis.

This difference in chemical reactivity can be converted into a differential etch rate by treating the exposed film with an appropriate silylating reagent. For example, chlorotrimethylsilane, hexamethyldisilazane and bis(trimethylsilyl)acetamide are well know to react with phenolic hydroxyls, carboxylic acids and other nucleophilic species to form the corresponding trimethylsilyl derivatives.

Correspondingly, during the subsequent silylation step, silicon was also incorporated down to the substrate. However, several groups have shown that oxygen plasma only converts the top surface of an organosilicon polymer to the corresponding oxide, which then functions as the etch barrier (7,8). This observation implies that it is only necessary to add silicon to the top surface of a resist to obtain oxygen RIE resistance. In this study, a 1.0μ thick film was exposed at 3 KV such that reactive functionality was only generated near the resist-air interface. As a result, silicon is only incorporated into this thin layer when the film is treated with the silylating agent. The final relief image is developed with oxygen RIE. Figure 5 shows a typical example of this process. In this case a 1.0μ thick coating of resist was exposed at $1.25\ \mu C/cm^2$ at 3 KV (main column at 15 KV, wafer bias of 12 KV), silylated for 10 minutes, and developed with 10 minutes of oxygen RIE (50 mTorr, 40 SCCM oxygen, 150W RF, and -270 V bias). In this example, the final film thickness was found to be 0.96μ.

The scanning electron micrograph in Figure 5 shows several interesting features. This micrograph is of a 1.0μ line-space pattern, and the resist line at the end of the array is the same size as a resist line within the array. This observation is consistent with reduced proximity effects due to limited penetration of the incident electrons. While the wall profile of this image is reasonable, it could be improved by altering the RIE conditions. In any dry develop scheme where the etch barrier is only incorporated into the surface, the specific RIE parameters will determine the wall angle.

Spectroscopic Evidence of Limited Penetration. The calculated curves in Figure 3 predict that at 3 KV, all of the beam energy should be deposited into the top 0.25μ of the resist, while at 10 KV energy is deposited in a reasonably uniform fashion through a 1.5μ thick film. We have examined this proposal experimentally by overcoating a 1.0μ resist film with a 0.4μ film of carbowax, and exposing this 2-layer system at both accelerating potentials. Carbowax was chosen as the top layer because it is an organic polymer that does not contain carbonyl groups and can be spin coated onto the resist layer. The calculations in Figure 3 predict that exposing this 2-layer system at 3 KV should produce no chemical changes in the resist layer as all of the electrons should be contained within the 0.4μ carbowax layer. On the other hand, exposure at 10 KV should result in significant exposure of the resist layer. In this experiment, the extent of exposure was determined by following the loss of carbonyl by IR; results are shown in Figure 6. This figure shows that when the 2-layer system is exposed with $1.25\ \mu C/cm^2$ at 10 KV, there is a significant reduction in carbonyl intensity. However, when the same 2-layer system is exposed with $1.25\ \mu C/cm^2$ at 3 KV, no radiation chemistry occurs within the resist film. This experiment clearly demonstrates that at 3 KV net landing energy, the electron penetration depth must be less that 0.4μ.

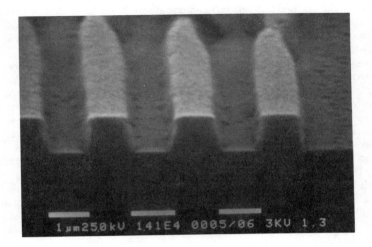

Figure 5. Scanning electron micrograph of relief image generated by the process described in this paper. The resist line at the right of the photograph is the last image in a long line–space array. There are several more resist lines to the left of the photograph.

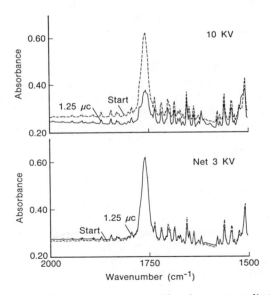

Figure 6. IR spectra of 2-layer test structure that demonstrate limited penetration of 3 KV e-beam.

Acknowledgments

The authors thank M. G. Rosenfield for generating the data in Figure 3.

Literature Cited

1. Lin, B. J. In Introduction to Microlithography: Theory, Materials and Processing; Thompson, L. F.; Willson, C. G.; Bowden, M. J., Eds.; ACS Symposium Series No. 219, American Chemical Society: Washington DC, 1983; pp. 287-350.
2. Jackel, L. D.; Howard, R. E.; Mankiewich, P. M.; Craighead, H. G.; Epthworth, R. Appl. Phys. Lett. 1984, 45, 698.
3. Yau, Y. W.; Pease, R. F.; Iranmanesh, A. A.; Polasko, K. J. J. Vac. Sci. Technol. 1981, 19, 1048.
4. MacDonald, S. A.; Ito, H.; Hiraoka, H.; Willson, C. G. "Photopolymers: Principles, Processes, and Materials," in Proceedings of the Society of Plastic Engineers, 1985, p. 177.
5. Taylor, G. N.; Stillwagon, L. E.; Venkatesan, T. J. Electrochem. Soc. 1984, 131, 1658.
6. Venkatesan, T.; Taylor, G. N.; Wagner, W.; Wilkens, B.; Barr, D. J. Vac. Sci. Technol. 1981, 19, 1379.
7. MacDonald, S. A.; Ito, H.; Willson, C. G. Microelectronic Eng. 1983, 1, 269.
8. Chou, N. J.; Tang, C. H.; Paraszczak, J.; Babich, E. App. Phys. Lett. 1984, 46, 31.

RECEIVED April 8, 1987

Chapter 30

Oxygen Ion Etching Resistance
of Organosilicon Polymers

H. Gokan, Y. Saotome, K. Saigo, F. Watanabe, and Y. Ohnishi

Microelectronics Research Laboratories and Fundamental Research Laboratories, NEC Corporation, 4-1-1, Miyazaki, Miyamae-ku, Kawasaki 213, Japan

Etch resistance for 12 organosilicon polymers has been studied under oxygen ion-beam etching (O_2-IBE) and oxygen reactive ion etching (O_2-RIE) conditions. Under O_2-IBE conditions, the etching rate for organosilicon polymers is found to be proportional to N/N_{Si}, where N and N_{Si} denote the number of total atoms in a monomer unit and the number of silicon atoms in a monomer unit, respectively. This means that the etching rate is determined only by the silicon content in the polymer. Under O_2-RIE conditions, the etching rate depends not only on the silicon content, but also on the polymer structure. Etched surface observation shows that polymers, having N/N_{Si} values over 40, are sensitive to temperature during etching. Low T_g polymers show higher etch resistance than high T_g polymers, even if the silicon contents are the same. The result suggests that the surface mobility for the silicon atoms during O_2-RIE is also an important factor for forming an excellent SiO_2 barrier over the polymer surfaces.

Organosilicon polymers are potentially useful in two level resist processes. Since these polymers generate highly resistant SiO_2 films on the polymer surfaces in O_2-RIE (reactive ion etching), they are utilized as a top imaging layer to obtain steep profiles in the bottom planarizing layer for device fabrication processes. The effect of silicon on oxygen plasma resistance were first reported by Taylor and Wolf ([1]). The use of organosilicon polymers in two level resist processes was initiated by Hatzakis et al. ([2]), who employed commercially available polysiloxanes as an electron or deep UV sensitive imaging layer. To meet with the requirement for use in practical device fabrication processes, various kinds of

0097–6156/87/0346–0358$06.00/0

organosilicon polymers have been specially designed and evaluated by many researchers (3 - 11). Suzuki et al. stated that polymers containing the order of 1×10^{16} silicon atoms/cm^2 practically stop the etching in O$_2$-RIE (3). Chou et al. investigated the etched organosilicon polymer surfaces by Auger and XPS analyses and indicated that an about 10Å thick oxide layer is formed on the polymer surface and that the layer proceeds the etching front in a steady state manner (7). Babich et al. compared O$_2$-plasma etch rates of various organosilicon polymers (12) and Paraszczack et al. reported etch rates of organosilicon polymers in various plasmas (13).

It is generally accepted that the increase in silicon content in the polymer units improves the etching resistance in O$_2$-RIE. However, since the O$_2$-RIE resistance for polymers strongly depends on etching conditions such as bias voltage, pressure, excitation frequency and flow rate, it is not easy to quantitatively compare the etch resistance for various polymers from reported data.

This paper discusses the etching resistance for various organosilicon polymers under O$_2$-IBE (ion-beam etching) and O$_2$-RIE conditions. Under O$_2$-IBE, ion bombardment and ion-assisted chemical reaction predominate and the etching due to radical species is negligibly small. The results will provide a fundamental understanding of the etching due to energetic species. Under O$_2$-RIE, although the effect of ion bombardment and the radical species cannot be separated, in comparison with O$_2$-IBE, the results will provide practical guideline for the organosilicon polymer etch resistance.

Experiments

The polymers evaluated are shown in Fig. 1. These twelve polymers cover typical types of organosilicon polymers, such as polysiloxane, ladder type polysiloxane, polysilane, polymethacrylate, polysilylstyrene, and novolak. Polymers 1 and 2 were supplied by Shin-etsu Chemical Co., Ltd. Actually, polymer 2 evaluated contains very small amount of vinylsiloxane for crosslinking by deep UV irradiation after coating, because the pure polysiloxane is a fluid at room temperature and is impossible to measure the etching depth. These polymers have Si-O bonds in the main chain structure. Polymer 3 was purchased from Shin Nisso Kako Co., Ltd. and was purified in house. Polymers 4 to 12 were synthesized in house. Polymers 11 and 12 have Si-O bonds in side chain structure. Thermal SiO$_2$ 13 was evaluated as a reference.

For the polymer evaluations, O$_2$-IBE and O$_2$-RIE systems were used. The O$_2$-IBE system is comprised of a 3 inch diameter Kaufman ion-gun and an exhausting chamber installing a water cooled rotary stage. The temperature rise in the samples during etching was supressed to below 40°C, using a heat sink material between the stage and the back of the sample substrates. Etching was performed at 2.7×10^{-2} Pa oxygen pressure. The O$_2$-RIE system comprises 280 mmφ parallel plate electrodes excited at 13.56 MHz. A Teflon insulator ring between the cathode electrode and the ground chamber walls was completely covered by a 318 mmφ fused quartz table in order to prevent fluorine release during etching. Since the

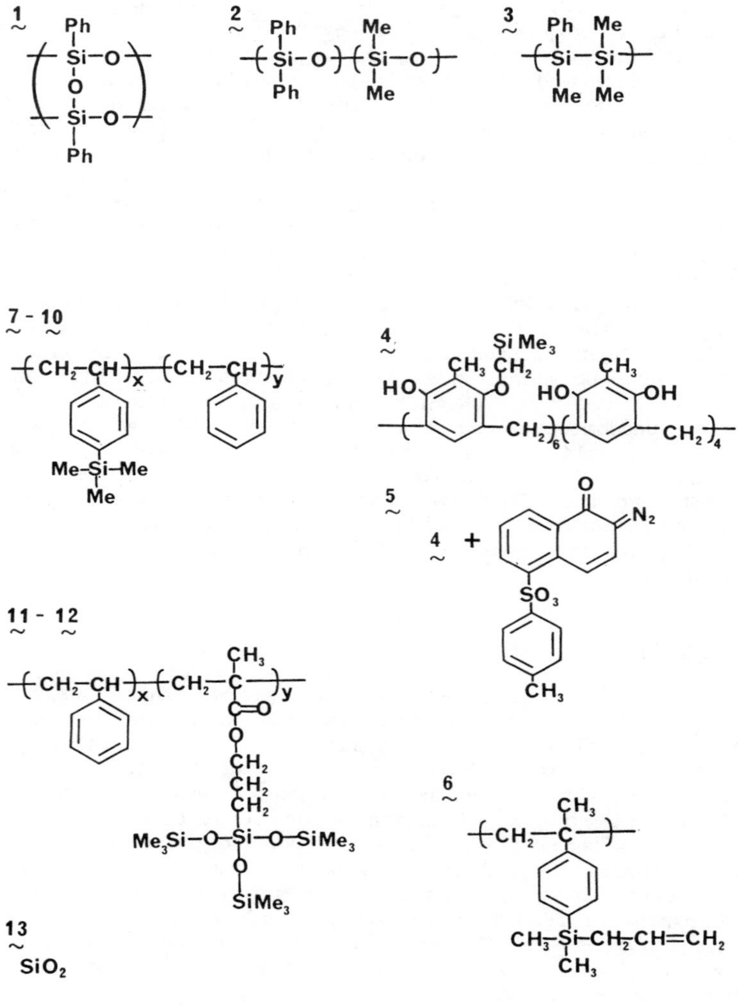

Figure 1.   Organosilicon polymers examined in this study.

organosilicon polymers etching was significantly enhanced by the fluorine release, a chamber cleaning process by O_2 discharge was employed before etching. This cleaning process is necessary, when fluorine-containing gas, such as CF_4, is introduced into the chamber before the run. Unless otherwise stated, etching was performed at 50 W (0.08 W/cm^2) power, 350 V induced self-bias potential, 5 sccm flow rate and 1.6 Pa chamber pressure. Under this condition, anisotropic etching is realized in the present etching system. Etching rates were calculated by measuring the thickness change, before and after the etching, using a Talystep.

Results and Discussion

Etch rate of metal-free polymers was reported to be inversely proportional to the carbon atom content in the polymer (14). For organosilicon polymers, role of silicon atoms under O_2-IBE is discussed as an analogy of that for the carbon atoms in carbonacious polymers. The number of silicon atoms in a unit volume is given by

$$\rho \cdot A \cdot N_{Si}/M = A(\rho/\overline{M}) \cdot (N_{Si}/N),$$

where ρ is the density of the polymer, A, Avogadro's number, M, the molecular weight in a monomer unit, $\overline{M}$, the average molecular weight in a monomer unit, N_{Si}, the number of silicon atoms in a monomer unit and N, the total number of atoms in a monomer unit. Since the $\overline{M}/\rho$ is nearly constant, etching rate is proportional to the factor N/N_{Si}. Figure 2 shows etching rate for organosilicon polymers versus N/N_{Si} under O_2-IBE. The results show that the rate determining step for organosilicon polymers under O_2-IBE is the sputtering of silicon atoms in the polymers. Since the sputtering yield of silicon decreases with ion energy, higher etching rate ratio between metal-free polymers and the organosilicon polymers is realized with decreasing ion energy (11). Under low ion energy condition, the chemical structure may be reflected on the etching resistance. However, no significant difference in chemical structure dependence is observed down to 100 eV. This may be because the 100 eV is still much greater energy than the chemical bond energy.

Under O_2-RIE, contrary to the O_2-IBE, etching depth does not linearly increase with etching time, as is typically shown in Fig. 3. Etching begins with a relatively high rate and the rate slows down towards the end with a steady state. This is mainly because, at an initial stage, atoms other than silicon are consumed by oxygen radical species, until the oxidized silicon protection layer is formed on the polymer surface. The phenomenon was analytically explained by Watanabe and Ohnishi in pertinent literature (15). Figure 4 shows O_2-RIE rate versus N/N_{Si}. The etching rate for some polymers is shown as bars showing a range, because the rate gradually varies with etching time and cannot be defined exactly. Polymers 3, 11 and 12, which degrade by main chain scission, showed a crumpled surface during etching and the etching rates could not be defined. Polymer surfaces for 4 and 10 were, more or less, roughened during etching, suggesting that the protective SiO_2 layer was not formed uniformly on the polymer surfaces. In the Figure,

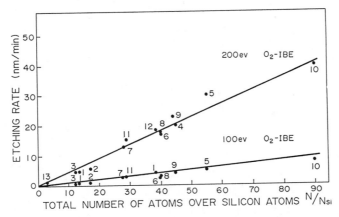

Figure 2.   Etching rate versus N/N$_{Si}$ under O$_2$-IBE.   Etching was
performed at 2.7×10^{-2} Pa for 100 eV and 200 eV beam
energies.

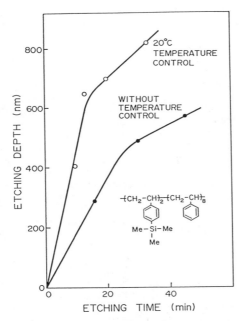

Figure 3.   Etching depth versus etching time typically observed
for organosilicon polymes under O$_2$-RIE.

polymers accompanied by the surface roughening are marked with "$\sim\!\!\sim\!\!\sim$". Although polymer 5 contains less Si atoms than does polymer 4, no surface roughening on polymer 5 was observed. This fact can be attributed to enhancement of dense SiO_2 film formation on the surface of polymer 5, because softning point of polymer 5 is lower than that of polymer 4.

Major differences between O_2-IBE and O_2-RIE are considered to be the degree of radical species contribution and the temperature difference during etching.   In order to clarify the effect of temperature increase, care was taken to suppress temperature rise during O_2-RIE.   The cathode electrode was cooled by circulating temperature controlled ethanol inside it and heat sinks were used. Figure 5 shows the rate versus N/N_{Si}, measured under 20°C temperature control.   Etching resistance for polymers 3, 11 and 12 was greatly improved. This is because these polymers are not stable at high temperature. For the other polymers, the etching rates were increased under the low temperature condition. The unusual etching rate dependence on temperature for organosilicon polymers may be because the reactivity between Si and O decreases due to low surface temperature, or, more likely, the mobility of Si may have substantially decreased.   Etched surface observation reveals that polymers having N/N_{Si} values over 40 are sensitive to temperature rise.

To clarify the origin of the rough surface formation, polymer 5, which shows relatively low T_g value, was further investigated. Polymer 5 was coated on the cured polyimide (PI-2555, DuPont) 1 μm thick and was subjected to the etching process.   Figure 6 to 8 show SEM photographs of the polymer surface after the etching process. For each etching process, the exposure time was adjusted to the period required to etch 1.5 μm thick cured polyimide. Figure 6 shows surfaces under different O_2-RIE power and also with and without temperature control. Apparently, surfaces become porous under low temperature conditions.   To check the surface SiO_2 quality, samples were subjected to a slight BHF (buffered hydrofluoric acid) treatment as are shown in Fig. 7. Figure 8 shows surfaces after O_2-IBE, and after O_2-RIE (50 W, without temperature control) with and without deep UV treatment. It is noted that, under O_2-RIE, deep UV treatment before etching and supressing the temperature rise during etching cause a similar effect.   Deep UV exposure makes the polymer surface crosslinked and reduces the polymer flow during etching. This leads to reducing substantial silicon mobility on the polymer surface and retards the reaction between Si and O, which makes the SiO_2 layer porous.

Conclusion

Oxygen ion-etching resistance of organosilicon polymers has been investigated under O_2-IBE and O_2-RIE conditions. The rate determining step for etching under ion bombardment is the sputtering of silicon atoms in the polymers. Polymer flow and/or surface temperature rise enhances non porous SiO_2 film formation under O_2-RIE, as long as the polymers are thermally stable.

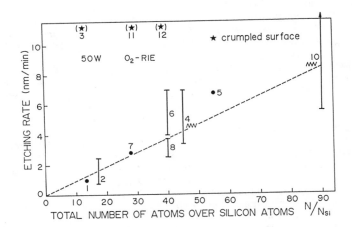

Figure 4.   Etching rate versus N/NSi under O2-RIE.   Etching was
performed under 50 W power, 5 sccm flow rate and 1.6
Pa pressure conditions without temperature control.

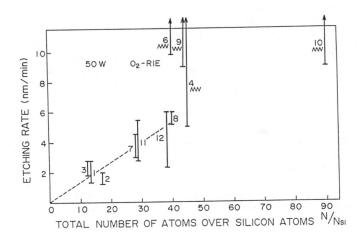

Figure 5.   Etching rate versus N/NSi under O2-RIE.   Etching
condition was the same as that in Fig. 4, except that
the temperature was controlled to 20°C.

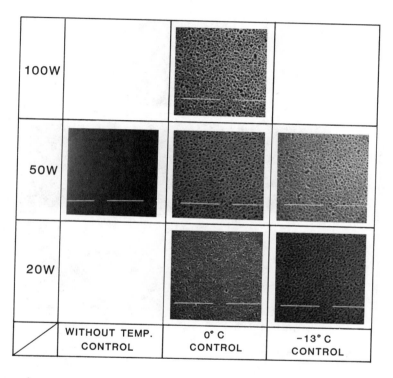

Figure 6.   SEM observation for the etched organosilicon polymers after O_2-RIE.

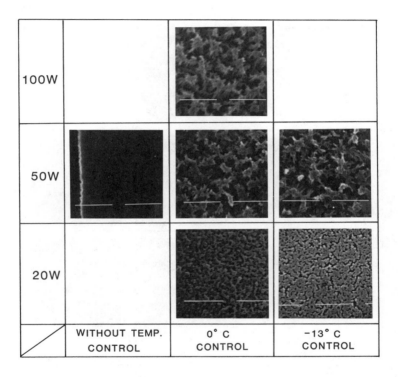

Figure 7.   SEM photos of surfaces corresponding samples to Fig.
             6, after BHF treatment.

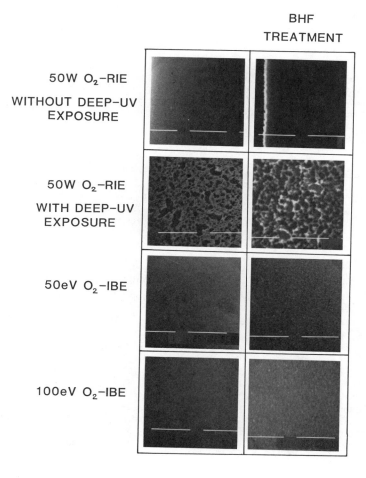

Figure 8.   SEM photos of surfaces after O_2-IBE, O_2-RIE without temperature control, O_2-RIE without temperature control but with deep UV treatment, and corresponding samples after BHF treatment.

Acknowledgments

The authors would like to thank K. Matsumi, T. Matsubara, K.
Yoshimi, S. Esho and H. Makino for their encouragement. They also
thank K. Tanigaki for discussion and M. Mukainaru for technical
assistance.

Literature Cited

1. G. N. Taylor and T. M. Wolf, Proc. Technical Conf.,
    Photopolymers Principles-Processes and Materials, p.175,
    Ellenville, N.Y. 1979.
2. M. Hatzakis, J. Paraszczak and J. Shaw, Proc. Microelectronics
    Engineering (Lausanne), p.386, 1981.
3. M. Suzuki, K. Saigo, H. Gokan and Y. Ohnishi, J. Electrochem.
    Soc., vol. 130, p.1962, 1983.
4. K. Morita, A. Tanaka, S. Imamura, T. Tamamura and O. Kogure,
    Japan. J. Appl. Phys., vol. 22, p. L659, 1983.
5. D. C. Hofer, R. D. Miller and C. G. Willson, SPIE vol. 469,
    Advances in Resist Technology, p. 16, 1984.
6. Y. Ohnishi, M. Suzuki, K. Saigo, Y. Saotome and H. Gokan, Proc.
    SPIE vol. 539, Advances in Resist Technology and Processing II,
    p. 62, 1985.
7. N. J. Chou, C. H. Tang, J. Paraszczak and E. Babich, Appl.
    Phys. Lett., vol. 46, p. 31, 1985.
8. M. A. Hartney, A. E. Novembre and F. B. Bates, J. Vac. Sci.
    Technol., vol. B3, p. 1346, 1985
9. E. Reichmanis and G. Smolinsky, J. Electrochem. Soc., vol. 132,
    p. 1178, 1985.
10. A. Tanaka, M. Morita and K. Onose, Japan. J. Appl. Phys., vol.
    24, p. L112, 1985.
11. Y. Saotome, H. Gokan, K. Saigo, M. Suzuki and Y. Ohnishi, J.
    Electrochem. Soc., vol. 132, p. 909, 1985.
12. E. Babich, J. Paraszczak, M. Hatzakis and J. Shaw,
    Microelectronic Engineering, vol. 3, p.279, 1985.
13. J. Paraszczak, E. Babich, M. Hatzakis and J. Shaw, J. Vac. Sci.
    Technol., vol. B3, p.358, 1985.
14. H. Gokan, K. Tanigaki and Y. Ohnishi, Solid State Technology,
    vol. 28, p. 163, 1985.
15. F. Watanabe and Y. Ohnishi, J. Vac. Sci. Technol., vol. B4, p.
    422, 1986.

RECEIVED April 8, 1987

POLYMERS IN PHOTONIC APPLICATIONS
AND DEVELOPMENTS

POLYMERS IN PHOTONIC APPLICATIONS AND DEVELOPMENTS

The use of light to accomplish many of the functions conventionally performed by electronics is not only the focus of much current research and development, but is also a growing reality in the world outside the laboratory. The potential exists for photonic applications involving direct replacement of conductive circuitry, as well as many new technologies which offer increased capabilities, convenience and savings. Optical-fiber transmission and optical-disk data recording/storage, for example, represent technologies that have already been commercialized. Specialized optical signal processors and switching units are available in small quantities, with greater production and utilization being visible on the horizon. Optical bistability, optical computing and active optical circuitry are not only "hot" research topics, but are also the object of several research consortia and cooperatives around the world.

We are currently in the midst of a transition from electronics to photonics, particularly in technologies related to communication and information handling. The reasons for this evolution are varied and include high data storage density associated with optical technology, low transmission losses, high transmission or processing bandwidths, immunity from EMI (electromagnetic interference), immunity from eavesdropping, limited crosstalk, potential for massive parallel processing capabilities and low cabling mass.

Organic materials (including polymers) offer a wide variety of properties that may make them the materials of choice for many of these applications. Obviously the diversity of properties, structures, and compositions of organic materials coupled with the diversity and wide flexibility of fabrication techniques presently available has stimulated their utilization in photonic processes and devices. As described in the following papers, protective coatings/barriers are essential for silica optical fibers. Materials with special properties can minimize thermal difficulties associated with fiber-optic cables for long-haul transmission which requires extremely low attenuation constants. Cabling and connections rely heavily on structural polymers while polymeric optical fibers are useful for local communication loops and links. In optical disk technology, polymers not only find application as inexpensive substrates, but are also used as passive layers (e.g., antiscratch, defocusing overcoats) and, feasibly, as recording layers. The latter two applications stem from the ability to accurately and reproducibly deposit organic and polymeric films. Beyond this, organics have the potential to function as active media in active (i.e., nonlinear) photonic components.

Organics have been used as active optical components, at least in the laboratory. Examples include dye lasers, saturable absorbers for passive Q-switching and mode-locking, polymeric films containing the dyes and saturable absorbers, photochromics and thermal media for switching, etc., and liquid crystal devices. There is considerable interest in their potential for the extremely fast nonlinear optical processes needed for a variety of photonic applications such as very high-rate electro-optical modulation and all-optical switching. As described in the following papers, electron delocalization can greatly enhance microscopic nonlinear polarizability. An understanding of these effects and the means to fabricate or molecularly engineer useful structures are still fairly limited, but are actively under study in a growing number of industrial and academic laboratories. Research activities reported in this section include a study of the nonlinear properties of conjugated polymers and their fabrication into films or platelets, nonlinear properties of dye-in-polymer orientational-electret films and inclusion complexation/crystallization of nonlinearly polarizable molecules. The latter two papers address the frustrating problem of how to prevent the orientational cancellation of the second-order nonlinear polarizability.

Gerald R. Meredith
Central Research and Development Department
E. I. duPont de Nemours & Co., Inc.
Experimental Station - 356
Wilmington, DE 19898

Chapter 31

Nonlinear Excitations and Nonlinear Phenomena in Conductive Polymers

A. J. Heeger, D. Moses, and M. Sinclair

Department of Physics and Institute for Polymers and Organic Solids,
University of California, Santa Barbara, CA 93106

Semiconductor polymers such as polyacetylene and polythiophene have experimentally demonstrated nonlinear optical processes with characteristic time scales in the sub-picosecond range. Fast transient photoconductivity measurements on trans-$(CH)_x$ as a function of temperature and photon energy indicate a relatively high quantum efficiency for the photoproduction of mobile, charged, nonlinear excitations, consistent with the Su-Schrieffer mechanism for the photogeneration of charged solitons. The major shifts in oscillator strength due to these nonlinear photoexcitations lead to relatively large resonant third-order nonlinear optical processes ($\chi^{(3)}$) on time scales of order 10^{-13} s. A direct measurement of $\chi^{(3)}$ in polyacetylene has been carried out by third harmonic generation. The measured nonresonant value of $\chi^{(3)}(3\omega = \omega+\omega+\omega) = 5\times10^{-10}$ esu. The implied value for $\chi_{\shortparallel}^{(3)}$ is comparable to the corresponding value for polydiacetylene.

I. PHOTOEXCITATION: PHOTO-INDUCED ABSORPTION, PHOTO-INDUCED BLEACHING AND PHOTOCONDUCTIVITY

Photo-excitation studies of conjugated semiconductor polymers were stimulated by the calculations of Su and Schrieffer [1] which demonstrated that in trans-$(CH)_x$ an e-h pair should evolve into a pair of solitons within an optical phonon period or about 10^{-13} sec. Thus, the absorption spectrum was predicted to shift from $\hbar\omega_l$ to $\hbar\omega_S$ (see Figure 1) on a time scale of 10^{-13} seconds after photoexcitation.

The photo-generation of soliton-antisoliton pairs implies formation of states at mid-gap. Time resolved spectroscopy [2] has been used to observe the predicted absorption due to photo-generated intrinsic gap states in trans-$(CH)_x$. Moreover, the time scale for photo-generation of these gap states has been

0097–6156/87/0346–0372$06.00/0

investigated (3,4). Using sub-picosecond resolution, these studies demonstrated that the gap states and the associated interband bleaching are produced in less than 10^{-13} seconds, consistent with the theoretical predictions.

Vardeny et al (5) and Blanchet et al (6) have observed the photoinduced absorption arising from both the mid-gap electronic transition and the associated infrared active (IRAV) modes introduced by the local lattice distortion. Infrared spectroscopy of lightly doped trans-$(CH)_x$ has demonstrated that the same spectroscopic features arise upon doping (7). Moreover, these doping-induced absorptions are independent of the dopant and are therefore identified as intrinsic features of the doped trans-$(CH)_x$ chain (7). These important results demonstrate that both the photo-induced spectroscopic features and those induced by doping are associated with the same charged state. The observed frequencies and line shapes are consistent with those expected for charged soliton excitations (8). Moreover, these excitations have the reversed spin-charge relation (9) predicted for solitons (10). Thus, the photoinduced IRAV modes can be used as a signature of soliton formation.

Recent measurements of fast transient photoconductivity (11) in trans-$(CH)_x$ have demonstrated that the photogenerated solitons are mobile and contribute to the electrical conductivity. Figure 2 shows the transient photoconductivity following a 1 μJ pulse at 2.1 eV with a bias voltage of 300 V. The charge carriers are produced within picoseconds of optical excitation. The fast rise is followed by (approximately exponential) decay with a time constant of ~ 300 ps. The magnitude and time decay of $\sigma_{ph}(t)$ are temperature independent (for $t < 10^{-9}$ s) from 10 K to 300 K.

The photoinduced change in conductivity is large (11). For an absorbed photon flux of 10^{15} cm^{-2} per pulse, the photocurrent (at ~50 ps) is $I_p = 4 \times 10^{-4}$ A (i.e., current density ~ 10^3 A/cm^2) with good reproducibility from sample to sample. This corresponds to a conductivity of about 0.3 S/cm with a possible error of a factor of two due to uncertainty in αd in the top-illumination measurement geometry. The increase in σ_{ph} over the dark room temperature value is five to six orders of magnitude!

The relationship between the photocurrent and the incident photon flux is given by

$$\sigma_{ph} = (E/\hbar\omega)e\eta\phi\mu \qquad (1)$$

where $(E/\hbar\omega)$ is the number of absorbed photons per unit volume, η is the quantum efficiency, ϕ is the probability to escape geminate (or early time) recombination and μ is the mobility. The picosecond (3,4) and sub-picosecond (12) $\delta\alpha(t)$ data indicate that the number of photoexcitations has already decayed to less than 10^{-2} of the initial value at 50 ps. Thus, assuming $\eta \sim 1$, the carrier yield at 50 ps is $\phi \sim$ 0.01. Thus using $\eta = 1$ and σ_{ph} (50 ps) ~ 0.3 S/cm results in a mobility of approximately 1 cm^2/V-s. With this value, the net distance drifted in

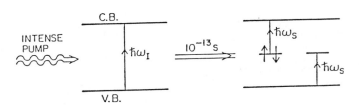

Fig. 1.   A photo-pump makes e-h pairs which evolve in 10^{-13} seconds
to soliton pairs with states at mid-gap. The oscillator strength
shifts accordingly.

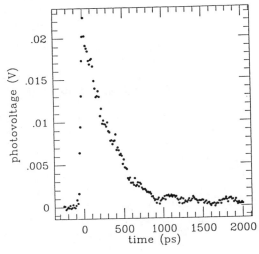

Fig. 2. Transient photovoltage across 50 Ω of <u>trans</u>-$(CH)_x$ in the
Auston switch configuration.

the measured decay time (300 ps) is about 400 Å, in good agreement with that inferred from the picosecond decay of photoinduced dichroism (4).

The similar excitation profiles for the photoconductivity (11) and charged soliton photogeneration (13), the conclusion that the initial quantum efficiency for photogeneration of charged photoexcitations is relatively high (11) and the observation of photoinduced bleaching (implying nonlinear excitations) on the same time scale imply that the photocurrent is carried by mobile charged solitons.

Although the branching ratio of charged to neutral excitations has not been measured, the close agreement between the spectral onset of absorption and high η photoconductivity is traditionally interpreted as ruling out the generation of neutral excitons as the primary excitations. We suggest, therefore, that the photoinduced absorption at 1.4 eV due to neutral excitations (2,5) is generated as a secondary process during the rapid initial recombination of the photoinduced charged excitations.

II.   MEASUREMENT OF THE THIRD ORDER SUSCEPTIBILITY OF TRANS-(CH)$_x$ BY THIRD HARMONIC GENERATION (14)

The existence of these fast nonlinear processes is not only important as confirmation of the proposed mechanism for the photogeneration of charged solitions but also establishes this class of conjugated polymers as extremely fast nonlinear optical materials with relatively large third-order susceptibilities (15). These potentially important nonlinear optical properties arise directly from the shifts in oscillator strength which result from the novel nonlinear photoexcitations (solitons in the case of a degenerate ground state and polarons or bipolarons when the ground state degeneracy is lifted (16)).

The demonstration of third harmonic generation using thin films of polyacetylene, (CH)$_x$, as the nonlinear optical medium was recently reported (14,17,18). The third harmonic was generated by focusing the pulse train of a mode locked Nd:YAG laser (λ = 1.06 μm) on the sample. The transverse mode structure of the fundamental beam (TEM$_{00}$) and the mode-locked pulse width (100 ps FWHM) were carefully characterized, and the input power was measured with a calibrated photodiode. A half wave plate and a polarizing cube were used as a variable attenuator so that the intensity dependence of the third harmonic could be measured. With a spot size of 15 μ, peak intensities of 1-100 MW/cm^2 could be achieved. In order to reduce the risk of sample damage due to heating, the laser was chopped with a duty cycle of 1.2×10^{-2}. The generated third harmonic was collimated with a lens, spectrally separated from the fundamental and detected with a photomultiplier tube. The responsivity of the photomultiplier at 355nm was carefully determined by direct comparison to a calibrated photodiode. Several samples were measured with good reproducibility from sample to sample.

Figure 3 is a log-log plot (14) of the third harmonic intensity (in W/cm^2) as a function of fundamental intensity (in

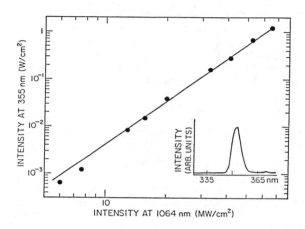

Fig. 3. Log-log plot of the third harmonic intensity (W/cm^2) as a function of the fundamental intensity (MW/cm^2). The solid line indicates a cubic dependence of the third harmonic intensity on the fundamental intensity. The inset shows that the spectral content of the third harmonic is resolution limited at 355 nm; Ref. 14.

MW/cm^2). The solid line is a best fit to the data assuming a cubic dependence of the third harmonic intensity on the fundamental intensity. The constant of proportionality is 4.3 x 10^{-6}. The inset shows the spectral content of the third harmonic intensity showing a resolution limited peak at 355 nm. Using Q-switched, modelocked pulses, we have extended the measurement to peak pump powers in excess of 10 GW/cm^2 without damage to the sample.

In the presence of strong absorption at ω and 3ω, the intensity of the third harmonic can be written in terms of the third order polarization as (19):

$$I(3\omega) = \frac{K|P^{(3)}(3\omega)|^2}{[(\Delta k)^2 + \frac{(\alpha_3 - 3\alpha_1)^2}{4}]} \tag{2}$$

with

$$K = \frac{2\pi(3\omega)^2}{c\sqrt{\varepsilon_3}} [\exp(-3\alpha_1 L) + \exp(-\alpha_3 L) - 2\cos(\Delta kL)\exp-(3\alpha_1 + \alpha_3)L/2]$$

and $\tag{3}$

$$|P^{(3)}(3\omega)|^2 = \left(\frac{2\pi}{c\sqrt{\varepsilon_3}}\right)^3 |\chi^{(3)}(3\omega)|^2 I^3(\omega) \tag{4}$$

where $P(3\omega) = \chi^{(3)}(3\omega = \omega+\omega+\omega)E(\omega)E(\omega)E(\omega)$, α_1 and α_3 are the absorption coefficient at ω and 3ω, Δk is the phase mismatch and ε_3 is the dielectric constant at 3ω. This equation is a plane wave result which is valid when the sample thickness (0.1 μ) is much less than the confocal beam parameter of the focused gaussian beam (300 μ). Since $\alpha_3 \gg \alpha_1$ and $\alpha_1 L < 1$, we can rewrite equation 2 in the simplified form

$$I(3\alpha) = \frac{2\pi(3\omega)^2}{c\sqrt{\varepsilon_3}} \frac{|P^{(3)}(3\omega)|^2}{[(\Delta k)^2 + 1/4(\alpha_3)^2]} \tag{5}$$

Using equations 3, 4 and 5, one can relate the experimentally measured intensity at 3ω to the third order susceptibility:

$$\chi^{(3)}(3\omega) = \frac{c^2}{12\pi^2\omega} \{\sqrt{\varepsilon_1^3\varepsilon_3} \, [(\Delta k)^2 + \frac{(\alpha_3)^2}{4}] \frac{I(3\omega)}{I^3(\omega)}\}^{1/2} \qquad (6)$$

Using $I(3\omega)/I^3(\omega) = 4.3 \times 10^{-38}$ (in esu), and the following ([20]) optical constants, $\alpha_3 = 1 \times 10^5$ cm^{-1}, $\Delta k = 3 \times 10^5$ cm^{-1}, $\varepsilon_1 = 4$ and $\varepsilon_3 = 14$, we obtain $\chi^{(3)} = 5 \times 10^{-10}$ esu. Although there are many possible sources of inaccuracy in the absolute determination of the susceptibility, we have taken care in all aspects of the measurement; we estimate that the result is accurate to within a factor of two. This value of $\chi^{(3)}$ compares favorably to the $\chi^{(3)}$ values measured in the polydiacetylenes ($\chi^{(3)} \sim 10^{-10}$ esu in PTS) ([21,22]) and in inorganic semiconductors ($\chi^{(3)} \sim 10^{-10}$ for Ge) ([21])

Since the optical properties of polyacetylene are highly anisotropic, the measured $\chi^{(3)}$ is related to $\chi_{\parallel}^{(3)}$ through the relation ([23]):

$$\chi^{(3)} = \chi_{\parallel}^{(3)} \, [<\cos^6\theta>]^{1/2} \qquad (7)$$

where the angle θ is the angle between a given $(CH)_x$ chain and the E field and the brackets denote an angular average. Assuming the polymer chains are oriented randomly in the plane of the thin film, $<\cos^6\theta> = 5/16$. The implied value of $\chi_{\parallel}^{(3)}$ is therefore comparable to the corresponding value for polydiacetylene.

III. CONCLUSION

Although the importance of the effect of the delocalized π-electron system on the nonlinear susceptibilities has long been realized, the role of the strong electron-phonon coupling in these systems has been largely overlooked. In resonant processes, the photoproduction of the nonlinear excitations with associated structural distortions which characterize these systems (i.e., solitons, polarons and bipolarons) is a direct consequence of the electron-phonon interaction ([3,11]). These nonlinear excitations are responsible for the large shifts of oscillator strength which have been observed with resonant pumping. This interaction can be expected to affect nonresonant processes as well. Direct photoproduction of solitons has been demonstrated in polyacetylene for incident photons in the energy range $4\Delta/\pi < \hbar\omega < 2\Delta = E_g$. This direct coupling to the nonlinear excitations leads to a nonresonant mechanism analogous to that proposed for excitons in the polydiacetylenes by Greene et al (24). The direct coupling to soliton excitations leads to the formation of a soliton-polariton so that for $\hbar\omega$ below the absorption threshold "photons" propagating in the polymer are part solitons. As a result, the oscillator strength shifts described above can be effective in virtual processes, giving rise to large nonresonant $\chi^{(3)}$.

In summary, semiconductor polymers such as polyacetylene and polythiophene have experimentally demonstrated nonlinear optical processes (photo-induced absorption, photo-induced bleaching and photo-luminescence) with characteristic time scales in the picosecond range or faster. These phenomena are intrinsic and originate from the instability of these conjugated polymers toward structural distortion.

ACKNOWLEDGMENT: Supported by the Office of Naval Research.

REFERENCES

1. Su, W. P.; Schrieffer, J.R. Proc. Nat. Acad. Sci. U.S.A. 1980, 77, 5626.
2. Orenstein, J.; Baker, G. Phys. Rev. Lett. 1982, 49, 1043.
3. Shank, C. V.; Yen, R.; Fork, R. L; Orenstein, J.; Baker, G. L. Phys. Rev. Lett. 1982, 49, 1660.
4. Vardeny, Z.; Strait, J.; Moses, D.; Chung, T.-C.; Heeger, A. J. Phys. Rev. Lett. 1982, 49, 1657.
5. Vardeny, Z.; Orenstein, J.; Baker, G. L. J. Phys. Colloq. 1983, 44 C3-325; Phys. Rev. Lett. 1983, 50, 2032.
6. Blanchet, G. B.; Fincher, C. R.; Chung, T.-C.; Heeger, A. J. Phys. Rev. Lett. 1983, 50, 1938.
7. a) Heeger, A. J. Polymer Journal, 1985, 17, 201 and references therein.
    b) Fincher, C. R.; Ozaki, M.; Tanaka, M.; Peebles, D.; Lauchlan, L.; Heeger, A. J.; MacDiarmid, A. G. Phys. Rev. B, 1979, 20, 1589.
    c) Etemad, S.; Heeger, A. J.; MacDiarmid, A.G. Ann. Rev. Chem. Phys. 1982, 33, 433.
8. Horovitz, B. Solid State Commun. 1982, 41, 729.
9. Flood, J. D.; Heeger, A. J. Phys. Rev B, 1983, 28, 2356; Moraes, F.; Park, Y.-W.; Heeger, A. J. Syn. Mtls. 1986, 13, 113.
10. Su, W. P.; Schrieffer, J. R.; Heeger, A. J. Phys. Rev. Lett. 1979, 42, 1698; Phys. Rev. B, 1980, 22, 2209.
11. Sinclair, M.; Moses, D.; Heeger, Solid State Commun. (in press).
12. Shank, C.V.; Yen, R.; Orenstein, J.; Baker, G. L. Phys. Rev. B, 1983, 28, 6095.
13. Blanchet, G. B.; Fincher, C. R.; Heeger, A. J. Phys. Rev. Lett. 1983, 51, 2132.
14. Sinclair, M.; Moses, D.; Heeger, A. J.; Vilhelmsson, K; Valk, B.; Salour, M. Solid State Commun. (in press).
15. Heeger, A. J.; Moses, D.; Sinclair, M. Syn. Mtls. 1986, 15, 95.
16. Heeger, A. J. Phil. Trans. R. Soc. London A, 1985, 314, 17.
17. Gookin, D. M.; Hicks, J. C. SPIE Advances in Materials for Active Optics, 1985, 567, 41.
18. Etemad, S.; this volume.
19. Thalhammer, M.; Penzkofer, A. Appl. Phys. B, 1983, 32, 137.

20.   Fujimoto, H.; Kamiya, K.; Tanaka, J.; Tanaka, M. Syn. Mtls. 1985, 10, 367.
21.   Flytzanis, C. In Nonlinear Optical Properties of Organic and Polymeric Materials, Williams, D. J., Ed.; American Chemical Society Symposium Series 233; American Chemical Society: Washington, DC, 1983; Ch. 8.
22.   Carter, G. M.; Thakur, M. K.; Chen, Y. J.; Hryniewica, J. V. Appl. Phys. Lett. 1985, 47, 457.
23.   Sauteret, C.; Hermann, J. P.; Frey, R.; Pradere, F.; Ducuing, J.; Baughman, R. H.; Chance, R. R. Phys. Rev. Lett. 1976, 36, 956.
24.   Greene, B. I.; Orenstein, J.; Millard, R. R.; Williams, L. R. (preprint).

RECEIVED May 13, 1987

Chapter 32

Dipolar Alignment for Second Harmonic Generation: Host–Guest Inclusion Compounds

David F. Eaton, Albert G. Anderson, Wilson Tam, and Ying Wang

Central Research and Development Department, Experimental Station,
E. I. du Pont de Nemours and Company, Wilmington, DE 19898

ABSTRACT. Solid, polycrystalline inclusion complexes between β-cyclodextrin, tris-*ortho*-thymotide, or thiourea hosts and a variety of polarizable organic and organometallic guests have been prepared. The inclusion complexes generate second harmonic light at 530 nm when irradiated with high power 1.06 μ light from a Nd-YAG laser. The relative SHG ability of the complexes has been measured relative to urea by powder techniques. The solid state structure of several of the complexes has been determined. The structure and activity of the materials is correlated using a simple two state model for the SHG process.

Nonlinear optics is the study of the interaction of electromagnetic fields with materials to produce new fields which are different from the input field in phase, frequency or modulation. Second harmonic generation (SHG) is a nonlinear optical process which results in the conversion of an input optical wave into an output wave of twice the input frequency. The process ocurrs within a nonlinear medium, usually a crystal such as potassium dihydrogen phosphate (KDP) or potassium titanyl phosphate (KTP). Chemists are familiar with SHG through its widespread use in laser chemistry to frequency double, e.g., a Nd-YAG laser, operating at 1.06 μ, to the green region (530 nm). However, such processes will undoubtedly see their major application in optical communications, laser medicine and in the emerging field of integrated optics. Recent activity in many laboratories has been directed toward understanding and enhancing second and third order nonlinear effects in inorganic, organic and polymeric materials. Several recent reviews attest to the high interest in this area.[1-5]

NOTE: This chapter is Contribution No. 4217 in a series.

Effective second harmonic generation is attained in materials which have both high second-order molecular hyperpolarizability, β, and high second-order bulk susceptability, $\chi^{(2)}$. Molecular polarization is described by the field dependent molecular dipole moment, μ (eq 1), expanded as a function of the applied field strength, **E**, which may be electric or optical (that is, electromagnetic) in nature. The field strength **E** is a vector, and μ_0 is the intrinsic dipole

$$\mu = \mu_0 + \alpha \cdot \mathbf{E} + \beta \cdot \mathbf{E} \cdot \mathbf{E} + \gamma \mathbf{E} \cdot \mathbf{E} \cdot \mathbf{E} + ... \quad (1)$$

moment of the material. Many organic materials are highly polarizable and are thus inherently good candidates for second harmonic generation. However, because the field strength is a vector, the equation for induced molecular polarization is not scalar, and the constants in (1) are tensors. Second harmonic generation is in fact measured on bulk samples which consist of ensembles of individual molecules. It is therefore necessary to consider the bulk polarization of the material, which again can be expanded with the applied field, as in eq. 2:

$$\mathbf{P} = \mathbf{P}_0 + \chi^{(1)} \cdot \mathbf{E} + \chi^{(2)} \cdot \mathbf{E} \cdot \mathbf{E} + \chi^{(3)} \cdot \mathbf{E} \cdot \mathbf{E} \cdot \mathbf{E} + ... \quad (2)$$

It can be shown[6] that the odd-order terms in (1) are orientationally independent, but that the even terms depend critically on the symmetry projections of the individual molecular components of the polarizability on the field orientation. One effect is that bulk materials for second harmonic generation must be noncentrosymmetric. That is they must be members of an acentric space group. An ancillary effect is that, while there is not a one-to-one mapping between the molecular second order term (β) and the bulk coefficient [$\chi^{(2)}$], the most advantageous relationship between β and $\chi^{(2)}$ can be shown [6] to exist for polar molecules which crystallize in point groups 1, 2, m or mm2. This is both an advantage and a disadvantage for chemists interested in optimum materials for second harmonic generation. The advantage is that it allows one to search crystal structure literature seeking molecules which are members of only several of many possible symmetry groups. This approach has been taken by Tweig[7] in a systematic search for suitable materials. The disadvantage of this approach is that it prevents the physical organic chemist from choosing the optimum molecule (highest molecular polarizability), because it may or may not crystallize in an appropriate space group. The technique we have developed is one of dipolar alignment, in which a general, polarizable guest, is inserted into an inclusion matrix host in a manner that ensures bulk acentric orientation. We have reported on our intitial work in this area previously.[8-10] That work involved cyclodextrin complexation of nitroaniline derivatives as a method of inducing SHG activity in organic materials with inherently high β, but vanishing $\chi^{(2)}$ beacause of the guests' centric crystal habits. A similar approach was reported by a Japanese group[11] independent of our own work.

In the present work we describe considerable progress towards the generalization of the principle of inclusion induced dipolar alignment for SHG. We have discovered several new classes of host matrices which can be effective, including cage-, channel- and lattice-forming materials. We have also extended the range of possible guests to include organometallics in addition to organic materials.

We first describe the experimental methods used in this work. Then the specific example of the inclusion of *p*-nitroaniline derivatives in β-cyclodextrin[8-10] will be briefly reviewed. Nitroanilines were chosen for our initial study because of their high inherent nonlinear polarizability[3] and analogy to known effective SHG active materials (2-methyl-4-nitroaniline, **MNA**[12]), and the fact that the *p*-nitroaniline crystal is centrosymmetric and therefore inactive for SHG itself. We will then expand the range of guests and hosts by explication of our results with inclusion complexes of the two hosts tris-*ortho*-thymotide (**TOT**) and thiourea with a variety of organic and organometallic guests.

EXPERIMENTAL SECTION

Method of Measuring SHG.

The ability of powdered complex to generate second harmonic radiation was demonstrated in the apparatus shown in Scheme 1. The technique is based on the method of Kurtz and Perry.[14] A Nd-YAG laser is directed through an

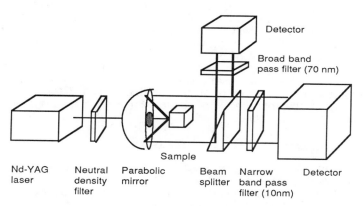

| Nd-YAG laser | Neutral density filter | Parabolic mirror | Beam splitter | Narrow band pass filter (10nm) | Detector |

Scheme 1. Apparatus for powder second harmonic generation measurements

optical neutral density filter to adjust the light intensity. The beam then passes through a hole in a parabolic mirror and illuminates a sample. Light emerging from the sample is collected by the mirror and passed through a beam splitter. One portion of the signal is passed through an optical narrow band-pass filter (FWHM 10 nm) which passes light at the second harmonic frequency only (532 nm). The second portion passes through another filter which is a broad band filter (FWHM 70 nm). Each portion of the split beam is detected with photomultipliers. The detected signals are compared electronically. Thus in each experiment, two channel detection is employed to discriminate against spurious signals (fluorescence or scattered light) which can be generated. Polycrystalline urea having an average particle size of 90-125 µm is used as a reference material.[3,15] No effort was made during this study to examine any of the materials for phase-matchability. Random, as prepared, distributions of crystallites were examined; no particle size classification was done. Errors in SHG signal intensities can in principle be quite large since grain size differences are not taken into account. In general, for these series of similar complexes, we suspect that those differences are not exceptional. We consider differences of SHG intensities of the order of 2-3x to be real and presume them to be related to molecular and material differences in the materials.

Preparation of Complexes.

In general, complexes were prepared by mixing a solution of the guest, in an appropriate solvent, with a solution of the host in an appropriate solvent, and allowing the complex to precipitate as crystals from the mixture. For cyclodextrin (**CD**) complexes, distilled water was used to dissolve the **CD**, and the guest was added in a water miscible, but incompatible solvent, such as ether, and slow mixing was used to distribute the complex-forming reagents (Method A). For thiourea, and **TOT** complexes, methanol was the solvent of choice. For thiourea complexes, about 200-500 mg of the guest was added to 15-20 ml of a 50% saturated solution of thiourea in methanol. The mixture was heated if necessary to dissolve the guest, and the warm solution filtered and then allowed to stand. Crystals formed on cooling to 0°C or near -20° C in extreme cases (Method B). Alternatively, thiourea and the guest in a ratio of near 3:1 molar were dissolved in methanol, followed by removal of solvent slowly under reduced pressure (Method C). For **TOT** complexes, similar procedures were followed, except that **TOT** (150 mg) was dissolved in 15 ml of warm methanol, filtered and cooled to room temperature before addition of a methanolic solution of the guest. The solution was filtered again, and crystals formed on slow evaporation of solvent at room temperature or on cooling (Method D). These methods gave polycrystalline powders that were suitable for SHG measurement. Specific examples are given below.

β-Cyclodextrin--p-Nitroaniline Complex (**CD-PNA**).

Method A was used. 2.0 g of β-**CD** (1.8 mmol) was dissolved in 100 ml of distilled water, and the solution was filtered. p-Nitroaniline (**PNA**), 0.25 g (1.8 mmol), was dissolved in 40 ml of diethyl ether, and added all at once to the aqueous **CD** solution. The mixture was stirred, open to the atmosphere to allow evaporation of the ether, overnight, and then filtered. The yellow precipitate was washed with ether and dried. Yield, 1.4 g, for which CH and N analysis indicated the inclusion of 4-6 moles of water per mole of complex. The actual degree of hydration varied from preparation to preparation. Melting occured with decomposition near 288-289 °C.

Thiourea--Benzenechromium Tricarbonyl.

The procedure of method B was followed using 15 ml of 50% thiourea solution and 250 mg of benzenechromium tricarbonyl. After two days at 0° C, yellow needles were isolated: mp 158-170 ° (decomp.); IR (KBr): 1948 (s), 1892 (m), 1979 (s), 1848 (w), 1632 (w), 1615 (m), 1490 (w) cm^{-1}; anal., calc. for 3 thiourea to one organometallic residue $C_{12}H_{18}CrN_6O_3S_3$: C, 32.57, H 4.10, Cr, 11.75; found, C, 32.72, H, 4.18, Cr, 11.29.

Tris-*ortho*-thymotide--*p*-Dimethylaminocinnamaldehyde(**TOT-PDMAC**).

According to method D, **TOT** (150 mg) was dissolved in 50 ml boiling methanol, and filtered hot into a solution of PDMAC (200 mg) in 10, ml of methanol. After 6 days at 0°, yellow rhomboid crystals, 90 mg, were isolated, mp 217-218 (decomp), which was found to be 2TOT:1**PDMAC** by x-ray analysis: calc. for $C_{77}H_{85}NO_{13}$: C, 75.03, H, 6.95; found, C, 74.81, H, 6.82.

X-ray Structural Analyses.

X-ray analyses were performed by Dr. J. Calabrese and Dr. I. Williams of this department.   Data were collected on either a Syntex R3 or an Enraf-Norius CAD4 diffractometer, using graphite-monochromated MoKα radiation. Crystals, roughly 0.4-0.5 mm along the longest dimension, were prepared by solvent difussion or slow evaporation techniques. Structures were refined by Patterson (PHASE) methods or by direct methods (MULTAN). Final R-factors were in the range 2.9-5% depending on the sample, except for **TOT-PDMAC**, which refined to R=10.6 %, and exhibited considerable disorder. Each structure will be discussed fully in the text. Graphics used in the Figures was produced using CHEM-X, designed and distributed by Chemical Design Ltd., Oxford, England. The crystallographic coordinates and symmetry parameters were used to generate the packing diagrams. Full details of the structures will be reported separately.

RESULTS AND DISCUSSION

The SHG efficiencies at 1.06 μ of the materials examined in this study are shown in Tables 1-3. All SHG values are listed relative to urea as a standard. The data show that a wide variety of inclusion complexes can exhibit substantial SHG. The range of activity observed is nearly 100-fold, from a low value of 0.015x urea to a high value of 4x urea. Interestingly, both the high and low examples come from the same class of inclusion compound: **PNA-CD** is the most active solid prepared, and the **CD** complex with *p*-dimethylaminobenzonitrile is the least active (Table 1). However, with the exception of **CD** itself, which is very slightly SHG active (ca. 0.1% of urea) as a result of its chirality, none of the pure hosts or guests listed in the tables are themselves SHG active. Therefore the net SHG activity observed for the complexes must be considered to be remarkable.   The inclusion phenomenon has resulted in a dipolar orientation of the polar guests within the inclusion matrices. We also note that the observation of substantial SHG activity among organometallic inclusion complexes is new.

The **CD** examples all involve direct inclusion of the guest within the cavity of the **CD**. The polar orientation of the solid arises as a result of the crystal habit of the **CD**. For the thiourea and **TOT** complexes, the inclusion complexation is a cooperative phenomenon between host and guest. The nature of the polar orientation, and its apparent commonality, deserve special comment. We have prepared nearly 25 examples of inclusion complexes of polar molecules in various matrices as part of this work. Only

Table 1. Inclusion Compounds with β-Cyclodextrin

Guest	Host:Guest	SHG Rel. to Urea
p-Nitroaniline	1:1	2.0-4.0
p-(N,N-Dimethylamino)cinnamaldehyde	1:1	0.4
N-Methyl-*p*-nitroaniline	1:1	0.25
2-Amino-5-Nitropyridine	1:1	0.07
p-Dimethylaminobenzonitrile	1:1	0.015

Table 2. Inclusion Complexes with Thiourea

Guest	Host:Guest	SHG Rel. to Urea
Benzenechromium tricarbonyl	3:1	2.3
(Fluorobenzene)chromium tricarbonyl	3:1	2.0
(Cyclopentadienyl)rhenium tricarbonyl	ND	0.5
(1,3-Cyclohexadiene)iron tricarbonyl	3:1	0.4
(1,3-Cyclohexadienyl)manganese tricarbonyl	3:1	0.4
(Trimethylenemethane)iron tricarbonyl	3:1	0.3
(Cyclopentadienyl)manganese tricarbonyl	3:1	0.3
(1,3-Butadiene)iron tricarbonyl	ND	1.0
Pyrrolylmanganese tricarbonyl	ND	0.2
(Cyclopentadienyl)chromium dicarbonyl nitrosyl	3:1	0.1

ND = Not Determined

Table 3. Inclusion Complexes with Tris-*ortho*-thymotide

Guest	Host:Guest	SHG Rel. to Urea
p-Dimethylaminocinnamaldehyde	2:1	1.0
p-Dimethylaminobenzonitrile	2:1	0.3
(*p*-Cyanobenzoyl)manganese pentacarbonyl	ND	0.2
(Indane)chromium tricarbonyl	1:1	0.1
(Anisole)chromium tricarbonyl	ND	0.1
(Tetralin)chromium tricarbonyl	1:1	0.1
[Benzenemanganese tricarbonyl] tetrafluoroborate	ND	0.1

ND = Not determined

three complexes have failed to exhibit SHG. They were that between thiourea and pyridinetungsten pentacarbonyl, the **TOT**-*cis*-stilbene-chromium tricarbonyl complex, and a CpRh(olefin)$_2$CO complex with thiourea. A crystal structure of the PyW(CO)$_5$-(**TOT**)$_2$ complex showed that indeed the inclusion environment was centrosymmetric and therefore incapable of SHG. The other two complexes have not been examined by x-ray, so no firm conclusion can be made for the failure to exhibit SHG. However, we suspect that the hyperpolarizability of these materials may be too low to exhibit substantial SHG even if the crystals were acentric. All of the other complexes prepared and examined to date have proved, by virture of their SHG activity, to be acentric and polar materials. A plausible explanation for the ubiquity of dipolar alignment in these structures will be described later. We first examine the solid state structures of some of the complexes solved by x-ray analysis.

Thiourea--Benzenechromium Tricarbonyl.

The x-ray data was solved within the space group R3c. The structure confirms the channel nature of the inclusion. The guest molecules stack along the axes of the channels with a separation of 2.84 Å between the plane of the carbonyl oxygens and the adjacent benzene ring. Each guest is staggered by 13.3° from neighboring Cr(CO)$_3$C$_6$H$_6$ molecules across the channel. The channel network resembles a honeycomb as a result of extensive hydrogen bonding within the thiourea channel structure. Each sulfur is associated with four adjacent amino groups of neighboring thioureas, with S-N distances in the range 3.4-3.5 Å. The channels formed by the thiourea assembly runs parallel to the crystal c-axis and has an internal diameter of 9.3Å. The closest contact between guests is 3.1Å, between atoms O1 and N1. The molecular structure is shown in Figure 1, and the honeycomb channel structure is shown in Figure 2. Figure 3 is a side view perpendicular to the c-axis, which clearly shows that all the Cr(CO)$_3$ moieties reside on one side of the channel. The crystal is enantiomorphic, and polar. We ascribe the SHG activity to this orientational arrangement, which places the transition moment vector of the lowest-lying charge transfer excited state (CTML) along a single axis. The simple two-state model[16] for SHG activity states that the change in dipole moment on optical excitation is partially responsible for second-order polarization effects in the molecule. We consider this effect to be primarily responsible for the SHG in all the systems we describe here, and suggest that differences in the SHG activity of these systems is caused, at the molecular level, by differences in the CT character of the lowest lying excited states of the molecules incorporated into the inclusion matrices.

Thiourea--(1,3-Cyclohexadiene)manganese Tricarbonyl.

This structure was also solved within the space group R3c, and its structure is entirely isomorphous to the benzenechromium tricarbonyl structure. Figures 4 shows the x-ray solution, and Figure 5 shows the polar nature of the bulk structure. The view shown in Figure 5 is identical to that shown in Figure 3 for the isomorphous Cr complex. In this system, the staggering of adjacent Mn guest molecules along the channel is 35.1Å, and the interplanar CO-to-cyclohexadienyl intermolecule spacing is 3.19Å. The internal cavity diameter is 9.48Å. Thus the Mn structure is slightly larger in volume than the isomorphous Cr structure, but the polar orientation is preserved. It is therefore of interest to compare the observed SHG activity

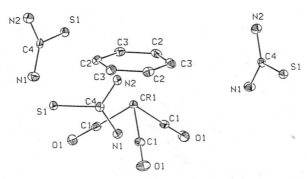

Figure 1.   Molecular structure of benzenechromium tricarbonyl complex with thiourea as determined by x-ray analysis.   There are three thiourea molecules per organometallic unit.

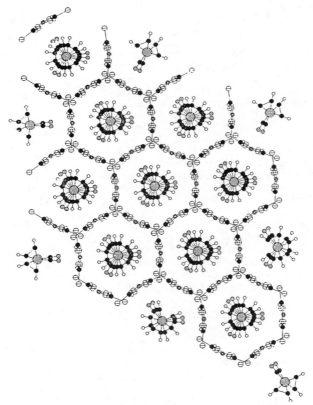

Figure 2.   View of molecular packing arrangement in benzenechromium tricarbonyl--(thiourea)$_3$ complex.   The view is down the crystal c-axis, looking down the honeycomb formed by the thiourea lattice.

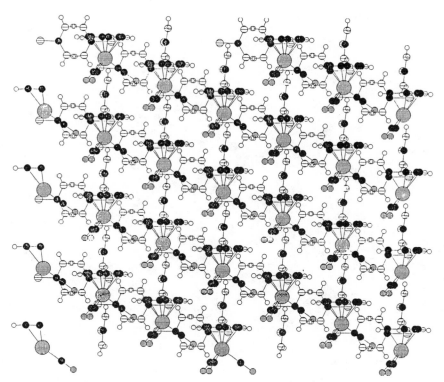

Figure 3.   A view of benzenechromium tricarbonyl--thiourea perpendicular to the c-axis.   This view highlights the polar arrangement of the organometallic moieties within the thiourea lattice: note that all molecules are aligned.

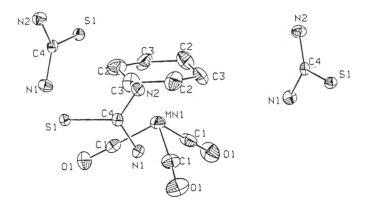

Figure 4.   Molecular structure of 1,3-(cyclohexadienyl)manganese tricarbonyl complexed with three units of thiourea.

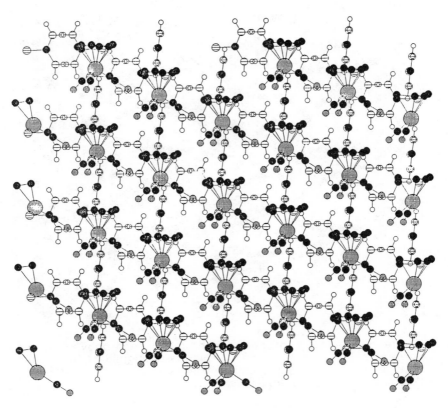

Figure 5.   Crystal packing of the 1,3-CHDMn(CO)$_3$--(thiourea)$_3$ complex illustrating the polar alignment.   The cystal structure is isomorphous with the chromium complex shown in Figures 2 and 3.

of the two isomorphous systems, and to ask why the Mn activity (0.4x urea) is lower than the Cr activity (2.3x urea). At the present stage of our understanding of the SHG phenomenon in this class of compounds, we attribute this difference to a lower degree of CT character in the lowest lying excited state of the Mn compound compared to the Cr species. We can not, however, eliminate differences such as domain size effects or correlation size effects at this time. The question is being addressed both theoretically, by calculation of the excited state CT character of the materials, and experimentally by measuring the second-order polarizabilties using electric field induced SHG techniques.

Thiourea--(1,3-Cyclohexadienyl)iron Tricarbonyl.

An arrangement different than the previous two structures is encountered in this complex. The data were solved within the orthorhombic space group Pna2$_1$. The inclusion structure, 1,3-CHDFe(CO)$_3$ ·(thiourea)$_3$ is confirmed for this material by the structure solution. In this case, however, the organometallic guest stacks within the channels sideways, as opposed to the perpendicular stacking observed in the previous structures. The packing is again influenced by hydrogen bonding. The iron to iron distance within a channel is 6.29Å, and it is 9.46Å across channels. The arrangement is again polar, but the the director of one layer of Fe-CHD units is canted with respect to the next layer by some 60°. That is, there is not a single polar direction common to all CHD-Fe units, but two, which average to an overall resultant polar axis. These conclusions are illustrated in Figures 6-8. In spite of the difference in the arrangement of the organometallic units in the CHDFe- and CHDMn (CO)$_3$ inclusion complexes, the observed SHG is identical for both (0.4x urea).

Tris-*ortho*-thymotide--(Tetralin)chromium Tricarbonyl.

This complex is a 1:1 complex between the guest and the host, unlike most other **TOT** complexes encountered in this study. It is orthorhombic, space group Pca2$_1$. The structure of the inclusion complex is shown in Figure 9 (the **TOT** has been removed for clarity). The Figure illustrates the polar environment and indicates that there are two tetralinchromium tricarbonyl units staggered within the channel and slightly tilted toward one another within the channel, and a second pair pointing toward the first, but not in an inversion relationship. The structure is thus a weakly polar one, and the observed SHG activity is low (0.1x urea).

Tris-*ortho*-thymotide--*p*-Dimethylaminocinnamaldyhyde.

The solution of this structure gave extreme difficulty, and the observation of SHG proved to be a key observation which allowed solution of the structure. The structure was solved as a triclinic, centrosymmetric crystal habit, P1 containing two **TOT** molecules per aldehyde. A strong centric domination of the **TOT** unit was noted that forced a disorder onto the cinnamaldehyde units. The solution is illustrated in Figure 10 and more clearly in Figure 11, in which the **TOT** moieties are removed. The SHG activity dictates an acentric arrangement. The cinnamaldehyde molecules occupy two perpendicular channels. While it is impossible to assign the enantiomorphic structure of a particular crystal exactly, it is permissible, according to the SHG activity, to arbitrarily assign a sense of polarity to the aldehyde units in

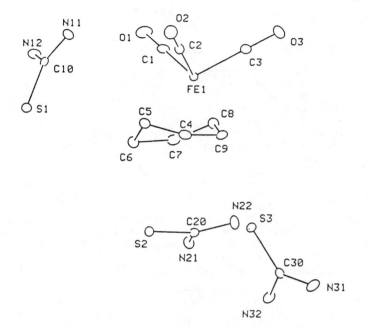

Figure 6.   Molecular structure of 1,3-(cyclohexadiene)iron tricarbonyl--(thiourea)$_3$.

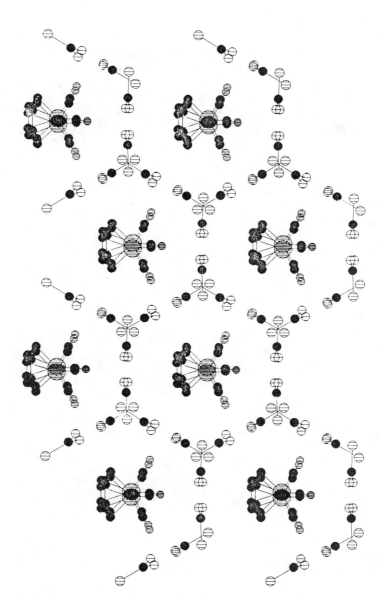

Figure 7. View of the crystal packing in 1,3-CHDMn(CO)$_3$--(thiourea)$_3$ illustrating the sideways arrangement of the organometallic in the thiourea channels.

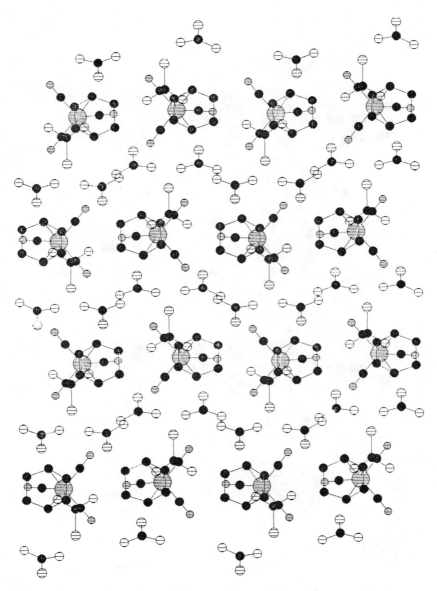

Figure 8. A view of 1,3-CHDMn(CO)$_3$--(thiourea)$_3$ illustrating the polar alignment. The view is down the honeycomb axis of the crystal, and it shows the sideways arrangement of the organometallic in the crystal lattice, unlike the arrangements shown in Figure 2 for the chromium complex.

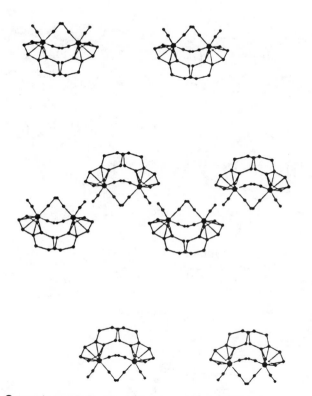

Figure 9. Crystal packing arrangement of tetralinchromium tricarbonyl complexed with **TOT**. The **TOT** units are not shown for clarity. The view illustrates the polar arrangement of the organometallic in the bulk structure.

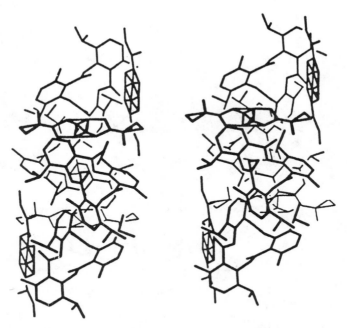

Figure 10. Crystal structure of *p*-dimethylaminocinammaldehyde complexed with two molecules of **TOT**. The **PDMAC** units are shown disordered about a center of symmetry in the solution, an arrangement proved to be impossible by the SHG result.

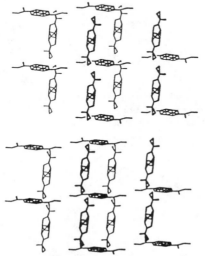

Figure 11. The **PDMAC-TOT** structure solution shown with removal of the **TOT** units.   The figure shows the perpendicular channel arrangement for occupation by the aldehyde units. The SHG result allows arbitrary solution of the structure. See text.

each of the channels. Thus the structure can be solved by the combination of the x-ray and SHG data.

β–Cyclodextrin--*p*-Nitroaniline.

The preparation and characterization of **CD-PNA** has been described in some detail previously.[10] All analytical data (CHN, NMR, IR, Raman) support the assignment as a 1:1 incluson complex, but it has to date defied attempts to prepare crystals adequate for x-ray determination. Powder x-ray data indicate that it is not a simple admixture of the two ingrediants, and ultraviolet spectroscopy shows that a strong 1:1 complex can exist in solution (Benesi-Hildebrand analysis), and that the complex is oriented such that the nitro group entered the **CD** cavity first (circular dichroism and NMR).

The picture shown below best describes the structure of the complex, according to our data:

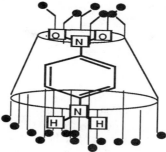

Solid State Properties of **PNA--β-CD.**

Microcrystalline **PNA--β-CD** complex was shown to produce SHG with efficiencies of 2-4 times that of urea. Other nitroaniline derivatives, and several other polarizable small organics, have also been included within β-**CD** and their cyrstalline powders shown to generate SHG. Table 1 lists observed SHG efficiencies relative to urea.

Crystals of the complex were grown by slow diffusion of ethereal **PNA** into **CD** solutions or by slow evaporation. Water of hydration is retained in varying amounts by the crystals depending on method of crystal growth. Data was collected, but solution and refinement of the structure could not be effected because of disorder in either the **CD** or the **PNA.**

For the many cyclodextrin inclusion structures that have been solved,[17] two main structrual types are normally observed, channels and cages. Among the channel types, head to head or head to tail arrangement of **CD** units within the channels can occur. For the cage structures, herringbone and brick-work arrangements are noted. Scheme 2 illustrates these common arrangements. Only the head to tail channel structure or the cage structure is capable of exhibiting SHG. Crystals of those types would be polar.

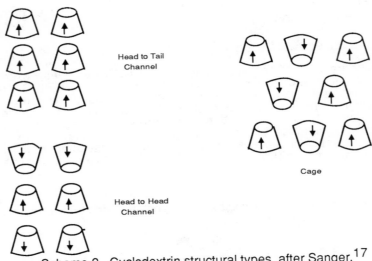

Scheme 2. Cyclodextrin structural types, after Sanger.[17]

CONCLUSIONS

The work reported here has shown that inclusion complexation of organic and organometallic chromophores by thiourea, **TOT** and cyclodextrins can induce second harmonic generation capability in the polar crystals which result, even when the original bulk materials are themselves incapable of SHG. Structural evidence has been presented to show tht the solid state inclusion structures are acentric, and a simple electronic picture for the polarization response of these materials within the two-state model[16] has been discussed. In an earlier section we remarked that of the many complexes we have made, only one has NOT been acentric. This result was not anticipated. We postulate that it is a natural tendancy in such materials, rather that an exception. If we consider a dipolar molecule in isotropic solution, we can imagine that if it were to aggregate, it would do so in a head to tail fashion in order to minimize electrostatic repulsion. The situation is illustrated in Scheme 3. The arrangement that would result is centrosymmetric.

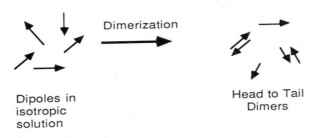

Scheme 3. Dipolar species tend to dimerize head to tail.

In a channel or cage inclusion matrix however, a polar molecule can only dimerize if two molecules can be accomodated within the matrix. If the thickness of the molecular guest is greater than one-half the diameter of the channel or cage, then the head to tail dimer structure observed in solution will be disfavored by large steric repulsion factors. The inclusion will preferentially occur in an end to end fashion along the growing channel. That arrangement is energetically (electrostatically) the preferred one, in which unlike polarity ends of the guest are nearest to one another. Such an arrangement is depicted in Scheme 4. A final requirement to assure polar alignment

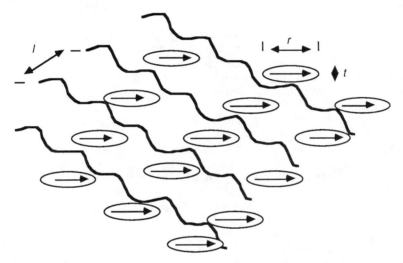

Scheme 4. Here, l = lattice diameter, r = length of guest, t = guest thickness, and the guest is depicted as a dipole oriented within the lattice matrix.

within the cavity, after the structure is established, is that the guest is longer than the matrix channel is wide. If that is the case, then the dipolar molecule can not rotate within the channel into a centrosymmetric arrangement. The concept we espouse here is one of complex growth which minimizes electrostatic repulsion. We expect, but can not yet prove without more detailed analyses of the available crystal structures, that the most stable arrangement of guests between channels may be a staggered one as shown in Scheme 4. Such an arrangement again minimizes head to head (like-charge to like-charge) electrostatic contacts. We are continuing to pursue this and other questions in our search for new materials for nonlinear optics.

ACKNOWLEDGMENTS

The authors thank Drs. Joseph Calabrese and Ian Williams for determining all the crystal structures reported here. Drs. Gerald Meredith and David

Thorn provided thoughtful comments and suggestions during the course of this work. Technical assistance was provided by Mrs. Sarah Harvey , Mr. John Lockhart, and Mr. Barry Johnson.

LITERATURE CITED

1. Gedanken, A., Robb, M. B., and Keubler, N. A., *J. Phys. Chem.*, **1982**, *86*, 4096.
2. "Nonlinear Optical Properties of Organic and Polymeric Materials," Williams, D. J., ed., *ACS Symposium Ser. No. 233*, Washington, D. C., 1983.
3. Williams, D. J., *Angew. Chem. Int. Ed. Eng.*, **1984**, *23*, 690.
4. Basu, S., *Ind. Eng. Chem. Prod. Res. Dev.*, **1984**, *23*, 183.
5. Glass, A. M., *Science*, **1984**, *226*, 657.
6. Oudar, J. L., and Zyss, Z., *Phys. Rev.* , **1982**, *A26,* 2016, 2028.
7. a.) Jain, K., Crowley, J. J., Hewig, G. H., Cheng, Y. Y., and Tweig, R., *Opt. Laser Tech.* , **1981**, *13*, 297.
   b.) Twieg, R., Azema, A., Jain, K., and Cheng, Y. Y., *Chem. Phys. Letts.*, **1982**, *92*, 208.
8. Wang, Y., and Eaton, D. F., U. S. Ser. No. 732, 652 (May 10, 1985).
9. Wang, Y., and Eaton, D. F., *Proc. Conf. Lasers and Electroopt. (CLEO '85)*, Paper THM44, p 212, Baltimore, Md., May 20-24, 1985.
10. Wang, Y., and Eaton, D. F., *Chem. Phys. Letts.*, **1985**, *120*, 441.
11. Tomaru, S., Zembutsu, S., Kawachi, M., and Kobayashi, M., *J. Chem. Soc. ,Chem. Commun.*, **1984**, 1207.
12. Levine, B. F., Bethea, C. G., Thurmond, C. D., Lynch, R. T., and Bernstein, J. L., *J. Appl. Phys.*, **1979**, *50*, 2523.
13. Bergeron, R., and Rowan, R., III, *Bioorg. Chem.*, **1976**, *5*, 425.
14. Kurtz, S. K., and Perry, T. T., *J. Appl. Phys.*, **1968**, *89*, 8798.
15. Halbout, J. M., Blit, S., and Tang, C. L., *IEEE J. Quantum Electron.*, **1981**, *QE-17*, 513.
16. Oudar, J. L., *J. Chem. Phys.*, *67*, 446 (1977).
17. Saenger, W., "Structural Apsects of Cyclodextrins and their Inclusion Complexes," in **Inclusion Compounds**, Vol. 2, J. L. Atwood, J. E. Davies and D. D. MacNicol, eds., Academic Press, New York, 1984, Chapter 8.

RECEIVED May 18, 1987

Chapter 33

Polymers for Integrated Optics

John E. Sohn, Kenneth D. Singer, and Mark G. Kuzyk

Engineering Research Center, AT&T, Princeton, NJ 08540

The potential applicability of organic and polymeric materials to integrated optics is large owing to both their microscopic and bulk properties. Two of the advantages of using such materials are flexibility in the fabrication of optical structures and the tailoring of optical properties through material engineering. For application in guided-wave nonlinear optical devices high optical quality and low dielectric constant are but two of the requisite properties. Polymer glasses have been shown to possess these properties, and recently have been rendered optically nonlinear using electric field poling of nonlinear optical molecular-doped polymer glasses. Organic and polymeric materials are discussed in the context of the requirements of integrated optics, including development of doped poled polymer glasses.

Presently, in order for optical information to be processed, the information must first be converted into electrical information before it can be operated upon by control electronics. Once the processing is complete, conversion to optical information is done before transmission. The transmission medium of today is optical (optical fiber) and the switching medium is electronic (integrated circuit).

Electro-optic devices, *e.g.* titanium indiffused lithium niobate ($Ti:LiNbO_3$), will shortly see commercial application. With these devices, an electronic control is used directly on an optical signal, obviating the need for optical to electrical conversion and reconversion. Further down the road, all-optical processing (optical control of an optical data stream) may be realized. These developments require materials with large optical nonlinearities and the ability to be processed and integrated with optical sources, detectors and drive electronics, and where necessary, the ability to be formed into structures capable of supporting guided waves.

Numerous investigations into the nonlinear optical properties of certain organic and polymeric materials have shown that these materials possess the largest observed optical nonlinearities.[1] [2] [3] [4] These nonlinearities, whose physical mechanisms were elucidated by basic studies over the last two decades, arise from charge correlated features present in conjugated π-electron moieties constituting these organic materials.[5] [6] [7] [8] A variety of approaches to the fabrication of bulk organic nonlinear optical materials have been pursued including molecular crystals, crystalline polymers, Langmuir-Blodgett films, liquid crystals and liquid crystal polymers.[1] [3] Significant efforts in crystal growth [3] and single crystal thin film fabrication[9] [10] as well as our recent demonstration of optically nonlinear polymer glasses[11] show promise for using organic and polymeric materials in applications using both second and third order optical nonlinearities.

Nonlinear Optics

The polarization response of a material to external electromagnetic fields is given by (in the electric dipole approximation)

$$P(E_{total}) = \overset{\leftarrow(0)}{\chi} + \overset{\leftarrow(1)}{\chi} \cdot \vec{E}_{total} + \overset{\leftarrow(2)}{\chi} : \vec{E}_{total}\vec{E}_{total} + \overset{\leftarrow(3)}{\chi} \vdots \vec{E}_{total}\vec{E}_{total}\vec{E}_{total} + \cdots \quad (1)$$

where the χ's are the susceptibilities and E_{total} is the sum of electric fields of various frequencies and polarizations. $\chi^{(0)}$ is the permanent polarization, $\chi^{(1)}$ the linear susceptibility, and the higher order χ's are the nonlinear optical susceptibilities. For simplicity, tensor notation is ignored. Linear processes, such as refraction and absorption, arise from the linear susceptibility $\chi^{(1)}$, and nonlinear processes arise from the higher order terms.

When electromagnetic fields, ω_1, ω_2, traverse a material whose higher order susceptibilities are zero or negligible, no interaction of the two fields occurs. However, with an optically nonlinear material ($\chi^{(2)}$ and/or higher order terms nonzero), these fields do interact and fields of different frequency can be generated.

For example, for a material possessing nonzero $\chi^{(2)}$, $\omega_1 = \omega_2$ results in second harmonic generation, that is, a field of twice the incident frequency ($2\omega_1$) is produced. Frequency conversion or parametric mixing occurs when $\omega_1 \neq \omega_2$ and a wave at $\omega_3 = \omega_1 \pm \omega_2$ is generated. The linear electro-optic (Pockels) effect, $\omega_2 = 0$, yields a change in the index of refraction of the material, thus changing the optical path length. Parametric amplification occurs when energy is exchanged between incident beams. These effects lead to potential applications such as frequency doublers, optical mixers, optical amplifiers, switches, and modulators. Processes arising from the third order susceptibility $\chi^{(3)}$ include optical bistability, the intensity dependent index of refraction, and optical phase conjugation.

The odd order susceptibilities are nonzero in all materials. However, owing to the fact that $\chi^{(2)}$ is a third rank tensor, the second order susceptibility is nonzero only in noncentrosymmetric materials, that is, materials possessing no center of symmetry. The focus of this paper is on second order processes, and the relationships between the bulk susceptibility, second harmonic generation, and the linear electro-optic effect. For second harmonic generation, $\chi_{ijk}^{(2)}$ is symmetric in i, j, leading to the relationship between the second harmonic coefficient d_{ijk} and the bulk second order susceptibility $\chi^{(2)}$ [12]

$$\chi_{ijk}^{(2)}(-2\omega;\omega,\omega) = 2d_{ijk}(-2\omega;\omega,\omega). \quad (2)$$

Since the electro-optic coefficient $r_{ij,k}$ is defined by the electric field dependence of the optical indicatrix, $r_{ij,k}$ is related to the second order bulk susceptibility through[12]

$$\chi_{ijk}^{(2)}(-\omega;\omega,0) = -\frac{1}{2}\epsilon_{ii}(\omega)\epsilon_{jj}(\omega)r_{ij,k}(-\omega;\omega,0). \quad (3)$$

The intrinsic nonlinearity of certain organic materials is substantially higher than the nonlinearity of inorganic and semiconducting materials.[4] The origin of the linear electro-optic effect in organic compounds arises from their electronic structure. An understanding of the molecular origins of the nonlinearity is essential in the development of optically nonlinear organic and polymeric materials. The electronic contribution to the bulk second order susceptibility is related to the microscopic susceptibility β_{IJK} by the van der Waals sum[13]

$$\chi_{ijk}^{el}(-\omega_3;\omega_1,\omega_2) =$$

$$N f_i^{\omega_3} f_j^{\omega_1} f_k^{\omega_2} \sum_{I,J,Ks=1}^{n} \cos(i,I(s)) \cos(j,J(s)) \cos(k,K(s)) \beta_{IJK}(s) \quad (4)$$

where s is summed over molecules in the unit cell, N is the number of unit cells per unit volume, f's are local field factors, the cosines transform the molecular to bulk axes, and β_{IJK} is the electronic molecular nonlinear optical susceptibility. The major contributions to the molecular susceptibility, for $2\omega < E_1$ (below the first excited state), are the transition moment μ_{1g}, the change in dipole moment $\Delta\mu$ between the ground state and first excited state of the molecule, and the excited state energy E_1, and are related by

$$\beta = \frac{e^3}{4h^2} \frac{6\mu_{1g}^2 E_1^2 \Delta\mu}{(E_1^2 - \omega^2)(E_1^2 - 4\omega^2)}.$$ (5)

   Thus, the larger the transition moment and change in dipole moment, the larger the microscopic susceptibility. Approaches to increasing this susceptibility that have proved successful include the placement of strong electron-donor and electron-acceptor groups at opposite ends of a conjugated π-electron system and increasing the conjugation length (See Table 1).[14] [11] [15]

TABLE 1. Values of $\beta\mu$ measured at $\lambda = 1.9 \ \mu m$

Molecule	$\beta\mu$ $(10^{-30} cm^5 D/esu)$
2-methyl-4-nitroaniline (MNA)	120[13]
Disperse Red 1*	525[11]
MPC merocyanine**	-2600[14]

* 4-[N-ethyl-N-(2-hydroxyethyl)]amino-4'-nitroazobenzene
** N-methyl-[4(1H)-pyridinylidene ethylidene]-2,5-cyclohexadien-1-one

Construction of a bulk material possessing a large bulk susceptibility $\chi^{(2)}$ requires not only molecular constituents with large microscopic susceptibilities but a noncentrosymmetric system where the orientation of the molecular species results in additivity of the molecular susceptibilities.

Doped Poled Polymer Glasses

A variety of approaches to macroscopic structures have been reported.[1] [3] [9] [10] The majority of these approaches involve the engineering and use of crystalline materials. We have chosen to use amorphous polymer glasses, and recently demonstrated that molecularly-doped poled polymer glasses possessing reasonably large second order nonlinear optical susceptibilities can be produced.[11] This approach has the advantages of material processability, the capability of integration with sources, detectors, and drive electronics, and material properties required for application to integrated optics, namely high optical quality, and low dielectric constant and dielectric loss.

   The concept involves the preparation of a solid solution of an optically nonlinear organic molecule dissolved in a polymer glass, processing of the solid solution into a thin film, and electric filed poling to align the dipolar dye molecules, thus removing the inversion center inherent to amorphous glasses, and allowing the nonlinear optical properties to be additive. Orientationally ordered films are prepared by heating the film above the glass-rubber transition temperature where molecular motion is enhanced. A strong electric field is applied which aligns the nonlinear optical species in a Boltzmann distribution. Cooling the sample and removing the electric field below the glass-rubber transition temperature results in locking in the induced polarization, yielding a noncentrosymmetric material.

The nonlinear optical susceptibility can be calculated by assuming a one-dimensional molecule and a non-interacting molecular ensemble at the poling conditions, and is given by [11] [4]

$$\chi_{zzz}(-\omega_3;\omega_1,\omega_2) = N\beta_{333}(-\omega_3;\omega_1,\omega_2) \, f^{\omega_3} f^{\omega_1} f^{\omega_2} \left\{ \frac{q}{5} - \frac{q^3}{105} + \cdots \right\}, \tag{6}$$

where

$$q = \left[\frac{\epsilon(n^2 + 2)}{n^2 + 2\epsilon} \right] \frac{\mu E_p}{kT}, \tag{7}$$

N is the number density, β_{333} is a component of the molecular nonlinear optical susceptibility, μ is the static dipole moment, E_p is the poling field, ϵ is the static dielectric constant, n is the index of refraction, and where the $f's$ are local field factors at the appropriate frequencies $(f^\omega = (n_\omega^2 + 2)/3)$.

The model embodied in Equations (6) and (7) can be used to evaluate the potential of doped poled polymer films in nonlinear optics. Using reasonable values for the parameters of the model and molecules with large values of $\beta\mu \sim \beta_{333}\mu_3$, susceptibilities comparable to those measured in crystals can be realized (See Table 2).

TABLE 2. Potential second harmonic coefficient for various molecular dopants in poled poly(methyl methacrylate) (PMMA) using Eqs. (6) and (7) with $N = 3 \times 10^{20}/cm^3$, $E_p = 0.5 \, MV/cm$, $n = 1.52$, and $\epsilon = 3.6$. Values of $\beta\mu$ are measured separately at $\lambda = 1.9 \, \mu m$

Molecule	$\beta\mu$ $(10^{-30} cm^5 D/esu)$	d_{33} $(10^{-9} \, esu)$
2-methyl-4-nitroaniline (MNA)	120	1.6
Disperse Red 1	525	6.6
MPC merocyanine	-2600	-32

To demonstrate the concept, we used the organic dye Disperse Red 1 (4-[N-ethyl-N-(2-hydroxyethyl)]amino-4'-nitroazobenzene) and the polymer glass poly(methyl methacrylate) (PMMA). Solid solutions of 0-12wt% of the dye in PMMA were prepared and, using standard coating techniques, rendered into thin films on indium tin oxide coated glass. The thin indium tin oxide layer is transparent and acts as one electrode for the poling process. The other electrode is transparent gold deposited on top of the film. The film is heated above its glass-rubber transition temperature $(T_g \sim 100°C)$. An intense electric field of $0.2-0.6MVcm^{-1}$ is applied to align the nonlinear dopant. The field is maintained until the sample is well below T_g.

The second-order nonlinear optical properties of the poled film are investigated using second harmonic generation in transmission. The sample preparation technique results in a poled glassy film possessing a unique axis in the direction parallel to the field direction within the point group of ∞mm. Thus the symmetry operations are an infinite-fold rotation axis and an infinity of mirror planes. There are then five non-zero tensor components, three of which are independent.[16] The second harmonic polarization has the form

$$
\begin{aligned}
P_x^{2\omega} &= 2d_{15}E_xE_z \\
P_y^{2\omega} &= 2d_{15}E_yE_z \\
P_z^{2\omega} &= d_{31}E_x^2 + d_{31}E_y^2 + d_{33}E_z^2,
\end{aligned}
\tag{8}
$$

where the standard contracted notation is used.[16] Kleinman symmetry gives $d_{15}=d_{31}$,[17] while the thermodynamic model yields $d_{31}=d_{33}/3$. Thus the second harmonic coefficient d_{33} is determined directly from the second harmonic intensity.

Second harmonic generation was measured in the films at a fundamental wavelength of $\lambda=1.58\mu m$. The d_{33} second harmonic coefficient, determined by measuring the second harmonic intensity as a function of angle, is: $d_{33} = 6.0\pm1.3 \times 10^{-9}esu$, which compares favorably with that of KDP ($d_{36} = 1.1 \times 10^{-9}esu$). KDP, potassium dihydrogen phosphate, is a commonly used second harmonic generating crystal. Using the thermodynamic model and the properties of the measured film, namely that $N=2.74\times10^{20}$, $E_p=0.62MV/cm$, $\beta\mu=525\pm100\times10^{-30}cm^5/esu$[11], $n=1.52$, and $\epsilon=3.6$, d_{33} is calculated to be $d_{33} = 7.5\pm1.5 \times 10^{-9}\ esu$. The agreement between the experimentally determined value of d_{33} and the calculated value indicate that Equations (6) and (7) adequately describe the poled polymer solution. This implies that at the poling conditions the system behaves as a simple solution, where there is negligible interaction between the solute dye molecules and the polymer solvent.

The dependence of the second harmonic coefficient on the number density is found to be linear as predicted by the model. However, the measured values of d_{33} are lower than those predicted by the model. This result can be attributed to several factors. The third order susceptibility γ_{ijkl} may not be negligible in the dc second harmonic generation used to determine the molecular susceptibility, but γ_{ijkl} does not contribute to the film susceptibility. Trapped charges in the sample may screen the poling field, reducing alignment. The strong dependence of the model on local fields (to high powers) may also account for the differences observed. also, polymer glasses are not in thermodynamic equilibrium and molecular relaxation occurs. Such relaxation processes may have reduced the polarization.

It has been established experimentally that the origin of the electro-optic effect in organic materials is largely electronic.[4] This implies that the linear electro-optic coefficient can be estimated from the second harmonic coefficient.[4] By properly accounting for the dispersion (using a two level model), the electronic contribution to the electro-optic coefficient is calculated to be $r_{33}^{el} = 2.4\pm0.6 \times 10^{-12}\ m/V$ at $\lambda=0.8\mu m$. Measured values of the electro-optic coefficient are in agreement within experimental uncertainty. These values compare favorably with that of GaAs ($r_{41} = 1.2 \times 10^{-12}\ m/V$).

In addition to nonlinear optical properties, these polymer glasses possess other properties important for integrated optics, including high optical quality and low dielectric constant and dielectric loss.

Optical Properties

Polymer glasses are widely used in the optics industry as bulk and micro optical components.[18] The most commonly used materials are polyacrylates, polycarbonates, and polystyrenes. Low optical losses are found with these materials due to their amorphous nature. Crystalline regions in a material will cause scattering, thus reducing the optical quality.

Kaino has discussed the processes contributing to optical loss in polymer optical fibers.[19] The intrinsic processes are absorption, due to electronic transitions and high harmonics of carbon-hydrogen (C-H) absorption, and various scattering processes. Extrinsic processes, those that can be controlled through fabrication and processing, are also twofold - absorptive and scattering. Absorption arises from contaminants (organic and transition metal) and absorbed water; those contributing to scattering are dust and microvoids, orientational birefringence, core-cladding boundary imperfections, core diameter fluctuations, and Rayleigh scattering.

Using poly(methyl methacrylate), PMMA, perdeuterated PMMA, and polystyrene, it was found that absorptive losses are very low (<0.1 dB/cm) at wavelengths between 0.5 and 1.0 μm, and that losses due to C-H overtones can be made lower than 0.1 dB/cm between 1 and 2 μm by deuteration.[20] [21]

High optical quality polymer waveguide structures were first reported in 1972 by scientists at Bell Laboratories.[22] Light-guiding films from photoresist were used in the fabrication of strip guides of width 5 μm. An argon ion laser was used to write the guide in the photoresist. Subsequently, several other methods were employed to produce low loss waveguides. The results have been reviewed elsewhere,[23] [24] and are listed in Table 3.

TABLE 3. Propagation losses in glassy polymer waveguides fabricated by various techniques

Fabrication Technique	Propagation Loss (dB/cm)
RF discharge polymerization	0.04
photolocking	0.2-0.4
mechanical	0.5
fiber	0.1<

The high optical quality of these glassy polymer structures compares favorably with the propagation loss for devices fabricated in $Ti:LiNbO_3$ ($\sim$ 0.1 dB/cm). In addition, Kurokawa et al investigated the use of selective photopolymerization to fabricate optical dividers and low loss couplers. [25]

Loss measurements for organic crystals have also been reported. Tomaru et al have produced optical channel waveguides of single crystal m-nitroaniline. [26] Scattering losses for a 5 mm long waveguide were reduced to 5 dB/cm with a laser zone-melting technique. Nayar has prepared single-crystalline, void-free benzil fibers (diameter 2-10μm) with a propagation loss estimated to be 2 dB/cm at 633 nm.[27] The quoted value was about 1 dB/cm greater than the attenuation in the bulk crystal. These values for crystalline materials are considerably higher than those measured in amorphous polymer glasses and in $Ti:LiNbO_3$; if crystals are to see application to guided-wave devices, these losses must be reduced substantially.

Dielectric Properties

The speed of electro-optic devices is greatly determined by the dielectric properties of the electro-optic material. For instance, the bandwidth per modulating power ($\Delta f/P$) of a guided-wave lumped-element modulator is given by[28]

$$\frac{\Delta f}{P} \propto \frac{1}{1 + \epsilon/\epsilon_0}.$$ (9)

The rise time of a traveling wave modulator is given by $t_r \propto \sqrt{\epsilon/\epsilon_0} - n$,[29] thus a lower dielectric constant results in a faster device.

The dielectric properties of organic and polymeric materials are one of their more attractive properties. Table 4 gives the dielectric constants of PMMA and polystyrene which are typical polymer glasses.[30]

TABLE 4. Dielectric properties of PMMA and Polystyrene, where ϵ/ϵ_0 is the dielectric constant at frequency f, and $\tan\delta$ is a measure of the dielectric loss

Material	ϵ/ϵ_0	$\tan\delta$	f
PMMA	3.6	0.062	50 Hz
	3.0	0.055	1 kHz
	2.6	0.014	1 MHz
	2.57	0.007	30 GHz
Polystyrene	2.49-2.55	0.0015	flat to 1 GHz

As is also seen, the dielectric constant does not exhibit substantial dispersion in the modulating frequency ranges of interest.

Dielectric loss will also degrade device performance and affect the switching power through its dissipative effects. As can be seen in Table 4, the dielectric losses are also low and nondispersive. The decrease in $\tan\delta$ in PMMA with frequency indicates that the losses are conductive, and probably due to impurity ions.

Processing

Recent advances in semiconductor processing, especially in areas of microstructure fabrication and optoelectronic integration, have led to the reproducible fabrication of optical waveguide structures in polymeric materials. Optical microstructures have been fabricated in PMMA, doped PMMA, and other amorphous polymer glasses using techniques compatible with semiconductor processing, such as conventional lithography with a positive photoresist, deposition of an etch mask, and anisotropic etching methods.

The ability to integrate an electro-optic material with other optical devices, *e.g.* light sources and detectors, and with electronic drive circuits is important. Integrability implies that the electro-optic materials and the processing of these materials are compatible with the other components, and that electrical and optical interconnects can be fabricated. Polymer glasses are widely used in the fabrication of electronic devices and device interconnects. Polymers are also used as photoresists and as dielectric interlayers for electrical interconnects. As a result, a body of knowledge already exists concerning planarization methods of polymers on substrates, the definition of microscopic features, and the fabrication of microstructures in planar polymer structures.

Refractive index differences between the guiding region and the surrounding medium in waveguide structures can be tailored depending on the fabrication process, materials, and structures fabricated. These differences can be used to control bending in guided wave configurations. The index differences obtained from several techniques can range from 0.01 to 0.5. The refractive indices of typical glassy polymers range from 1.48 to 1.6. These values are relatively close to the index of optical fibers ($\sim$ 1.45), thus mode mismatch between fibers and optical structures comprised of glassy polymers can be minimized. The large numerical aperture (dependent on the difference in refractive indices between the guiding region and the surrounding medium) of polymers can be used to advantage in fabrication and integration of guided wave structures.

Summary

The potential application of organic and polymeric materials, especially glassy polymers, is promising as evidenced by the pertinent material properties. A model nonlinear optical glassy polymer system, combining the large nonlinear susceptibilities of the dopant and favorable optical properties of the host polymer glass, was fabricated and evaluated. The nonlinear optical properties compare favorably with those of other materials under development for nonlinear optics. Further, the results are consistent with a thermodynamic model based on noninteracting dopant molecules. Future optimization of nonlinear optical properties, along with the integrability , and high optical and dielectric quality of polymer glasses, make this material class a promising one for use in integrated optics.

Acknowledgments

The authors wish to thank H. M. Gordon, L. A. King, H. E. Zahn, A. G. Kerr, S. J. Lalama, and R. D. Small for their contributions.

Literature Cited

1. Williams, D. J. *Angew. Chem. Int. Ed. Engl.*, **1984**, *23*, 690.

2. Williams, D. J., Ed.; *Nonlinear Optical Properties of Organic and Polymeric Materials*; ACS Symp. Ser. No. 233; American Chemical Society: Washington D. C.; 1983.

3. Zyss, J. *J. Molec. Electron.*, **1985**, *1*, 25.

4. Singer, K. D.; Lalama, S. J.; Sohn, J. E.; Small, R. D. In *Nonlinear Optical Properties of Organic Molecules and Crystals*; Chemla, D. S.; Zyss, J., Eds.; Academic Press: New York, 1987, p. 437.

5. Davydov, B. L.; Derkacheva, L. D.; Dunina, V. V.; Zhabotinskii, M. E.; Zolin, V. F.; Koreneva, L. G.; Sarmokhina, M. A. *JETP Lett.*, **1970**, *12*, 16.

6. Levine, B. F.; Bethea, C. G. *J. Chem. Phys.*, **1977**, *66*, 1070.

7. Oudar, J. L.; Chemla, D. S. *J. Chem. Phys.*, **1977**, *66*, 2664.

8. Lalama, S. J.; Garito, A. F. *Phys. Rev. A*, **1979**, *20*, 1179.

9. See *inter alia* Ledoux, I.; Josse, D.; Vidakovic, P.; Zyss, J. *Opt. Eng.*, **1986**, *25*, 202.

10. See *inter alia* Carter, G. M.; Chen, Y. J.; Tripathy, S. K. *Opt. Eng.*, **1985**, *24*, 609.

11. Singer, K. D.; Sohn, J. E.; Lalama, S. J. *Appl. Phys. Lett.*, **1986**, *49*, 248.

12. Kaminow, I. P. *An Introduction to Electrooptic Devices*; Academic Press: New York, 1974.

13. Zyss, J; Oudar, J. L. *Phys. Rev. A*, **1982**, *26*, 2028.

14. Teng, C. C.; Garito, A. F. *Phys. Rev. Lett.*, **1983**, *50*, 350.

15. Singer, K. D.; Sohn, J. E.; Cai, Y. M.; Man, H. T.; Garito, A. F., to be published.

16. Nye, J. F. *Physical Properties of Crystals*; Clarendon: Oxford, 1957.

17. Kleinman, D. A. *Phys. Rev.*, **1962**, *126*, 1977.

18. Musikant, S. *Optical Materials: An Introduction to Selection and Application*; Marcel Dekker, Inc.: New York, 1985.

19. Kaino, T. *Jpn. J. Appl. Phys.*, **1985**, *24*, 1661.

20. Kaino, T.; Fujiki, M.; Jinguji, K. *Rev. Elect. Commun. Lab.*, **1984**, *32*, 478, and references cited therein.

21. Avakian, P.; Hsu, W. Y.; Meakin, P.; Snyder, H. L. *J. Polym. Sci. Polym. Phys. Ed.*, **1984**, *22*, 1607, and references cited therein.

22. Weber, H. P.; Ulrich, R.; Chandross, E. A.; Tomlinson, W. J. *Appl. Phys. Lett.*, **1972**, *20*, 143.

23. Lalama, S. J.; Sohn, J. E.; Singer, K. D. *SPIE Proc.*, **1985**, *578*, 168, and references cited therein.

24. Tomlinson, W. J.; Chandross, E. A. In *Advances in Photochemistry*; John Wiley & Sons: 1980, p. 201, and references cited therein.

25. Takato, N.; Kurokawa, T. *Appl. Opt.*, **1982**, *21*, 1940, and references cited therein.

26. Tomaru, S.; Kawachi, K.; Kobayashi, M. *Opt. Commun.*, **1984**, *50*, 154.

27. Nayar, B. K. In *Nonlinear Optical Properties of Organic and Polymeric Materials*; Williams, D. J., Ed.; ACS Symp. Ser. No. 233; American Chemical Society: Washington D. C.; 1983, p 153.

28. Kaminow, I. P.; Carruthers, J. R.; Turner, E. H.; Stulz, L. W. *Appl. Phys. Lett.*, **1973**, *22*, 540.

29. White, G.; Chin, G. M. *Opt. Commun.*, **1972**, *5*, 374.

30. Brandrup, J.; Immergut, E. H. Eds.; *Polymer Handbook*, 2nd Ed.; John Wiley & Sons: New York, 1975.

RECEIVED July 6, 1987

Chapter 34

Polymer Materials for Optical Fiber Coating

L. L. Blyler, Jr., F. V. DiMarcello, A. C. Hart, and R. G. Huff

AT&T Bell Laboratories, Murray Hill, NJ 07974

In order to protect their surface from damage caused by abrasion, optical fibers must be coated with one or more polymers as they are drawn. The properties of the coating are crucial to the optical and mechanical performance of the fiber in a lightwave transmission system. The role played by the coating in determining fiber performance is described.

The performance properties of optical fibers used in telecommunications applications are strongly influenced by the polymer coating applied during fiber drawing. The properties most affected by the coating are fiber strength and transmission loss. In addition manufacturing economy and productivity are directly linked with the coating process, which usually imposes the major limitation to the attainment of high drawing speeds (1).

Fiber Drawing and Coating

A fiber drawing and coating tower is shown schematically in Figure 1. The glass preform rod, manufactured in a separate process, is fed into the top of a high-temperature (2200°C) furnace at a controlled rate. The preform softens in the cylindrical furnace and the fiber is freely drawn from the molten end of the preform rod by means of a capstan located at the base of the tower. The fiber passes through a diameter monitor and control apparatus immediately below the furnace. The device provides a high resolution, e.g., 0.1 μm, measurement of the fiber diameter at a rapid update rate, e.g., 500 Hz, and generates a signal for use in a feedback control loop which adjusts capstan speed to control fiber diameter. Typically, fiber diameter can be held to a standard deviation of 0.2% (2), an important concern for fiber splicing and connectorization.

Because the surface of the silica glass fiber is very susceptible to damage caused by abrasion, it is necessary to coat the fiber in-line, as it is drawn, before it comes into contact with any solid surface. Prior to coating, however, the fiber must be allowed to cool. If its temperature is too high, it cannot be coated in a stable fashion (1,3). Hence, in present practice, tall draw towers, 7 meters or more in height, are employed in order to provide sufficient distance for convective cooling of the fiber (1). Forced convective cooling with appropriate gaseous heat transfer media such as helium, is also practiced (4).

Once the fiber is sufficiently cool, e.g. below 80°C (1), it can be coated, generally with one or two layers of organic polymers. Because the application method must not damage the glass surface, the polymers are applied in the liquid state, commonly as reactive prepolymer or hot melt formulations. The coating diameter and concentricity are monitored and controlled via suitable techniques (5). Once applied the coating must solidify very rapidly, before the fiber reaches the capstan at the base of the tower. Photocuring polymer formulations are therefore very advantageous and drawing speeds of greater than 10 m/sec have been achieved with these systems (6,7).

As mentioned earlier, the optical fiber performance properties most affected by the polymer coating are strength and transmission loss. The relationship between the coating properties and fiber properties will now be described.

<u>Fiber Strength</u>

The integrity of optical fibers deployed in the field depends upon the assurance that their strength is maintained over their entire length, which may be several kilometers. While silica glass possesses a very high intrinsic tensile strength, its strength is severely degraded by the presence of bulk or surface defects which act as stress concentrators (8). Such defects act as local weak points along the fiber. It is essential to eliminate sources of these defects in the materials used to fabricate optical fibers and in the fiber manufacturing processes as well.

Low strength fiber defects may arise from a number of sources. Surface flaws such as scratches or abrasions on the glass preform surface, which may not be healed by the fiber drawing process, can be removed by fire polishing (9). Solid particles emanating from the drawing furnace, or convected into the furnace from the drawing environment, may become lodged in the molten glass at the end of the preform. These particles are drawn into the fiber as surface particles whose size and shape determine the severity of the flaw produced. Thus the drawing environment must be as particle-free as possible, and the drawing furnace must be scrupulously maintained to avoid particle generation.

Another source of particles is the polymer coating material itself. Particles present in the material may locate at the coating-glass interface, where they may produce flaws by contact or abrasion. Therefore it is necessary to filter fiber coating materials to remove particles with sizes in the micron range. As an illustration of the relationship between coating particles and fiber strength, Figure 2 shows strength data for fibers coated with materials containing intentionally-added alumina particle contamination (10). Tensile strengths are high (~ 6 GPa) for fiber coatings with no added particles (control) as well as for those with particles in the $1-3\,\mu$m size range. Larger particles, however, degrade the strengths to 1.4–3.5 GPa (200-500 kpsi). Similar results are obtained for coatings having high or low modulus values, providing they have relatively high adherence to the fiber. If the adhesion is poor, however, relative movement can occur between the fiber and coating, leading to increased abrasion between the fiber surface and any particles located at the interface. Figure 3 displays a plot of the number of tensile failures per kilometer of a fiber in a prooftest (continuous tensile test of a fiber over its entire length) versus the force required to pull a 1 cm length of the fiber out of its coating. This force is a function of fiber-to-coating adhesion. The number of prooftest failures increases with decreasing adhesion for coatings containing particles in the $6-10\,\mu$m range. Significant failure frequencies also occur for fiber coatings with very low adhesion containing particles in the $1-3\,\mu$m size range. This range lies below the critical size for highly adherent coatings, typified by those used to produce the data of Figure 2. Therefore to ensure against fiber failure due to particle contamination of the coating, highly adherent coatings, filtered to remove all particles larger than $3\,\mu$m in size, should be suitable for most applications.

Coating defects which may expose the fiber to subsequent damage arise primarily from improper coating application. Defects such as large bubbles or voids, highly non-concentric coatings with unacceptably thin regions, or intermittent coatings, can be overcome through proper applicator design and process controls. Pressurized applicator designs have been successful in eliminating bubble incorporation in the coating (11). Coating concentricity can be monitored by optical techniques, e.g., laser forward scattering, and the applicator positioned so as to apply the coating concentrically (5). Intermittent coating is overcome by insuring the fiber is suitably cool at the point of its entry into the coating applicator so as to avoid coating flow instabilities (1).

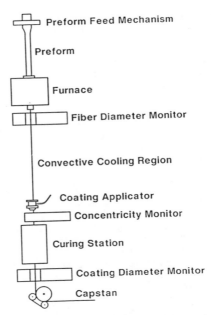

Figure 1   Schematic of an optical fiber drawing tower.

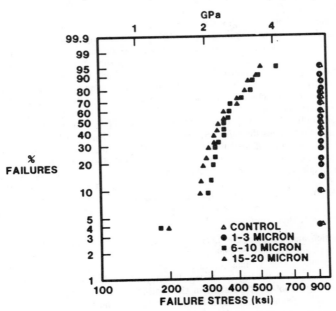

Figure 2   Failure stress distribution for fibers with coatings having various particle size ranges. (Reproduced with permission from Reference 10. Copyright 1985 IEEE).

Transmission Loss - Microbending

Optical fibers are susceptible to a transmission loss mechanism known as microbending (12,13). Since the fibers are thin and flexible, they are readily bent when subjected to mechanical stresses, such as those encountered during placement in a cable or when the cabled fiber is exposed to varying temperature environments or mechanical handling. If the stresses placed on the fiber result in a random bending distortion of the fiber axis with periodic components in the millimeter range, light rays, or modes, propagating in the fiber may escape from the core. These losses, termed microbending losses, may be very large, often many times the intrinsic loss of the fiber itself. Thus the fiber must be isolated from stresses which cause microbending. The properties of the fiber coating play a major role in providing this isolation, with coating geometry, modulus and thermal expansion coefficient being the most important factors (3).

Two types of coating geometries, displayed in Figure 4, are commonly used. Single coatings, employing a high modulus, e.g. 10^9 Pa, or an intermediate modulus, e.g. 10^8 Pa, are most easily produced and are used in applications requiring high fiber strengths or in cables which employ units, e.g., buffer tubes (15), where fiber sensitivity to microbending is not a significant problem. Dual coatings (12,14), consisting of a low modulus ($10^6 - 10^7$ Pa), elastomeric primary coating surrounded by a high modulus secondary coating, are used for more sensitive applications. This structure isolates the fiber very well from external stresses which would tend to cause local bending. Such stresses may be imposed in two distinct ways. First, non-uniform lateral stresses, imposed by the cable structure surrounding the fiber, may cause bending with periodic components in the microbending regime. The dual coating serves to cushion the fiber via the primary layer and to distribute the imposed forces via the secondary layer, so as to isolate the fiber from bending moments. Second, axial compressive loading of the fiber occurs when the surrounding cable components contract relative to the fiber. Such contraction results from both the differential thermal contraction of the cable materials relative to the glass fiber and from the viscoelastic recovery of residual orientation present in the cable materials. If the axial compressive load imposed on the fiber becomes large enough, the fiber will respond by bending or buckling (16). The low modulus primary coating is effective in promoting long bending periods for the fiber which are outside the microbending range.

A variety of techniques have been employed to induce microbending losses in fibers so that they may be studied. Some of the methods used include tensioning the fiber on a microscopically rough drum (9,13), pressing the fiber between blocks (17), and winding the fiber over itself under tension in a multilevel fashion on a spool (18). The microbending losses induced thereby decay with time owing to viscoelastic stress relaxation of the fiber coating. Figure 5 illustrates the time-dependence of the transmission losses of an optical fiber determined in the spool test. The loss changes parallel the time-dependence of the relaxation modulus of the fiber coating, also shown in Figure 5. These and other studies (9) reveal that the microbending losses decay to their minimum value when the coating modulus relaxes to a value of approximately 10 MPa (1500 psi) or below. Thus the equilibrium modulus of the primary coating used in a dual coating structure is desirably kept below 10 MPa. The most effective range is generally taken to be 0.5 to 5.0 MPa (75 to 750 psi), with moduli below 0.5 MPa regarded as too low for adequate physical protection.

The primary coating should possess a low glass transition temperature interval as well as a low modulus. The low glass transition temperature range insures that the primary coating will provide adequate resistance to microbending at the lowest temperatures encountered in field environments, e.g. $-40°C$ or lower. The secondary coating forms a tough, elastic shell around the primary coating which provides the fiber with suitable handling characteristics, e.g. abrasion resistance, low friction. Its glass transition (or melting temperature for crystalline thermoplastics) should lie well above normal field environment temperatures.

Coating Materials

Several types of primary coating materials have been employed commercially. The earliest materials used in large scale production were thermally-cured silicones (14). They are notable

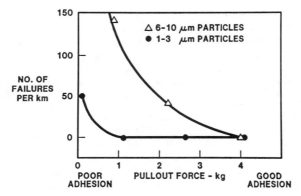

Figure 3   Effect of coating-to-fiber adhesion on prooftest failure frequency. (Reproduced with permission from Reference 10. Copyright 1985 IEEE).

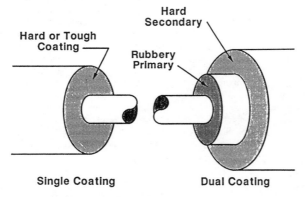

Figure 4   Schematic of single and dual coatings for optical fibers.

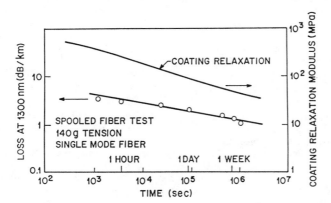

Figure 5   Relaxation of microbending losses and coating modulus for an optical fiber spooled under tension.

for their exceptionally low glass transition temperatures, ranging from approximately $-50°C$ for methyl-phenyl silicones to $-120°C$ for poly(dimethyl siloxane). These materials are supplied as two-part systems, usually consisting of a vinyl-terminated siloxane resin, a multifunctional silane crosslinker and a platinum catalyst. After mixing, the pot life of these materials ranges from a few to several hours, but they can be forced to crosslink very rapidly at elevated temperatures (200–400°C). Hence they can be applied to a fiber and cured in-line at draw speeds up to a few meters per second.

A serious problem has arisen with the curing chemistry used in these thermally-curable siloxanes. Residual silane groups remaining after the crosslinking reaction can generate hydrogen gas (19) which can then diffuse into the silica fiber (20). Within the glass network this interstitial molecular hydrogen possesses a vibrational spectrum which results in increased fiber attenuation in the wavelength regions used for optical transmission i.e., 1300 and 1550 nm (21,22). Furthermore the diffused hydrogen is capable of chemically reacting with the glass constituents to form OH groups and other chemical species which introduce irreversible absorption losses into the fiber (21,22). Consequently, the use of thermally curable silicones as primary coatings for optical fibers is declining.

Ultraviolet (UV)-radiation curable formulations have become the most extensively used primary coating type. In large measure the drive toward high fiber draw speeds (5 m/sec or greater) in production processes has led to UV-curable coatings as the most viable materials. Many of these materials will cure to an adequate degree at UV-doses as low as a few tenths Joule/cm^2. Such rapid curing is required to assure suitable property development during the short UV exposure times encountered in high-speed fiber drawing.

UV-curable primary coatings are most commonly based on urethane acrylate oligomers which have been modified to yield polymers with low glass transition temperatures, in some cases as low as $-40°C$ (23). Formulations based on acrylated polybutadienes (24) and silicone oligomers (25) have also been developed which yield even lower glass transition temperatures. In many cases, however, these materials with improved low temperature properties have deficiencies in other areas, such as low thermal stability, poor mechanical toughness, or low adhesion to glass. Thus, despite many years of intense research effort, improved UV-curable primary coatings for optical fibers are still needed.

Hot-melt thermoplastic primary coatings for optical fibers have also been employed to a limited extent (26,27). They may be formulated with reasonably low glass transition temperatures, but particle contamination is a difficult quality control problem with these materials. UV-curable coating formulations have largely supplanted hot melt coatings owing to the increased draw speed which they offer.

In general two types of secondary coatings are used in conjunction with the primary coatings described above. UV-curable epoxy acrylates or urethane acrylates are commonly used for in-line application, as the fiber is drawn. These same materials are also used extensively as single coatings. Extruded thermoplastics, such as nylon or polyester elastomers, are frequently used for off-line application processes.

Future Needs

The development of the high performance optical communication systems which are now being implemented would not have been possible without the advances in the polymer coating materials described herein. Despite the successes enjoyed to date, substantial progress in coating technology will be required as optical fiber applications in new areas are pursued. For example specialized coating development is needed for optical fiber sensors and for fibers intended for hostile chemical or high temperature environments. An understanding of the relationships between the materials and processes used in the fabrication of optical fibers and fiber performance is essential to the success of this effort.

Literature Cited

1. Paek, U. C.; Schroeder, C. M. Appl. Opt., 1981, 20, 1230.
2. Smithgall, D. H. Top. Meet. Opt. Fiber Transm., 3rd Tech. Dig., 1979, WF-4.
3. Blyler, L. L., Jr.; Aloisio, C. J. in Applied Polymer Science; Tess, R. W.; Poehlein, G. W.; Eds.; ACS Symposium Series No. 285; American Chemical Society: Washington, D.C. 1985; Chapter 38.
4. Jochem, C. M. G.; Van der Ligt, J. W. C. J. Lightwave Tech., 1986, LT–4, 739.
5. Smithgall, D. H.; Frazee, R. E. Bell Syst. Tech. J., 1981, 60, 2065.
6. Paek, U. C.; Schroeder, C. M. Electron. Lett., 1984, 20, 304.
7. Sakaguchi, S.; Kimura, T. OFC/OFS '85, Tech. Dig., 1985, MG-2.
8. DiMarcello, F. V.; Kurkjian, C. R.; Williams, J. C. in Optical Fiber Communications, Vol. 1, Li, T.; Ed., Academic Press: Olando, FL, 1985; chapter 4.
9. Blyler, L. L., Jr.; DiMarcello, F. V. Proc. IEEE, 1980, 68, 1194.
10. Huff, R. G.; DiMarcello, F. V. J. Lightwave Tech., 1985, LT–3, 950.
11. Lenahan, T. A.; Taylor, C. R.; Smith, J. V. OFC '82, Tech. Dig., 1982, WCC6.
12. Gloge, D. Bell Syst. Tech. J., 1975, 54, 245.
13. Gardner, W. B. Bell Syst. Tech. J., 1975, 54, 457.
14. Naruse, T.; Sugawara, Y.; Masuno, K. Electron. Lett., 1977, 13, 153.
15. Bark, P. R.; Oestreich, U.; Zeidler, G. Proc. Int. Wire & Cable Symp., 1978, 27, 379.
16. Blyler, L. L., Jr.; Gieniewski, C.; Quan X.; Ghoneim, H. Proc. Int. Wire & Cable Symp., 1983, 32, 144.
17. Katsuyama, Y.; Mitsunaga, Y.; Tanaka, C.; Waki, T.; Ishida, Y. Appl. Opt., 1982, 21, 1337.
18. Kaiser, P.; French, W. G.; Bisbee, D. L.; Shiever, J. W. IOOC '79, 1979, 7.4-1.
19. Kimura, T.; Sakaguchi, S. Electron. Lett., 1984, 20, 315.
20. Stone, J.; Chraplyvy, A. R.; Burrus, C. A. Optics Lett., 1982, 7, 297.
21. Uesugi, N.; Murakami, Y.; Tanaka, C.; Ishida, Y.; Mitsunaga, Y.; Negishi, Y.; Uchida, N. Electron. Lett., 1983, 19, 762.
22. Rush, J. D.; Beales, K. J.; Cooper, D. M.; Duncan, W. J.; Rabone, N. H. Telecom. Technol. J., 1984, 2, 84.
23. Broer, D. J.; Mol, G. N. IOOC–ECOC '85, 1985, 523.
24. Ohno, R.; Kikuchi, T.; Matsumura, Y. Proc. Int. Wire & Cable Symp., 1985, 34, 76.
25. Cush, R. J.; Lutz, M. A. Plastics in Telecommunications IV, 1986, p6.
26. Miller, T. J.; Hart, A. C., Jr.; Vroom, W. I., Jr.; Bowden, M. J. Electron. Lett., 1978, 14, 603.
27. Blyler, L. L., Jr.; Hart, A. C., Jr.; Levy, A. C.; Santana, M. R.; Swift, L. L. Proc. 10th European Conf. on Opt. Comm., 1982, 245.

RECEIVED May 29, 1987

Chapter 35

Oriented Polymers Obtained by UV Polymerization of Oriented Low Molecular Weight Species

D. J. Broer and G. N. Mol

Philips Research Laboratories, P.O.B. 80000, 5600 JA Eindhoven, Netherlands

Anisotropic polymer filaments could be produced by in-situ photopolymerization of oriented acrylate monomers. Ordering of the monomers was achieved
by an elongational flow prior to the polymerization process. The produced
polymers showed a high elastic modulus and a low thermal expansion coefficient in the direction of the orientation.

To reduce optical transmission losses under lateral forces, silica optical fibers are provided
with a soft primary coating and a hard secondary coating [1,2,3]. Both coatings are applied
on-line with the fiber drawing process and are cured by UV-irradiation. However, the
choice of conventional UV-curing coatings gives rise to enhanced optical attenuation at low
temperatures [4,5,6]. The large difference in linear thermal expansion coefficient between
the secondary coating ($\approx 10^{-4}$ °C^{-1}) and silica ($5 \cdot 10^{-7}$ °C^{-1}) causes microbending of
the silica fiber within the soft primary coating due to compression stresses in axial direction.
Ideally, the secondary coating should have a linear expansion coefficient equal to that of
silica. This requirement prompted us to investigate new polymeric materials with a low thermal expansivity which can be processed within the boundaries as fixed by the optical fiber
drawing process. The latter for instance means high extrusion rates at relatively low
pressures and ultra-fast solidification.

As known [7,8], the thermal expansion coefficient is reduced in the direction of the
molecular orientation obtained by stretching of a thermoplastic polymer during or directly
after its processing. In special cases thermotropic polyesters are applied to facilitate the process of molecular orientation [9]. However, in all these cases solidification must proceed
either by cooling down from the melt or by evaporation of the solvent. These relatively slow
processes are not suited for on-line optical fiber coating.

In this paper we will demonstrate that molecular orientation can also be achieved by
starting with a low molecular weight species (M.W. < 2000) which is oriented in an elongational flow and subsequently cured under UV-irradiation. The orientation of the monomer
is frozen-in by the ultra-fast polymerization and crosslinking. The advantages are low-
pressure, low-temperature processing, fast extrusion and solidification within 0.1 seconds.
Moreover a three dimensional polymer network is formed, which should maintain the
molecular ordering over a wide temperature range.

To get a better understanding of the material properties, the present investigations have
been carried out on polymeric filaments, i.e. there are no silica fibers involved as a carrier
for the investigated polymer.

0097–6156/87/0346–0417$06.00/0

Experimental

The present studies have been concentrated on the molecular ordering and photopolymerization of the polyesterurethane acrylate shown in Figure 1, to which 4 w% 2,2-dimethyl-2-hydroxyacetophenone was added as photoinitiator. Figure 2 schematically shows the preparation of oriented filaments. The molten monomer is extruded through a die while temperature and pressure are controlled. The resulting liquid thread is stretched and subsequently cured by irradiation with a UV lamp (Fusion Systems, electrodeless mercury lamp, 0.3 W.cm^{-2} in the 365 nm region). After leaving the die, the liquid filament phases passes a room temperature zone where some cooling takes place before it enters the irradiation zone. The temperature during irradiation was difficult to measure exactly but was estimated to be between 80 and 150°C, depending on the extrusion rate. The data given in fig. 2 serve as an example and were varied during our experiments.

Results and Discussion

The polyesterurethane acrylate monomer is a semi-crystalline solid at room temperature with a glass transition of the amorphous phase at −6°C and melting of the crystalline phase at 52°C. When cooled from the isotropic liquid phase to room temperature the monomer remains undercooled for several hours. The rate of photopolymerization and the ultimate conversion of the acrylate groups strongly depends on the polymerization temperature. The overall activation energy for the rate of polymerization above the melting temperature is 10 kJ/mole which is a normal value for photoinitiated acrylate polymerization [10,11]. Below this temperature the activation energy increases to 89 kJ/mole, indicating that the polymerization kinetics are strongly influenced by the mobility of the monomer. After polymerization under isotropic conditions, the glass transition becomes very broad with a temperature range of − 20 to + 90°C as determined by DSC and DMTA. DSC and TMA measurements show melting of a small crystalline phase at 55°C. Crystallinity could not be detected by polarization microscopy.

To obtain filaments the monomer was extruded from the melt at 80°C, well above the melting temperature (fig. 2). Figure 3 shows the extrusion flow rate as a function of pressure for two different die geometries, indicating a linear behaviour in both cases. It was observed that there is only a limited flow rate region where stable liquid filaments are formed. This region decreases with increasing die length. At low flow rates the stability of the filaments is controlled by the surface tension, i.e. the liquid thread breaks up into droplets. At high flow rates the stability of the filaments is controlled by the process of molecular orientation in the die and relaxation after leaving the die. At too high flow rates melt fracture phenomena can be observed in the liquid thread. The flow rate region for stable filament production can be broadened by the addition of small amounts of surfactant to the reactive precursor.

The ultimate molecular orientation in the filament is induced by an elongational flow prior to polymerization. The stresses needed for stretching the liquid threads are small (0.1 to 10 MPa). As orientation is reduced by fast relaxation of the molecules, the UV irradiation was already started during stretching. Since there is a continuous equilibrium between molecular orientation and relaxation, the degree of orientation along the fiber axis (z-axis) can be considered to be a function of the strain rate. The strain rate at position $z, \dot{\epsilon}(z)$, is given by Equation 1 [12]:

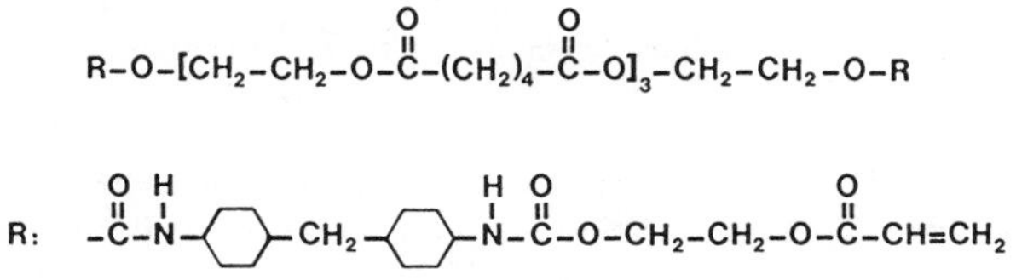

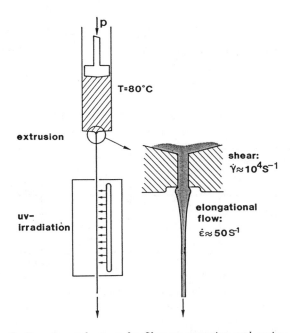

Fig. 1. *Polyesterurethane diacrylate reactive precursor.*

Fig. 2. *Experimental set up for filament extrusion and curing.*

$$\dot{\epsilon}(z) \ = \ \frac{F}{S(z) \cdot \kappa(z)}$$

where F is the axial tensile force, S(z) the cross section of the filament and $\kappa(z)$ the elongational viscosity at position z. F remains constant during the drawing process, however S(z) and $\kappa(z)$ change along the z-axis. S(z) decreases due to the stretching process, whereas $\kappa(z)$ increases due to the cooling of the fiber. As a result, $\dot{\epsilon}(z)$ decreases gradually with the distance z. On the other hand the relaxation rate of the oriented molecules also decreases due to cooling of the filament. The total effect of all these phenomena is that the die-lamp distance is not very critical. A suitable distance appeared to be 0.2 m. As soon as the polymerization starts, $\kappa(z)$ increases greatly while F and S(z) remain constant. From the resulting large decrease of $\dot{\epsilon}(z)$ it can be concluded that elongation only takes place in the liquid, non-polymerized phase.

Polymeric filaments produced in this way have a diameter of 100 to 300 μm and show anisotropy in their properties, indicating molecular orientation. For instance the birefringence of this fiber is 0.0073. Table 1 compares the modulus measured both in axial and in lateral direction, with the modulus of the same material cured under isotropic condition at 80°C. The increase in modulus in axial direction is obvious. The decrease of the modulus at higher temperatures can be ascribed to both the glass transition and melting of crystalline areas.

Table 1.

Modulus in GPa of polyesterurethane acrylate, respectively cured under isotropic and anisotropic conditions

Temperature (°C)	Isotropic	Anisotr. axial direction	Anisotr. lateral direction
− 40	1.8	34.2	—
25	0.6	14.6	0.6
80	0.02	0.6	—

As already stated in the introduction the main interest of our investigation is the influence of the molecular orientation on the axial linear thermal expansion coefficient. Figure 4 shows the linear expansion of an oriented filament measured in the two main directions and compares these measurements with the linear expansion of an isotropically cured sample. Again the differences are obvious. Overall, the thermal expansivity of the oriented fiber in axial direction is one order of magnitude lower than that of the isotropic material. In addition, the increase of the thermal expansion coefficient at the glass transition temperature has become less pronounced. The thermal expansion coefficient measured in lateral direction is higher than that of the isotropically cured sample. This can be understood from the consideration that the volume expansion coefficient is not subjected to large changes. Table 2 compares the volume thermal expansion coefficients as calculated from the linear thermal expansion coefficients. Above the melting temperature the values of the isotropic and the oriented samples are identical. Below this temperature there are some small differences indicating a somewhat higher degree of crystallinity in the case of the oriented sample.

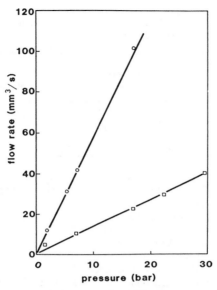

Fig. 3. *Flow rate vs. pressure during extrusion of the monomer at* 80°C. *The die length is 5 mm* (o) *and 20 mm* (□) *resp.*

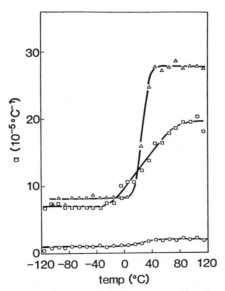

Fig. 4. *Linear thermal expansion coefficients as a function of temperature:* (□) *isotropically cured at* 80°C; (o) *and* (△) *oriented polymer filament* (200 μm diameter) *measured in the axial and lateral direction, resp.*

Table 2.
Volume thermal expansion coefficients in $°C^{-1}$ as calculated from the linear
thermal expansion coefficients

Temperature (°C)	Isotropic	200 μm Oriented filament
− 40	$2.1 \cdot 10^{-4}$	$1.7 \cdot 10^{-4}$
25	$3.9 \cdot 10^{-4}$	$3.6 \cdot 10^{-4}$
80	$5.8 \cdot 10^{-4}$	$5.8 \cdot 10^{-4}$

The question may arise whether the same changes in material properties can also be obtained by stretching the filaments after the polymerization has taken place. In our experience the anisotropy was never so great in such cases, whereas the isotropic state was almost completely recovered when the samples were heated above the melting temperature. At this temperature the post-polymerization drawn filaments retained the original dimensions they possessed before stretching. Apparently some strain-induced crystallization yielded a metastable anisotropy which was lost under the combined action of entropy and strained crosslinks when the crystalline areas were melted.

The investigation on oriented polymeric networks obtained by the photopolymerization of oriented low molecular weight species, as presented in this paper, has been carried out with a more or less conventional acrylate monomer. Already with this material an anisotropy in properties could be demonstrated. It is to be expected that even more pronounced effects can be obtained with monomers which have a strong tendency to alignment. Based on this idea we are now investigating liquid crystalline monomers in our laboratory.

Conclusions

It has been demonstrated that molecular orientation can be achieved starting with a low molecular weight species which is oriented in an elongational flow and subsequently cured under UV-irradiation. The orientation of the monomer is frozen-in by the ultra-fast process of polymerization and crosslinking. Both extrusion and stretching can be carried out at relatively low temperatures and pressures. Polymer filaments produced in this way are definitely anisotropic as is evidenced by their birefringence and by a strong increase of the tensile modulus and a decrease of the thermal expansion coefficient in the axial direction.

References

1. L.L. Blyler, A.C. Hart, A.C. Levy, M.R. Santana and L.L. Swift, 8th European Conf. Opt. Commun., Cannes, France, 1982.
2. D. Gloge, Bell Syst. Techn. J. 1975, **54**, 245.
3. D.J. Broer and G.N. Mol, 5th Int. Conf. Int. Optics Opt. Fibre Comm. and 11th European Conf. Opt. Comm., Venezia, Italy, 1985.
4. Y. Katsuyama, Y. Mitsunaga, Y. Ishida and K. Ishihara, Applied Optics 1980, **19(24)**, 4200.

5. N. Yoshizawa, M. Ohnishi, O. Kawata, K. Ishihara and Y. Negishi, J. Lightwave Techn. 1985, **LT-3**, 779.
6. T.A. Lenahan, AT&T Techn. J. 1985, **64**, 1565.
7. R.S. Porter, N.E. Weeks, N.J. Capiati and R.J. Krzewki, J. Thermal Anal. 1975, **8**, 547.
8. M. Jaffe, In Thermal Characterization of Polymeric Materials, E.A. Turi, Ed.; Academic Press, New York, 1981.
9. S. Yamakawa, Y. Shuto and F. Yamamoto, Electr. Letters, 1984, **20**, 199.
10. G.R. Tryson and A.R. Schultz, J. Pol. Sci., Pol. Phys. Ed., 1979, **17**, 2059.
11. G. Odian, In Principles of Polymerization, 2nd ed., Wiley Interscience, New York, 1981.
12. C.D. Han, In Rheology in Polymer Processing, Academic Press, New York, 1976.

RECEIVED May 18, 1987

HIGH-TEMPERATURE POLYMERS FOR DIELECTRIC APPLICATIONS

HIGH-TEMPERATURE POLYMERS FOR DIELECTRIC APPLICATIONS

The use of polymers as dielectrics in electrical applications can be traced back to the early twentieth century when Bakelite was employed as an insulator. Since that time, the dielectric properties required of materials have become increasingly stringent to the point that polymers have been virtually excluded from such applications in microelectronic devices. However, as chip geometries shrink and chip manufacturing embraces multilevel technology, the need for new or improved dielectric polymers has assumed new importance.

In addition to their use as dielectrics in multilevel devices, high-temperature polymers are also finding use as passivation layers, barrier layers, adhesives, encapsulants, and fabrication aids. Most of the high-temperature polymers in use today are polyimides and polyimide-like materials. From 1983 to 1985, sales of polyimides for electronics showed an annual growth of 30 to 35%, and it is estimated that this growth will accelerate to 40 to 50% per year (through 1989). This is particularly noteworthy given the prejudice and belief of many in an industry which has historically been very resistant to change, that organic materials do not belong in microelectronic devices.

The major driving force to replace inorganic materials such as silicon oxide and silicon nitride as interlevel dielectrics is the need for improved planarization in multilevel devices. The plastic properties of polymers, coupled with their ability to form highly uniform, integral films via spin-coating from solution, make them extremely attractive candidates as planarizing, dielectric materials. The degree of planarization depends on the type of polymer, molecular weight and other polymer properties. Polyimides are viewed by some as inferior planarizing materials and consequently work continues on other classes of high-temperature polymers such as polyacetylenes, which have been claimed to give better planarization.

An organic polymer must also possess other properties if it is to be useful in high-temperature dielectric applications. It must withstand temperatures as high as 450°C, which is encountered in sintering aluminum. Most high-temperature polymers currently available can meet this criterion. However, there is considerable effort underway to develop polymers that will extend the useful range to 500°C and higher. The polymer must also possess the required electrical characteristics along with physical and chemical properties needed to withstand temperature cycling and other processing steps. In order to prevent corrosion and related problems, the polymer must adhere to a wide variety of metal, semiconductor, and ceramic substrates. Most polymers require an adhesion promoter, but some silicon-containing polymers are inherently adherent. The polymer must meet strict purity requirements in order to prevent ion migration and related electrical problems. Typically, impurity levels no greater than 1ppm can be tolerated. Although most, if not all of these properties can be met by many of today's high-temperature polymers, there are still some areas of concern regarding device application. These include problems such as water adsorption by the film and film shrinkage during processing, both of which continue to be addressed.

Once the polymer film is applied, it must be patterned. This can be accomplished via multilevel techniques in which the pattern is lithographically defined in a resist layer coated on top of the polymeric dielectric layer and then etched into the bottom layer. Wet-etching may be used, although in most very large scale integrated circuit (VLSI) applications film removal is more commonly performed by oxygen plasma techniques such as reactive-ion etching. The most active area of research today is direct patterning using photosensitive polyimides, although concerns about photospeed, film shrinkage, and stability have limited the use and availability of this approach. Recent research has been directed toward laser patterning.

The following papers give a flavor of the type of research being conducted on high-temperature polymers for electronic applications. The initial paper gives an overview of the area followed by three papers describing chemistry associated with the preparation of improved polymers. The next paper deals with the leading edge in research on ultrathin dielectric films. This section concludes with a paper describing a working device which employs a high-temperature dielectric polymer, the goal of most research in this area.

It is only a matter of time before widespread acceptance of the use of organic materials as an integral part of a final device is reached. With that acceptance, the era of high-temperature polymers for electronic applications will have arrived.

Gary C. Davis
General Electric
Research & Development Center
Schenectady, NY

Chapter 36

Polyimides in Microelectronics

Stephen D. Senturia

Microsystems Technology Laboratories, Department of Electrical Engineering and Computer Science, Massachusetts Institute of Technology, Cambridge, MA 02139

Polyimides are finding increased use in microelectronics [1-6]. There are four primary areas of application: 1) as fabrication aids; 2) as passivants and interlevel insulators; 3) as adhesives; and 4) as components of the substrate or circuit board. In each application, the requirements on properties and performance differ. This paper addresses primarily the first two applications.

Polyimides for microelectronics use are of two basic types. The most commonly used commercial materials (for example, from Dupont and Hitachi) are condensation polyimides, formed from imidization of a spin-cast film of soluble polyamic acid precursor to create an intractable solid film. Fully imidized thermoplastic polyimides are also available for use as adhesives (for example, the LARC-TPI material), and when thermally or photo-crosslink able, also as passivants and interlevel insulators, and as matrix resins for fiber-reinforced-composites, such as in circuit boards. Flexible circuits are made from Kapton polyimide film laminated with copper. The diversity of materials is very large; readers seeking additional information are referred to the cited review articles [1-3, 6] and to the proceedings of the two International Conferences on Polyimides [4,5].

We now examine how applications and properties interact, by examining the uses of polyimides as fabrication aids and as passivants and interlevel insulators.

Fabrication Aids

Fabrication aids include such applications as photoresists, planarization layers in multi-level photoresist schemes, and as ion implant masks. In these applications, the polymer is applied to the wafer or substrate, is suitably cured and/or patterned, but is removed after use.

In the case of a photoresist, the ultimate definable feature size together with the ability of the material to withstand either chemical etchants or plasma environments determines the domain of utility. The feature size is in turn determined by the wavelength required for exposure, the sensitivity and contrast of the resist, and the dimensional stability of the material during exposure, development, and subsequent processing. Adhesion of the resist to the substrate is critical both for patterning and use, and adhesion can be affected by surface preparations, and by residual stresses developed during deposition and cure. While photo-imagable polyimides have been introduced, their principal intended application is as a component of the finished part, either as passivant or interlevel dielectric (see below).

In multi-level resist schemes, the polyimide itself does not need to be photo-imageable. Instead, it is used to planarize the substrate topography, as illustrated in Figure 1. The planarization process resembles that which occurs with ordinary paints and varnishes. Because polyimides are applied as liquids, the surface tension keeps the surface flat (Figure 1a) until the cure or drying process proceeds sufficiently far to prevent flow. Thereafter, shrinkage due to solvent loss or cure results in partial recovery of the underlying feature (Figure 1b). The achievable degree of planarization is controlled by shrinkage [7]. This, in turn, is controlled by the specific polyimide chemistry, by the molecular weight, and by the percent solids in the film at the point where flow ceases. In general, polyimides intended only for use as temporary

0097–6156/87/0346–0428$06.00/0

planarization layers can be of lower molecular weight than those intended for interlevel dielectric application, hence, can achieve better planarization. After the polyimide is applied and cured, the photoresist is then placed over the polyimide, in some cases with an intervening aluminum masking layer. The relatively flat surface provides for maximum resolution of the exposed pattern in the resist, while the aluminum serves as an etch mask for creating deep vertical cuts in the polyimide using reactive ion etching in oxygen.

The ability to create high-aspect-ratio etched structures in polyimide using an aluminum mask in an oxygen plasma operated in reactive-ion-etch mode permits polyimide to be used as a high-resolution ion-implant mask [8,9]. Of critical importance in this application is the stopping power of the polyimide at the desired implant energy, the dimensional stability of the structure under high implant doses, and the possible effects of temperature rise during the implant.

In all of these fabrication-aid applications, the polyimide is removed after it has performed its task. Removal of polyimides is normally accomplished in an oxygen plasma.

Passivants and Interlevel Dielectrics

In order for a polyimide to be useful as an interlevel dielectric or protective overcoat (passivant), additional demanding property requirements must be met. In the case of the passivant, the material must be an excellent electrical insulator, must adhere well to the substrate, and must provide a barrier for transport of chemical species that could attack the underlying device. It has been demonstrated that polyimide films can be excellent bulk barriers to contaminant ion motion (such as sodium) [10], but polyimides do absorb moisture [11,12], and if the absorbed moisture affects adhesion to the substrate, then reliability problems can result at sites where adhesion fails. However, in the absence of adhesion failure, the bulk electrical resistance of the polyimide at ordinary device operating temperatures and voltages appears to be high enough to prevent electrochemical corrosion [13].

When used as an interlevel dielectric, even greater demands are placed on the polyimide. Because integrated circuit processing includes as a final step a metal sinter at 400 °C, the interlevel insulator film must withstand such exposures without degradation of electrical, chemical, or mechanical properties. In addition, the deposition, cure, and etch process must provide for reliable interconnection between the metal layers above and beneath the film (the "via contact") [8]. Issues of ion motion, moisture uptake, and electrical conduction both in bulk and at interfaces must also be considered carefully.

Ion motion can be studied using the method orignally suggested by Brown [10], and subsequently improved by Neuhaus [14]. The sample consists of a metal-polyimide-oxide-silicon (MPOS) parallel plate capacitor (Figure 2). The device is driven with a slow triangle-wave voltage, and the total charge on the capacitor is measured with an electrometer. By measuring the hysteresis in the charge-voltage characteristic of these capacitors at elevated temperature (Figure 3), the concentration and mobilities of mobile ions can be determined. The method can detect ion concentrations per unit area on the order of 2×10^{10} cm^{-2}. A typical polyimide has about 1 ppm sodium, which for a 1 μm thick film corresponds to about 10^{12} ions/cm^2. A critical issue is whether the sodium in polyimide will spontaneously transfer to oxides under normal device fabrication conditions. Studies at cure temperatures show no spontaneous transfer [14].

Extension of such measurements to ordinary device operating temperatures (below 100 °C) is complicated by the problem of moisture absorption. Moisture uptake can be monitored gravimentrically, but is more readily monitored using parallel-plate capacitors with perforated upper electrodes (Figure 4) [12]. The capacitance of these devices is observed to depend almost linearly on ambient relative humidity in the temperature range 20 - 80 °C (see Figure 5). An explanation for this effect is that the absorbed water molecule contributes its permanent dipole moment to the overall molar polarizability of the medium. Quantitative analysis of moisture absorption in a PMDA-ODA polyimide based on the Clausius-Mosotti equation has determined that, indeed, the dipole moment of the absorbed water molecule is equal to 0.9 times that of the free water molecule, suggesting that absorption is into microvoids created by the release of water during the imidization step [15].

Because of both moisture absorption and ion motion, it has proved difficult to carry out reproducible studies of bulk conduction at ordinary device temperatures. There is a large and somewhat contradictory literature which has been reviewed in [6] and in [13]. When a DC

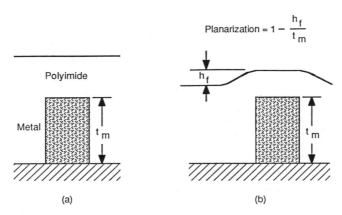

$$\text{Planarization} = 1 - \frac{h_f}{t_m}$$

Figure 1. Illustrating planarization over a metal line

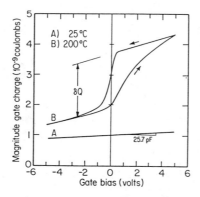

Figure 2. Cross section of metal- polymer-oxide silicon test structure.

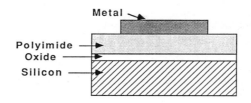

Figure 3. Charge–voltage data from MPOS device at two temperatures, with hysteresis due to mobile ions. (Reproduced with permission from Ref. 14. Copyright 1985 The Metallurgical Society.)

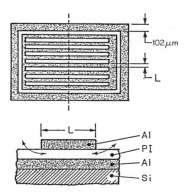

Figure 4. Top view and cross-section of device used for moisture uptake and conduction studies. (Reproduced with permission from Ref. 12. Copyright 1985 The Metallurgical Society.)

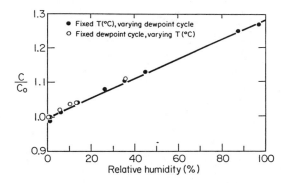

Figure 5. Relative capacitance of a polyimide capacitor versus moisture. (Reproduced with permission from Ref. 12. Copyright 1985 The Metallurgical Society.)

voltage is applied to a device structure like that in Figure 4, the current initially decays with time following a power law t^{-n} with n @ 0.8, then appears to level out at what is called the "transport current," although it, too, may actually be due to a slow internal transient. With careful attention both to contaminant ions and moisture, however, quantitative and reproducible conductivity data have been obtained, as illustrated in Figure 6. From data of this type at a wide range of temperatures, one can extract the temperature dependence of the transport current, as shown in Figure 7 for two device-grade polyimides. At temperatures below 100 °C, moisture has a strong effect on the current transient, as shown in Figure 8 [13].

There are fewer published measurements of mechanical properties and adhesion [16-19]. Generally, residual stresses have been measured with curvature of inorganic-polymer bimorphs, either by coating silica reeds or an entire silicon wafer. Mechanical properties have been measured both statically and dynamically. Adhesion has typically been measured with 90° peel force for a controlled geometry, although such measurements are primarily for comparative purposes, being difficult to interpret quantitatively in terms of the work of adhesion. Recently, we have developed new methods for in-situ measurement of residual stress, modulus, tensile stress, and work of adhesion using a suspended membrane technology microfabricated using anisotropic etching of silicon (see Figure 9) [20,21]. This work is still in its preliminary stages, but is expected to help clarify the role of process conditions on stress and adhesion.

Of great ultimate interest in the application of polyimides to microelectronics is a fundamental understanding of structure-property relations, that is, an understanding of how to relate critical performance properties to details of composition and structure. As new experimentas becomes better documented, and as insights into how to achieve reproducible measurements are obtained, the prospect of achieving that goal improve.

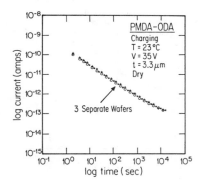

Figure 6. Demonstration of reproducibility in current versus time measurement. (Reproduced with permission from Ref. 13. Copyright 1987 The Metallurgical Society.)

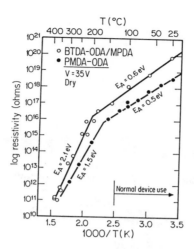

Figure 7.   Apparent resistivity versus 1/T for two polyimides.
(Reproduced with Ref. 13.   Copyright 1987 The Metallurgical
Society.)

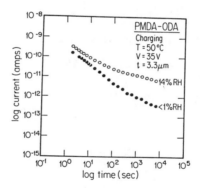

Figure 8. Effect of moisture on current versus time transient.
(Reproduced with permission from Ref. 13.   Copyright 1987 The
Metallurgical Society.)

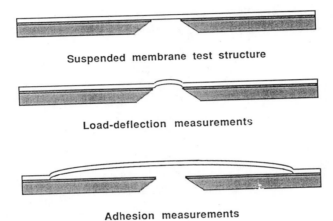

Suspended membrane test structure

Load-deflection measurements

Adhesion measurements

Figure 9. Schematic test site for measurement of residual stress, modulus, tensile stress, and adhesion of polyimide.

References

1. Wilson, A. M., "Polyimide Insulators for Multilevel Interconnections", Thin Solid Films, 83, 145-163 (1981).

2. Jenson, R. J.,Cummings, J. T., and Vora, H., "Copper Polyimide Materials System for High Performance Packaging", 1984 IEEE Electronic Components Conference, pp. 73-81.

3. Moghadam, F. K., "Development of Adhesive Die Attach Technology in Cerdip Packages; Material Issues", Solid State Technology, 27, 149-157 (1984).

4. Proc. 1st International Conference on Polyimides, Ellenville, NY, 1982; published as: K. L. Mittal, editor Polyimides: Synthesis, Characterization and Applications, New York, Plenum Press, 1984; Vols. 1,2.

5. Proc. 2nd International Conference on Polyimides, Ellenville, NY, 1985, sponsored by Mid-Hudson Section, Society of Plastics Engineers.

6. Senturia, S. D., Miller, Rebecca A., Denton, Denice D., Smith, Frank W., III, and Neuhaus, Herbert J., "Polyimides and VLSI: A Research Perspective", Ref. 5, pp. 107-118.

7. Day, D. R., Ridley, D., Mario, J., and Senturia, S. D., "Polyimide Planarization in Integrated Circuits", Ref. 4, pp. 767-782.

8. Herndon, T. O., Burke, R. L., and Landoch, W. J., "Inter-Metal Polyimide Via Conditioning and Plasma Etching Techniques", Ref. 4, pp. 809-826.

9. Herndon, T. O., Burke, R. L., and Yasaitis, J. A., "Polyimide for High Resolution Ion Implant Ion Masking", Solid State Technology, 27, 179 (1984).

10. Brown, G. A., "Reliability Implications of Polyimide Multilevel Insulators", IEEE 1981 Reliabilility Physics Symposium, pp. 282-286.

11. Sacher, E., and Susko, J. R., "Water Permeation of Polymer Films: III, High Temperature Polyimides", J. Appl. Polymer Sci., 26, 679-686 (1981).

12. Denton, D. D.,Day, D. R., Priore, D. F., Senturia, S. D., Anolick, E. S., and Scheider, D., "Moisture Difusion in Polyimide Films in Integrated Circuits", J. Elec. Mat., 14, 119-136 (1985).

13. Smith, F. W., Feit, Z., Neuhaus, H. J., Day, D. R., Lewis, T. J., and Senturia, S. D., "Conduction in Polyimide between 20 and 300°C", J. Electronic Mat., 16, 93-106 (1987).

14. Neuhaus, H. J., Day, D. R., and Senturia, S. D., "Sodium Transport in Polyimide-SiO_2 Systems", J. Elec. Mat., 14, 379-404 (1985).

15. Denton, D., Camou, J. B., and Senturia, S. D., "Effects of Moisture Uptake on the Dielectric Permittivity of Polyimide Films", Proc. 1985 ISA Conf. on Moisture and Humidity, pp. 505-513.

16. Rothman, L. B., "Properties of Thin Polyimide Films", J. Electrochem. Soc., 127, 2216-2220 (1980).

17. Lacombe, R. H., and Greenblatt, J., "Mechanical Properties of Thin Polyimide Films", Ref. 4, pp. 647-670.

18.   Kochi, M., Isoda, S., Yokota, R., Mita, I., and Kambe, H., "Mechanical Properties and Molecular Aggregation of Aromatic Polyimides", Ref. 4, pp. 671-682.

19.   Geldermans, P., Goldsmith, C., and Bedetti, F., "Measurement of Stresses Generated during Curing and in Cured Polyimide Films", Ref. 4, pp. 695-711.

20.   Mehregany, M., Allen, M. G., and Senturia, S. D., "Use of Micromachined Structures for the Measurement of Mechanical Properties and Adhesion of Thin Films", IEEE 1986 Workshop on Solid-State Sensors, Hilton Head NC, June 1986.

21.   Mehregany, M., Allen, M. G., Howe, R. T., and Senturia, S. D., "Novel microstructures for the study of residual stress in polyimide film", 1986 Electronic Materials Conf., Amherst MA, June 1986.

RECEIVED April 24, 1987

Chapter 37

Soluble Aromatic Polyimides for Film and Coating Applications

Anne K. St. Clair and Terry L. St. Clair

Materials Division, Langley Research Center, National Aeronautics and Space Administration, Hampton, VA 23665-5225

Because of their toughness, flexibility and remarkable thermal stability, linear all-aromatic polyimides are excellent candidate film and coating materials for advanced electronic circuitry and wire coating applications. In past years, however, the inherent insolubility (1-2) of these polymers has somewhat limited their usefulness for electronic applications.

The classic approach of incorporating aromatic pendant groups along the polymer backbone has been used successfully to improve the solubility of linear polyimides.(3-5) Variation in the isomeric points of attachment of bridging groups in the diamine monomers has also proved effective at enhancing solubility of polyimides in common organic solvents.(6) More recently, the combined effects of incorporating bulky ($-CF_3$ and $-SO_2$) groups, linking or bridging groups, and meta-linked diamines to reduce charge transfer complexing in aromatic polyimides and thereby facilitate solubility have been studied.(7) Other researchers have also reported the solubility of polyimides prepared with biphenyltetracarboxlic dianhydrides in a high-boiling solvent, N-methylpyrrolidone.(8-9)

The purpose of this investigation was to observe the effects on solubility of changing isomeric points of attachment of phenoxy units in the diamine portion of several all-aromatic polyimide systems. Hexafluoropropane- and oxygen-containing dianhydrides were used in this study because of their known value at contributing to polyimide solubility.(6-7)

EXPERIMENTAL

MATERIALS. The 4,4'-oxidiphthalic anhydride (ODPA) was recrystallized from anisole and sublimed at 200°-210°C/1 mm prior to use (m.p. 224°C). The 2,2-bis(3,4-dicarboxphenyl) hexafluoropropane dianhydride (6F) was obtained from American Hoechst and recrystallized from toluene/acetic anhydride (m.p. 241°C). The 4,4'-oxydianiline (4,4'-ODA) was obtained

commercially, recrystallized and sublimed (m.p. 188°C). The 3,4'-
and 3,3'-oxydianiline diamines (3,4'-ODA and 3,3'-ODA) are
experimental materials obtained from Mitsui Toatsu, Inc. The 3,4'-
ODA was recrystallized from chloroform/hexane (m.p. 74°C); and the
3,3'-ODA was vacuum distilled at 125°C/0.1 mm (m.p. 77°C). Source
and purity of the 2,4'-ODA has been reported (10) elsewhere. The
1,4-bis(4-aminophenoxy)benzene, 1,4(4)-APB; 1,3-bis(4-aminophenoxy)
benzene,1,3(4)-APB; and 1,4-bis(3-aminophenoxy)benzene, 1,4(3)-APB,
were experimental diamines obtained and used as received from
Mitsui Toatsu, Inc., with melting points of 171°, 115°, and
127° respectively. The 1,3-bis(3-aminophenoxy)benzene, 1,3(3)-APB,
was obtained commercially (m.p. 105°C). Dimethylacetamide (DMAc)
used as a solvent for polymerization was vacuum distilled at 107°C
from calcium hydride.

PREPARATION of POLYMERS. Polyamic acids were prepared at 15%
solids (w/w) by adding the diamine and DMAc to a flask flushed with
dry nitrogen. An equimolar amount of solid dianhydride was then
added to the dissolved diamine. After stirring 8-24 hours at room
temperature, the resulting polyamic acid solutions were
refrigerated. Films were prepared by casting the amic acid resins
onto soda-lime glass plates in an enclosed dust-free chamber at 10%
relative humidity. Films were cured unless otherwise indicated by
heating in a forced air oven for one hour each at 100°, 200° and
300°C. Resulting polyimide films were approximately 1 mil (.0025
cm) in thickness.

CHARACTERIZATION. Melting points were determined on an E. I.
DuPont Series 99 Thermal Analyzer at 20°C/min. Inherent
viscosities of polyamic acid solutions were obtained at a
concentration of 0.5% (w/w) in DMAc at 35°C. Glass transition
temperatures (T_g) of the fully cured polymer films were measured
by thermomechanical analysis (TMA) on a DuPont 943 Analyzer in air
at 5°C/min. Films fully-cured at 300°C were tested for
solubility at 3-5% (w/w) solids concentration in DMAc,N,N-
dimethylformamide (DMF), and chloroform ($CHCl_3$). Solubilities at
room temperature were noted after periods of 3 hours, 1 day and 5
days. Refractive indices of 1 mil thick films were obtained at
ambient temperature by the Becke line method (11) using a
polarizing microscope and standard immersion liquids obtained from
R. P. Cargille Labs.

RESULTS AND DISCUSSION

The polyimide films prepared from the monomers shown in Figures 1-3
are listed in Table I. Reaction of monomers yielded pale yellow to
colorless polyamic acid solutions with inherent viscosities ranging
from 0.34 to 1.82 dl/g. Tough, flexible, transparent films were
produced by thermally converting the polyamic acids to polyimides
at 300°C in air. Films ranged in color from a light yellow color
to essentially colorless depending on thickness. The film colors
described in Table I were for nominally 1 mil thick films. Glass
transition temperatures of films increased in value within each
series of polymers as the linkages of the aromatic diamines were

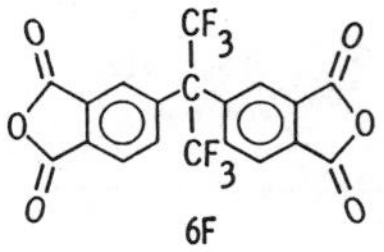

6F

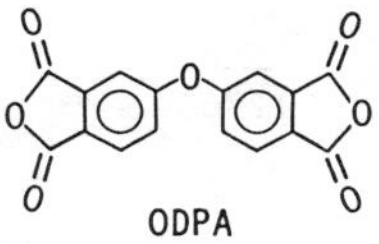

ODPA

Figure 1. Dianhydride Monomers

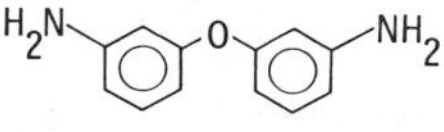

3,3'-ODA

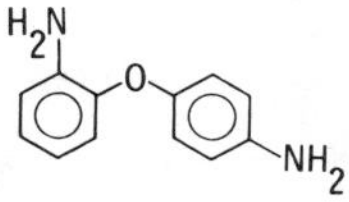

2,4'-ODA

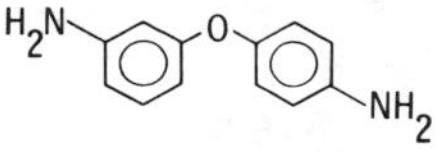

3,4'-ODA

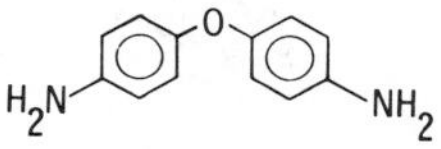

4,4'-ODA

Figure 2. Oxydianiline Diamines

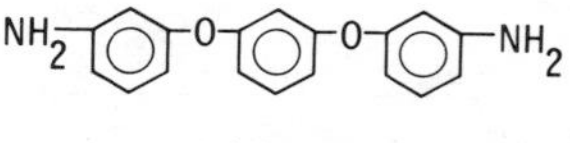

1,3(3)-APB

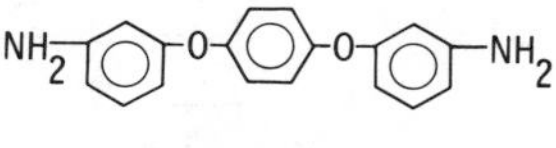

1,4(3)-APB

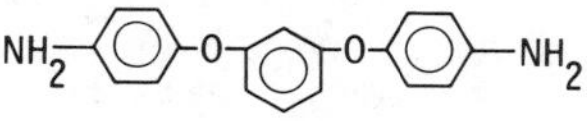

1,3(4)-APB

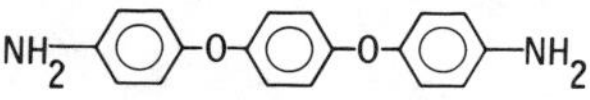

1,4(4)-APB

Figure 3. Aminophenoxybenzene Diamines

TABLE I. Properties of Polyimide Films

Polymer	n_{inh} (dl/g)	Tg, oC	Refractive Index (n)	Film Appearance
6F + 3,3'-ODA	1.00	244	1.60	Pale to Colorless
6F + 2,4'-ODA	0.75	276	--	Pale to Colorless
6F + 3,4'-ODA	0.79	280	1.60	Pale Yellow
6F + 4,4'-ODA	1.11	326	1.60	Pale Yellow
ODPA + 3,3'-ODA	1.09	186	1.69	Pale to Colorless
ODPA + 2,4'-ODA	0.77	264	1.67	Pale to Colorless
ODPA + 3,4'-ODA	0.61	245	1.69	Pale Yellow
ODPA + 4,4'-ODA	0.34	273	1.69	Light Yellow
6F + 1,4(4)-APB	1.82	281	1.60	Light Yellow
6F + 1,3(4)-APB	1.58	255	1.62	Light Yellow
6F + 1,4(3)-APB	1.19	230	1.61	Pale to Colorless
6F + 1,3(3)-APB	1.02	209	1.61	Pale to Colorless
ODPA + 1,4(4)-APB	1.46	245	1.67	Light Yellow
ODPA + 1,3(4)-APB	1.29	217	1.69	Light Yellow
ODPA + 1,4(3)-APB	1.06	204	1.68	Pale Yellow
ODPA + 1,3(3)-APB	0.98	182	1.68	Pale to Colorless

varied from all-meta to all-para. Polyimides prepared with 6F dianhydride consistently displayed higher T_g than did ODPA dianhydride-containing polymers with the same diamine. Refractive index measurements were obtained on each of the polyimide films because of their potential use as transparent/optical coatings. The ODPA films had a higher refractive index (n = 1.67-1.69) than did the 6F polymers n = 1.60-1.62). Refractive indices for these series of films was governed by the dianhydride portion of the polymer chain. These values of refractive index are all lower than the reported values of commercial polyimide film (1.78).

The solubilities of polyimide films cured at 300°C are presented in Tables II and III. Polymers prepared with ODPA dianhydride were less soluble overall in the solvents listed than those prepared with 6F. This phenomenon is not surprising since the 6F dianhydride, because of its bulky $-CF_3$ groups, is most effective in preventing charge transfer complexing (CTC) between polymers chains through steric hindrance.(7) ODPA containing oxygen as a "separator" or "linking atom" is also effective (although less so than 6F) in producing a reduction in CTC when compared to pyromellitic (PMDA) or benzophenone tetracarboxylic (BTDA) dianydrides. When coupled with the diamines in Figure 2, these latter two dianydrides produce totally insoluble polymers.

Table II.   Solubilities of Polyimide Films Prepared with 6F
Dianhydride[a,b]

Diamine	DMAc			DMF			CHCl$_3$		
	3 hr	1d	5d	3 hr	1d	5d	3 hr	1d	5d
3,3'-ODA	s	s	s	s	s	s	s	s	s
2,4'-ODA	s	s	s	s	s	s	s	s	s
3,4'-ODA	i	s	s	i	s	s	i	s	s
4,4'-ODA	i	i	i	i	s	s	i	s	s
1,4(4)-APB	i	i	i	i	s	s	i	s	s
1,3(4)-APB	s	s	s	s	s	s	s	s	s
1,4(3)-APB	ps	s	s	ps	ps	ps	ps	ps	ps
1,3(3)-APB	s	s	s	s	s	s	s	s	s

[a]Solubilities were tested after 3 hours, 1 day and 5 days at room temperature

[b]s = totally soluble; ps = partly soluble; i = insoluble

Table III. Solubilities of Polyimide Films Prepared with ODPA
                        Dianydride

Diamine	DMAc			DMF			CHCl$_3$		
	3 hr	1d	5d	3 hr	1d	5d	3 hr	1d	5d
3,3'-ODA	i	s	s	i	i	i	s	s	s
2,4'-ODA	s	s	s	s	s	s	-	-	-
3,4'-ODA	i	i	i	i	i	i	i	i	i
4,4'-ODA		i	i	i	i	i	i	i	i
1,4(4)-APB	i	i	i	i	i	i	i	i	i
1,3(4)-APB	i	i	i	i	i	i	i	i	i
1,4(3)-APB	i	i	i	i	i	i	i	i	i
1,3(3)-APB	ps	ps	ps	ps	ps	ps	ps	ps	ps

The polyimides prepared with 6F dianhydride (Table II) were
exceedingly soluble in the solvents studied. Solubility increased
with incorporation of meta or ortho isomerism which serves to
create more "kinks" and dissymmetry in the polymer chains. The
same trend was observed for ODPA-containing films (Table III)
except to a lesser degree. Ortho isomerism appeared to have a
greater effect on the solubility of ODPA films than did meta
isomerism as had been noted previously.(6)

Several polymers were tested for solubility at high
concentrations of polymer in the solvent. Previous studies had
shown that ODPA + o,p'-ODA and 6F + o,p'-ODA films had a high
degree of solubility in DMF at room temperature (>40% w/w).(6)
The polyimide powder of 6F + 3,3'-ODA was likewise tested for its
solubility limit in DMAc. The imide powder was prepared by
chemically imidizing the 6F + 3,3'-ODA polyamic acid with
pyridine/acetic anhydride, precipitating in distilled water,
thoroughly drying at 60°C and heating for 2 hours at 200°C.
This powder was gradually added to DMAc while stirring. After
dissolving amounts greater than 30% (w/w) stirring became so
difficult, the experiment was stopped.

Another method for solubilizing polyimide film by curing the
polyamic acid for longer times at lower temperatures was
attempted. The 6F + 4,4'-ODA film cured 1 hour at 300°C was
found to be insoluble in DMAc at ambient temperature. The same
material cured for 5 hours at 200°C was totally soluble in DMAc
upon stirring for several hours. The enhanced solubility of this
polymer could be due to a possible lowering in molecular weight or
incomplete imidization of the polyamic acid. This method for
obtaining solubility is not a preferred method but is mentioned

here only as an example of how solubility can be achieved in some systems. Although the T_g of this polymer was 10-15°C lower than that of the polymer cured at 300°C, the infrared spectrum of the film showed only a slight hint of anhydride -C=O peak.

CONCLUSIONS

Soluble all-aromatic polyimides have been produced by coupling hexafluoropropane-(6F) and oxygen-(ODPA) containing dianhydrides with oxydianiline and bis(aminophenoxy)benzene diamines. Solubility was enhanced by the presence of meta and ortho isomer links in the diamine portion of the molecule. The polymers prepared with 3,3'-ODA and 2,4'-ODA were found to be readily soluble at 30-40% solids at room temperature in amide solvents. These polyimides are also readily soluble in low-boiling chlorinated solvents. They can therefore be spray-coated onto desired substrates in the fully-imidized form and thus eliminate the need for taking the substrate to elevated temperatures. These soluble phenoxy-linked polyimides yield tough, flexible, colorless to pale yellow transparent films from amide or cholorinated solvents. Their potential for use in electronic applications should be excellent.

ACKNOWLEDGMENTS

The authors are indebted to Mr. Robert Ely for his expert technical assistance and Mr. Edward Shockey for refractive index measurements.

LITERATURE CITED

1. Adrova, N. A., Bessonov, M. I., Laius, L. A. and Rudakov, A. P. Polyimides, A New Class of Thermally Stable Polymers; Technomic Publishing Co.: Stamford, CT, 1970, p 89.
2. Sroog, C. E., Endrey, A. L., Abramo, S. V., Berr, C. E., Edwards, W. M., and Olivier, K. L. J. Polym. Sci., 1965, A, 3, 1373.
3. Korshak, V. V., Vinogradova, S. V. and Vygodskii, Y. S. J. Macromol. Sci. - Rev. Macromol. Chem., 1974, C-11 (1), 45.
4. Harris, F. W., Feld, W. A. and Lanier, L. H. Polymer Letters Edition, 1975, 13, 283.
5. Harris, F. W., Norris, S., Lanier, L., Reinhardt, B., Case, R., Varaprath, S., Padaki, S., Torres, M., and Feld, W. In Polyimides; Mittal, K. L., Ed.; Plenum Press: New York, 1984; Vol. 1, p 3.
6. St. Clair, T. L., St. Clair, A. K., and Smith, E. N. In Structure-Solubility Relationships in Polymers; Harris, F. W. and Seymour, R. B., Ed.; Academic Press, Inc.: New York, 1977; p 199.
7. St. Clair, A. K., St. Clair, T. L., Slemp, W. S. and Ezzell, K. S. Proc. 2nd International Conference on Polyimides, Ellenville, New York, 1985, p 333; NASA-TM-81650.
8. Nakano, T. Proc. 2nd International Conference on Polyimides, Ellenville, New York, 1985, p 163.

9. Yamane, H. *Proc. 2nd International Conference on Polyimides*, Ellenville, New York, 1985, p 86.

10. Bell, V.. L., Stump, B. L. and Gager, H. *J. Polymer Sci.: Polym. Chem. Ed* 1976, A-1, 14, 2275.

11. Faust, R. C., *Proc. Phys. Soc.*, 1955, 68B, 1081.

RECEIVED July 6, 1987

Chapter 38

Cocyclotrimerization of Aryl Acetylenes: Substituent Effects on Reaction Rate

Daniel J. Dawson, Janice D. Frazier, Phillip J. Brock, and Robert J. Twieg

Almaden Research Center, IBM, San Jose, CA 95120-6099

Cyclotrimerization of polyfunctional aryl acetylenes offers a unique route to a class of highly aromatic polymers of potential value to the micro-electronics industry. These polymers have high thermal stability and improved melt planarization as well as decreased water absorption and dielectric constant, relative to polyimides. Copolymerization of two or more monomers is often necessary to achieve the proper combination of polymer properties. Use of this type of condensation polymerization reaction with monomers of different reactivity can lead to a heterogeneous polymer. Accordingly, the relative rates of cyclotrimerization of six *para*-substituted aryl acetylenes were determined. These relative rates were found to closely follow both the Hammett values and the spectroscopic constants $\Delta\delta_H$ and $\Delta\delta C_\beta$ for the *para* substituents. With this information, production of such heterogeneous materials can be either avoided or controlled.

Since its inception, microelectronics has been evolving towards denser configurations of smaller circuit elements. Critical to these devices is the concurrent development of insulating materials capable of isolating these elements to allow the construction of multi-level circuitry. The chemical and physical semiconductor processing requirements are aggressive enough to eliminate most polymers from this application. Due to their high thermal stability, polyimides have been widely used for this purpose but have two major drawbacks: 1) their polar imide structure permits water absorption and release, which are potential sources of corrosion of nearby metals; and 2) the high melt viscosity of most polyimides limits the degree of planarization that can be achieved.

One class of materials that minimizes both of these problems without compromising thermal stability is poly(aryl acetylenes). Over the past several years, two types of these polymers have been investigated in our laboratories: diethynyl oligomers and cyclotrimerized multi-ethynyl aromatics (1). Examples of these two types of oligomers (1 and 2, respectively) are illustrated below.

0097–6156/87/0346–0445$06.00/0
© 1987 American Chemical Society

1   2

The highly aromatic structure of these materials provides good thermo-oxidative stability, extremely low water absorption, and a low dielectric constant. The branched structure of these oligomers, together with their low polarity, affords a low melt viscosity (which leads to good planarization) and high solubility in most organic solvents. Because the cyclotrimerized materials (2) undergo a slower, more easily controlled thermal cure, they have been the target of our recent work.

Due to the branching inherent in the cyclotrimerization reaction, use of a multi-ethynyl aromatic monomer alone inevitably leads to gel formation if the reaction is driven to completion. Stopping the reaction before the onset of gelation gives low yields and difficulty in molecular weight reproduction. To circumvent these problems, co-cyclotrimerizations can be conducted with a mono-ethynyl aromatic co-monomer, such as phenylacetylene, that terminates the growth of many of the polymer branches. This strategy results in a more controlled reaction, which can be pushed to high conversion with good reproducibility.

Key to the success of this co-cyclotrimerization procedure is the selection of the appropriate monomers. A co-cyclotrimerization in which one monomer reacts much more rapidly than the other will result in a heterogeneous product as the monomer ratio changes. Moreover, if the mono-ethynyl capping agent reacts much more slowly than the multi-ethynyl monomer, gel formation can occur early in the reaction. Alternatively, if the mono-ethynyl material reacts much more rapidly, it can be exhausted early in the reaction, having produced a non-reactive trimer, and the multi-ethynyl monomers will gel later in the reaction. These problems can be avoided by using monomers of comparable reactivity or by adjusting the feed ratio to compensate for unequal reactivities. With either approach, it is necessary to determine the relative cyclotrimerization rates of each ethynyl monomer. In this paper, we report the initial results of our measurements of these rates.

Discussion

In order to avoid complications in kinetics and analysis due to polymerization, multi-ethynyl monomers were not used during the initial rate studies. This study focussed instead on *para*-substituted phenylacetylenes.

The substituted phenylacetylenes (6) were all prepared by a modified (2-6) Stephens-Castro coupling (7) of an aryl halide (3) with a monoprotected acetylene [2-methyl-3-butyn-2-ol (4, R' = -C(CH$_3$)$_2$OH) or trimethylsilylacetylene (4, R' = -Si(CH$_3$)$_3$)] in a refluxing dialkylamine solvent, followed by a deprotection step (Scheme I).

The catalyst system for the coupling reaction was a Pd(II)-tri-phenylphosphine complex, usually prepared *in situ*, with excess triphenyl-phosphine and either cuprous iodide or cupric acetate as a co-catalyst. Al-ternatively, a preformed catalyst mixture prepared from these reagents may be utilized (see Experimental Section). When 2-methyl-3-butyn-2-ol was used as the protected acetylene, the intermediates 5a–d were converted to the corresponding aryl acetylenes 6a–d by a retro-Favorskii-Babayan (8) reaction utilizing potassium *t*-butoxide in toluene under conditions of slow distillation. In the case of *p*-iododimethylaniline (3e), trimethylsilylacetylene was used as the ethynyl source. The intermediate (5e) was treated with hydroxide to generate the free aryl acetylene 6e. The syntheses of 6d and 6e are described in the Experimental section below.

Scheme I

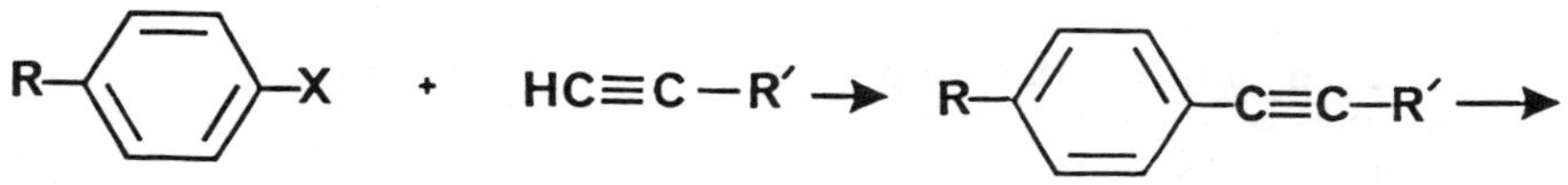

	R	R'	X
a	-CN	-C(CH₃)₂OH	Br
b	-CF₃	-C(CH₃)₂OH	Br
c	-C₆H₅	-C(CH₃)₂OH	Br
d	-OC₆H₅	-C(CH₃)₂OH	Br
e	-N(CH₃)₂	-Si(CH₃)₃	I

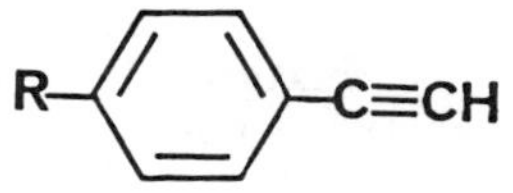

The cyclotrimerization reactions were conducted in dioxane, using a nickel acetylacetonate/triphenylphosphine catalyst system at 90°C. This catalyst system produces cyclotrimerized products in preference to linear polyenes (9–11). An effort was made to minimize the rate-affecting variables in the kinetics runs. Use of an oil bath and a stirred reactor provided good heat transfer and the assurance that the desired temperature was maintained de-spite the exothermic nature of cyclotrimerizations. This reaction is quite sensitive to catalyst concentration and temperature; it is drastically inhibited or terminated by oxygen contamination. Accordingly, care was taken to use the same catalyst formulation in each kinetics run and the reaction was thor-oughly deoxygenated and conducted under argon. To compensate for any unforeseen variables that might have affected the overall reaction rate, each aryl acetylene 6a–e was cyclotrimerized with an equimolar amount of pheny-lacetylene, allowing a direct ranking of reaction rates. HPLC was used to monitor the concentration of each of the starting materials. An inert com-pound (naphthalene) was included as an internal HPLC standard.

As a rule, each phenylacetylene derivative was evaluated at an initial concentration of 125 mM; reactions using phenylacetylene alone were conducted at concentrations of 125, 250, and 500 mM in order to establish the effect of ethynyl concentration on the measured rate. Because more than one material is produced in even the simplest cyclotrimerization reaction, all reactions were followed only by measuring the disappearance of the starting material(s). Although attempts were made to fit the resulting data into the expected second- or third-order kinetics plots, it was finally concluded that the reactions were better described as zero-order. Accordingly, data were plotted on linear concentration and time scales.

Figures 1 to 3 illustrate the disappearance of phenylacetylene, starting at three different concentrations: 125, 250, and 500 mM. From these plots, it is clear that there were two different reactions taking place. At high concentrations (>200 mM ethynyl), the reaction was rapid; below 150 mM, a second reaction took over with a rate approximately an order of magnitude less. The fact that both rates appeared to be zero-order suggests that catalyst turnover was rate-limiting, with the effective catalytic species altered by the ethynyl concentration.

In order to obtain information on both the fast and slow reactions, the kinetics runs on the substituted aryl acetylenes were conducted at an initial total ethynyl concentration of 250 mM. The resulting plots are shown in Figures 4 to 8. Several conclusions were drawn from these plots:

1) The fast/slow pair of reactions occurred with each of the five aryl acetylene mixtures.

2) With one exception (6b, R = p-CF$_3$), the rate of disappearance of the aryl acetylenes shifted from fast to slow at a total ethynyl concentration of 120-140 mM. The mixture of p-trifluoromethylphenylacetylene and phenylacetylene changed rates at a total concentration of ~195 mM.

3) There was a definite substituent effect on the high-concentration (fast) reaction rates. The low-concentration (slow) rates were almost always identical for each pair of reactants.

Because preparative cyclotrimerization reactions are usually conducted at high concentration, the initial, faster rates in this study were considered more important. For each run, the rates of disappearance of the substituted aryl acetylene and phenylacetylene, along with the reaction ratio, are listed in Table I.

TABLE I. Cyclotrimerization Rates for p-Substituted Phenylacetylenes (Sub) and Phenylacetylene (PA)

Substituent	Reaction Rates ($\times 10^{-6}$ mole L^{-1} sec^{-1})		Rate Ratio
	Sub	PA	Sub/PA
CN[a]	4.5	2.0	2.25
CF$_3$	0.98	0.69	1.4
C$_6$H$_5$	4.2	3.6	1.2
OC$_6$H$_5$	2.2	2.1	1.05
N(CH$_3$)$_2$	0.48	1.1	0.43
H	-----1.7-----		--

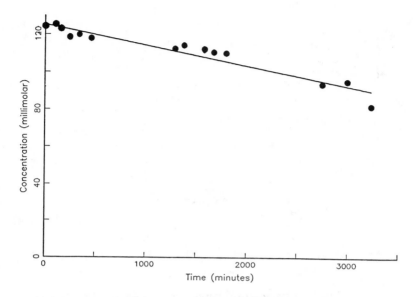

Figure 1. Concentration of Phenylacetylene (125 mM) vs. Time.

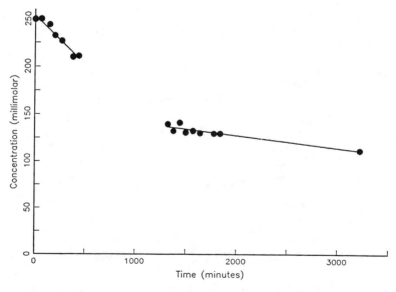

Figure 2. Concentration of Phenylacetylene (250 mM) vs. Time.

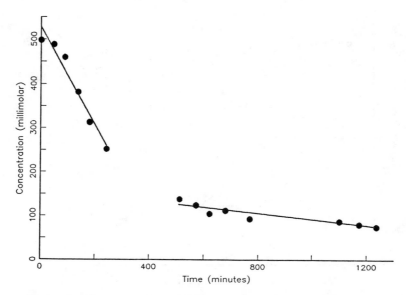

Figure 3. Concentration of Phenylacetylene (500 mM) vs. Time.

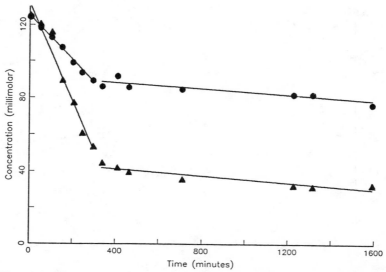

Figure 4. Concentration of *p*-Cyanophenylacetylene (▲) and Phenylacetylene (●) vs. Time.

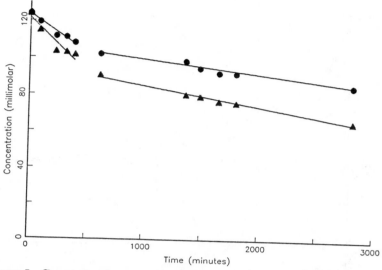

Figure 5. Concentration of *p*-Trifluoromethylphenylacetylene (▲) and Phenylacetylene (●) vs. Time.

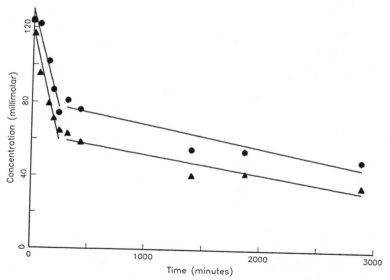

Figure 6. Concentration of *p*-Phenylphenylacetylene (▲) and Phenyl-acetylene (●) vs. Time.

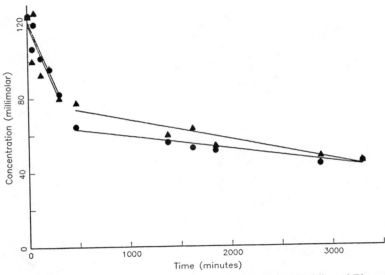

Figure 7. Concentration of p-Phenoxyphenylacetylene (▲) and Phenylacetylene (●) vs. Time.

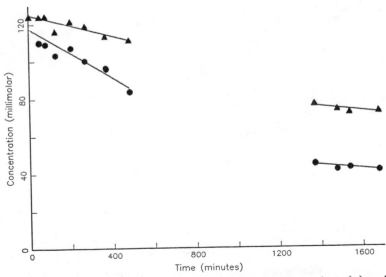

Figure 8. Concentration of p-Dimethylaminophenylacetylene (▲) and
Phenylacetylene (●) vs. Time.

) As a control experiment, benzonitrile and phenylacetylene were subjected .o the standard cyclotrimerization conditions. Benzonitrile did not react.

Two different interpretations of this data are possible. If the direct reaction rates of the substituted aryl acetylenes are compared, they would be ranked as follows:

$$CN > C_6H_5 > OC_6H_5 > H > CF_3 > N(CH_3)_2$$

However, because the reaction rate for phenylacetylene in these same runs varied by a factor of five, this type of ranking is suspect. If, as originally intended, the rate ratios are compared, a different ranking is produced:

$$CN > CF_3 > C_6H_5 > H \sim OC_6H_5 > N(CH_3)_2$$

which is very close to the order of the Hammett values for the *para*-substituents:

$$CN(.66) > CF_3(.54) > C_6H_5(.01) \sim H(0) > OC_6H_5(-.32) > N(CH_3)_2(-.83)$$

It is interesting to note that the rate ratios follow the spectroscopic $\Delta\delta_H$ (chemical shift difference of the acetylene hydrogen on the substituted vs. parent phenylacetylene) and the $\Delta\delta C_\beta$ (chemical shift difference of the terminal acetylene carbon in the substituted vs. parent phenylacetylene) (12).

$\Delta\delta_H$:
$$CN(.251) > CF_3(.159) > C_6H_5(.0515) > H(0) > OC_6H_5(-.0453) > N(CH_3)_2(-.147)$$

$\Delta\delta C_\beta$:
$$CN(4.02) > CF_3(2.43) > C_6H_5(0.56) > H(0) > OC_6H_5(-0.65) > N(CH_3)_2(-2.28)$$

Because the exact mechanism of the cyclotrimerization reaction is not adequately understood, it is useless to conjecture on exactly how the substituent influences the reaction rate. However, it is useful to know that spectroscopic data correlates with the observed rates and this may prove advantageous in the prediction of cyclotrimerization rates for other substituted phenylacetylenes.

Conclusion

The results of this study suggest that diethynyl monomers such as diethynylbiphenyl, diethynylterphenyl, and diethynyldiphenylether could be cyclotrimerized with equimolar amounts of phenylacetylene to give homogeneous polymers in high yield. Bis(*p*-ethynylphenyl)X monomers where X = -SiR$_2$- and -P(R)- with Hammett constants close to 0 would be expected to behave similarly. In contrast, compounds in which X = -SO$_2$-, -C(=O)-, -CF$_2$-, and -N=N- would be expected to cyclotrimerize two to three times faster than phenylacetylene. Use of these materials to prepare analogous homogeneous polymers in high yield might well require constant adjustment of reaction stoichiometry or the use of an electron-deficient mono-ethynyl component such as the *p*-cyano or *p*-CF$_3$ materials described above.

Experimental

4-Phenoxyphenylacetylene (6d). A 250-mL round-bottomed flask equipped
with a heating mantle, magnetic stir bar, thermometer, and condenser with a
nitrogen/vacuum source, was charged successively with 6.27 g (74.5 mmol,
1.626 equiv per Ar-Br) of 2-methyl-3-butyn-2-ol, 11.42 g (45.8 mmol) of 4-
bromodiphenylether, 0.35 g (1.33 mmol, 0.029 equiv) of triphenylphosphine,
84.2 mg (0.27 mmol, 0.0059 equiv) of cupric acetate monohydrate, 38.6 mg
(0.218 mmol, 0.0048 equiv) of palladium(II) chloride, and 100 mL (73.8 g,
729 mmol, 15.9 equiv) of di-n-propylamine. The light purple suspension was
deoxygenated and then heated to reflux. By the time the internal temperature
had reached 80°C the reaction mixture had become a clear, light yellow solution. Di-n-propylamine hydrobromide began to precipitate before the reaction mixture reached reflux temperature (115°C). After 2 hours at reflux,
TLC (silica gel, 1:1 ether/hexane) indicated that the starting material had
been consumed. After a further 10 minutes at reflux, the heating mantle was
removed, and the reaction mixture was cooled to room temperature with a
water bath. The thick crystalline slurry was vacuum filtered and the cake was
carefully washed with two 16-mL portions of toluene. The combined filtrate
was filtered through 16 g of silica gel (40-140 mesh) prepared in toluene. The
column was then rinsed with two 16-mL portions of toluene and blown dry
with nitrogen. The resultant clear yellow solution weighed 103 g and was used
immediately in the next step.
    A 250-mL round-bottomed flask equipped with a heating mantle, mag-
netic stir bar, thermometer, and a Claisen head/distillation system, was
charged with the yellow solution prepared above. Potassium t-butoxide (0.85
g, 7.87 mmol, 0.164 equiv) was added to the stirred solution, which was then
immediately deoxygenated, left at 250 mm Hg vacuum, and heated to boiling
in order to drive off the acetone formed in the reaction. The internal temperature during distillation was initially 94°C, dropped to 92° over a 12-min-
ute period, and then slowly rose to 96°C after 22 minutes of distillation time.
TLC (silica gel, 10% ether/hexane) indicated that the intermediate had been
consumed. The brown reaction mixture was transferred to a 1-liter flask and
the solvents stripped off at reduced pressure. Hexane (150 mL) was added to
the residue, followed by 0.8 mL of acetic acid in 10 mL of hexane. Norite (4
grams) was introduced next. This suspension was stirred and warmed to
60°C over a 20 minute period, 2.1 g of Celite was added, and the hot mixture
was filtered through GF/A filter paper and then through a column containing
32 g of 40-140 mesh silica gel in hexane. The eluant was stripped to dryness
to afford 3.06 g (34%) of 4-phenoxyphenylacetylene (6d) as a pale violet oil:

TLC (silica gel, 10% ether/hexane) R_f 0.66; ^{1}H NMR (CDCl$_3$) δ 2.95 (s, 1H,
C≡CH), 6.9-7.3 (m, 9H, ArH).

Preformed Ethynylation Catalyst. Into a 500-mL round-bottomed flask
equipped with a magnetic stir bar, reflux condenser and nitrogen bubbler was
placed palladium chloride (1.77 g, 10 mmol), di-n-propylamine (100 mL) and
triphenylphosphine (15.74 g, 60 mmol). The resulting slurry was boiled for
two hr after which time the brown palladium chloride had been consumed and
the yellow bis(triphenylphosphine) palladium(II) chloride had formed. The
slurry was cooled and cupric acetate monohydrate (1.99 g, 10 mmol) was added in one portion and the slurry boiled for one hr longer. After cooling, the
solvent was removed by rotary evaporation and then finally under high vac-

uum to give the ethynylation catalyst mixture as a yellow-brown solid in quantitative yield.

Each 19.5 mg of the solid catalyst mixture contained about 0.01 mmol of palladium. Samples of this mixture lost little if any activity over a one-year period although the material gradually darkened with age.

4-N,N-Dimethylamino(trimethylsilylethynyl)benzene (5e). A 500-mL round-bottomed flask equipped with an oil bath, magnetic stir bar, condenser and nitrogen bubbler was charged successively with 4-iodo-N,N-dimethylaniline (13) (24.8 g, 100 mmol), diisopropylamine (100 mL, 72.2 g, 0.714 mole, 7.14 equiv), ethynyltrimethylsilane (12.30 g, 125 mmol, 1.25 equiv) and preformed catalyst mixture (1.0 g). The resulting slurry was heated at a gentle reflux with stirring for 16 hr after which time TLC analysis indicated that the starting iodide had been consumed. The slurry was concentrated by rotary evaporation to remove excess solvent and then taken up in ethyl acetate and water and transferred to a separatory funnel. The organic phase was washed repeatedly with water, then dried ($MgSO_4$), filtered through a short column of silica gel in ethyl acetate and concentrated by rotary evaporation to a brown solid. Two recrystallizations from ethanol afforded pure product, 15.01 g (69%) of 5e, mp 88-9°C: ^{1}H NMR δ 0.24 (s, 9H, Si(CH$_3$)$_3$), 2.94 (s, 6H, N(CH$_3$)$_2$), 6.50 (d, J=9, 2H, ArH o to N), 7.26 (d, J=9, 2H, ArH m to N). Additional crude product, 2.30 g (10%), could be obtained from the mother liquors.

4-N,N-Dimethylaminophenylacetylene (6e). A 250-mL round-bottomed flask equipped with a stir bar and nitrogen bubbler was charged successively with 4-N,N-dimethylamino(trimethylsilylethynyl)benzene (10.86 g, 50 mmol), tetrahydrofuran (THF) (50 mL), methanol (25 mL), and 45% KOH solution (6.25 g). After 2.0 hr of stirring at ambient temperature, the reaction was checked by TLC and found to be complete. The THF and methanol were stripped off by rotary evaporation and the residue was taken up in hexane and water and transferred to a separatory funnel. The phases were separated and the hexane phase washed with water, dried ($MgSO_4$) and filtered through a pad of silica gel. After concentration by rotary evaporation, the crude product was crystallized from methanol to afford 6.35g (87%) of 6e: mp 51-2°C [lit. (14) mp 52-3°C].

Kinetics Experiments. A 100-mL three-necked, round bottomed flask equipped with a magnetic stir bar, serum cap, thermocouple temperature sensor, and reflux condenser with an argon/vacuum source, was positioned in an oil bath which was carefully controlled to provide a reaction temperature of 90.0±0.3°C. The flask was charged with 50 mL of a stock solution of p-dioxane (OmniSolv, EM Science) containing nickel acetylacetonate (15 mM), triphenylphosphine (45 mM), and naphthalene (0.5 wt%), deoxygenated, and allowed to warm to 90°C under argon. At time $t = 0$, a solution of 6.5 mmol (664 mg) of phenylacetylene and 6.5 mmol of one of the substituted phenylacetylenes in 2 mL of dioxane was added by syringe. Samples (0.5-mL) were removed for analysis by syringe over a period of 1 to 3 days; samples not analyzed immediately were stored at 2°C. Sample analysis was performed on a Waters HPLC (Model 6000 pump, Model 660 solvent programmer) using a C_{18} reverse-phase column (12-inch, Analytical Sciences, Inc.) and a solvent program running from 40% acetonitrile/water to 100% acetonitrile with a flow rate of 1.0 mL/min. A Waters (Model 450) UV detector (254 nm) was connected to an HP Model 87 minicomputer for peak integration. The areas of each of the aryl acetylene peaks were divided by that of the naphthalene peak, then normalized to 0.125 M (the starting concentration) at $t = 0$ for plotting purposes.

Acknowledgments

The authors would like to thank Ms. Heidi Bauer for her many invaluable contributions to the work described above.

Literature Cited

1. Dawson, D. J.; Fleming, W. W.; Lyerla, J. R.; Economy, J. In Reactive Oligomers; Harris, F. W.; Spinelli, H. J., Eds.; ACS Symposium Series No. 282; American Chemical Society: Washington, DC, 1985; pp 63-79.
2. Sonogashira, K.; Tohda, Y.; Hagihara, N. Tetrahedron Lett. 1975, 4467.
3. Sabourin, E. T.; Selwitz, C. M. U.S. Patent 4 223 172, 1980.
4. Austin, W. B.; Bilow, N.; Kelleghan, W. J.; Lau, K.S.Y. J. Org. Chem. 1981, 46, 2280.
5. Takahashi, S.; Kuroyama, Y.; Sonogashira, K.; Hagihara, N. Synthesis 1980, 627.
6. Ames, D. E.; Bull, D.; Takundwa, C. Synthesis 1981, 364.
7. Stephens R. D.; Castro, C. E. J. Org. Chem. 1963, 28, 2163, 3313.
8. Shchelkunov, A. V.; Muldakhmetov, Z. M.; Rakhimzhanova, N. A.; Favorskaya, T. A. Zhurnal Organicheskoi Khimii 1970, 6, 930.
9. Jabloner, H. Ger. Offen. 2 235 429, 1973.
10. Cessna, L. C. U.S. Patent 3 926 897, 1975.
11. Jabloner, H.; Cessna, L. C. Polym. Prepr., Am. Chem. Soc. Div. Polym. Chem. 1976, 17(1), 169.
12. Dawson, D. A.; Reynolds, W. F. Can. J. Chem. 1975, 53, 373.
13. Reade, T. H.; Sim, S. A. J. Chem. Soc. 1924, 157.
14. Barbieri, P. Compt. Rend. 1950, 231, 57.

RECEIVED May 1, 1987

Chapter 39

Photo-Cross-Linking and Imidization of Poly(amic acid) Methacrylate Esters

H. Ahne, W.-D. Domke, R. Rubner, and M. Schreyer

Corporate Research and Development Laboratories, Siemens AG D-8520 Erlangen, Federal Republic of Germany

Polyamic acid methacrylate esters are the first technically used self-patternable polyimide precursor to give polyimide relief patterns in a direct process. Photo-crosslinking during UV-irradiation was investigated by means of FT-IR spectroscopy. Results of the efficiency of photocross-linking in layers of variable thickness are given. Thermal conversion of the partially photo-crosslinked polyimide precursor into a highly heat resistant polyimide was followed up by means of FT-IR spectroscopy and DSC measurements. The splitting off of the crosslinking bridges and depolymerization was monitored with thermal gravimetric and mass spectroscopic methods. More than 95 % of the volatile product is the monomer hydroxyethylmethacrylate.

Polyamic acid methacrylate esters are the first self-patternable, pure organic polyimide precursors to be described. They are the polymer basis of the first technically applied resist to produce polyimide patterns in a direct process. They are synthesized simply by the addition of hydroxyethylmethacrylate to aromatic acid dianhydride, and subsequent polycondensation of the intermediate tetracarboxylic acid diester with aromatic diamines. These polyimide precursors give rise to a number of special photoresist properties which lead to important applications, such as photolithographically produced protection layer against α-radiation on memory

0097–6156/87/0346–0457$06.00/0
© 1987 American Chemical Society

chips[1]. Figure 1 shows the chemical principle and the two
main processing steps for direct production of polyimide
patterns:
1. Photo-crosslinking during UV-radiation, yielding
    intermediate products with high solubility diffe-
    rences between the exposed and non-exposed re-
    gions even in thick layers, and
2. Thermal conversion of the partially photo-crosslinked
    polyimide precursor into a highly heat-resistant
    polyimide.
    In this paper we want to discuss the quantitative
analytical results of the studies concerning these two
processing steps. With regard to photo-crosslinking we
will discuss the percentage decrease in photo-crosslinking with increasing layer thickness, the percentage of
photo-crosslinking necessary as a function of layer
thickness for the production of fine and normal patterns
and the improvements in photosensitivity which can be
achieved with the aid of new photo-reactive additives.

Photo-crosslinking of Polyimide Precursors

We have used the sensitizer system of Michler's ketone/
azidosulphonylphenylmaleic imide, which was first developed by Siemens, for the fundamental study of photo-
crosslinking in a wide range of layer thickness. The absolute values of the radiation dose necessary in these
studies are high as this is not a very sensitive system.
However, based on Siemens' fundamental technology new
photoresist systems which are significantly more photosensitive are now commercially available[2]. The
fundamental studies discussed here apply in principle
also to these new systems.
    Due to the high natural absorption of the polyamic
acid methacrylate ester in the region below 400 nm only
light with a wavelength > 400 nm is primarily available
for photopolymerization, particularly with the higher
layer thicknesses. In order to determine the degree of
photocrosslinking we have quantitatively investigated the
reduction in the bending modes of the methacrylate double
bond at 945 cm^{-1} by FT-IR spectroscopy. The decay of
methacrylate (i. e. relative crosslinking) in dependence
on photoresist thickness was investigated on 5 - 50 µm
layers of photoresists which were spin coated on Si-
wafers. The exposure was made using a 350 W ultra high
pressure mercury vapour lamp with a radiant power of 21
mW/cm^2 (measured with 365 probe from the OAI Company) and
a time of 120 s (dose 2,5 J/cm^2). Since the light intensity per wavelength decreases exponentially with the
layer thickness (in accordance with Lambert-Beer's law),
similar behaviour was also expected for the crosslinking

of double bonds. Figure 2 depicts a very good concurrence of the experimental values with such a curve.

To determine the degree of photo-crosslinking which is necessary for the production of patterns with sharp contours we used gray wedge exposure on different layers up to 50 μm thick. The percentage of photo-crosslinking of the individual gray wedge stages was determined using FT-IR spectroscopy and the layers were subsequently developed. In this way, it was possible to determine the photo-crosslinking required for the production of contour-sharp patterns. The percentage of photo-crosslinking necessary depends on the finess of the structure and lies between 50 % for fine structures and 35 % for normal structures (see gray area in Figure 2). This can easily be achieved by an proper radiation dose or by using an adequate sensitizer system.

The Photoresist systems now commercially available require a dose of 50 - 70 mJ/cm² for thin layers of up to about 6 μm and approximately 500 - 750 mJ/cm² for thick layers of 100 μm. These dose values account for highly crosslinked patterns which show no reduction in film-thickness after developing.

Thermal Conversion to Polyimide

The second significant stage in the direct production of polyimide structures involves the thermal conversion of the patterned crosslinked film to the patterned polyimide film. It is important to understand how and under what condition the photo-crosslinked polyimide precursor is converted into polyimide as well as how completely. Mechanistically it is intriguing to determine wether the crosslinking fractures are split into small pieces or escape as pure hydroxyethylmethacrylate comparable to the zip-off depolymerization of polymethylmethacrylate.

For our investigation we used several analytical methods: FT-IR spectroscopy, differential scanning calorimetry (DSC), thermogravimetric analysis (TGA) and mass-spectroscopy.

In thermal treatment of the partially crosslinked polyamic acid ester, the conversion of the amide structures to imide structures starts at about 100°C. This conversion was investigated by means of FT-IR spectroscopy (Bruker IFS 85, reflexion measurement) on annealed films on Al-wafers, where the amide-carbonyl stretching vibrations at 1680 cm⁻¹ and 1540 cm⁻¹ and the corresponding imide band at 1780 cm⁻¹ were quantitatively evaluated.

Figure 3 shows these results in comparison to DSC measurements (heating rate always 10 K/min). An uniform exothermic reaction with a maximum at 160°C is indicated by DSC (DuPont 1090). This correlates with the main imidization process. The FT-IR results show that at this

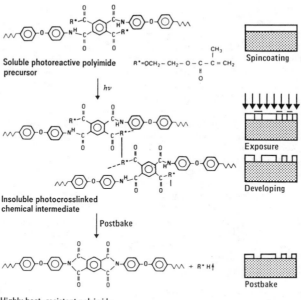

Soluble photoreactive polyimide precursor

Insoluble photocrosslinked chemical intermediate

Highly heat-resistant polyimide

Chemical principle

Spincoating

Exposure

Developing

Postbake

Processing steps

Figure 1.   Chemical principle and processing steps for direct production of polyimide patterns

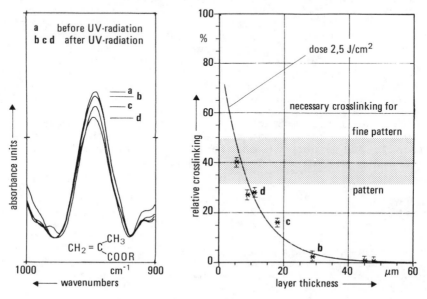

Figure 2.   Decay of relative crosslinking with increasing layer thickness

temperature already 60 % of the amide structures are converted to imide. Up to 200°C the imide amount rises to
about 85 %. This imide level remains constant up to
340°C. The residual imidization proceeds above 340 to
400°C. This may be due to an increased mobility of the
polymer chain and may explain the exothermic rise in heat
flow. After heating to 400°C for 1 h no amide-carbonyl
stretching vibrations but an intensive polyimide band can
be observed (Figure 4).

These results can be correlated with the weight loss
of the sample during annealing. Weight loss was followed
by TGA (DuPont 1090) and correlated with the amount of
residual methacrylate double bonds and amide structures
(Figure 5). According to TGA and IR-measurements inci-
pient cleavage of volatile products associated with loss
of amide and methacrylate structures is evident above
100°C. Above 160°C the decay of double bonds is faster
than the weight loss. That means that the residual double
bonds begin to polymerize thermally. At higher temperatures the crosslinking bridges are split off. This is
indicated by the weight loss. The conversion of the residual amide to imide above 340°C and the measured weight
loss run parallel again and attain the theoretically
possible weight loss at 400°C after 2 h. It was established by mass-spectroscopy (Figure 6) that more than 95
% of the volatile product was hydroxyethylmethacrylate
so that a simultaneous depolymerization takes place. Only
to a minor extent are the dimer and some lower molecular
weight fragments found.

Conclusion

We investigated analytically the two main processing
steps of self patternable photoresists based on polyamic
acid methacrylate ester.

With regard to photo-crosslinking the main results
show that the photolithographically produced pattern is
highly crosslinked (35 - 50 %). Therefore we have no
reduction in layer thickness after developing.

With regard to thermal conversion to polyimide, we
evaluated the suitable reaction conditions for thermal
conversion of the photo-crosslinked patterns into
completely polyimide patterns and found, that the
crosslinked bridges are split off and depolymerization
takes place. More than 95 % of the volatile product is
the monomer hydroxyethylmethacrylate.

Figure 7 shows a typical photo-crosslinked test pattern before and after conversion to polyimide.

These results are also valid for the commercially
available photoresist system based on polyamic acid
methacrylate ester, which have an improved photospeed for
technical applications.

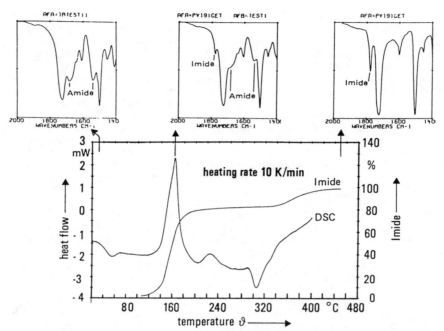

Figure 3. Differential scanning calorimetry of polyamic acid methacrylate ester: Structures and degree of imidization

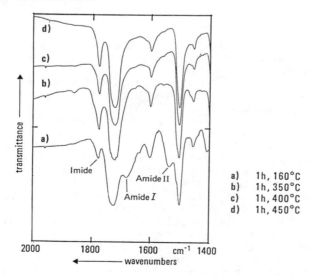

Figure 4. FT-IR spectroscopic determination of the thermal conversation of polyamic acid methacrylate ester to polyimide

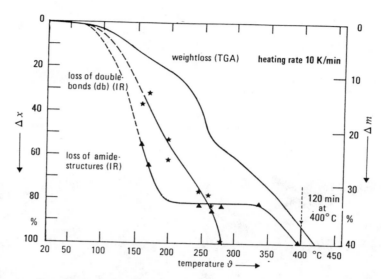

Figure 5.   Polyamic acid methacrylate ester: Correlation of weightloss to loss of methacrylate double bonds and amide structures by annealing from ambient to 400 °C

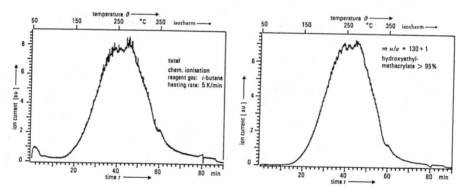

Figure 6.   Polyamic acid methacrylate ester: Mass-chromatogram from 50 °C to 350 °C and 60 minutes isotherm at 350 °C

├─────┤ 7 μm   ├─────┤ 7 μm

photo-cross-linked patterns   polyimide patterns after thermal
layer thickness: 7 μm   treatment
                                         layer thickness: 3.5 μm

Figure 7. SEM recordings of test patterns on silicon
wafers before and after thermal treatment

Acknowledgments

We thank Mr. Schmidt for his committed technological assistance.

References

/1/ Ahne, H.; Krueger, H.; Pammer, E. and Rubner, R.; Polyimides, Vol. 2, Plenum Publishing Corporation, 1984, 905-18
/2/ Photosensitive polyimides with improved photospeed are supplied by Asahi Chemical, Ciba Geigy, E.I DuPont and E. Merck.

RECEIVED April 8, 1987

Chapter 40

Polyimides as Interlayer Dielectrics for High-Performance Interconnections of Integrated Circuits

Ronald J. Jensen

Physical Sciences Center, Honeywell, Inc., Bloomington, MN 55420

Advancements in the speed and density of integrated circuits (ICs) have created a need for new packaging technologies that can provide high density interconnections with controlled electrical characteristics for propagating high speed signals between ICs. A packaging technology that meets these needs uses multiple layers of a thin film conductor and polymer dielectric to achieve high density interconnections. Polyimides are uniquely suited as an interlayer dielectric because of their high stability, processability, and low dielectric constant. This paper reviews the process technology and reliability issues associated with copper/polyimide thin film multilayer (TFML) interconnections, with emphasis on the deposition and patterning of polyimide and its stability under environmental stress. Some recent demonstrations of this technology will also be presented.

New technologies are needed to package and interconnect the latest generation of high-density and/or high-speed integrated circuits (ICs) such as VLSI, VHSIC, and GaAs ICs. Fine-line, multilayer conductor patterns are required to interconnect the large number of input/outputs (I/Os) on highly integrated circuits. Interconnections must be short and have well-controlled electrical characteristics in order to propagate high speed signals with minimum delay and distortion. Finally, the package must provide effective heat removal and environmental protection for the ICs.

Conventional single-chip packages have limited packing density on printed wiring boards (PWBs) and limit the system speed due to the large delay time for signals propagated between chips. Multichip packaging technologies are being developed to overcome the size and performance limitations of single-chip packaging. Compared to single-chip packaging, multichip packaging permits greater chip density, fewer external connections resulting in improved reliability, and reduced power consumption and delay time required to drive signals between chips.

The high interconnect density of multichip packaging requires technologies that can define high-resolution conductor patterns in multiple layers on relatively large substrates. Multilayer interconnect technologies that have been been used or proposed for high performance packaging include: (1) high density PWBs with plated Cu conductor and glass-reinforced polymer dielectrics (1), (2) thick film multilayer interconnects with screen-printed conductor pastes (e.g., Cu, Au) and ceramic/glass dielectrics (2,3), (3) multilayer co-fired ceramic with refractory metals (W or Mo) and alumina dielectric (3-5), (4) wafer scale integration using IC metallization processes on Si substrates (6,7), and (5) thin film multilayer (TFML) interconnections using Cu, Au, or Al conductors and polymer dielectrics such as polyimide (PI) (3,5,7-12).

The design requirements for multilayer interconnections are dictated by a combination of electrical, thermal, and interconnect performance requirements (13). Interconnections should be short to minimize signal propagation delay, resistive losses, and noise due to reflections and crosstalk. Interconnect lengths are minimized by densely packing the chips on a multichip package; the resulting high interconnect density requires a small conductor line pitch and multiple layers of interconnections. Ultimately, the chip packing density will be limited by the thermal density or the physical size of the chips, bonding pads, and off-package connections, rather than the interconnect density.

Conductor lines must be narrow to maximize the space/linewidth ratio and thus minimize the crosstalk between adjacent lines. The lines must also have a large cross-section, and thus a high aspect ratio (thickness/width), to minimize resistive losses which attenuate signals and increase signal risetimes. High conductivity conductor materials are also desirable for low resistive losses. The dielectric layers should be thick to achieve low interconnect capacitance (or high characteristic impedance), which reduces the power consumption of driver circuits and the RC delay of the interconnect. Finally, the dielectric material should have a low dielectric constant (ϵ_r) to minimize the propagation delay (which is limited by the speed of light in the dielectric), the interconnect capacitance and the crosstalk between signal lines.

A TFML interconnect technology based on Cu conductor and PI dielectric offers several advantages over other packaging technologies in meeting these design needs. TFML technologies are being developed at Honeywell and a number of other electronics and packaging companies. This paper will first describe a concept for multichip packaging with Cu/PI TFML interconnections and review the advantages and technical issues associated with this approach. The processes used to fabricate TFML structures will be described, with an emphasis on the deposition and patterning processes for polyimide. Several studies of PI stability and the reliability of Cu/PI interconnections will be summarized. Finally, some recent demonstrations of multichip packaging with TFML interconnections will be presented.

Thin Film Multilayer Packaging Approach

TFML interconnections can be fabricated on a variety of substrates, including ceramics, metals, or silicon wafers. An approach proposed by Honeywell (8), which uses a multilayer co-fired ceramic substrate, is illustrated in Figure 1. The co-fired ceramic substrate is 50-100 mm square, with internal metal layers for power and ground distribution and pins brazed to the bottom for connection to a PWB. Metallized strips on the bottom of the substrate contact the PWB to conduct heat away from the package. A metal seal ring around the perimeter of the substrate permits hermetic sealing to provide mechanical and environmental protection for the chips and interconnections.

The high density interconnections between chips are patterned in TFML Cu/PI. A typical configuration contains five metal layers: two layers of signal interconnections sandwiched between ground or voltage reference planes, and a top metal layer for chip attachment and bonding. This places the signal lines in an offset stripline configuration with constant, controlled impedance. For typical 50 ohm signal lines, the conductor lines are 25 μm wide and 5 μm thick, and dielectric layers are 25 μm thick. The pitch between line centers is typically 75-125 μm. The chips are attached to pads on the surface of the PI, with thermally-conductive Cu vias providing heat transfer through the PI layers to the substrate. The chips can be electrically bonded to the multichip package by wire bonding or tape automated bonding (TAB).

Advantages and Technical Issues

Cu/PI TFML interconnects offer a number of inherent advantages over other interconnect technologies. Table I compares the material properties, geometries, and electrical properties of TFML interconnects with the two primary competing technologies: co-fired ceramic and multilayer thick film. First, thin film patterning processes such as photolithography and dry etching can define higher resolution and higher aspect ratio features in the conductor and dielectric materials than the screen printing and punching processes used for co-fired ceramic or thick film. The TFML geometries result in high interconnect density and low interconnect resistance and capacitance. Secondly, thin film metal deposition processes such as sputtering permit the use of high conductivity metals (Cu, Al) and achieve nearly bulk resistivity in thin films, as compared to the lower conductivity W and Mo pastes used in co-fired ceramic technology or the Cu and Au pastes used for thick film.

From a manufacturing standpoint, TFML technology offers cost advantages over thick film and co-fired ceramic technology by replacing labor-intensive screen printing processes with automated semiconductor processes, and by replacing hard-tooled punches or screens with fast-turnaround photolithographic masks. Finally, TFML technology is extendible over a wide range of package geometries and performance requirements. It is currently being used for multichip packaging of VLSI, VHSIC, and GaAs ICs, and is extendible to even finer geometries required for wafer scale integration (7).

Table I.   Comparison of Multilayer Interconnect Technologies

	Co-fired Ceramic	Thick Film	Thin Film
Conductor			
Material (Alternative)	W (Mo)	Cu(Au)	Cu(Au,Al)
Sheet Resistance (mΩ/□)	10	3	3
Thickness (μm)	~25	15	5
Linewidth (μm)	100	100	25 → 10
Pitch (with vias)(μm)	750 → 250	250	125 → 50
Max. metal layers	7 → 30+	5 → 10	5 → 8 (?)
Dielectric			
Material	Al_2O_3	glass/ceramic	Polyimide
Dielectric constant	9.5	6-9	3.5
Thickness (μm)	250-500	35-65	10-25
Via diameter (μm)	200 100	200	25
Propagation delay (ps/cm)	102	90	62
Min. stripline cap. (pf/cm)	2.0	4.3	1.2
Line resistance (Ω/cm)	1.0	0.3	1.3

Favorable Properties of Polyimide.   Polyimides possess a
combination of physical properties and process characteristics that
make them uniquely suited as a dielectric material.   The synthesis,
characterization, and properties of PIs have been extensively
discussed in two conference proceedings (14,15).   PIs can be
obtained in very pure solution form as polyamic acid (PAA) or
soluble PI.   These solutions can be deposited using a variety of
techniques such as spin coating, spray coating, or screening to
cover a wide range of film thickness (1-100 μm).   The ability of
polyimide to flow before curing enhances the smoothing effect or
planarization of underlying conductor topography.   After driving
off most of the solvents and partially imidizing the PAA at 90-
140°C, the polyimide films can still be stripped or patterned.
   After complete imidization at 350-420°C, PI has high thermal
stability (decomposition temperature > 450°C) and is chemically
inert and insoluble, which prevents degradation during subsequent
processing steps involving high temperature (e.g., metal
deposition, soldering) or strong solvents and acids (e.g.,
photolithography, wet etching).   PIs have relatively high tensile
strength (100-200 MPa) and a large elongation at break (10-25
percent), making the films resistant to cracking despite the large
stresses created by the thermal expansion mismatch between PI and
substrates.   Finally, the dielectric properties of PI, particularly
its low dielectric constant (ϵ_r), are favorable for high speed
signal propagation.   PI has a low ϵ_r of 3.5 (typically), compared
to co-fired ceramic dielectrics with ϵ_r = 9-10 or thick film
glass/ceramic with ϵ_r = 6-9.   The low ϵ_r results in high signal
propagation speeds, low interconnect capacitance which requires
less power dissipation by output drivers, and low levels of
crosstalk between adjacent conductor lines.   PI also has a high
breakdown voltage ($V_B \geq 10^6$ V/cm), and a low dissipation factor
(tan δ < 0.01) resulting in low dielectric losses.

Technical Issues. A number of problems are presented by the processing of relatively thick layers on large substrates for TFML interconnects. Although the minimum feature sizes of TFML are large relative to IC dimensions (25 μm vs 1 μm), the thickness of the conductor and dielectric layers is substantially greater. Anisotropic etching processes such as plasma etching or ion milling are required to achieve high aspect ratios in thick films. These processes involve expensive vacuum equipment and require long processing times for thick films. High aspect ratio features also create large topographies that must be planarized. Finally, the strain energy due to thermal expansion mismatch increases with increasing film thickness, creating potential problems with cracking or loss of adhesion.

Large ceramic substrates such as the one shown in Figure 1 present additional problems for TFML processing. Typical IC fabrication equipment that is designed for the automatic handling of a single size wafer may be unable to accommodate the variety of substrate sizes and shapes used for package applications. Ceramic substrates have greater surface roughness and camber than silicon wafers, which limits the photolithographic resolution and minimum feature size. The large shrinkage tolerance (typically 1 percent) in co-fired ceramic substrates creates a problem with pattern registration between the thin film layers and conductor features metallized into the substrate. Finally, substrates may contain structures such as seal rings or pins that must be protected during processing.

Perhaps the greatest challenge for TFML processing arises from the large substrate area and its effect on yield. A variety of defects can cause critical faults in TFML structures; these include dielectric pinholes causing shorts between metal layers and photolithographic defects causing open or shorted conductor lines. If one assumes a Poisson distribution of defects on the surface, i.e., no clustering effects, the yield for a particular type of fault is given by

$$Y = \exp(-A_c D)$$

where A_c is the critical area in which a defect will cause a fault, and D is the defect area density. For faults that are dependent on defect size, it can be shown (5) that

$$Y = \exp(-k \ (L/x)^2)$$

where L is the side length of a square substrate and x is the minimum critical defect size, which is related to the miniumum feature size of the conductor pattern. Since the yield depends exponentially on the ratio $(L/x)^2$, the yield for patterning 10 μm lines on a 10 cm substrate is equivalent to patterning 1 μm lines on a 1 cm chip, which approaches the limits of current IC production technology. Because of this severe yield constraint, in-process testing methods such as high speed capacitance probes or voltage-contrast scanning electron microscopy, and repair techniques such as laser etching or wire bonding, may be required to obtain acceptable manufacturing yields.

Processing

TFML interconnections are fabricated using a repetitional sequence
of thin film processes to deposit and pattern the conductor and
dielectric layers. A variety of individual processes and process
sequences, including both additive and subtractive approaches, have
been used. The subtractive process sequence shown in Figure 2 has
been used at Honeywell for a variety of patterns (8,9) and is
offered as an example.

The conductor layers, consisting of 5 µm of Cu sandwiched
between thin (20-100 nm) Cr or TiW adhesion layers, are deposited
by dc and rf sputtering, respectively. Relatively thick (2-6 µm)
photoresist layers are deposited and patterned, using proximity
alignment to accommodate the large camber of ceramic substrates and
to prevent yield loss caused by contact between the mask and
substrate. The conductor materials are etched using wet processes
for aspect ratios less than 0.5 or ion beam milling for higher
aspect ratios.

Polyamic acid (PAA) solutions are deposited by either spinning
or spraying. They are cured by heating at a controlled rate from
50°C to $350\text{-}420^{\circ}$C to evaporate solvents and reaction products
(primarily H_2O) and convert the PAA to PI. Multiple coats are
deposited to achieve the thick (20-40 µm) planarized dielectric
layers required for high impedance TFML interconnects. Figure 3
shows a cross-section of 5 µm thick conductor lines planarized with
25 µm (three coatings) of spray-coated PI.

Features such as vias for connection between metal layers are
patterned in PI by reactive ion etching (RIE). To serve as a mask
for the RIE of thick PI layers, a slow-etching material such as
SiO_2 must first be deposited on the PI and patterned by
photolithography and RIE or wet etching. Vias are metallized by
sputtering a conformal metal layer and patterning both the vias and
next conductor layer with a single photolithography and etching
step. Figure 4 shows a 50 µm via covered with a 100 µm pad after
metallization and patterning. The sequence of deposition and
patterning steps is repeated for additional metal layers, using
staggered vias to connect through more than one metal layer.

A number of alternative processes may be used for depositing
and patterning the conductor and dielectric layers. Some
investigators have used additive processes such as selective
electroplating (12) or lift-off (7) for defining the conductor
patterns. These additive approaches have the important advantage
of allowing vertical stacking of vias through several dielectric
layers, however, they require additional processing steps because
the vias and conductor patterns are defined in separate
photolithography steps. Options for depositing and patterning
polyimide dielectric layers are discussed in more detail below.

Polyimide Deposition. The most accurate processes for depositing
thin, uniform coatings of PAA or PI solutions are spin coating,
spray coating, and a recently reported screening process (12).
Spin coating is a commonly available process in the IC industry and
has been characterized and modeled for a variety of polymer
solutions (16). The thickness of spin coated PI films depends

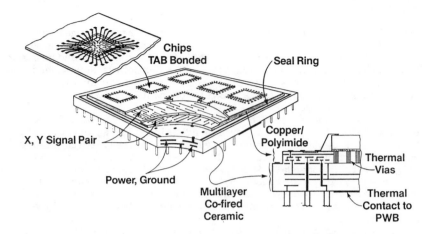

Figure 1.   Proposed approach for multichip packaging using thin
            film multilayer Cu/polyimide interconnections.

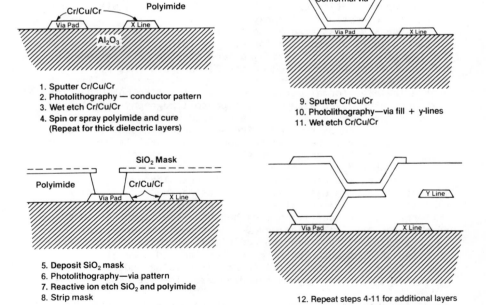

Figure 2.   Process sequence for fabricating multilayer
            Cu/polyimide structures.

Figure 3.   Cross-section of 5 um thick conductor lines on
ceramic, coated with 25 um of polyimide.

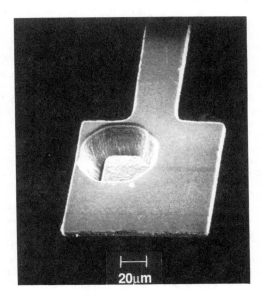

Figure 4.   Conformally-filled via hole for connection between
metal layers.

strongly on solution viscosity and solids content and can be
accurately controlled over a wide range (1-8 μm for DuPont 2555 PI)
by varying the time and speed of spinning. However, in order to
deposit thick films, high viscosity solutions must be spun at low
speeds for short times. All of these factors adversely affect the
uniformity of coating thickness. Furthermore, the size of
substrate which can be uniformly spin coated is limited.
        To overcome these limitations, thicker PI films can be
deposited on larger substrates by spray coating. Spray coating has
the additional advantages of high throughput, reduced formation of
edge bead, and the ability to coat non-square substrates with large
topography. PI can be accurately spray coated in commercial
conveyorized systems, where the uniformity and thickness of the
films depend on a number of process parameters such as solution
viscosity and concentration, solution flow rate, spray nozzle
diameter, atomization pressure, nozzle-to-substrate distance, and
conveyor speed. The selection of a diluting solvent that will
rapidly evaporate from the freshly sprayed film is also crucial.
Statistically designed experiments have proven to be essential in
developing optimized spray coating processes. Processes developed
at Honeywell are capable of spray coating PI films 2-15 μm thick,
to accuracy of ± 1 μm, with 20-30 percent planarization of conductor
lines per coating (60 percent planarization after three coats).

Polyimide Patterning. PI films can be patterned by a variety of
techniques including wet etching, plasma etching or RIE, direct
photopatterning of photosensitive polyimide (PSPI), and laser
ablation. Wet etching can pattern only partially-cured PI films
and is limited to aspect ratios less than 1. RIE involves more
steps and more expensive equipment than wet etching, but is capable
of patterning high aspect ratio features with nearly vertical
sidewalls in thick, fully-cured PI films. By varying process
parameters, the sidewall angle can be controlled to produce the
taper required for good metal step coverage. RIE processes
developed at Honeywell are being used to etch vias as small as 12
μm in diameter in PI films 25-50 μm thick at an etch rate of 1
μm/min, with a controlled sidewall angle of $65\pm2^{\circ}$ from horizontal
and less than 5 percent standard deviation in via diameter across 3
inch square substrates. Figure 5 shows a 50 μm-diameter via hole
in a 25 μm-thick polyimide layer after RIE.
        PSPIs are polyamic acids which have been esterified with
photoreactive alcohols and sensitized with monomeric additives to
form a PI precursor that will crosslink under near-UV exposure
(17,18). The PSPI can thus be patterned like a negative
photoresist; after dissolving out the unexposed material with a
developer, the exposed pattern is cured to convert the crosslinked
PAA to PI and drive off the crosslinking groups. This
photopatterning process involves significantly fewer steps and
fewer material interactions than either wet or plasma etching.
However, it is currently limited to aspect ratios of about 1:2 in
thin films and 1:3 in thick films. Thick films are difficult to
pattern because of the large shrinkage (about 50 percent) during
cure and the high UV absorption of the PSPI. PSPI is still under
intensive development at several companies (DuPont, Merck, Ciba-
Geigy, Toray, Hitachi) and is a promising alternative for future

processing. At Honeywell we have patterned line gaps 2 μm wide in
PSPI films 1 μm thick (hard cured), and via holes 50 μm in diameter
in multiple-coated films 25 μm thick. The latter result is shown
in Figure 6, where three coatings and developments were used to
pattern a 50 μm bottom-diameter via in PSPI (Toray Photoneece) 25
μm thick.

A number of investigators have reported the ablative etching of
PI using eximer lasers at different UV wavelengths (19,20).   So
far most of these studies have been concerned with the mechanism of
the process, particularly the relative roles of thermal and
photochemical dissociation, rather than the practical problems of
masking, optics, throughput, and control of feature geometries.
However, laser ablation offers the exciting possibility of a
maskless, programmable technique for directly writing via holes or
other features in polyimide films.

Reliability of TFML Packaging

The high performance systems which require TFML interconnects
generally have high reliability requirements due to the high cost
of the ICs and the adverse environments under which they must
perform. The use of an organic material such as PI in a system
demanding high reliability raises a number of concerns. Compared
to inorganic dielectrics, most polymers have poor thermal and
mechanical stability, high moisture absorption, and poor adhesion
to metals and substrates. Although PI has exceptional chemical and
thermal stability and can be obtained in semiconductor-grade
purity, there are a number of other issues relating to the
stability of PI during package fabrication, assembly, and long term
use in adverse environments. Critical concerns for packaging
applications include: (1) the effects of PI chemistry, cure
conditions, and surface treatments on its electrical, mechanical,
and adhesion properties, (2) the effects of environmental stress,
i.e., humidity, thermal shock, temperature excursions, and
radiation, on the electrical and mechanical properties of PI and
the adhesion of PI/PI, PI/metal and PI/ceramic interfaces; and (3)
the species outgassed by PI, their rate of desorption, and their
effect on package and IC performance. The military qualification
procedures specified in MIL-STD-883C provide a well-standardized
and severe set of environmental tests for packaging structures.

A number of studies addressing these reliability issues have
been conducted at Honeywell (8,21) and elsewhere (11,12,22-24).
The reader is referred to these references for greater detail. A
brief summary of results from studies performed at Honeywell is
given here:

(1) The dielectric constant of PI increases linearly with humidity,
    ranging from 3.1 to 4.1 over 0-100 percent relative humidity
    (r.h.) at room temperature;
(2) The dissipation factor of PI increases and breakdown voltage
    decreases with increasing humidity, but both remain within
    electrical design requirements for dielectric loss;

Figure 5. Reactive ion etched via hole, 50 um diameter x 25 um
           deep.

Figure 6. Via hole in 25 um-thick photosensitive polyimide.

(3) The fracture strength of PI (200 MPa) is about five times greater than the internal stress developed due to thermal expansion mismatches with Si or Al_2O_3 substrates; thus internal stress alone will not cause cracking of the PI film;

(4) The fracture strength, dielectric constant, dissipation factor and adhesion of PI are unaffected by cummulative gamma ray doses of up to 10^8 rads (Si) (unpublished results);

(5) The adhesion of PI to Cr as measured by peel strength is excellent and does not degrade with temperature cycling, thermal shock, or high humidity; adhesion to TiW and Mo is about 1/3 to 1/2 that of Cr; and an amino-silane adhesion promoter is required to obtain acceptable adhesion of PI to Al_2O_3 and to prevent degradation of adhesion after temperature shock and humidity;

(6) There is no evidence of delamination, corrosion, or other visual changes for TFML structures subjected to MIL-STD-883C tests for temperature cycling ($-65°C$ to $150°C$, 100 cycles), thermal shock ($-55°C$ to $125°C$, 15 cycles), moisture resistance (Method 1004.5), and accelerated aging at $85°C$/85 percent r.h. for 1000 hours (unpublished results);

(7) Temperature cycling and thermal shock cause no change in via resistance and no open vias for strings of 245 vias, 25 μm or larger in diameter (unpublished results);

(8) The major outgassed species from fully-cured PI (as measured by mass spectroscopy) is H_2O, which is rapidly desorbed upon heating and reaches background levels of detection at temperatures above $150°C$; residual solvents are observed above $170°C$ and CO_2 is observed above $240°C$, however, these temperatures are well above MIL-STD requirements;

(9) Cu/PI TFML substrates that have been vacuum baked for 16 hours at $150°C$ prior to hermetic sealing in an N_2 atmosphere satisfy MIL-STD moisture levels of less than 5000 ppm H_2O at $100°C$.

Some general conclusion from these studies are: (1) Cu/PI TFML structures have excellent thermal and mechanical stability under extremes of temperature, humidity, and radiation; (2) the adhesion of polyimide is highly dependent on interface chemistry and surface preparation; (3) PI rapidly absorbs and desorbs water, which has an appreciable effect on its dielectric properties and thus the electrical charactersitics of TFML interconnections; the electrical design tolerances must accommodate these variations or the package must be hermetically sealed; (4) properly baked and sealed TFML packages can maintain MIL-STD internal moisture levels of less than 5000 ppm at $100°C$.

Demonstrations of Cu/PI TFML Packaging

The Cu/PI interconnect technology described above has been demonstrated in a number of multichip packages and test vehicles (8-12,25). A demonstration package that was designed and fabricated at Honeywell is shown in Figure 7. Nine bipolar LSI gate array ICs (HE-2000) are interconnected on an 80 mm square substrate. Each chip has 174 bonding pads and there are 420 off-package connections around the edge of the substrate. There are six metal layers and five dielectric layers of Cu/PI. Figure 8

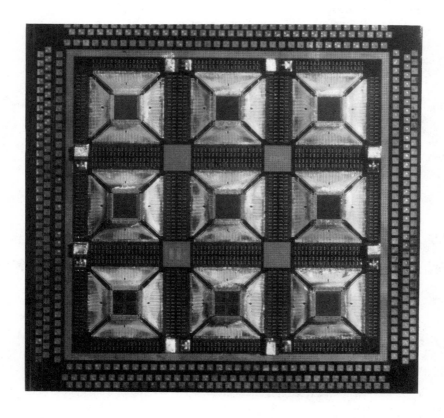

Figure 7. Nine-chip LSI gate array module with Cu/PI
           interconnects.

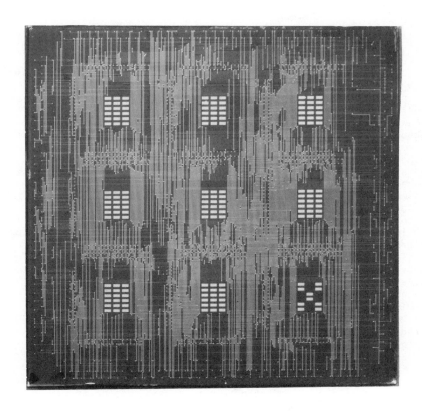

Figure 8. Nine-chip module processed through two layers of signal lines.

shows the substrate processed through the second layer of signal
lines, which are 37 μm wide on a 250 μm pitch. Figure 9 shows a
close-up view of a tape-automated-bonded chip and the fan-out
pattern to rework pads.

Honeywell has also demonstrated TFML interconnections in a
multichip package for digital image processing applications (9).
The completed package, shown in Figure 10, contains 18 chips (two
8000-gate CMOS gate arrays, twelve 16K x 4 static RAMs, and four
SSICs) on a 57 mm square substrate housed in a metal flatpack. The
high density interconnections (50 μm wide lines on a 250 μm pitch),
power and ground planes, and top layer features are patterned in
four metal layers of copper conductor and polyimide dielectric on a
blank ceramic substrate. The chips are wire bonded to the TFML
substrate, which is then epoxied and wire bonded inside an 84-lead
commercial metal flatpack that can be hermetically sealed. TFML
technology was selected for this application primarily to reduce
the size and weight of the current system which uses single-chip
packages surface-mounted to a PWB. A board area of 94 cm^2 was
reduced to 36 cm^2 by using the multichip TFML package.
Calculations show that power dissipation will be reduced by a
factor of three, due to the shorter length and lower capacitance of
the interconnects. Functional testing of this package is currently
underway.

The most advanced application of TFML technology has been
reported by NEC for their SX-1 and SX-2 supercomputers, which are
currently in production (10). The logic module for the computer
contains 36 1000-gate CML chips packaged in individual ceramic
carriers that are soldered face-down to the multichip substrate
through an area array of metallic bumps. The 100 mm square co-
fired ceramic substrate contains internal metal planes for power
and ground distribution and a grid of 2177 I/O pins on 2.5 mm
centers for connection to a PWB. The high density interconnections
(25 μm lines on a 75 μm pitch), ground planes, and chip bonding
layer are patterned in five metal layers of TFML gold/polyimide on
the ceramic substrate. Pistons contact the back side of the chip
carriers to transfer heat to a liquid-cooled cap. The multichip
module permits the system to achieve a performance of 1300
megaflops (million floating point operations per second) with a
machine cycle time of 6 ns.

Conclusions

High-density, high-speed interconnections between ICs can be
achieved using thin film multilayer (TFML) Cu/polyimide structures.
The physical and electrical properties of polyimide, combined with
its processability in thin film form, make it uniquely suited as an
interlayer dielectric. We have demonstrated that Cu/PI TFML
structures can be patterned to the geometries required for high
performance interconnections. The primary issues for this
technology revolve around the high cost and severe yield
constraints of processing thick layers on large substrates of
varying shape, flatness, and roughness. Reliability studies have
shown that PI has excellent thermal, mechanical, and chemical
stability under environmental stress; however, the electrical
properties of TFML structures depend on moisture absorption, which

Figure 9. Close-up view of TAB-bonded chip on nine-chip module.

Figure 10. Multichip package with TFML interconnections housed
in a metal flatpack.

can be controlled by hermetically sealing the package. The
feasibility of TFML Cu/PI interconnects has been demonstrated in a
variety of multichip packages, including a 9-chip LSI gate array
module for computer applications and an 18-chip hermetically-sealed
module for image processing applications.

Acknowledgments

Many individuals at Honeywell's Physical Sciences Center have
contributed to this technology development, in particular, D.
Saathoff, M. Propson, B. Ihlow, F. Belcourt, R. Douglas, and T.
Moravec. N. Griffin at Honeywell's Microelectronics Technology
Center designed the 9-chip microprocessor module and D. Kompelien,
now at Defense Systems Division, designed the 18-chip image
processor module.

Literature Cited

1. Bupp, J. R.; Challis, L. N.; Ruane, R. E.; Wiley, J. P. IBM J.
    Res. Dev. 1982, 10, 306-317.
2. Handbook of Thick Film Hybrid Microelectronics; Harper, C. A.,
    Ed.; McGraw-Hill: New York, 1974.
3. Teresawa, M.; Minami, S.; Rubin, J. Int. J. Hybrid
    Microelectronics 1983, 6, 607-615.
4. Schwartz, B. J. Phys. Chem. Solids 1984, 45, 1051-1068.
5. Ho, C. W. In VLSI Electronics: Microstructure Science;
    Einspruch, N. G., Ed.; Academic: New York, 1982; Vol. 5,
    Chapter 3.
6. Spielberger, R. K.; Huang, C. D.; Nunne, W. H.; Mones, A. H.;
    Fett, D. L.; Hampton, F. L. IEEE Trans. Components, Hybrids,
    Manuf. Technol. 1984, CHMT-7, 193-196.
7. McDonald, J. F.; Steckl, A. J.; Neugebauer, C. A.; Carlson,
    R. O.; Bergendahl, A. S. J. Vac. Sci. Technol. 1986, A4, 3127-
    3138.
8. Jensen, R. J.; Cummings, J. P.; Vora, H. IEEE Trans.
    Components, Hybrids, Manuf. Technol. 1984, CHMT-7, 384-393.
9. Kompelien, D.; Moravec, T. J.; DeFlumere, M. Proc. Int. Symp.
    Microelectronics, 1986, 749-757.
10. Watari, T.; Murano, H. IEEE Trans. Components, Hybrids, Manuf.
    Technol. 1985, CHMT-8, 462-467.
11. Tsunetsugu, H.; Takagi, A.; Moriya, K. Int. J. Hybrid
    Microelectronics 1985, 8, 21-26.
12. Takasago, H.; Takada, M.; Adachi, K.; Endo, A.; Yamada, K.;
    Makita, T.; Gofuku, E.; Onishi, Y. Proc. 36th Electronic
    Components Conf., 1986, 481-487.
13. Jensen, R. J. submitted for publication in Chemical
    Engineering in Electronic Materials Processing; Hess, D. W.;
    Jensen, K. V., Eds.; Advances in Chemistry Series; American
    Chemical Society: Washington D.C.
14. Polyimides: Synthesis, Characterization and Applications;
    Mittal, K. L., Ed.; Plenum: New York, 1984; 2 vols.
15. Proc. 2nd International Conference on Polyimides, Society of
    Plastics Engineers, 1985.

16.   Jenekhe, S. A. In Polymers for High Technology ; Bowden, M. J.; Turner, R. S., Eds.; ACS Symposium Series; American Chemical Society: Washington, D.C., (this volume, in press).

17.   Rubner, R.; Ahne, H.; Kuhn, E.; Kolodziej, G. Phot. Sci. Eng. 1979, 23, 303-309.

18.   Deutsch, A. S.; Schulz, R. Proc. Sixth Int. Elec. Packaging Conf., 1986, 331-339.

19.   Srinivasan, V.; Smrtic, M. A.; Babu, S. V. J. Appl. Phys. 1986, 59, 3861-3867.

20.   Brannon, J. H.; Lankard, J. R.; Baise, A. I.; Burns, A. I.; Kaufman, J. J. Appl. Phys. 1985, 58, 2036-2043.

21.   Douglas, R. B.; Smeby, J. M. Proc. 37th Electronic Components Conf., 1987 (in press).

22.   Homa, T. R. Proc. 36th Electronic Components Conf., 1986, 609-615.

23.   Samuelson, G.; Lytle, S. Proc. 1st Tech. Conf. Polyimides, 1984, 2, 751-766.

24.   Senturia, S. D.; Miller, R. A.; Denton, D. D.; Smith, F. W.; Neuhaus, H. J. Proc. 2nd Int. Conf. Polyimides, 1985, 107-118.

25.   Lane, T. A.; Belcourt, F. J.; Jensen, R. J. Proc. 37th Electronic Components Conf., 1987 (in press).

RECEIVED April 8, 1987

Chapter 41

Preparation of Polyimide Mono- and Multilayer Films

Masa-aki Kakimoto[1], Masa-aki Suzuki[3], Yoshio Imai[1], Mitsumasa Iwamoto[2], and Taro Hino[3]

[1]Department of Textile and Polymeric Materials, Tokyo Institute of Technology, Meguro-ku, Tokyo 152, Japan
[2]Department of Electrical and Electronic Engineering, Tokyo Institute of Technology, Meguro-ku, Tokyo 152, Japan
[3]Department of Physical Electronics, Tokyo Institute of Technology, Meguro-ku, Tokyo 152, Japan

Mono- and multilayer films of polyimides were successfully prepared using Langmuir-Blodgett technique.
Monolayer films of polyamic acid long alkylamine salts
were prepared at the air-water interface. The monolayer films were deposited on appropriate plates to
produce multilayer films of the precursor to polyimide
films. Finaly, the polyimide multilayer films were
obtained by treatment of the multilayer films of the
polyamic acid amine salts with acetic anhydride and
pyridine. The polyimide multilayer films had excellent
coating ability giving a very smooth surface. They also
exhibited insulating characteristics as reliable as
polyimide thick films.

Wholly aromatic polyimides are highly thermally stable engineering
plastics, and have been widely used as the reliable insulating materials in microelectronics. Recent developments in this field toward
higher integration of devices required ultra thin films of polyimides.
Minimum thickness of polyimide films cast by spin coating was
about 0.1 μm.
    Since polyimides 5 are essentially infusible and insoluble in
organic solvents, they are processed into films at the stage of polyamic acids 3 which are readily synthesized from tetracarboxylic dianhydrides 1 and diamines 2. Thermal treatment of polyamic acid films
to 300 °C affords polyimide films through cyclodehydration (Scheme 1).
Alternatively, chemical treatment of polyamic acid films with a mixture of acetic anhydride and pyridine is also effective in obtaining
polyimide films.(1)
    Langmuir-Blodgett technique (LB technique) is one of the most
promising methods for the preparation of ultra thin ordered multilayer
films with uniform thickness.(2) Typical monomeric multilayer films
of amphiphilic substances such as long alkyl carboxylic acids are
thermally and mechanically unstable. Improvements in the stability
have been achieved using polymeric multilayer films which have been

0097-6156/87/0346-0484$06.00/0

prepared by polymerization of amphiphilic multilayer films having polymerizable functions such as diacetylenes,(3) olefins,(4) and aminoacid esters.(5) A few examples of direct preparation of multi-layer films of preformed polymers have also been reported, including long alkyl polyacrylates,(6) alternate copolymers of polymaleic an-hydride,(7) and cellulose esters with long alkyl acids.(8) These polymeric multilayer films still possess thermally unstable long alkyl chains.

In this article, we describe the first successful preparation of polyimide mono- and multilayer films without any pendant long alkyl chains using LB technique.

RESULTS AND DISCUSSION

The present method consists of three steps as illustrated in Scheme 1. In the first step, monolayer films of polyamic acid long chain alkyl-amine salts <u>4</u> at the air-water interface are prepared.(9) Unexpected-ly, the polyamic acid <u>3</u> itself, which possess hydrophilic carboxyl functions in the polymer backbone, did not afford a stable monolayer at the air-water interface. Introduction of a hydrophobic long alkyl chain into <u>3</u> was performed by mixing polyamic acids and longchain alkylamines. Polyamic acid salts <u>4</u>, thus obtained, afforded very stable monolayer films at the air-water interface. In the second step,(10) the polyamic acid salt monolayer films are successfully deposited on appropriate plates such as glass, quartz, or silicon wafer. Finally polyimide multilayer films are obtained by treatment of polyamic acid salt multilayer films on the plates with a mixture of acetic anhydride and pyridine.

<u>Preparation of Monolayer Films of Polyamic Acid Alkylamine Salts at Air-water Interface</u>:

The structure and the code of various polyamic acids <u>3</u> are shown in Table 1. A solution of <u>3</u> in a mixture of N,N-dimethylacetamide (DMAc) and benzene (1:1) prepared to a concentration of 1 mmol/L and a solu-tion of dimethylhexadecylamine (DMC16) in the same mixed solvent with the same concentration were combined to produce a solution of polyamic acid salts <u>4</u>. The solutions were then spread on deionized water. The measurement of surface pressure-area relationships (π–A isotherms) was carried out at about 1.8 cm/min compression speed at 20 °C using Wilhelmy type film balance (Kyowa Kaimenkagaku Co., Japan). Figure 1 shows π–A curves of polyamic acid salts <u>4a</u> where the molar ratio of polyamic acid <u>3a</u> to DMC16 was varied. Each curve rose steeply and the collapsed pressures fell by increasing the ratio of <u>3a</u> to DMC16. Polyamic acid <u>3a</u> could not be spread into a stable film as illustrated on curve A. When the ratio of <u>3a</u> to DMC16 was 1:2 (curve D), the collapse point reached 40 dyne/cm. In this case, an equimolar amount of carboxylic function and amine function was present. Figure 2 illustrates the molecular model of the repeating unit of polyamic acid <u>3a</u> in which the aromatic ring coming from the pyromellitic acid lies flat on the water surface. The area of the square line around the model was calculated as 1.28 nm^2, which was in good agreement with a surface area of 1.38 nm^2 obtained by extrapolation of the steep rise

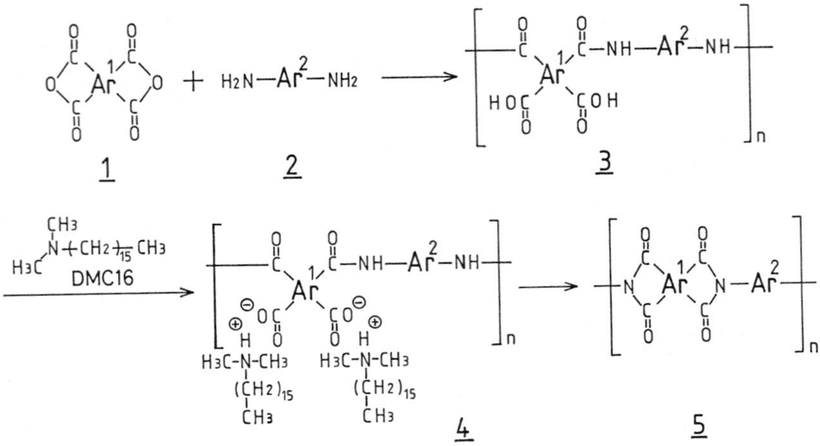

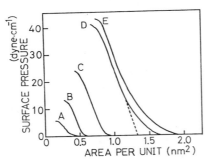

Scheme 1.

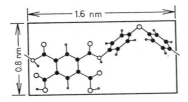

Figure 1 π-A Curves of polyamic acid salts <u>4a</u> varying the ratio of N,N-dimethylhexadecylamine (DMC16) to polyamic acid <u>3a</u>.
A; polyamic acid <u>3a</u>,   B; <u>3a</u>:DMC16=2:1,   C; <u>3a</u>:DMC16=1:1,
D; <u>3a</u>:DMC16=1:2,   E; <u>3a</u>:DMC16=1:4.

Figure 2   Molecular model of polyamic acid <u>3a</u>.

Table 1. Extrapolated Surface Area of
Various Polyamic Acids Salts

Polyamic Acid	Surface Area (nm²/unit)
a	1.38
b	1.65
c	1.39
d	1.60
e	1.30
f	1.50
g	1.10

of curve D to zero pressure. This fact strongly supports that the
spread film at the air-water interface has the monolayer structure.
When an excess of DMC16 was employed (curve E in Figure 1), the extrapolated surface area did not increase markedly compared with the case
of curve D.

A variety of polyamic acids 3b-3f were used for the preparation
of monolayers to examine the behavior of monolayers at the air-water
interface in more detail. Two equivalent of alkylamine DMC16 were
used to produce polyamic acid salts 4b-4f, which were spread onto the
pure water under the same conditions described above. In all cases,
steeply rising π-A curves were obtained. The values of each extrapolated surface area are summarized in Table 1. The difference between the area of 4a and 4b was almost equivalent to the difference
between 4c and 4d, corresponding to the surface area of benzoyl function which was calculated to be about 0.20 nm^2 by the use of a molecular model. Similarly, the observed area of the phenoxy group from the
difference between 4a and 4g, 4e and 4f as well as that of the phenyl
group from the difference between 4a and 4f were in good agreement
with the corresponding calculated values.

Preparation of Multilayer Films of Polyamic Acid Alkylamine Salts:

Deposition of polyamic acid salt 4a was carried out at a surface
pressure of 25 dyne/cm onto a appropriate plate by drawing down and up
through the air-water interface at a rate of 3-5 mm/min at 20 °C
(vertical dipping method).

The plots of absorbance at 258 nm in the UV spectra of multilayer
films of 4a on the quartz plate against the number of layers exhibited
a linear relationship. This fact suggested that the films had an
ordered multilayer structure.

In the transmission FT-IR spectrum of the 4a film (200 layers)
deposited on a silicon wafer, typical absorptions were observed at
around 2920 cm^{-1}, and 1675 and 1610 cm^{-1} due to the long chain hydrocarbon and carbonyl groups, respectively.

Observation by scanning electron microscopy (SEM) showed that no
voids or cracks were present on the film surface.

The extrapolated surface area of the repeating unit of 4a was observed to be 1.38 nm^2 as described above. If the two long alkyl
chains included in each repeating unit were condensed like in the
typical fatty acid type monolayer films, the surface area of the
repeating unit of 4a should be about 0.50 nm^2, which was far smaller
than the observed value. This fact led to the conclusion that the
present monolayer film was not a solid condensed film, where the
extrapolated surface area indicated the area of the repeating unit of
polyamic acid backbone. Furthermore, the X-ray diffraction analysis
suggested that the multilayer films had no crystalline structure. As
we are interested in the conformation of the alkyl chain in such a
liquid film, the tilting angle of alkyl chain from the plane of the
plate was measured using polarized FT-IR spectroscopy. The modified
method reported by Fukuda et al. was applied for the measurement of
the tilting angle of the long alkyl chain of the polyamic acid salt
multilayer films of 4a, 4e, and 4g.(11) The results are summarized
in Table 2. The tilting angles θ increased with decreasing extrapolated surface area to a maximum value of 90° for 4g. The calcu-

lated tilting angle θ_{cal} was obtained using the observed monolayer thickness values d measured by ellipsometry. The d values increased with decreasing extrapolated surface area, and 4g had a value of 1.9 nm which is assumed as the maximum thickness of the monolayer. The tilting angles θ_{cal} were smaller than the observed θ.

Table 2. Tilting Angles of Alkyl Chain
and Monolayer Thickness of Polymeric
Acid Amine Salts Multilayer Films 4

Polyamic Acid Salt	θ (°)	d (nm)	θ_{cal} (°)
4a	80	1.65	60
4e	83	1.80	71
4g	90	1.90	90

θ: Observed tilting angle of alkyl chain.
θ_{cal}: Calculated tilting angle of alkyl chain using d value.
d: Monolayer thickness of 4 measured by ellipsometry.

Preparation of Multilayer Films of Polyimides:

A multilayer film of 4a obtained as above was immersed overnight in a mixture of acetic anhydride, pyridine, and benzene (1:1:3) to afford a polyimide film of 5a.(1)

A linear relationship was again obtained between the number of layers and the absorbance at 284 nm of 5a films, which indicates that the layers did not come off during the imidation step.

In the IR spectrum of polyimide films of 5a, the absorption due to hydrocarbon group of the 4a film disappeared, and new characteristic absorptions corresponding to the imide carbonyl groups appeared at 1780 and 1720 cm^{-1}. This suggested that the cyclization of the polyamic acid salt 4a to polyimide 5a proceeded almost completely with the removal of the long chain alkylamine.

The elemental analysis of multilayer films (50 layers) of polyamic acid salt 4a and polyimide 5a was performed by ESCA (electron spectroscopy for chemical analysis), with the following results:
4a; calculated for $(C_{58}H_{92}N_4O_7)_n$, C: 1.00, O: 0.12, N: 0.07. Found C: 1.00, O: 0.21, N: 0.08.
5a; calculated for $(C_{22}H_{10}N_2O_5)_n$, C: 1.00, O: 0.23, N: 0.09. Found C: 1.00, O: 0.24, N: 0.06.
The films of 4a seemed to be converted into the polyimide under the high energy of X-ray irradiation, while the analysis value of 5a was in good agreement with the calculated value.

Characterization of Polyimide Multilayer Films:

The chemical resistance of the polyimide multilayer film of 5a was

determined by the change of UV spectra before and after immersing it
in various solvents for 1 h.   Although the film readily decomposed in
alkaline   solution,   it   was quite stable in strong acid and organic
solvents such as benzene, alcohols, and dimethylformamide.

     The surface of the polyimide film of 5a was observed by SEM to be
as flat as that of 4a.   The multilayer film of 5a was prepared on a
silicon wafer which contained evaporated aluminum lines in order to
examine the coating ability of the present LB films.   Figure 3(a)
exhibits the mechanical stylus probe (Talystep) pattern of aluminum
lines (30 nm high and 2000 nm wide) and (b) shows the lines after 120
layers of 5a were deposited. Since the depth was observed as 30 nm in
both cases, it is concluded that the multilayer film can conformally
coat the uneven surface. This feature can be directly observed in the
photographs of Figure 4 where a 60 layer film of 5a is coated on an
aluminum line (200 nm of high and 2000 nm wide).

     Decomposition of the multilayer film of 5a was observed to be
around 250 °C which was determined by using SEM as well as by measurement of conductivity at elevated temperature.

     The X-ray interference pattern of the multilayer film of 5a (100
layers) suggested that the film has uniform thickness of around 40 nm.
Furthermore, the thickness of the same film was measured directly to
be 42 nm using a Talystep.   According to the ellipsometry results,
the monolayer thickness was 0.45 nm.

     Figure 5 shows the reciprocal capacitance plotted versus the
number of layers for polyimide 5a.   The linear relationship obtained
suggested that the number of layers were directly proportional to the
thickness of the films.   If the monolayer thickness was assumed to be
0.4 nm, the dielectric constant of the 5a film calculates to a value
of 3.3 which is in good agreement with that of the corresponding
polyimide resin.   It was remarkable that only a monolayer film of the
polyimide performed as a good dielectric.

     Figure 6 shows the conductivities, determined from current-
voltage relationships at room temperature, of multilayer films of 5a
against the number of layers.   The conductivities were almost a constant value of about 8×10^{-15} Scm^{-1} for the samples possessing over
10 layers.   Moreover, the electrical breakdown strength of the polyimide multilayer film is larger than 10^7 Vcm^{-1}.   These results indicate that polyimide multilayer films have insulating characteristics
comparable to normal polyimide films.   The electrical properties of
multilayer films of 5a as well as other insulating materials are
summarized in Table 3.

     The   orientation of the aromatic rings of the polyimide was es-
timated by using the dichroic ratio obtained form polarized UV
spectra.   The samples had 20 layers on one side of a quartz plate.
The measurement was carried out as shown in Figure 7; thus the incident angle α of UV light was 0° or 45° and the UV electric field was
polarized parallel to the transfer direction in spectrum V and was
polarized perpendicular in spectrum H.   The dichroic ratio D_{α} (Absorbance at V/Absorbance at H) at 289 nm was 1.07 at 0° and 0.67 at
45°, respectively. The tilting angle θ of the aromatic rings from the
plane of the plate could be calculated using the modified equation
(1) reported by Yoneyama et al..(12) As the result of the calculation, it was found that θ was 14°, that is, the aromatic rings should
lie flat on the plate. Thus, it was found that the monolayer structure

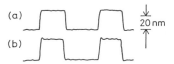

Figure 3 Talystep pattern. (a) Aluminum line (30 nm depth and 2000 nm of width). (b) Multilayer films of 5a (120 layers) are deposited on the aluminum line of (a).

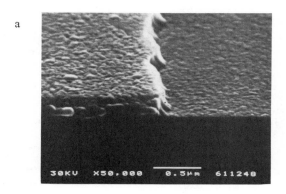

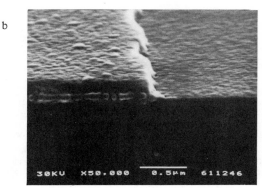

Figure 4 Feature of coating of the plate. (a) Aluminum line (200 nm depth). (b) Multilayer films of 5a (60 layers) are deposited on (a).

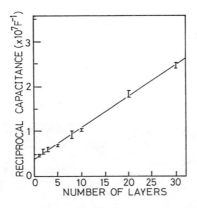

Figure 5   Plots of reciprocal capacitance against number of layers of polyimide 5a.

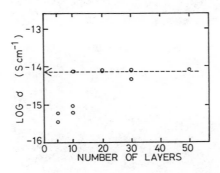

Figure 6   Conductivity against number of layers of polyimide 5a.

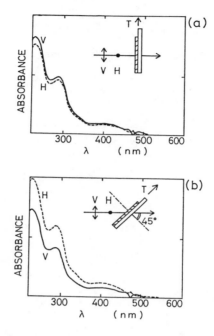

Figure 7 Polarized UV spectra of multilayer films of 5a (20 layers). (a) Incident angle is 0°. (b) Incident angle is 45°.

in which the aromatic rings lay flat at the air-water interface was maintained in the final polyimide multilayer films. Furthermore, this results implies that the present thin films should have an ordered multilayer structure.

$$\cos^2\theta = \frac{(1/D_0)-\{1+(1/D_0)\sin^2\beta\}\cdot(1/D_{45})}{(1-2\sin^2\beta)\cdot(1/D_{45})-\{1+(1/D_{45})\sin^2\beta\}\cdot(1/D_0)} \qquad (1)$$

β:refraction angle when α is $45°$, $\beta=38°$ was used in this case.(12)

Table 3. Electrical Properties of Multilayer Films of 5a and Other Insulating Materials

	5a-LB[a)]	Kapton H[b)]	C_{20}Cd-LB[c)]	SiO_2
Monolayer Thickness (nm)	0.4	---	2.77	---
Dielectric Constant	3.3	3.5	2.7	3.9
Resistivity (Ωcm)	10^{14}-10^{15}	10^{18}	10^{12}-10^{13}	10^{14}-10^{16}
Breakdown Strength (Vcm^{-1})	$>10^7$	2.76×10^6	$>10^6$	10^7
Refractive Index	1.85	1.78	1.51	1.46
Thermal Stability (°C)	>200	500	100	---

a) Multilayer Films of 5a. b) Polyimide film (25 μm thickness).
c) Cadmium arachidate LB film.

It was observed that the liquid crystalline substances incorporated on the polyimide films which were only 0.8 nm (2 layers) thick were successfully oriented to the transfer direction without a surface rubbing treatment of the polyimide film. This observation can be explained by orientation of the polymer molecule to the film transfer direction as observed by polarized UV and IR spectroscopy.

To our knowledge, the present polyimide monolayer films possess the minimum thickness ever produced. We believe that this simple and efficient method for the preparation of polyimide mono- and multilayer films will bring about new technology especially in microelectronics.

REFERENCES
1)   Sroog, C. E. In Macromolecular Synthesis; Moore, J. A., Ed., John Wiley & sons: New York, 1977; Coll. Vol. 1, P 295.

2)   For special issue of Langmuir-Blodgett films: Thin Solid Films 1980, 68, 1983, 99, and 1985, 132, 133, and 134.

3)   Akimoto, A., Dorn, K., Gros, L., Ringsdorf, H., and Schupp, H., Angew. Chem. Int. Ed. Engl. 1981, 20, 90.

4)   Fukuda, K., and Shiozawa, T., Thin Solid Films 1980, 68, 55.

5)   Fukuda, K., Shibasaki, Y., and Nakahara, H., J. Macromol. Sci., Chem. 1981, A15, 999.

6)   Mumby, S. J., Swalen, J. D., and Rabolt, J. F., Macromolecules 1986, 19, 1054.

7)   Tredgold, R. H., Vickers, A. J., Hoorfar, A., Hodge, P., and Khoshdel, E., J. Phys. D: Appl. Phys. 1985, 18, 1139.

8)   Kawaguchi, T., Nakahara, H., and Fukuda, K., J. Colloid Interface Sci. 1985, 104, 290.

9)   Suzuki, M., Kakimoto, M., Konishi, T., Imai, Y., Iwamoto, M., and Hino, T., Chem. Lett. 1986, 395.

10) Kakimoto, M. Suzuki, M., Konishi, T., Imai, Y., Iwamoto, M., and Hino, T., Chem. Lett. 1986, 823.

11) Nakahara, H., and Fukuda, K., J. Colloid Interface Sci. 1979, 69, 24.

12) Yoneyama, M., Sugi, M., Saito, M., Ikegami, K., Kuroda, S., and Iijima, S., Jpn. J. Appl. Phys. 1986, 25, 961.

RECEIVED May 11, 1987

POLYMERS FOR ELECTRONICS PACKAGING AND INTERCONNECTION

POLYMERS FOR ELECTRONICS PACKAGING AND INTERCONNECTION

Plastic packaging of microelectronic devices continues to grow in importance as processing costs decline and the use of chips in consumer products increases. Both trends promote the acceptance of lower cost, non-hermetic packages. Packaging is the culmination of a long and complicated manufacturing process that may entail 100 separate steps. Defective chips are removed in evaluation tests so that those selected for packaging are the small residual of a much larger number of chips on the wafer. Any further reduction in yield on account of packaging problems, such as flow-induced stresses and thermal shrinkage stresses associated with the plastic molding material, is particularly costly because much value has already been added to each chip by carrying it so far along the manufacturing sequence. High productivity of packaging operations is therefore essential to remaining competitive in the global market.

Packaging also has an important impact on device performance and reliability. Molding compounds with improved crack resistance (often achieved through polymerization and morphology control) allow large chips to be packaged reliably in small-outline and surface-mount packages. The correlation between low ionic impurities in the molding compound and survival rates in accelerated aging tests is also well established. There is, in addition, a growing awareness that higher yield translates to better reliability. Low yield means many marginal packages that are prone to premature failure.

Interconnection of devices is also an important area that many consider to be an impediment to continued performance improvements. Continued miniaturization has resulted in geometric increases in the number of logic elements on a processor chip, and the number of input/output pins is expected to increase to more than 200 in the near future. Accommodating such large numbers of I/O connections requires closer attention to integrating the device into the next higher level of interconnection, typically the epoxy laminated circuit board. Technologies such as wafer-scale integration, hybrid circuits, optical interconnection, and pin grid arrays are under investigation, but the material issues of interconnection, particularly the disparate electrical and thermal expansion properties, have lagged behind.

For these reasons, polymers used in packaging and interconnection of microelectronic devices have begun to receive their due attention among industrial and academic researchers. The following collection of papers represents a cross section of efforts devoted to understanding and improving the productivity and performance of plastic packaging and interconnection.

Louis T. Manzione
AT&T Bell Laboratories
600 Mountain Avenue
Murray Hill, NJ 07974

Chapter 42

Prediction of Lay-Up Consolidation During the Lamination of Epoxy Prepregs

H. M. Tong[1] and A. S. Sangani[2]

[1]Thomas J. Watson Research Center, IBM, Yorktown Heights, NY 10598
[2]Department of Chemical Engineering and Materials Science, Syracuse University, Syracuse, NY 13210

A model has been developed for the prediction of lay-up consolidation (i.e., the lay-up thickness, the inter-spacings between prepreg fabrics and the resin flow field) during the squeeze-flow lamination of circular prepreg lay-ups for the manufacture of printed circuit boards. This model allows for time-dependent lamination temperature and pressure schedules. The input parameters required by this model are both the fabric parameters (such as the yarn diameter and the pitch distances) and knowledge of resin viscosity as a function of time and temperature. While the former is usually available from the fabric manufacturer, the later can be obtained from separate squeeze-flow experiments for the neat resin using a thermomechanical analyzer. The lay-up thickness predictions provided by this model have been found to be in good agreement with experimental results obtained using a thermomechanical analyzer for the lamination of up to five epoxy prepreg layers.

Multilayer printed circuit boards are typically manufactured by laminating circuitized and fully cured epoxy/glass laminates using epoxy prepreg (i.e., a glass fabric impregnated with partially cured epoxy; see Figure 1) at elevated pressures and temperatures. The temperature and pressure schedules during lamination significantly affect the duration and magnitude of the resin flow which can be normal to plate surfaces of the press as in the autoclave process where a porous material collects the excess resin (1-3), or parallel to plate surfaces as in the squeeze-flow lamination process where the excess resin is squeezed off the plate circumference (4,5). In both lamination processes, the resin flow results in lay-up consolidation which manifests itself with a decrease in lay-up thickness until the resin gel point or a sufficiently high resin viscosity is reached. Knowledge of the lay-up consolidation process is essential for the selection of lamination conditions to meet many ultimate board requirements such as the dielectric spacing and void content.

During lamination, the resin flow follows a complex tortuous path both around the yarns of the prepreg fabric and through the spaces or pores that separate the individual glass filaments in each yarn (Figure 1). This is complicated by a resin viscosity which can vary both with time and position by orders of magnitude due to the resin curing reaction and the temperature non-uniformities in the lay-up. In addition, the yarns may deform from their initial shapes and begin to carry part of the applied pressure, particularly, when the press platens and the glass yarns are touching during the later stages of lamination (5). This, in effect, reduces the resin pressure that drives the flow.

0097–6156/87/0346–0499$06.00/0

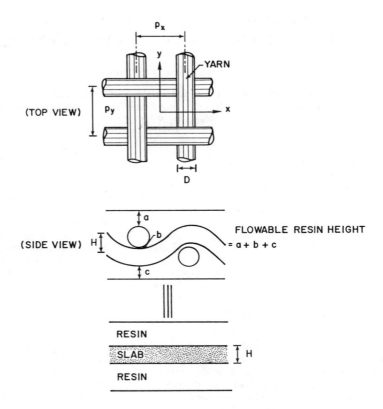

Figure 1.   Top and side views of a resin/glass prepreg showing
the definition of flowable resin height.   (Data are from Ref. 4.)

The purpose of this study is to develop a simple model which retains some of the features of the above complex process to predict the lay-up thickness as a function of time during the squeeze-flow lamination of circular prepreg lay-ups. The prepreg of interest is of the type commonly adopted in the board manufacturing industry. It is composed of two outer resin layers and a fabric core constructed of interlaced yarns oriented in two directions perpendicular to each other (Figure 1). The fabric core is treated as a porous slab characterized by a constant Darcy permeability coefficient (see k in Darcy's law, i.e., Equation 2 below) which can be estimated from fabric parameters such as the yarn diameter and the pitch distances. The lay-up thickness predictions provided by this model have been found to be in reasonable agreement with experiment for the lamination of up to five epoxy prepreg layers.

Prior to the present study, various approaches have been suggested (2-5) for the flow modeling of either circular or rectangular prepreg lay-ups during lamination. For the autoclave process, Springer (2) assumed that the whole lay-up is homogeneously porous and the resin flow is represented empirically by Darcy's law. A more detailed analysis by Lindt (3) involves macroscopic force balances on all glass yarns set in relative motion by the resin flow. As for the squeeze-flow process, Aung (4) reduced the problem to the lamination of a neat resin with a thickness equaling the flowable resin height defined as the sum of the heights of individual resin layers in the lay-up (Figure 1). In contrast to Aung's model, Bartlett (5) took into account the tortuosity of the fabric by estimating the sizes of various channels established between the yarns in adjacent prepreg layers. In these treatments, the prepreg lay-up is considered as either homogeneously porous throughout (2) or consisting of impermeable yarns embedded in the resin matrix (3-5). Moreover, the yarn deformation or the part of the pressure carried by the yarns is ignored except in Bartlett's approach where it is considered on an empirical basis. Recently, the yarn deformation was modelled by Gutowski (6) assuming that the yarns make up a deformable, elastic network. In addition to allowing for the flow through the yarns which may be important in determining the fate of the voids trapped in the yarns, we present a detailed calculation for the inter-spacings between prepreg fabrics and the resin flow field in the lay-up. As an approximation, the yarn deformation which is more important towards the later stages of lamination is ignored in the present study.

Analysis

Problem Formulation.

Consider the squeeze-flow lamination of a circular prepreg lay-up of radius R, consisting of N ($\geq$ 1) prepreg layers under a pressure P. The porous slab (with a constant thickness, H, equaling the yarn diameter) in each prepreg layer is assumed to be characterized by a constant permeability coefficient k for the flow in the lateral direction, r, and a large permeability for the flow in the thickness direction, z (Figure 2). The origins of the r and z coordinates are fixed at the center of the lay-up.

It is known that the resin behaves as an incompressible and Newtonian fluid (7), at least for a significant portion of the lamination process during which resin flow occurs. The Reynolds number of the flow is usually so small that the inertia terms in the equations of motion can be neglected. Also, because the aspect ratio, R/h (h being the thickness of the lay-up), is much greater or greater than unity in our experiments, we assume $v_r >> v_z$ and $(\partial p/\partial r) >> (\partial p/\partial z)$ where v_r is the lateral velocity (in the r direction), v_z is the axial velocity (in the z direction) and p is dynamic pressure defined as the pressure above the ambient. Under these conditions, the resin flow satisfies

$$\frac{1}{\eta} \frac{dp[r]}{dr} = \frac{\partial^2 v_r[r,z]}{\partial z^2} \tag{1}$$

in the resin layers and Darcy's law

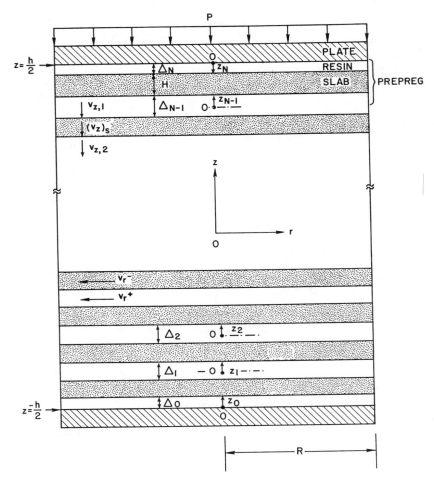

Figure 2: Definition of parameters for the lamination of N prepreg layers.

$$\frac{dp[r]}{dr} = - \left(\frac{\eta}{k}\right) v_r^-[r] \tag{2}$$

in the porous slabs where η is the resin viscosity, the superscript "-" above is used to differentiate lateral velocities in the resin layers and the slab, and the term (η/k) represents the fabric resistance to flow. The validity of Darcy's law for the flow of common viscous liquids through porous media at low Reynolds numbers has been established (8). The two velocities, v_r and v_z, are coupled via the continuity equation:

$$\frac{1}{r}\frac{\partial}{\partial r}(rv_r) + \frac{\partial v_z}{\partial z} = 0 \tag{3}$$

In addition to the usual initial condition for the resin layer thicknesses (Figure 2):

$$2\Delta_0 = \Delta_1 = \Delta_2 = \cdots = \Delta_{N-1} = 2\Delta_N \ @ \ t = 0 \tag{4a}$$

the boundary and other initial conditions associated with Equations 1 - 3 are

$$h = h_0 \ @ \ t = 0 \tag{4b}$$

$$v_r = 0, v_z = \mp \frac{1}{2}\frac{dh}{dt} = \mp \frac{1}{2}h \ @ \ z = \pm \frac{h}{2} \tag{4c}$$

$$\alpha v_r^- = v_r^+ \tag{4d}$$

$$p = 0 \ @ \ r = R \tag{4e}$$

$$\pi R^2 P = 2\pi \int_0^R prdr \tag{4f}$$

$$(v_z)_s = \frac{v_{z,1} + v_{z,2}}{2} \tag{4g}$$

where t is the time, h_0 is the initial lay-up thickness, v_r^+ is the lateral velocity in the resin layer at the resin/slab interface, α is the slip coefficient, $(v_z)_s$ is the speed of the slab in the z direction, and $v_{z,1}$ and $v_{z,2}$ are the two axial velocities just outside the slab (Figure 2). It is well known that the lateral velocity (v_r) may not remain continuous across the fluid/slab interface. Beaver and Joseph (9) have, in addition to examining the validity of Darcy's law, determined the slip coefficients for various porous materials. For the present case, we expect the slip coefficient, α, to be nearly unity due to the large area occupied by the resin at the resin/slab interface. Hence, although we will derive the model using an arbitrary slip coefficient, we shall specialize our calculations to $\alpha = 1$. While Equations 4b-4f are typical of squeeze-flow situations (10), Equation 4g which implies that the resin drags the porous slab with the average velocity in the z direction is assumed to uniquely specify the mathematical system when more than one prepreg layer is involved. The mathematical system of Equations 1-3 and 4a-4g describes the squeeze-flow lamination process for circular prepreg lay-ups.

As mentioned earlier, the viscosity, η, of the resin typically changes by orders of magnitude during a lamination cycle. Equation 5 below is a dual-Arrhenius type expression often used to describe the viscosity changes:

$$\ln \eta[t, T] = \ln \eta_\infty + \frac{\Delta E_\eta}{R'T} + k_\infty \int_0^t \exp\left[\frac{-\Delta E_k}{R'T}\right] dt \tag{5}$$

where T is the temperature, R' is the universal gas constant and η_∞, ΔE_η, k_∞ and ΔE_k are constant parameters (11). This expression accounts for both an exponential viscosity decay with temperature due to the softening of the resin (first two terms on the right-hand side) and a linear viscosity increase with the number density of the cross-links occurring as a result of a first-order curing kinetics (the last term on the right-hand side). It must be noted that the last two terms on the right-hand side requires knowledge of the thermal history of the resin for its evaluation. Thus, interestingly enough, even two fluid elements having the same temperature at some time could have very different viscosities if their temperatures at previous times were not identical. Whenever a temperature non-uniformity exists in the lay-up, the trajectory of each fluid element must be evaluated, in general, to determine the spatial distribution of the resin viscosity. This can be done, in principle, by solving the energy conservation equation together with the equations of motion for a given set of initial and boundary conditions. The extent of temperature non-uniformity would depend on the Biot number which is small for the experiments described in this work involving small circular samples (0.147 inch in diameter). Hence, the variations of temperature with r and z are neglected and the viscosity is assumed to be a function of time alone. In much larger samples typically employed during manufacturing, however, the spatial variation of temperature could play an important role and in such cases it would be necessary to carry out detailed trajectory calculations. The effect of the non-uniform spatial distribution of temperature on the flow field is currently being studied and the results will appear in a future article.

Single Prepreg Layer. For the sake of illustration, we shall first confine ourselves to the lamination of one prepreg layer for which the two resin layers are of equal initial thickness ($\Delta_0 = \Delta_1 = \Delta$).

Assuming that v_z is a function of z alone (10), it is clear from Equation 3 that

$$v_r[r,z] = rf[z] = \frac{-r}{2} \frac{dv_z[z]}{dz} \tag{6}$$

whereupon Equations 1 and 2 yield

$$\frac{1}{\eta r} \frac{dp[r]}{dr} = -\frac{v_r^-[r]}{rk} = f'' = \frac{d^2f[z]}{dz^2} \tag{7}$$

Applying Equations 7, 4c and 4d, we obtain

$$\dot{h} = f'' \left[\frac{\Delta^3}{3} + 2k(\alpha\Delta + H) \right] \tag{8}$$

where

$$\Delta = \frac{h - H}{2} \tag{9}$$

$f''[z]$ in Equation 8 can be determined from Equations 4e, 4f and 7:

$$f'' = -\frac{4P}{R^2 \eta[t]} \tag{10}$$

It should be noted that the time dependence of the resin viscosity, η, appears only through Equation 10. This is because of the quasi-steady-state approximation implicit in the present analysis. Equation 8 reduces to an expression similar to that derived by Aung (4) when $k = 0$

cm^2, i.e., when the slab is impermeable. Using Equations 4c, 4d and 7, we obtain the lateral velocity field:

$$v_r = rf[z_j] = rf'' \left[\frac{z_j^2}{2} - \left(\frac{\Delta}{2} + \frac{k\alpha}{\Delta} \right) z_j \right], \quad j = 0, 1 \tag{11}$$

in the resin layers and

$$v_r^-[r] = -rkf'' \tag{12}$$

in the slab where z_j (Figure 2) is the thickness-direction coordinate whose origin resides either at the plate surfaces (j=0, N) or at the centers of inner resin layers (j=1, 2, ..., N-1).

Once the resin viscosity (η) and the permeability coefficient (k) are known, Equations 8 - 12 allow calculation of h, Δ, v_r and v_r^- as a function of time during lamination.

N Prepreg Layers. The above analysis can be generalized to the case of N prepreg layers (Figure 2). The lateral flow within each slab is again given by Equation 12. Using Equation 4g, it can be shown that

$$\dot{\Delta}_0 = \frac{d\Delta_0}{dt} = f'' \left[\frac{\Delta_0^3}{6} + k(\alpha\Delta_0 + H) \right] \tag{13}$$

$$\dot{\Delta}_j = f'' \left[\frac{\Delta_j^3}{6} + 2k(\alpha\Delta_j + H) \right], \quad j = 1, 2,..., N - 1 \tag{14}$$

$$\dot{\Delta}_N = f'' \left[\frac{\Delta_N^3}{6} + k(\alpha\Delta_N + H) \right] \tag{15}$$

$$\dot{h} = \sum_{j=0}^{N} \dot{\Delta}_j \tag{16}$$

The lateral velocity distribution in the resin layers can be determined from

$$v_{r,j} = rf_j, \quad j = 0, 1, 2, 3, ... , N \tag{17}$$

with

$$f_0 = f'' \left(\frac{z_0^2 - \Delta_0 z_0}{2} - \frac{k\alpha z_0}{\Delta_0} \right)$$

$$f_j = f'' \left(\frac{z_j^2}{2} - \frac{\Delta_j^2}{8} - \alpha k \right), \quad j = 1, 2, ... , N - 1$$

$$f_N = f'' \left(\frac{z_N^2 - \Delta_N z_N}{2} - \frac{k\alpha z_N}{\Delta_N} \right)$$

Given k and η, Equations 4a, 4b and 12-17 describe the lay-up consolidation process during the lamination of N prepreg layers.

Estimation of Parameters. The resin viscosity, η, as a function of time and/or temperature can be obtained using either a generalized dual-Arrhenius rheology model (Equation 5) or the thickness - time relationship for the neat resin from a separate squeeze-flow experiment (7).

In order to estimate k, we first determine the pressure drop for the flow through a yarn of diameter H and represented by a bundle of glass filaments. We assume that this pressure drop corresponds approximately to that for the flow through an assembly of impermeable cylinders. Sangani and Acrivos (12) have obtained the numerical results for the case where the cylinders are arranged in either square or hexagonal arrays. It can be shown that when the volume fraction (c) of the cylinders is less than 0.4, the pressure drop is nearly independent of the geometrical arrangement of the cylinders. Using their expression for the pressure drop and correcting for the empty space between two successive yarns, we arrive at the following estimates of the permeability coefficients in the two directions (x, y) perpendicular to each other (Figure 1):

$$k_x = \frac{d_f^2}{c} \frac{p_x}{H} \frac{p_y - H}{p_y} \kappa \tag{18}$$

$$k_y = \frac{d_f^2}{c} \frac{p_y}{H} \frac{p_x - H}{p_x} \kappa \tag{19}$$

$$\kappa = \frac{-2 \ln[c] - 2.98 + 4c - c^2}{16\pi} \tag{20}$$

where d_f is the diameter of the glass filament, and p_x and p_y are the pitch distances in the x and y directions, respectively (Figure 1). Thus, we see that the resistance to the flow in the plane of the prepreg depends on the orientation of the flow with respect to its weave. In this study, we shall use an average pitch distance $\tau = (p_x + p_y)/2$ and

$$k = \frac{d_f^2}{c} \frac{\tau - H}{H} \kappa \tag{21}$$

to calculate the permeability coefficient k.

Experimental

The resin used in this study was a tertiary amine catalyzed, dicyandiamide cured FR-4 epoxy system. The prepreg was fabricated by impregnating the glass fabric (industry designation: 108; H=D=1.15 mils) with the FR-4 epoxy varnish and heating to remove the casting solvents and partially cure the resin. From the number of glass filaments in the yarn and the diameter of the glass filament supplied by the manufacturer as well as the diameter of the yarn obtained using scanning electron microscopy, the volume fraction of glass in the yarn was estimated to be 0.39. By Equation 21, the permeability coefficient k for the 108 fabric was then calculated to be 1.28×10^{-7} cm^2. Both the prepreg and the neat resin (i.e., without the glass fabric) prepared under identical conditions were kept in vacuum before use.

Using a thermomechanical analyzer (TMA) as the parallel-plate rheometer, the neat resin was laminated under identical conditions as did the prepreg to determine the viscosity history. Throughout this study, the quartz probe (0.145 in. in diameter) which is attached to a linear variable differential transformer for thickness monitoring exerted a constant force of 2 grams on the sample sitting on a non-moving quartz platform in the TMA. A detailed description of the TMA experiment can be seen elsewhere (7).

Results and Discussion

Figure 3a shows the good agreement between theory and experiment in the lay-up thickness - time relationship for the lamination of a prepreg layer. Also shown in Figure 3a are the predicted thickness-time relationships at other k values between 0 (impermeable, i.e., $c \rightarrow 1$) and 10^{-6} cm^2 (arbitrary). Although the predicted thickness history does not vary significantly with k when $k \leq 1.28 \times 10^{-7}$ cm^2, the k value may play an important role in determining the magnitude of the lateral flow in the fabric layer (see Equation 12).

A typical lateral velocity profile corresponding to the sample in Figure 3a is plotted in Figure 3b at an arbitrary time (t=20 sec). It is seen that the lateral velocity is negligible at r=0 and increases at a constant z (in either the resin layers or the slab) with increasing r, reaching its maximum at r=R. Thus, the resin removed from the prepreg comes mainly from regions confined to the vicinity of plate edges. That the maximum velocity in the resin layers (2.87 x 10^{-3} cm/sec) is an order of magnitude higher than that in the slab (1.87 x 10^{-4} cm/sec) reflects the high drag imposed by the glass yarns on the resin flow.

For the lamination of two and five prepreg layers, the lay-up thickness - time relationships are given in Figures 4 and 5 along with the calculated resin layer thicknesses (Δ_j's). In both cases, reasonably good agreement between theory and experiment is obtained with a maximum deviation of the order of 10%. The agreement is notably better especially if the theoretical curves are given a slight displacement to the right corresponding to a lag of a few seconds. The discrepancies here may be due partly to the assumption of a position-independent viscosity (or temperature) used in the theoretical calculations.

Summary

A model has been developed for the prediction of the lay-up consolidation process during the lamination of prepreg lay-ups for the manufacture of printed circuit boards. This model can be used advantageously in prepreg material screening and in the optimization of lamination schedules for high-performance structural composites.

Legend of Symbols

c: volume fraction of glass in a yarn.
d_f (cm): diameter of the glass filament.
h (cm): lay-up thickness.
h_0 (cm): initial lay-up thickness.
H (cm): slab thickness or yarn diameter.
k (cm^2): permeability coefficient.
k_x, k_y (cm^2): permeability coefficients in the x and y directions (Figure 1a).
p (dyne/cm^2): dynamic pressure.
p_x, p_y (cm): pitch distances in the x and y directions (Figure 1a).
P (dyne/cm^2): applied pressure.
r (cm): lateral coordinate.
R (cm): radius of lay-up.
R' (erg/gmole/$^\circ$K) : universal gas constant.
t (sec): time.
T ($^\circ$C): lay-up temperature.
v_r (cm/sec): lateral velocity.
v_r^+ (cm/sec): lateral velocity in the resin layer at the resin/slab interface.
v_r^- (cm/sec): lateral velocity in the slab.
v_z (cm/sec): axial velocity.

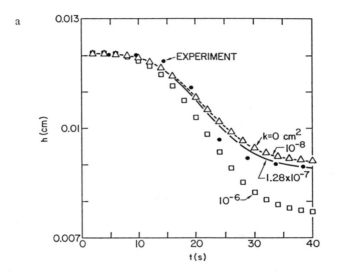

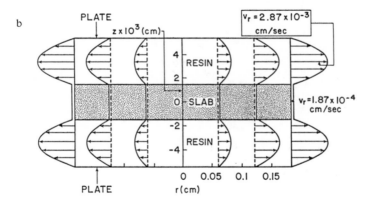

Figure 3: (a)The h-t relationship and (b)v_r and v_r^- at t = 20 sec for the lamination of one prepreg layer (k=1.3 x 10^{-7} cm^2). Also shown in (a) are the predicted thickness histories at k=0, 10^{-8} and 10^{-6} cm^2 .

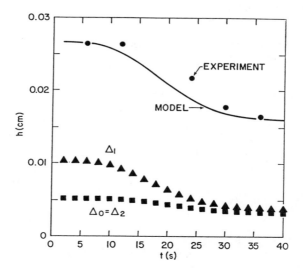

Figure 4: The h-t relationship for the lamination of two prepreg layers.

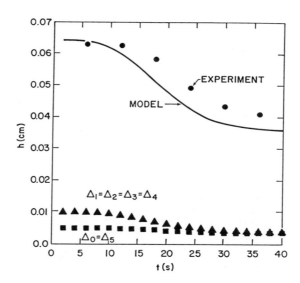

Figure 5: The h-t relationship for the lamination of five prepreg layers.

$(v_z)_s$ (cm/sec): slab speed in the z-direction.
$v_{z,1}$, $v_{z,2}$ (cm/sec): axial velocities just outside the slab (Figure 2).
z (cm): axial coordinate.
z_j (cm): thickness-direction coordinate defined in Figure 2.
η (poise): resin viscosity.
Δ_i (cm): resin layer thickness (Figure 2).
α : slip coefficient.
η_∞, ΔE_η, k_∞, ΔE_k: parameters defined in Equation 5.
τ (cm): average pitch distance.

Acknowledgments

We are indebted to Dr. J. J. Ritsko for his continuous support throughout this study. We would also like to thank Dr. G. Martin for suggesting some of the ideas presented in this study and Dr. C. Feger for useful discussions. In addition, Mr. G. Thomas is gratefully acknowledged for providing the information on glass fabric.

Literature Cited

1. Kardos, J. L.; Dudukovic, M. P., McKague, E. L.; Lehman, M. W. In Composite Materials: Quality Assurance and Processing; Browning, C. E., Ed.; ASTM STP 797; American Society for Testing and Materials, 1983; pp 96-109.
2. Springer, G. S. In Progress in Science and Engineering of Composites; Hayashi, T.; Kawata K.; Umekawa, S., Eds.; ICCM-IV: Tokyo, Japan, 1982; p 28.
3. Lindt, J. T. SAMPE Quarterly, Oct., 1982, pp 14-9.
4. Aung, W. Proc. INTERNEPCON/EUROPA, 1973, pp 13-23.
5. Bartlett, C. J. J. Elast. and Plastics, 1978, 10, 369
6. Gutowski, T. G. SAMPE Quarterly, 1985, 16, 58.
7. Tong, H. M.; Appleby-Hougham, G. J. Appl. Polym. Sci., 1986, 31, 2509.
8. Wiegel, F. W. Fluid Flow Through Porous Macromolecular Systems; Springer-Verlag: New York, 1980.
9. Beavers, G. S.; Joseph, D. D. J. Fluid Mech., 1967, 30, 197.
10. Bird, R. B.; Armstrong, R. C.; Hassager, O. Dynamics of Polymeric Liquids; Wiley: New York, 1977; Vol. 1, pp 19-21.
11. Roller, M. B. Polym. Eng. Sci., 1975, 15, 406.
12. Sangani, A. S.; Acrivos, A. Int. J. Multiphase Flow, 1982, 8, 193.

RECEIVED March 19, 1987

Chapter 43

Effect of Room-Temperature-Vulcanized Silicone Cure in Device Packaging

Ching-Ping Wong

Engineering Research Center, AT&T, Princeton, NJ 08540

At AT&T, RTV silicone is widely used in all of our Bipolar, Metal Oxide Semiconductor (MOS) and Hybrid Integrated Circuitry (HIC) encapsulation. This RTV silicone has been proven to be one of the best encapsulants for alpha particle, moisture and electrical protection of these sensitive devices. The complete cure of the RTV silicone affects not only its chemical, physical and electrical properties, but also the packaging yield of these devices, especially in the RTV silicone-coated Gated Diode Crosspoint (GDX) - a super-high voltage, ultrafast switch used in AT&T No. 5 Electronic Switch System (ESS) and 256K Dynamic Random Access Memory (DRAM) device packaging. However, it is difficult, if not impossible, to detect the degree of cure of RTV silicone in the device manufacturing packaging process. Microdielectrometry, a recently developed technique which utilizes a miniature IC sensor and a wide range of frequencies to monitor the polymer cure, becomes a very attractive technique. This microdielectrometry, coupled with the time-dependent solvent extraction technique, provides a sensitive tool to quantify the degree of RTV silicone cure. This paper will describe the microdielectric measurement and solvent extraction experiments that we have used to investigate curing of the RTV silicone encapsulant systems to optimize the device packaging yields.

The rapid development of integrated circuit (IC) technology from small-scale integration (SSI) to very large-scale integration (VLSI) has had great technological and economic impact on the electronic industry.[1] [2] The exponential growth of the number of components per IC chip, the steady increase of the IC chip dimension and the exponential decrease of minimum device dimensions have imposed stringent requirements, not only on the IC physical design and fabrication but also in the IC device packaging. For bipolar, metal oxide semiconductor (MOS) and hybrid IC, Room Temperature Vulcanized (RTV) silicone has proven to be one of the best device encapsulants for protecting these devices.[3] It is also important that the device encapsulant receives proper cure during packaging to ensure the long-term device reliability. It has been known for some time that the dielectric method was successfully used in monitoring the cure of epoxy resins.[4] [5] The enormous change of the material dielectric properties during the transformation of the resin from a viscous liquid to a brittle solid reveals the crucial degree of cure of the polymeric material. These simple dielectric measurements have become widely used in material analysis and process control.[3] [4] [5] [6] [7] The miniaturized integrated circuit sensor, with the on-chip sensor and wide frequency range, makes microdielectrometry a very attractive technique to monitor the polymer cure. This is especially important in the RTV silicone system where the complete cure of the material is critical in the nonhermetic and hermetic packaging processes. The RTV silicone-coated Gated Diode Crosspoint (GDX), an ultrafast and super high voltage switch, and 256K Dynamic Random Access Memory (DRAM) devices are excellent current examples. Besides, we have developed a time-dependent solvent extraction technique that can couple with the dielectric measurement to quantify the degree of cure of the RTV silicone encapsulant. The combination of this new microdielectric measurement and time-dependent solvent extraction technique provides a definitive method which detects the degree of cure of this RTV material. In this paper, we will describe both of these dielectrometric and solvent extraction experiments that we have investigated for curing the RTV silicone system.

EXPERIMENTAL

1. Material Preparations

A. Dielectric Measurement of RTV Silicone

A thin layer (approximately 10 mil thick) of RTV silicone was coated on the Micromet mini-dielectric sensor. The RTV silicone was first cured at room temperature, then in an oven at 120°C for a period of time, similar to certain RTV production cure schedules. The permittivity and loss factor of the RTV silicone were recorded during this cure cycle.

B. Solvent Extraction of RTV Silicone

RTV silicone samples were obtained from a thin coating which had been cured on a Teflon-coated steel or aluminum plate according to the standard cure schedule (room temperature and final oven cure). For RTV diluted to 50% by weight of the xylene, 20 grams of the material coated on the 5"x5" Teflon-coated aluminum plate will result in a 25 mil uniformly thick sample.

2. Experimental Measurements

A. Dielectric Measurements

The microdielectric sensor which incorporates an IC on-chip emiconductor diode thermometer for combined dielectric and temperature measurements is only 0.5 mm thick, 5 mm wide, with a flat Kapton ribbon package 35 cm long. A small sample of the RTV silicone dispersion (approximately 20 mg) was placed on the sensor electrodes (with interelectrode spacing fixed at 12 um) and its degree of cure was followed by the Micromet Instruments System II Microdielectrometer. A newly designed Fourier Transform Digital Correlator of the Micromet System II, with an accessible frequency range of 0.005-10,000 Hz, was used to analyze the sensor response. Loss factors at frequencies as low as 0.01 Hz can be measured, representing a typical cured resin to tangent value on the order of 0.003[8]. For elevated temperature measurement, the sensor and socket assembly were placed into an oven preheated to 120°C. The sample temperature, as measured by the on-chip temperature sensor, could reach an equilibrium within 10 minutes. The dielectric permittivity (dielectric constant), loss factors and temperature were recorded during the sample cure cycle. Results are shown in Figures 1-4.

B. Time-Dependent Solvent Soxhlet Extraction Measurements

A sheet of the RTV silicone sample, which was prepared by the method mentioned above, was weighed after being rolled up in a stainless steel wire gauge (250 mesh

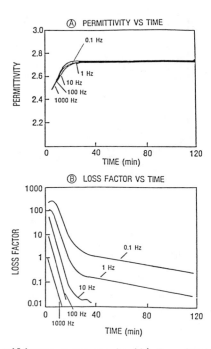

Figure 1.   RTV silicone cure study (A) Permittivity measurement, (B) Loss factor measurement with time at ambient temperature.

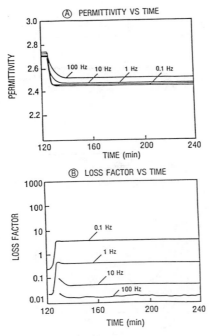

Figure 2. RTV silicone cure study (A) Permittivity measurement,
(B) Loss factor measurement with time at 120°C after 2 hr.
ambient cure.

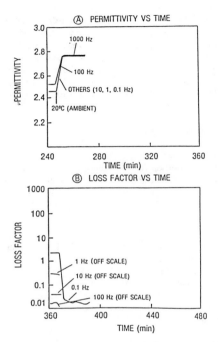

Figure 3. RTV silicone cure study (A) Permittivity measurement, (B) Loss factor measurement at ambient after 2 hr. ambient and 2 hr. at 120°C cure.

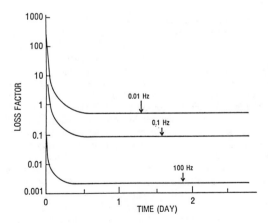

Figure 4. RTV silicone loss factor measurement at complete ambient cure.

size) to prevent the RTV from sticking to itself. This sample was placed inside the Soxhlet extractor which was attached below the water-cooled condenser. Freshly distilled Freon TA was continuously extracted and overflowed from the top of the side-arm of the extractor. RTV silicone extractables were collected on the bottom of the boiling flask. After the standard two-hour extraction, the RTV stainless steel package was removed and dried at 120°C for one hour to remove the excess Freon. The oven-dried sample was cooled to room temperature and reweighed. The percent extractables was calculated. Results of the percent extractable of cured RTV with Freon TA are shown on Figure 5.

C. FT-IR Analysis of Solvent Extractables

Soxhlet extractables which were recovered from the Freon TA solvent were eva-porated to a concentrated solution for the chemical analysis. A Nicolet Model 7199 FTIR was used to analyze the extractables. NaCl plates were used as sample hold-ers. Standard procedures were used to obtain the FT-IR spectrum. Results are shown in Figure 6.

D. GC/MS Analysis

Extractables were further identified by a Hewlett-Packard Model No.5993 GC/MS spectrometry. Solvent extractables were further diluted with Freon TA before GC/MS analysis. Six foot long, 0.3% SP2250 OV17 type material was used as the GC column, with 20-30 cc/min flow rate of helium as the carrier gas. Mass/charge (m/e) units were scanned from 30-800 atomic mass units (amu). Three-dimensional spectra GC/MS were printed out at the end of each run. Results of the GC/MS are shown in Figure 7.

RESULTS AND DISCUSSION

A. Dielectric Study of RTV Silicone

Figure 1 shows the first two-hour, room-temperature (ambient) cure, RTV silicone results. The permittivity (dielectric constant) and loss factor of the RTV silicone were recorded with the curing time and measured at frequencies of 0.1, 1, 10, 100, 1000 Hz. The frequency dependence of the loss factor with cure time indicates the superposition of two components, an ionic conductivity and a dipole relaxation[9]. The permittivity starts out with frequency at 0.1, 1, 10, 100 Hz around 2.5 and rapidly increases to about 2.7 within the first 30-minute room temperature cure. This phenomenon is due to the eva-poration of a low dielectric constant xylene solvent, leaving behind the higher dielectric constant silicone. During the same period (30 minutes) the loss factor of all frequencies (such as 0.1, 1, 10, 100, 1000 Hz) decreases by two orders of magnitude. Again, this is most likely due to loss of solvent, thus decreasing the ionic conductivity significantly. After 45 minutes ambient cure, the permittivity of the RTV is fairly steady, indicating no vitrification or loss of dipoles. However, the loss factor keeps decreasing but at a moderate rate. This is probably due to either slower system loss or a tightening of the network by crosslinking of the silicone matrix. Figure 2 shows the permittivity and loss factor change after 120 to 240 minutes. Upon heating at 125 minutes from room tem-perature to 120°C, the permittivity at all frequencies experienced a sudden decrease to approximately 2.5. This observation of decreasing permittivity with increasing tempera-ture is usually attributed to thermal randomization of the dipoles. Due to thermal activa-tion of ionic conductivity, the loss factor during the same period decreased, and then steadily decreases with further reaction with 120°C oven cure. A tightening of the silicone network is a reasonable explanation (please note the log scale of the loss factor).

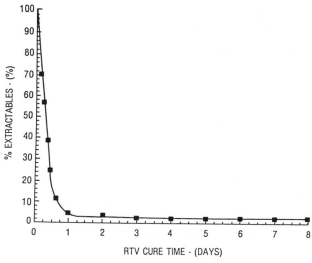

Figure 5.   RTV silicone percent solvent extractables in Freon TA.

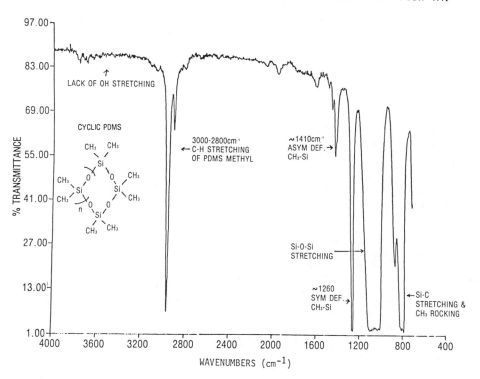

Figure 6.   FT-IR spectrum of RTV silicone solvent extractables.

Finally, in Figure 3, after two hours (240 minutes total curing) of heating at 120°C, the sample was cooled back to room temperature. The permittivities went back to approximately 2.7 due to the removal of the thermal randomization of dipoles in RTV. However, the loss factors were so low that only the 0.1 Hz was observed, indicating it was definitely cured further than before heating (compare Figures 1,2,3). The continuous decrease of loss factor at low frequency (such as 0.1 Hz) after 360 minutes of cure time is a good indication of the continuous further cure of the RTV silicone but at a much slower rate.

In order to further examine the RTV silicone cure, we have rerun the RTV silicone at 0.01 Hz and followed with a longer period of cure time. The RTV silicone was cured at room temperature for 3 days. During this cure process the permittivities show very little change. However, the loss factors in Figure 4 continue to show the decrease with curing time. This loss factor change indicates that the RTV silicone is still undergoing a curing process at room temperature prior to the first 16 hrs. The change is so small that we can only detect this change at the very low frequency (0.01 Hz) of the measurement. This result correlates with our solvent extraction data in estimating the complete RTV cure schedule. In normal RTV silicone material, under normal % relative humidity (50%) and room- temperature conditions, it takes approximately 1 day for the RTV to achieve the complete cure of a 20 mil thick sample (see Figure 6). Heating of the RTV could speed up the cure by removing the excess xylene solvent, unreactive cyclics which are present in the RTV. In addition, the heating of RTV increases the crosslinking rate of the RTV silicone. However, the moisture-initiated, catalyst-assisted RTV cure system does take a longer cure time than normal heat curable silicone. (See Figure 8).[10]

B.   *Time-Dependent Solvent Extraction Experiments*

The Freon TA Soxhlet extraction of the RTV silicone sample at different curing times with Freon reveals the degree of curing of the material. At time zero, when the RTV material was first coated for room temperature cure, almost 100% extractables were obtained (except residue of fillers). This 100% extractable indicates a 0% cure of the material. After 16 hrs. of room temperature cure at 50% relative humidity, most of the RTV material (>90%) was cured. After the second day, almost all of the RTVs studied were fully cured, and they seem to reach an extraction equilibrium. Further curing time shows no noticeable change in amount of extractables. For the fully cured RTV silicone, the level of extractables at their equilibrium extraction was a good indication of the unreactive cyclics present in this material. In Figure 5, the RTV silicone extractable is approximately 4%.

FT-IR measures the vibrational or rotational absorption for an organic molecule. It provides a fingerprint of regions of absorption for organic functional group identification. The collected extractables show a strong Si-O-Si stretching vibration between 1100 and 1000 cm^{-1}. This distinctive absorption is attributed to silicone compounds. Although the lack of absorption around the 3500 cm^{-1} region has indicated the absence of OH terminated silicone (silanol) compounds, it does not distinguish the type of extracted silicones.[11] Figure 6 clearly shows this characteristic feature of the silicone compounds.

GC/MS is one of the most useful analytical instruments to separate and identify mixtures of extracted organic compounds. Gas chromatography (GC) separates each of the components from the mixture by their differences in retention time. Mass spectrometry (MS) identifies each component by its fragmentation pattern. The fragmentation results from the ionization of the parent molecule into the radical cations. Each radical cation fragment has its own characteristic pattern and mass units which are used to identify the component. Low molecular weight volatiles were collected from the Soxhlet extraction using Freon and evaporated to a viscous oil. Figure 7 shows the components of

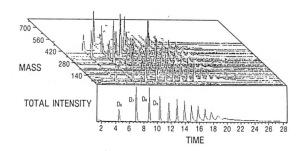

Figure 7.   GC/MS spectra of RTV silicone solvent extractables.

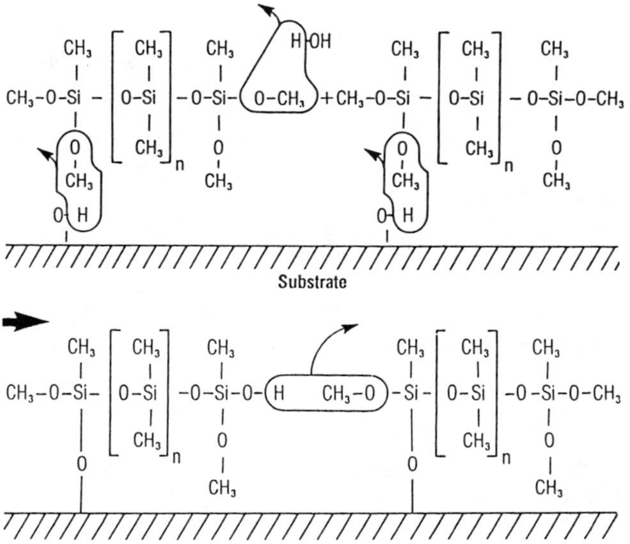

Figure 8.   RTV silicone cure mechanism.

the gas chromatography spectrum of the silicone extractables. Each extractable component was further identified by its mass spectrum. Cyclic siloxanes were identified by this MS analysis. The absence of peaks at m/e values of 58, 62, 76 or 78 suggested that there is no linear siloxane present. The observed mass spec fragments of m/e 73, 147, 133, 221, 286, 327, 355, 357, 415, 429, 503, 577 have structures of the silicone cyclic compounds and agree well with their GC/MS assignments. Prior to the completion of the RTV cure, there are unreactive OH fluids remaining in the solvent extractables, however, when the RTV silicone cure is completed, only unreactive cyclics are observed in the extractables.

CONCLUSIONS

Microdielectrometry, utilizing a miniature IC sensor to perform the dielectric measurement, is a sensitive technique to detect the RTV silicone degree of cure. This real time dielectric measurement technique and instrument, coupled with the time-dependent solvent extraction experiment, could be used to monitor incoming materials as well as packaging configurations using RTV when critical applications require a careful definition of "optimum cure." These methods appear to be very useful in monitoring curing or aging effects of the RTV silicone encapsulant in IC device packages.

ACKNOWLEDGMENT

The author would like to express his gratitude to David Day of Micromet Instrument Company for the helpful dielectric measurement discussion.

Literature Cited

1. Wong, C.P., "Polymeric Encapsulants", *Encyclopedia of Polymer Science and Engineering,* Vol. *5,* p. 638, Second Edition, John Wiley & Sons, Inc., NY, NY (1986).

2. Sze, S.M., ed., "VLSI Technology", McGraw-Hill Inc., NY, 1983 and references therein.

3. Wong, C.P., Rose, D.M., *IEEE Trans. Components Hybrids Manufacturing, CHMT* -6 (4), p. 485 (1983) and references therein.

4. Delmonte, J., *J. Appl. Polym. Sci., 2* (4), 108 (1959).

5. Warfield, R.W., Petree, M.C., *J. Polym. Sci., 37,* 305 (1959).

6. Baumgartner, W.E., Ricker, T., *SAMPE Journal, 19* (4), 6 (1983).

7. Dragatakis, L.K., Sanjana, Z.N., Insulation/Circuits, 27 (Jan. 1978).

8. Senturia, S.D., Sheppard, N.F., Lee, H.L., Marshall, S.B., *SAMPE Journal, 19* (4), 22 (1983).

9. Day, D.R., Lewis, T.J., Lee, H.L., Senturia, S.D., 1984 Adhesion Society, Jacksonville, Florida, Feb. (1984).

10. Wong, C.P., "Improved RTV Silicone Elastomers as IC Encapsulants", in Polymer Materials for Electronic Applications, edited by E.D. Feit, C. Wilkins, Jr., *ACS Symposium Series, Vol. 184,* 171 (1982).

11. Wong, C.P., "Thermogravimetric Analysis of Silicone Elastomers as IC Device Encapsulants," in Polymers in Electronics," edited by T. Davidson, *ACS Symposium Series, Vol. 242,* 285 (1984).

RECEIVED March 2, 1987

Chapter 44

Evolution of Epoxy Encapsulation Compounds for Integrated Circuits: A User's Perspective

H. J. Moltzan, G. A. Bednarz, and C. T. Baker

Texas Instruments, Dallas, TX 75265

During the last 15 years, the ever-changing integrated circuit has required continued improvement in epoxy encapsulation compound technology. These changes are due to performance requirements that have become increasingly demanding and are driven by the integrated circuit's complexity, feature size reduction, power dissipation, quality/reliability, package form, and manufacturing cost factors. The majority of these performance criteria have been met by parallel epoxy encapsulation compound technology improvements. Resin purity, modified resin structure, additives, fillers and compound manufacturing methods yield satisfactory compound for most of today's integrated circuit applications. This paper describes the epoxy encapsulation technology improvements that have been made to meet changing integrated circuit requirements; and the affect the everdemanding requirements have on analytical methodology and testing procedure. Finally, the requirements for encapsulation materials will be presented to help suppliers understand integrated circuit challenges for the near future.

The integrated circuit, which is basic to the electronics industry, has become the technical driving force in many parts of the world. It is now found in schools, cars, homes, offices, production areas, etc., and is used for almost any imaginable application. This has occurred because the integrated circuit can be designed to do many different things, and as time has progressed, it has been designed to perform increasingly difficult and complex tasks. This has been achieved by adding an increasing number of functions to the chip and doing this in a reliable and, most importantly, a cost-efective way.

The commercial integrated circuit (chip) is normally encapsulated in a thermoset plastic molding compound to protect the chip and permit handling and subsequent assembling without loss of circuit integrity. However, the increasing complexity of the chip has

required parallel improvements in the plastic encapsulant. Epoxy
molding compounds have generally been the encapsulant of choice and
this paper will focus on the improvements that have been made with
these materials to meet the ever-increasing demands of the chip. In
this paper, a specific type of chip, DRAM (Dynamic Randam Access
Memory), will be used as a reference point. The reason is that the
epoxy mold compound technology required by DRAMS is a driving force
which, in many cases, can be applied to other types of devices.
Several additional mold compounds will be discussed, along with
their advantages and disadvantages. Finally, this paper will address encapsulation challenges for the the near future.

Testing Procedure
Various testing procedures have been used to determine and ensure
that the epoxy encapsulant will meet the device requirements. This
testing can be divided into three categories: analytical characterization, functional testing and accelerated life tests. The
characterization includes the following:

1. Identification of resin and hardener.
2. Determination of thermal properties.
3. Determination of the flame retardants and their concentration.
4. Determination of impurities and their concentration.
5. Determination of the optimum cure cycle.
6. Determination of molding characteristics.

The accelerated life tests that are normally used are as follows:

1. 125° Operating Life
2. Temperature cycling (-65/150°C).
3. Temperature/Humidity with bias (85°C/85% RH).
4. Pressure cooker (121°C, 2 atm pressure, 100% RH).

These accelerated life tests include the proper electrical testing
at designated intervals.
     Many of the test procedures have remained the same, but there
have been some important changes. The development of the ICP (in-
ductively coupled plasma) and the ion chromatograph have given a
much more detailed and accurate picture of the impurities that are
present. For example, chloride used to be determined using turbidometric techniques, which had a detection limit of approximately
50 ppm. Now the use of the ion chromatograph allows the detection
and quantification of chloride at much lower levels (detection limit
<1 ppm). A new test for accelerated life is HAST (Highly Accelerat-
ed Stress Test). The exact conditions for this test are still
being determined, but it will probably be conducted with electrical
bias at 85% RH and at a temperature between 115°C and 140°C. This
test is similar to 85°C/85% RH, except that a higher temperature is
used. The purpose is to obtain additional acceleration of the reaction mechanisms responsible for device failures without inducing new
failures mechanisms.

Device Reliability

Device reliability is a combination of many different factors. These

factors include the device design, chip manufacturing processes, as-
sembly/packaging processes, packaging materials and the reliability
test criteria. There are many chip manufacturing processes that
will effect the reliability, but these will not be discussed here.
Generally, device reliability has been a "moving target". This is
because the chips are continually growing in size and complexity and
because the accelerated test requirements, as demanded by semiconductor users, are more stringent. For example, Figure 1 shows how
the DRAM chip size has changed from a 1K DRAM to a 256K DRAM. From
this figure, it is evident that the chip size is not increasing
linearly with the DRAM functionality. Instead, considerable
"shrinkage" of the chip is occurring as the functionality is
increased. This is evidenced by the fact that the 1K chip is actually larger than a later version of the 4K chip and the actual
256K chip size is only 6.5% of that extrapolated from a "linear" relationship curve. This chip "shrinkage" is made possible by the
size reduction of almost every design aspect of the chip. It
also means, however, that the chip is more susceptible to factors
such as contaminants and stress.

Figures 2 and 3 show that the DRAM chip performance has been
improved even though the chip functionality has increased for the
accelerated tests used by the semiconductor industry. The 85°C/85%
RH results are better because of a combination of improvements in
the chip design, the manufacturing procedures and the epoxy encapsulant. The temperature cycle test results, however, were primarily
improved by converting to a "low stress" epoxy encapsulant. The improvement in the pressure cooker and the 125°C operating life
(Figure 3) was also due to a combination of improvements, including
those in the epoxy encapsulant. These improvements in device reliability are especially remarkable when it is realized that the chip
susceptibility to contaminants and stress has increased tremendously
due to the 60-fold increase to functionality.

Corrosion. The initial epoxy encapsulants were anhydride-hardened
epoxies and they had several key reliability problems (1). First,
they had a low Tg (glass transition temperature) and this meant that
temperature cycling requirements (e.g., -65°C to 150°C) would be
difficult to meet. Second, they were not very "clean", in that they
contained a considerable amount of chloride which resulted in aluminum corrosion problem. Third, being anhydride-hardened epoxies,
they have poor moisture resistance, which allows the formation of
carboxylic acids by the hydrolysis of the ester that is formed and
by hydrolysis of the residual anhydride. The presence of these
carboxylic acids also results in aluminum corrosion. Anhydride-
hardened epoxies are still used on high power, high voltage and high
temperature applications because of their good thermal stability,
but their use is limited because moisture resistance is still not
adequate.

The introduction of the novolac epoxy corrected the Tg problem
and partially corrected the moisture resistance problem. Advances
in the novolac epoxy technology have continued to improve the moisture resistance, as noted earlier in Figures 2 and 3. Corrosion due
to chloride impurities, however, continued to be a very severe problem until the introduction of low chloride resins and hardeners by
the Japanese resin suppliers. Figure 4 (Bates, Wm., Morton-Thiokol,

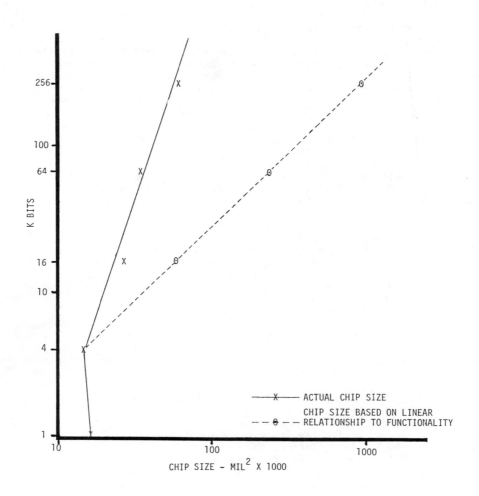

FIGURE 1
K BITS vs CHIP SIZE
DRAMS

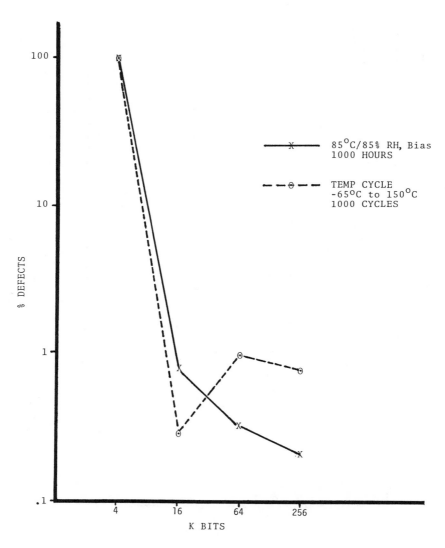

FIGURE 2
% DEFECTS vs K BITS
DRAMS

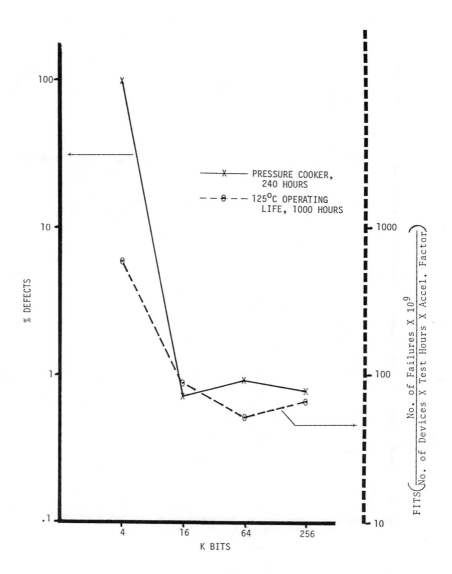

FIGURE 3
% DEFECTS vs K BITS
DRAMS

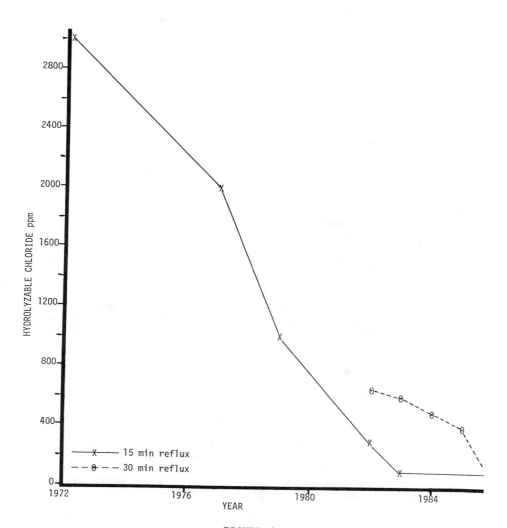

FIGURE 4
ECN RESIN PURITY
HYDROLYZABLE CHLORIDE - MAXIMUM LIMITS
Courtesy Morton Thiokol, Inc.

personal communication, 1986) which shows the maximum hydroxyzable
chloride limits of an ECN (epoxy cresolic novolac) resin used by a
U.S. mold compound supplier, illustrates this very well. In 1972
the maximum hydrolyzable chloride was 3000 ppm. By 1983, this was
reduced to 300 ppm and in 1986, the maximum is now 100 ppm using a
more stringent test condition. This is also illustrated in Table I,
which shows how the water extract conductivity and chloride of the
epoxy mold compound has improved with the use of cleaner resins and
hardeners.

Table I. Epoxy Mold Compound Purity

	Water Extract Conductivity micromhos	Cl, ppm
Anhydride epoxy, pre—1970	35	——
Novolac epoxy, 1972	7	125*
Flame—retarded novolac epoxy, 1975	2.5	10
Flame—retarded novolac epoxy, 1983	1.3	6
Flame—retarded novolac epoxy, 1985	1.2	3

* From turbidometric method.

In 1975 the flame retardant requirement for device encapsulants
was recognized, and by 1976 a flame—retardant novolac epoxy, which
met industry needs with regard to UL classification, was introduced.
This cured the flammability issue for plastic packaged chips, but it
aggravated the halide corrosion problem. The corrosion problem was
aggravated because most flame retardant systems are synergistic systems that contain a halogenated organic (either brominated or chlorinated) and an antimony oxide. Hydrated alumina was initially used
as a flame retardant, but it was found to be very abrasive and its
high alkalinity also caused aluminum corrosion problems. In the
past, both the halogenated organic and the antimony oxide have contained ionic impurities that caused corrosion. In addition, the
halogenated organic functions as a flame retardant by liberating the
halogen to eliminate flame propogation. Unfortunately, the liberation of any halide during high temperature storage or operating life
conditions can cause corrosion (2).
   Figures 5 and 6 show how the water extractable chloride and bromide change with storage at 175°C and 200°C for a 1983 and a 1985
vintage flame retarded novolac epoxy. In both of these figures, the
chloride, most of which comes from the ECN, changes from an initial
concentration of <10 ppm to a maximum concentration of 17 ppm after
1000 hours at 175°C and 23 ppm after 1000 hours at 200°C. The
amount of chloride extracted from both epoxies is similar. The
water extractable bromide, however, increases for the 1983 epoxy after an induction period of about 168 hours. The bromide from the
1985 epoxy also increases, but at a much slower rate. These results
show that the thermal stability problems of the brominated organic
can be minimized, provided the flame retardant system is carefully
selected.

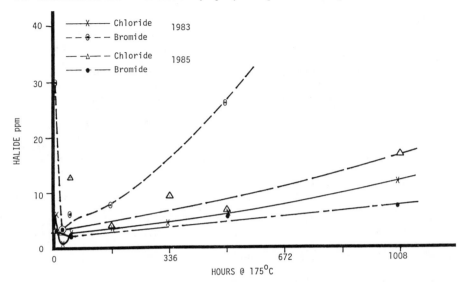

FIGURE 5
WATER EXTRACTABLE HALIDE
AFTER STORAGE AT 175°C
FLAME RETARDED NOVOLAC EPOXY

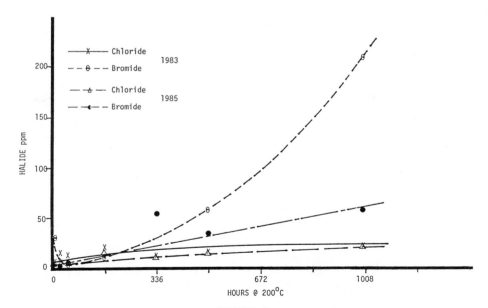

FIGURE 6
WATER EXTRACTABLE HALIDE
AFTER STORAGE AT 200°C
FLAME RETARDED NOVOLAC EPOXY

Older formulations were much worse, and Figure 7 shows a comparison of the water-extractable halide after 200°C storage for four flame-retarded novolac epoxy mold compounds and an early nonflame-retarded variety. Several items are evident from this figure. First, the early nonflame retarded novolac epoxy had a very high water extractable halide content, showing the high impurity levels of the early ECN resins. Second, the water extract halide concentration tended to "level out" with time. This may be indicative of volatilization of halide containing species after a certain halide level is reached. Third, the water extractable halide is reported as all chloride plus bromide. This is because the earlier analyses (up to and including the 1977 flame retarded novolac epoxy) were done using the turbidometric method, whereas the last two (the 1983 and 1985 samples) used the ion chromatograph method. Finally, this figure shows that both the purity and the thermal stability of the resins and flame retardant system has steadily improved. This is particularly dramatic when it is realized that the old method used only a 1.5-hour extraction time, whereas the new method has a 24 hour exraction time.

The water extractable halide (both chloride and bromide) content is important for several reasons. Control of aluminum corrosion, as mentioned earlier, has been the focus of attention for many years. Of more recent concern, are the many reports describing halide attack on the gold wire/aluminum contact bond (3). This attack appears to be due to halide evolution from halide containing impurities and from the flame retardant during high temperature storage (e.g., 200°C). This attack has been reported to occur through alteration of the growth of the Au–Al intermetallics, with the formation of a lamellar structure and Kirkendall voiding. This progressively weakens the gold wire/aluminum contact until failure occurs.

Stress. Initially, stress was concerned primarily with wire breakage and ballbond lifts during the temperature cycling. This was due to the thermal expansion mismatch of the chip, the gold wire and the encapsulant. It was corrected by the addition of a "suitable" filler system which reduced the linear expansion coefficients to a more acceptable level (from $35X10^{-6}$ to $28–30X10^{-6}/°C$), and by using an encapsulant that had a Tg of 150°C or greater. Because most temperature cycling tests use 150°C as the maximum temperature, a 150°C Tg was adequate to minimize or eliminate associated stress problems related to the expansion coefficient.

During the late 1970's, the chip size and stress sensitivity had increased to the point that the use of a filler like crystalline quartz was no longer satisfactory. This was typified by an excessive amount of cracked packages, cracked chips and other stress-related failures. The solution was to use a filler (fused silica) that had a lower expansion coefficient and which gave a "low stress" encapsulant having an expansion coefficient of $18–24X10^{-6}/°C$. This, however, was at the expense of thermal conductivity, because the use of the fused silica instead of the crystalline silica reduced the encapsulant's thermal conductivity from $30X10^{-4}$ cal/cm–sec–°C to $16X10^{-4}$ cal/cm/sec–°C.

As the chip size and functionality increased, the "low stress" epoxy encapsulant was no longer adequate and, in the early 1980's,

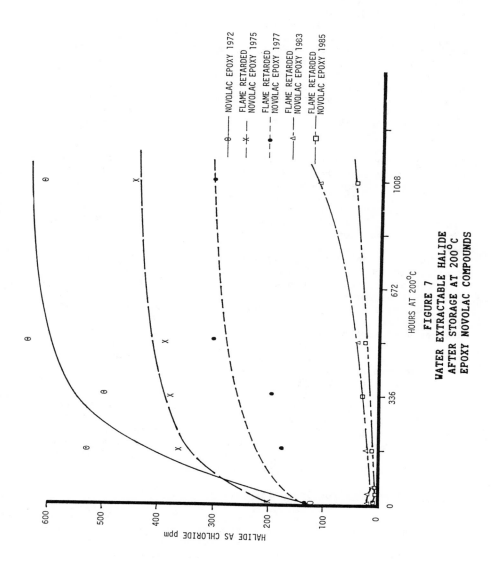

FIGURE 7
WATER EXTRACTABLE HALIDE
AFTER STORAGE AT 200°C
EPOXY NOVOLAC COMPOUNDS

the "ultra-low-stress" encapsulant was developed. This "ultra low stress" epoxy encapsulant contained various stress relieving agents in addition to the fused silica filler. These stress relieving a- gents are generally hydrocarbon rubbers or silicones (4) and their function is to relieve the stress before the stress becomes exces- sive and causes failure. Figure 8 illustrates how chip/surface stress has been reduced to match the higher circuit density require- ment in a comparable 16 pin, 300 mil plastic DIP DRAM package (5).

Another recent report (6) states that the filler particle size and shape also contributes to stress. In this case, a point of a large, irregularly shaped filler particle that contains the chip causes a "point stress" defect. The cure for this is the use of a filler having a smaller particle size or the use of a chip coat (to be discussed later).

Alpha Particles. Alpha particles are high energy particles that are emitted from various radioactive elements. These particles have the capability to change the state of the memory cell. As the size and functionality of the DRAMs has increased, the active cell size has decreased and the DRAMs have become more susceptible to alpha par- ticle damage. This damage became critical with the 16K DRAM and, since that time, action has been taken to eliminate alpha particle damage (7).

The main source of the alpha particles is trace quantities of uranium and thorium in the silica filler. Because silica fillers that did not contain these radioactive elements were not available, other methods for preventing alpha particles from reaching the ac- tive DRAM cells were devised. These early methods consisted of cov- ering the active cells with either a silicone or polyimide chip coat or with Kapton tape. These methods added extra steps to the manufacturing process which were cumbersome and labor intensive and, if not done precisely, had a negative reliability impact. These processes were not widely used once "low alpha fillers" became com- mercially available in 1982/1983. Initially, these "low alpha fill- ers", which contain <1 ppb uranium, were only available from one or two natural sources. Now, however, there are additional natural and synthetic sources of silica, all of which contain <1 ppb of uranium and have an alpha particle emission rate of less than .001 alpha particles/hr-cm^2. Figure 9 shows where the industry was in 1980 and where it stands today. An improvement by a factor of 30-50 has been achieved with "lower alpha" filler and compound manufacturing.

Additional Methods For Improving Device Reliability

Current leakage has also been a serious problem. It occurs by the migration of an electrical charge from one charged pole to an oppo- sitely charged pole and it can travel either on the surface of the chip or through the epoxy encapsulant. The presence of moisture, ionic impurities and/or polarizable species promote current leakage and current leakage has been controlled by improving the adhesion between the epoxy and the chip surface and by minimizing the ionic (or potential ionic) and polarizable species present in the epoxy. Again, purity is a key factor.

One method to minimize leakage and improve device reliability is the use of chip overcoats. Silicone or polyimide chip overcoats

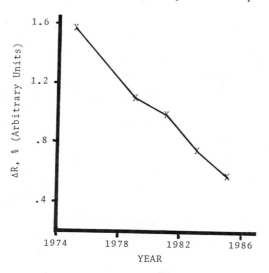

FIGURE 8
MOLD COMPOUND
CONTRIBUTION TO
PACKAGE STRESS

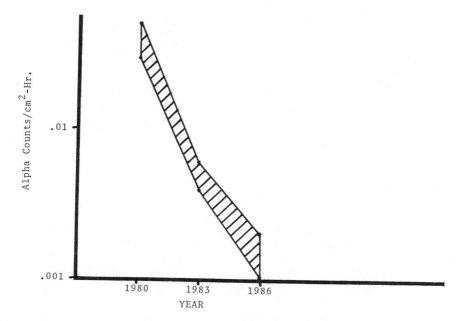

FIGURE 9
DRAMs MOLDING COMPOUND
ALPHA PARTICLE EMISSION

have been used for many years to protect the active surface of the
chip from moisture, halides and various other contaminants. They do
this by forming an adherent film on the active surface of the chip
which prevents the formation of a film of water and the migration of
halides and other impurities to the active chip surface. Chip overcoats are very effective in this, but they also have several disadvantages. Both the polyimide and the silicone materials require additional processing steps that can be labor intensive. In addition,
the polyimide requires a high cure temperature which can cause device degradation or other problems. The silicones must be selected
and applied carefuly, otherwise, wire breakage or bond lifts can occur during temperature cycling due to the large silicone thermal expansion coefficient.

Many devices, particularly high-power devices, require good
heat dissipation to prevent the device from overheating and degrading the device performance. This has generally been obtained
through the use of the proper leadframe material and/or the use of
an encapsulant that has a high thermal conductivity (e.g.,
$30-35X10^{-4}$ cal/cm-sec-^{o}C). High encapsulant thermal conductivity is
normally obtained by using as much filler as possible and by using a
filler that has good thermal conductivity, such as alumina or crystalline quartz. In conjunction with a higher thermal conductivity
filler, an anhydride hardened epoxy may offer a good alternative.
The anhydride hardened epoxies generally have better dielectric loss
properties above the Tg than do the novolac epoxies. Consequently,
a higher device junction temperature can be tolerated because of
less leakage at the plastic/chip interface and because the required
heat dissipation can be obtained via the filler. This is, unfortunately, obtained at the expense of moisture resistance because
the moisture resistance of the anhydride-hardened epoxies is not as
good as the novolac epoxies.

The addition of various corrosion inhibitors to the epoxy encapsulant formulation to prevent corrosion was also evaluated (8).
This procedure basically adds compounds or complexing agents to neu-
tralize or "getter" corrosion accelerating impurities or adds hydrophobic agents to reduce the moisture permeation of the plastic.
This is no longer attractive now that cleaner resins, fillers and
flame retardants are available.

Other Mold Compound Types

Several other types of encapsulants have been evaluated for use by
the semiconductor industry. Polyphenylene sulfide, a thermoplastic,
has the advantage of good thermal characteristics and a low viscosity for device encapsulation. In addition, since it is a thermoplastic, the molded runners can be reused. This improves the material utilization and reduces cost. Unfortunately, this material
also contains a significant amount of impurities that caused device
reliability problems and efforts to remove them have met with mixed
success.

Silicones and epoxy-silicone (9) hybrids have also been used.
The silicones are naturally flame retardant, have excellent electrical characteristics, good purity and good thermal stability. But
they have poor mechanical strength and poor adhesion to the lead-
frames. The poor strength has resulted in an excessive amount of

cracked packages with large bars during temperature cycling and the manufacturing trim and form operation. The poor leadframe adhesion resulted in poor salt spray performance. Consequently, this material is not widely used. The epoxy—silicone hybrid combines some of the properties of the silicone and the epoxy, but it is not inherently flame retardant and the addition of a flame—retardant system reduces the thermal stability. In addition, the mechanical strength, although better than that of the silicone, is not as good as the epoxy and is still inadequate.

Challenges For The Present And Future

Over the years there have been many improvements in the device plastic encapsulant, but the ultimate encapsulant has not yet been obtained. There are still many challenges remaining for the plastic encapsulant industry, both now and in the near future. These challenges are based on the fact that the chips are continuing to increase in both size and complexity. They are also being packaged in either the same size or smaller packages. The chips are, therefore, more sensitive to stress, thermal management, impurities, alphaparticle damage, etc. Further encapsulant improvements that are needed now and the reasons for this need are as follows:

1. Better low—stress properties to eliminate stress—related failures such as metal deformation, cracked passivation and parameter shifts. This must be done without sacrificing strength.
2. Higher thermal conductivity to more effectively dissipate the heat generated by higher power devices. This needs to be achieved without increasing the chip/surface or package stress.
3. Higher thermal stability so that the higher junction temperatures of today's and tomorrow's devices will not be limited by the thermal degradation of the plastic encapsulant.
4. Flame—retardant systems that have better purity and thermal stability .
5. Fillers that do not emit alpha particles so that the continually shrinking DRAM cell size will not be affected by loss charge and state change.

These challenges are substantial, but they are not unrealistic when considering the progress that has been made in the past and they are absolutely essential to the electronics industry.

Acknowledgments

The authors wish to thank Howard Ganden and Greg Getzan of Texas Instruments for DRAM reliability data, and Bill Bates of Morton—Thiokol for ECN resin purity information.

Literature Cited

1. Williams, R. E., <u>Proc. of the Symp. on Plastic Encapsulated/Polymer Sealed Semicon Devices for Army Equipment U.S. Army Electronics R&D Command, Ft. Monmouth, NJ,</u> 5/10—11/78, p. 184

2.   Thomas, R. E., et al., _Proc. IEEE Elect. Com. Conf._, 1977
     p. 182.
3.   Gale, R. J., _Proc. of 22nd Annual Reliability Phys. Symp._,
     1984, p. 37.
4.   Kuwata, K., et al., _Proc. IEEE, IRPS_ 1985, p. 18.
5.   Spencer, J. L., et al., _Proc. of 19th Annual Reliability Phys.
     Symp._, (1981), p. 74.
6.   Matsumoto, H., et al., _Proc. IEEE, ERPS_ 1985 p. 180.
7.   May, T. C., _Proc. of 17th Annual Reliability Phys. Symp._ 1979,
     p. 247
8.   Goosey, M. T., _Plessey Report No. CD6500140_, 1979.
9.   Trego, B., et al., _Electronic Production_, April 1978, p. 45.

RECEIVED March 19, 1987

Chapter 45

Stress Analysis of the Silicon Chip–Plastic Encapsulant Interface

S. Oizumi, N. Imamura, H. Tabata, and H. Suzuki

Electrotechnical Research Laboratory, Nitto Electric Industrial Company, Ltd., Shimohozumi-cho, Ibaraki-shi, Osaka 567, Japan

It is critical to increase the adhesion strength of the interface between the Si-chip and the molding compound. During temperature cycle testing, delamination starts from the chip edge. Once the delamination begins, it is facilitated toward the chip center. As the delamination spreads along the interface, the package cracking at the chip's edge increases because of high tensile stress and high maximum shear stress. The highest probability that package cracking will occur exists with complete delamination. Molding compounds which exhibit a higher thermal expansion coefficient and a greater flexural modulus display a higher tendency toward delamination due to the high shear stress at the chip edge.

Plastic molding encapsulants for IC and LSI contribute to failures caused by package cracks, deformation of the aluminum patterns, passivation layer cracks, etc., during temperature cycle tests. (1,2,3,4) These failures are due to differences between the thermal expansion coefficient (TEC) of the molding compound and that of the Si-chip. Solutions for the above failures are not simple. Lowering stress generated within the molding compound itself is one approach. (5,6) Silicone modified epoxy resins have been developed with this objective which display a lower TEC and flexural modulus. Redesigning the structure of the entire package is another method. (7) In this study, the mechanism of failure was investigated experimentally and mathematically. We estimated the relative stress, which is called experimental stress in this study, from TEC and flexural modulus data for the molding compound. In addition, we analyzed the stress at the interface between the molding encapsulant and the Si-chip during temperature cycling tests. The local stresses, which cause delamination between the molding encapsulant and Si-chip, were calculated by the Finite Element Method (FEM). The existance of the delamination was confirmed using Scanning Acoustic Tomography (SAT), a non-destructive evaluation technique used to detect voids and detached surfaces within a packaged device.

BACKGROUND

During temperture cycle testing from -80 °C to 200 °C, liquid to
liquid, we sometimes observed package cracking, deformation of the
aluminum pattern, and passivation layer cracking. Figure 1
displays SEM photographs that illustrate these phenomena. Example
A illustrates package cracking starting from the chip edge moving
upward at approximately a 120° angle. Example B shows deformation
of an aluminum pattern. The direction of the deformation is toward
the chip center and the degree of deformation is highest at the
outside of the chip. Example C is a passivation layer crack at the
bonding pad on the chip. The occurrence of these failures is often
used as an indication of the low stress characteristics of a molding compound. A correlation has been established that these
failures increase as the number of temperature test cycles
increases.

EXPERIMENTAL STRESS

The relative stress is calculated as experimental stress by using
the following Equation (1).

$$S = K \int_{T_1}^{T_2} E_{(T)} \cdot \alpha_{(T)} \, dT \qquad (1)$$

S: Experimental Stress
K: Constant
α(T):   Thermal Expansion Coefficient of Molding Compound
E(T):   Flexural Modulus of Molding Compound
T: Temperature

The TEC of the Si-chip is negligable because it is one order
of magnitude lower than that of the molding compound. The stress
values from Equation (1) are adequate for use in comparing the
stress levels generated between the molding compounds and the
Si-chip. Figure 2 shows the temperature dependance of the flexural
modulus and the TEC of a molding compound, which must be considered
when calculating the experimental stress. Flexural modulus
decreases slightly with increasing temperature and decreases
rapidly from around 160°C, which is the glass transition
temperature. In contrast, the thermal expansion coefficient
behavior is apparently opposite that of the flexural modulus, and
increases rapidly from approximately 160°C. We can calculate the
experimental stress using Equation (1) by measuring the temperature
dependance of these molding compound properties. Figure 3 exhibits
the experimental stress for three types of molding compounds. Compound A is a conventional molding compound. Compounds B and C are
low stress molding compounds. This data was generated from -50°C
to 175°C by using Equation (1). NOTE: The stress of the molding
compound to the Si-chip is assumed to be stress-free at the molding
temperature 175°C. This figure illustrates that the experimental
stress increases as temperature decreases. Constant K is 1 in this
calculation. Compound C apparently generates less than either

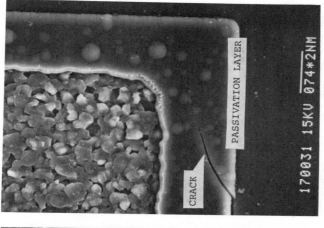

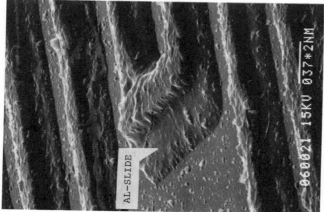

Figure 1: Package Failure

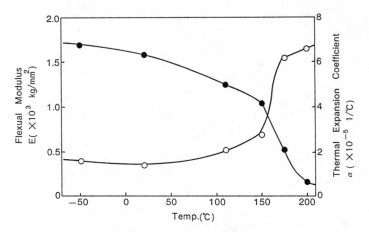

Figure 2: Temperature Dependency of Flexural Modulus and
Thermal Expansion Coefficient of Compound

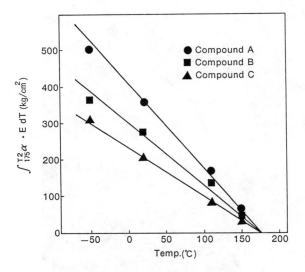

Figure 3: Experimental Stress

Compound A or B. Accordingly, Compound A has a higher probability of inducing package failures during temperature cycle testing. The experimental stress is an indication of a molding compound's low stress characteristics. Furthermore, it provides a satisfactory approximation of the stress levels that a molding compound would impart to the Si-chip. We define this experimental stress to be the total or "bulky" stress of a molding compound to a silicon chip.

Next, we focused on local stress in an entire package to determine which stress caused the package failure. To meet this objective, the Finite Element Method was chosen. However, its use required several assumptions to be defined prior to calculation.

ASSUMPTIONS

1. The temperature dependence of the material properties of the molding compound should be considered to calculate the stress accurately. The average TEC and average Flexural Modulus were calculated using Equation (2). These values are considered to be adequate for the comparison of the local stress of the molding compound (Table I).

$$
\begin{aligned}
\alpha &= \int_{T_1}^{T_2} \alpha(T)\, dT / \Delta T \\
E &= \int_{T_1}^{T_2} E(T)\, dT / \Delta T
\end{aligned}
\tag{2}
$$

α :   Average Thermal Expansion Coefficient
E: Average Flexural Modulus
$\alpha(T)$:   Thermal Expansion Coefficient
$E(T)$:   Flexural Modulus
T: Temperature
ΔT: $T_1 - T_2$ (Thermal Load)

TABLE I: MATERIAL PROPERTIES

MATERIAL	$(1/^\circ C)$	$E(kg/mm^2)$	ν
Compound A	2.38×10^{-5}	1290	0.25
Compound B	2.04×10^{-5}	1110	0.25
Compound C	1.81×10^{-5}	964	0.25
Si-Chip	2.60×10^{-6}	13000	0.28
Lead Frame	7.00×10^{-6}	20800	0.29

2. Analysis is 2-dimensional.
3. Analysis is plane stress condition.
4. Element is quadrilateral.

5.   Number of elements is 2107; number of nodal points is 2200.
                   Finite Element Model   Figure 4
6.   Thermal load is from $-80^{\circ}C$ to $175^{\circ}C$. The stress is assumed to
      be stress free at the molding temperature, $175^{\circ}C$.

RESULTS AND DISCUSSION

Figure 5 illustrates the shear stress distribution over a chip for
the three compounds whose experimental stress values are shown in
Figure 3. For all three compounds, the maximum shear stress occurs
at the chip edge, and this stress decreases rapidly toward the chip
center. It is believed that this maximum value of the shear stress
causes delamination at the interface between the Si-chip and
molding ecapsulant. As a result, it can be concluded that Compound
A has a higher possibility of delamination relative to Compounds B
and C. The occurrance of delamination was confirmed using SAT, a
non-destructive evaluation technique using sound waves to generate
optical images. Figure 6 shows the delamination advancement during
temperature cycle testing. Prior to temperature cycling, no
delamination is apparent between the Si-chip and molding
encapsulant interface. After 400 cycles, delamination originates
at the four corners of the silicon chip. Increasing to 800 cycles,
delamination advances half-way across the chip surface. At 1,000
cycles, there ceases to be any adhesion.
        Figure 7 graphically compares experimental stress versus
prinicpal stress at the chip's edge. The larger the experimental
stress, the larger the compressive stress, tensile stress, and
maximum shear stress. Compound A exhibits the largest experimental
stress value and, therefore, the highest possibility of package
cracking.
        Based on the fact that delamination advances during
temperature cycle testing, hypothetical changes were made in the
length of delamination during FEM calculations for Compound A, slip
was allowed between elements. Figure 8 shows the results of this
evaluation for delamination advancement. In addition to the chip's
edge, we see that there are also peaks of shear stress at the
hypothetical delamination points. These peaks of shear stress are
positioned higher than the shear stress for complete adhesion.
From this data it has been concluded that these peaks facilitate
the spreading of delamination from the edge of the chip towards the
chip's center.
        Figure 9 shows the principal stresses at the chip edge with
the delamination advancement. We understand that tensile stress
and maximum shear stress at the chip edge increases as the
delamination advances. When complete delamination occurs, tensile
stress and shear stress have reached their highest values. It can
be concluded then, that the maximum possibility of package cracking
occurs when there is complete delamination.
        This package cracking phenomena must also be related to
aluminum pattern deformation and passivation layer cracking.

CONCLUSIONS

1.   Lowering the thermal expansion coefficient and flexural mod-

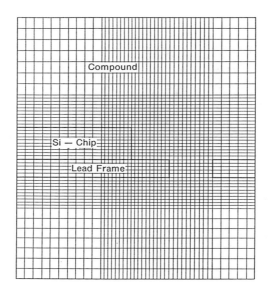

Figure 4: 2-Dimensional Finite Element Model

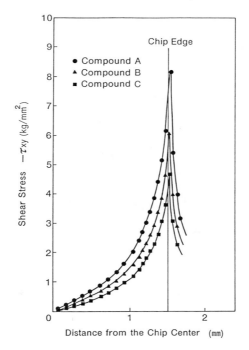

Figure 5: Shear Stress Distribution over a Chip

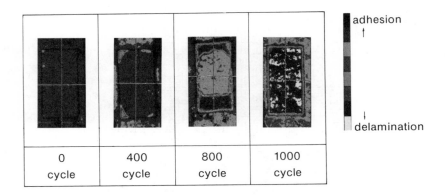

Figure 6: Delamination   Advancement   During   Temperature   Cycle
          Test

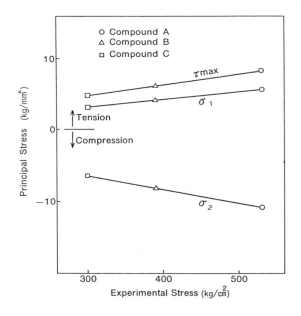

Figure 7: Principal Stress for 3 Compounds

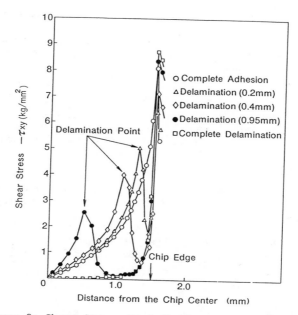

Figure 8: Shear Stress Distribution with a Delamination

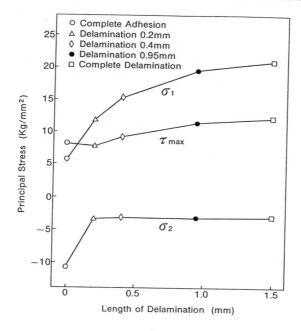

Figure 9: Principal Stress with a Delamination

ulus of a molding compound is one method to prevent package cracking at the chip edge.

2. Increasing the adhesion strength of a molding encapsulant and Si-chip interface is another method to prevent the package cracking at the chip edge.

ACKNOWLEDGMENTS

The authors would like to thank Dr. T. Moriuchi and Mr. K. Iko of Nitto Electric Industrial Co., Ltd., and Mr. K. Kuwada of Nitto Denko America, Inc. for their helpful comments and encouragements.

LITERATURE CITED

1. R.E. Thomas, Stress-Induced Deformation of Aluminum Metalization in Plastic Molded Semiconductor Devices, Proc. 35th ECC, 1985, pp. 37-45.

2. M. Isagawa, Y. Iwasaki, and T. Sutoh, Deformation of Al Metalization in Plastic Encapsulated Semiconductor Devices Caused by Thermal Shock, 18th Annual Proceedings, Reliability Physics, 1980, pp. 171-177.

3. S. Okikawa, M. Sakimoto, M. Tanaka, T. Sato, T. Toya, and Y. Hara, Stress Analysis of Passivation Film Crack for Plastic Molded LSI Caused by Thermal Stress, Proc. International Society for Testing and Failure Analysis, Oct. 1983, Los Angeles, CA.

4. K. Miyake et al., Thermal Stress Analysis of Plastic Encapsulated Integrated Circuits by FEM, IECE Proc., Japan, 1984, p. 2833.

5. K. Kuwata, K. Iko, and H. Tabata, Low Stress Resin Encapsulant for Semiconductor Devices, Proc. 35th ECC, 1985, pp. 18-22.

6. H. Suzuki, T. Moriuchi, and M. Aizawa, Low Mold Stress Epoxy Molding Compounds for Semiconductor Encapsulation, Nitto Electric Industrial Co., Ltd., 1979.

7. Steven Groothuis, Walter Schroen, and Masood Murtuza, Computer Aided Stress Modeling for Optimizing Plastic Package Reliability, 23rd Annual Proceedings, Reliability Physics, 1985, pp. 184-191.

RECEIVED April 8, 1987

Chapter 46

Patterning of Fine Via Holes in Polyimide by an Oxygen Reactive Ion Etching Method

Hiroshi Suzuki, Hiroyoshi Sekine, Shigeru Koibuchi, Hidetaka Sato, and Daisuke Makino

Yamazaki Works, Hitachi Chemical Company, Ltd., 4-13-1 Higashi-cho, Hitachi-shi, Ibaraki 317, Japan

A new silicon-containing positive photoresist having O_2 RIE (reactive ion etching) resistance was selected as an etching mask for the polyimide dry etch. The RIE selectivity of this resist to polyimide is more than five. By using this resist, 2 μm x 2 μm via holes are formed in 2 μm thick polyimide film. After the RIE treatment the resist width and thickness decreased, and the removal rates in horizontal and vertical direction are 5 and 10 nm/min., respectively. After the RIE treatment, a lawn-like scum was observed in a bottom of via hole. The amount of scum generated in silicon containing polyimide patterning is much more than in conventional polyimide patterning. However, the scum and the resist layer could be stripped by treatment with HF/NH_4F solution and a successive phenolic type resist stripper.

With the increase of the degree of integration of microcircuits, the multilevel interconnect technology becomes inevitable for future VLSI manufacture. Polyimide exhibits superior planarity over stepped structures and is expected to be one of the most promising materials for the dielectric insulation of VLSI's. However, since the smallest via holes so far achieved by wet etching is 3 μm (1), the formation of fine via holes by a dry etch process is needed for the application of polyimide to VLSI having fine metal wiring.

Usually, dry etching of polyimide is performed by RIE with O_2, CF_4, or their mixtures as an etchant gas, utilizing positive photoresist (2), metals such as aluminum (3), spin-on-glass (4), or SiN (5) as an etching mask. However, in the former case, it is difficult to define a fine via-hole as small as 2 μm or less because the resist thickness must be two or more times that of the polyimide as a result of the equal etching rates between photoresist and polyimide. In the latter case, though the fine pattern can be obtained the additional pattern transformation from the photoresist to the masking layer is necessary.

In this paper we report the results of our studies on the fine patterning of polyimide by the O_2 RIE method using silicon-containing photoresist as the masking material.

Selection of Photoresist

Many kinds of silicon-containing photoresists have been reported([6-11]) to have both O_2 RIE resistance and high resolution. Among them a positive photoresist called ASTRO([11]), a mixture of conventional positive photoresist and a silicon resin developed by Hitachi for the imaging layer of bi-level resist system, was selected as an etching mask in this experiment for the following reasons. First its positive tone imaging is more appropriate to forming via hole opening than negative photoresist. Also, the same development process as the conventional positive photoresist can be applied to ASTRO. Since the wettability of ASTRO to polyimide was not sufficient, a slight modification was made to ASTRO, and the resulting material is designated as X-8000K2.

Patterning of the Photoresist

Exposure characteristics of X-8000K2 and conventional positive photoresist OFPR-800 (Tokyo Ohka Co.) are shown in Figure 1. X-8000K2 shows the comparable sensitivity to conventional one. Line and space and via hole patterns of X-8000K2 are shown in Figure 2. 0.8 μm line and space patterns and 1.2 μm via hole patterns were obtained.

O_2 RIE Selectivity

The dependence of the etching selectivity of X-8000K2 relative to PIQ (a polyimide type resin from Hitachi Chemical Co.,) on O_2 RIE conditions was examined. RIE power, O_2 pressure, flow rate and the distance between the electrodes were selected as the parameters to determine the RIE condition. For these parameters, values of 100 W, 10 m Torr, 10 SCCM and 6 cm were selected as a standard condition. The changes in the etching rates are shown in Figure 3a, in which one parameter is varied and the remaining three parameters are fixed.
   Etching selectivity varies from 5 to 15 within the range of parameters investigated as shown in Figure 3b. When the RIE power increases the etching selectivity decreases and it increases when the O_2 pressure decreases. The optimum RIE condition determined from the selectivity and the pattern profile is shown in Table I. The etching selectivity of X-8000K2 under this condition is about 10.

Table I. Optimum RIE Conditions

Equipment	CSE-1110 (ULVAC, JAPAN)
Power	100 W
Flow rate	10 SCCM
Pressure	10 mTorr
Distance between electrodes	3 cm

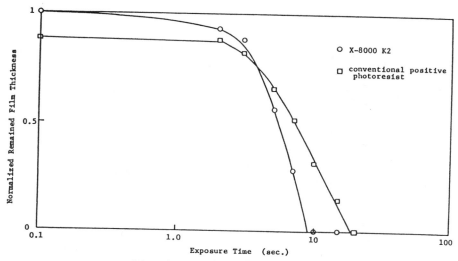

Lamp : 500 W Xe–Hg lamp
        1.65 mW/cm2 at 365 nm

Development : Immersion 23°C/60sec.
               by X–8000 K2 developer.

Film Thickness : 1.2μm

Figure 1.   Exposure characteristics of X–8000K2.

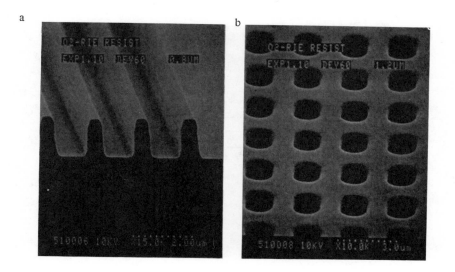

Figure 2.   Resist patterns of X–8000K2 printed by a RA–501
(g line) (a) 0.8μm L & S, (b) 1.2μm via-hole.

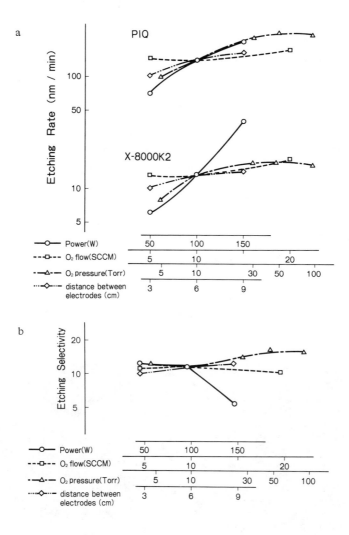

Figure 3. (a) Etching rate of X-8000K2. (b) Etching selectivity of X-8000K2.

Polyimide Patterning

Samples for polyimide patterning evaluations were prepared as follows. PIQ and a siloxane-containing polyimide PIX were spun on a silicon wafer followed by baking at 100°C for 30 minutes, 200°C for 30 minutes and 350°C for 60 minutes successively in a convection oven to get a 2 μm thick film. Then X-8000K2 was coated and prebaked at 90°C for 20 minute. X-8000K2 film with a thickness of 1.2 μm was exposed utilizing a Hitachi RA501 5:1 g-line reduction projection printer. Exposed film was developed at 23°C for 60 seconds by using a tetramethylanmonium hydroxide solution as a developer, followed by postbaking at 130°C for 20 minutes. Etching of PIQ and PIX films was performed by a ULVAC CSE-1110 RIE equipped with 250 mm diameter electrode.

Figure 4 shows a 2 x 2 μm via hole pattern formed in a PIX film under optimum RIE conditions shown in Table I. In this photograph the resist has not been stripped. The taper angle of the via hole is about 90°. Normally, taper angles required to prevent wire opening and to maintain high density wiring are around 70°. The dependence of the taper angle on the parameters mentioned above was examined, but it could not be reduced to 70° within the evaluation range shown in Figure 3.

After the RIE treatment, a lawn-like scum was observed in the bottom of via holes as shown in Figure 4. The generation of scum is enhanced with the longer etching period as clearly shown in the photographs of Figure 5. The change of resist pattern size also occurred as a result of the RIE treatment. The resist surface became rough and the resist width and thickness decreased. The resist errosion rates in the horizontal and vertical directions are 5 and 10 nm/min, respectively. In a optimum etching time of 24 minutes for 2 μm thick polyimide, the resist diminished about 120 nm in a horizontal direction.

The scum shown in Figures 4 and 5 is thought to be produced by the reaction of oxygen and silicon contained in both polyimide PIX and X-8000K2. In order to confirm this presumption, RIE experiments were performed for the various combinations of photoresist and polyimide, i.e.: (a) X-8000K2 (1.2 μm thick) on PIQ (2 μm thick) (b) X-8000K2 (1.2 μm thick) on a Si wafer (c) conventional positive photoresist (2 μm thick) on PIQ (0.5 μm thick) (d) conventional positive photoresist (2 μm thick) on PIX (0.5 μm thick). The results after RIE are shown in Figure 6. As can be seen in Figure 6a, when PIQ was patterned using X-8000K2 as a mask scum was also generated but in only slight amounts compared to PIX. When the X-8000K2 pattern was formed directly on a Si wafer, and was treated by O_2 RIE, no scum was observed as shown in Figure 6b. The absence of scum is also observed for sample 6c, in which conventional photoresist was formed on polyimide which contains no silicon atoms. However, in the case of the silicon containing polyimide, considerable scum appeared. These results are summarized in Table II.

The difference of the amount of scum among the samples is clearly explained assuming the scum to be silicon oxide produced by the reaction of oxygen and silicon atom. The difference of the amount of scum in Figures 6a and 6c comes from the different concentration of silicon contained in X-8000K2 and conventional positive photoresist. The amount of scum observed in Figure 5 is the

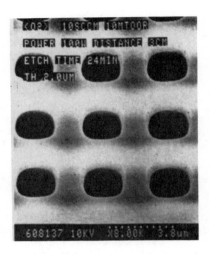

Figure 4. Via-hole pattern formed in PIX film under RIE
condition, 2μm via-hole (X-8000K2 1.2μm thick, PIX 2μm thick).

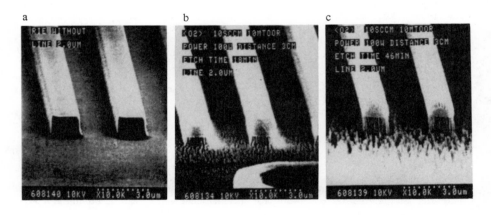

Figure 5. Line and space patterns of PIX after RIE treatment
(X-8000K2 1.2μm, PIX 2μm thick) (a) before etching, (b)
etching time 18min, (c) etching time 46min.

a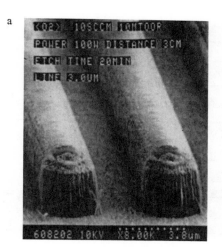
b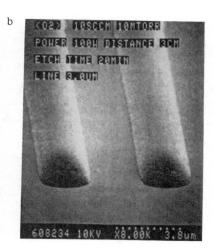

c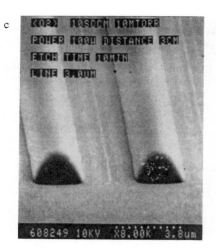
d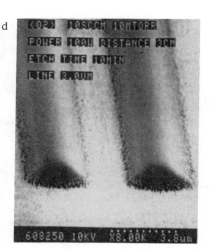

Figure 6.   Comparison of scum occurrence after RIE treatment, 3μm line and space (a) X-8000K2 (1.2μm thick) on PIQ (2μm thick), (b) X-8000K2 (1.2μm thick), (c) conventional positive photoresist (2μm thick) on PIQ (0.5μm thick), (d) conventional positive photoresist (2μm thick) on PIX (0.5μm thick).

largest.   In this case, the silicon oxide scum was generated from both the photoresist and the silicon containing polyimide.

Table II. Comparison of Scum Occurrence after RIE Treatment

| Substrate | Resist | |
	X8000K2 (Si Containing)	Conventional Positive Photoresist
Si Wafer	None	–
PIQ on Si Wafer	Slight Residue	None
PIX on Si Wafer	Moderate Residue	Moderate Residue

Resist Stripping

In order to remove the resist after the RIE treatment, wafers were immersed in phenolic-type resist stripper, but the resist as well as the scum could not be stripped. Since a thin silicon oxide layer is formed on the resist surface, and the composition of scum is thought to be silicon oxide as discussed above, it is necessary to remove this silicon oxide layer (scum) prior to the resist removal. Therefore, resist stripping was done in two steps. In a first step the wafer was immersed in buffered hydrofluoric acid solution to remove the silicon oxide and was then treated with conventional resist stripper.

When the wafer was dipped in a highly concentrated hydrofluoric acid solution or dipped for a long period, polyimide film tends to peel off the substrate.   From the experiment carried out for a various concentrations of HF/NH_4F solution, the most suitable conditions were found as follows: composition of buffered hydrofluoric acid $HF/NH_4F=1/20$; dipping time, $\leq$ 2 minutes at room temperature.

When the wafer was dipped into this buffer solution, the amount of scum decreased sharply during the first ten seconds, but after 30 seconds the removal rate became extremely low and residue remained as shown in Figure 7. However, the remaining scum and resist layer were removed perfectly in a second step by dipping into conventional resist stripper S-502(Tokyo Ohka Co.), at 110°C for 7 to 10 minutes as shown in Figure 8.

Conclusions

Positive working silicon-containing resist with 0.8 μm resolution and O_2 RIE selectivity greater than five was used for fine via hole formation in a polyimide film by O_2 RIE process.

     2 x 2 μm via holes were obtained in 2 μm thick polyimide by

Figure 7. Line and space pattern of PIX after HF/NH$_4$F treatment for 30 sec., 2μm L & S.

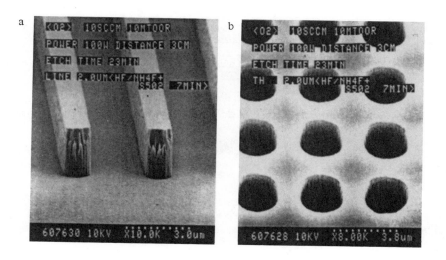

Figure 8. L & S and via-hole patterns of PIX after HF/NH$_4$F treatment and S-502 treatment, (a) 2μm L & S, (b) 2μm via-hole.

using this resist as an etching mask. Using RIE conditions to
achieve the etching selectivity of ten, the resist width and
thickness diminished after the RIE treatment. The removal rate of
the resist in horizontal and vertical directions was 5 and 10
nm/min., respectively. After the RIE treatment a lawn-like scum was
observed in a bottom of via hole. The amount of scum generated in
silicon-containing polyimide patterning is much greater than conventional polyimide patterning. However, the scum could be stripped
sequentially treatment with HF/NH$_4$F solution followed by a phenolic
type resist stripper.

Acknowledgments

We wish to thank Dr. T. Ueno and Dr. M. Toriumi of Hitachi Central
Research Laboratory for their invaluable advice about RIE equipment.
We also wish to thank T. Nishida of Hitachi Central Research
Laboratory for his helpful discussion.

Literature Cited

1. Saiki, A. Polyimides; Mittal, K. L., Ed.; Plenum Publishing
    Corp. 1984; Vol. 2, p 827.
2. Herndon, T. O. Kodak Interface Oct. 1979, 26.
3. Rivans, I. V. IEEE V-MIC Conf. 1984, 283.
4. Ting, C. H. IEEE V-MIC Conf. 1984, 106.
5. Samuelson, G. ACS Organic Coatings and Plastics Chemistry 1984,
    43, 446
6. Hazakis, M. Proc. Int'l. Conf. Microlithography 1981, 386.
7. Morita, M. Jpn. J. Appl. Phy. 1983, 22, L659
8. Wilkins, C. W. J. Vac. Sci. Technol. 1984, B3, 306.
9. Reichmanis, E. SPIE "Advences in Resist Technol. and Proc."
    1984, 469, 38.
10. Suzuki, M. J. Electrochem. Soc. 1983, 130, 1962.
11. Ueno, T. 4th Photo Polymer Conf. preprint 1985, 108.

RECEIVED May 13, 1987

CONDUCTING POLYMERS

CONDUCTING POLYMERS

Traditional thinking has always associated polymers with insulators. However, this perception was changed in the 1970s when it was discovered that certain conjugated polymers, of which polyacetylene is the archetype, could be oxidized or reduced (doped) by certain reagents yielding polymers with conductivity approaching that of metals. This work opened up entirely new avenues of potential application, and although it is unlikely such materials will directly replace metals, conducting polymers may very well find application as electrodes, EMF shielding materials, or as photoconductors. Even molecular-sized electronic devices might one day be possible if conductivity can be controlled on a molecular level.

Research on conducting polymers is distinguished from efforts in other fields of polymer science by the vast array of techniques that have been applied to the problem, making it perhaps the most interdisciplinary effort in contemporary materials science. Active research areas range from the synthesis of new polymers to the development of elegant theories to describe the excitations and conduction mechanisms in conducting polymers. Both chemist and physicist work in tandem; the chemist must learn which chemical structures are likely to lead to high conductivity, while the physicist must constantly adjust to new materials with different properties. The development of useful conducting polymers presents special problems, since both the electrical properties and the processability of the polymer need to be considered. This is a severe restriction since most conducting polymers are highly intractable, and many are environmentally unstable.

The following papers give a flavor of current research in conducting polymers. Three papers describe the search for new conducting polymers via classical organic chemistry, by electrochemical polymerization, and by pyrolysis reactions. In the fourth paper of this section, band structure calculations on several conjugated polymers are presented. These contributions illustrate the challenges and diversity of the field of conducting polymers.

Gregory L. Baker
Bell Communications Research
321 Newman Springs Rd.
Redbank, NJ 07701

Chapter 47

Electrochemical Synthesis and Characterization of New Polyheterocycles

M. Aldissi[1] and A. M. Nyitray[2]

[1]Los Alamos National Laboratory, Los Alamos, NM 87545
[2]Department of Chemistry, Colorado State University, Fort Collins, CO 80523

The direct synthesis by anodic oxidation of a new series of electrically conducting polymers is described. Our polymers derive from sulfur and/or nitrogen containing heterocycles such as: 2-(2-thienyl)pyrrole, thiazole, indole, and phthalazine. The anodic oxidation of these monomers is carried out in acetonitrile solutions containing tetrabutylammonium salts (TBA$^+$ X$^-$) with X$^-$ = BF$_4^-$, ASF$_6^-$, and the tetraethylammonium salt, TEA$^+$ H$_3$C-C$_6$H$_4$-SO$_3^-$. Characterization of the materials by electrical conductivity, electron spin resonance, uv-visible spectroscopy, and cyclic voltammetry is discussed.

New conducting polymers with conjugated double bonds, linear or cyclic, have attracted a wide interest due to the various interesting physical phenomena that they exhibit and to the imminent possibility of their use in technological applications. The electrochemical synthesis of conducting polyaromatics proved to be a convenient and efficient method which results usually in polymers in their conducting form as films or powders on the surface of the anode. Typical examples which have been widely studied are poly-(pyrrole) (1) and poly(thiophene) (2-4). The oxidation of other aromatic heterocycles followed, adding a large number of materials to the long list of conducting polymers. Examples of such compounds are derivatives of poly(pyrrole) and poly(thiophene) (5-10), poly(pyridazine) (11), and the bicyclic polymer, poly(thieno[3,2-b]pyrrole) (12). It has been found, however, that these polymers are mostly amorphous or have disordered structures (2,13). ^{13}C NMR (14) and XPS (15) studies have shown that polypyrrole consists of α,α'-bonding and α,β-bonding of the pyrrole ring, which have been suggested to be the origin of the disordered structures in the electrochemically prepared polymers.

We have focused in our study on newly synthesized polyheterocycles in which the repeating unit (Figure 1) consists of thiazole
(1), indole (2), phthalazine (3), and thienylpyrrole (4) for the
following reasons:
1) The decrease in disorder by reducing the number of coupling
sites (except in compound 4): α,α' coupling relative to the sulfur
atom in thiazole is the only type of coupling allowed; in phthalazine, coupling occurs on the α positions relative to both nitrogen
atoms, similarly to pyridazine (11); and in indole, the most likely
coupling to occur is α, β (on the pyrrole and fused phenyl rings
respectively). It should be noted that poly(indole) has been
previously prepared (16). Also, a decrease in disorder might
ultimately lead to decrease in the extent of the amorphous character. For example, when thiazole is synthesized by polycondensation
reactions the resulting material is highly crystalline with desirable mechanical properties. Therefore, the electrochemical technique could probably lead to a material with a crystalline character.
2) Compound 4 could, in principle, be polymerized to yield poly-
(thienylpyrrole) consisting of alternating pyrrole and thiophene
units. Such a polymer is analogous to conjugated diatomic polymers
which have been predicted (17) to give rise to excitations consisting of soliton pairs having either spin 0 or spin 1/2. We describe
in this paper the synthesis of the various materials and some of
their properties with an emphasis on poly(thienylpyrrole). A
detailed study of the various materials will be published in a
future paper.

EXPERIMENTAL SECTION

Synthesis: Monomers(1-3), solvents, and electrolytes, supplied by
Aldrich, were purified before use by distillation or recrystallization. Monomer 4 was synthesized by reacting methyl azidoacetate
with β-thienylacrolein, followed by cyclization then decarboxyla-
tion of the obtained derivative (18), which is a general method
employed for the synthesis of heteroarylpyrroles (19). The chemical structure of the four monomers is shown in Figure 1. Our
polymers were synthesized in a three compartment electrochemical
cell which contains In-Sn oxide conducting glass (ITO) or a platinum foil as the working electrode, a platinum counter electrode
(wire, foil, or mesh), and SCE or an $Ag/AgNO_3$ reference electrode.
The solutions from which the polymers were prepared consisted of
acetonitrile to which were added the supporting electrolyte and the
monomer to yield a concentration of 0.1 mole/l of both. The
polymers were grown on the platinum or the conducting glass electrode by applying a constant anodic current for periods of time
extending from a few minutes to several hours depending upon the
nature of the monomer and the electrolyte. $TBA^+ASF_6^-$ and TEA^+
$H_3C-C_6H_4-SO_3^-$ were used for monomers 1-4 and $TBA^+ BF_4^-$ was used for
monomer 4. The current densities were varied between 0.1 mA/cm^2
and 0.5 mA/cm^2. The anodic potentials during the synthesis were in
the range 0.25 V - 0.5 V vs. $Ag/AgNO_3$ reference electrode. The
reactions were carried out under argon at room temperature. The

synthesized materials were rinsed with acetonitrile to eliminate the supporting electrolyte from the interstices of the films or powders, then dried under vacuum.

Methods of Characterization: The polymers were characterized by four-probe electrical conductivity measurements between room temperature and liquid nitrogen, electron spin resonance (Varian E-line series), scanning electron microscopy (Hitachi 520), cyclic voltammetry (Princeton Applied Research Instruments), and uv-visible spectroscopy (Perkin Elmer 330).

RESULTS AND DISCUSSION

The various monomers underwent oxidation at the anode, yielding conducting films or powders with colors that depend on the monomer and which result from plasma reflection. In the oxidized form, the colors observed are dark blue and dark red, for poly(thienylpyrrole) and poly(thiazole) respectively, and dark brown for poly(phthalazine) and poly(indole). In the reduced form, which is accomplished by the reverse cathodic discharge, the colors are brown-yellow to yellow and result from the intrinsic interband absorption. The extent of reversibility of the oxidation/reduction processes depends upon the nature of the monomer or repeating unit in the chain for materials with comparable thicknesses. In general, an incomplete reversibility indicates that during the polymerization side reactions might have occurred. For example, crosslinking could take place particularly in the case of α-β coupling. Also, the formation of a slightly colored solution during the anodic oxidation accounts for the incomplete reversibility. Such an observation is supported by the fact that the coulombic efficiency of the various charge/discharge cycles, for compounds 1-3, is lower than what is usually observed for polyheterocycles, such as poly(pyrrole) and poly(thiophene), except in the case of few hundreds Å-thick films of poly(thienylpyrrole) for which the coulombic efficiency is comparable to that of the two latter compounds. By taking this assumption into account, an estimated composition of the various doped materials consists of 1 anion for every 3-4 repeating units. However, when the polymerization of thienylpyrrole is allowed to proceed for longer times a dark polymer solution is obtained simultaneously with an increase in the thickness of the film at the anode. The color of the solution or the film depends upon the electrolyte used. As for most electrochemically formed polyheterocycles, the morphology of the resulting materials exhibits, in the case of poly(thienylpyrrole) in its oxidized form (Figure 2), a fibrillar structure. The morphology of the other materials will be determined as well, including the effect of the nature of the electrolyte, specifically, the inserted ion on it. The nature of the inserted ion affects the mechanical properties and is also known to affect electrical conductivity of the oxidized polymer. For example, when poly(thienylpyrrole) is synthesized using TEA^+ CH_3-C_5H_4-SO_3^- as electrolyte, the material is more flexible than when synthesized using TBA^+ BF_4^-, which is in agreement with what was observed in the case of poly(pyrrole) (20). The cyclic voltammograms shown in Figure 3 were obtained for

(1) THIAZOLE

(2) INDOLE

(3) PHTHALAZINE

(4) 2-(2-THIENYL)PYRROLE

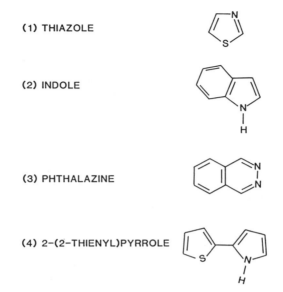

Fig. 1   Structures of the heterocyclic monomers.

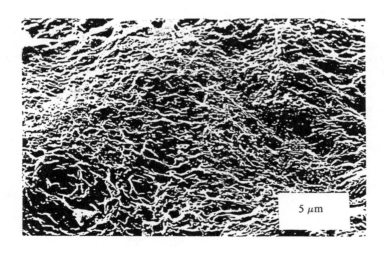

Fig. 2   Scanning electron micrograph of the surface of a
poly(thienylpyrrole) film.

poly(thienylpyrrole) formed on a platinum electrode using TBA$^+$ BF$_4^-$ in acetonitrile (0.1 mole/l) and an SCE as a reference electrode. The proportionality of the peak current for two sweep rates, 50 mV/sec and 100 mV/sec, is indicative, as for all conjugated polyheterocycles, of a charge injection reaction confined to the volume of the film and governed mostly by diffusion. In the conditions mentioned above, the oxidation and reduction peaks are observed at approximately 0.5 V and 0.25 V vs. SCE. It is important to note that the oxidation peak lies between those measured for poly(pyrrole) (21) (-0.1 V) and poly(thiophene) (22) (1.1 V).

The visible absorption spectrum of a poly(thienylpyrrole) film which is approximately 1000 Å thick on conducting glass (Figure 4) is characterized by an intense band with a maximum at 2.8 eV. Similar absorption characteristics are observed for the polymer solution. This value, which is the band gap of the conjugated polymer, lies between the published values (23) for poly(pyrrole) (2.9 eV) and for poly(thiophene) (2.6 eV). The absorption maxima are in the range which is expected for conjugated polyheterocycles (24). The intermediate maximum for poly(thienylpyrrole) suggests that the absorption is due to the transition $\pi-\pi^*$ of the conjugated system which consists of the thienylpyrrole repeating unit rather than a thiophene unit and a pyrrole unit. Such an assumption is supported by the fact the the oxidation/reduction processes of the polymer occur at different potentials than for the two polymers taken separately. This also indicates that an alternated structure might have been formed. A detailed study will be carried out to determine the exact structure. However, it should be mentioned that on one hand, theoretical calculations in conjunction with dipole moment data (25) showed that a planar cis conformation of the thienylpyrrole monomer is obtained. On the other hand, the proposed structure for i.e., poly(pyrrole) (26) consists of a trans conformation for the pyrrole units. This leads to the possible planar structure for the thienylpyrrole polymer shown in Figure 5. The width of the absorption band seems to be comparable to those of poly(thiophene) and poly(pyrrole) which suggests a comparable distribution for the length of the conjugated sequences in the three materials. The uv-visible absorption of the oxidized poly(thienylpyrrole) will be measured to detect the possible electronic states in the gap and whether they correspond to what was suggested for an AB polymer.

Electron spin resonance measurements at room temperature on poly(thienylpyrrole) showed an asymmetric line characteristic of conduction electrons, with a g value near that of the free electron indicating that the spin resonance is not due to the sulfur and/or nitrogen moieties.

The temperature dependence of conductivity of poly(thienylpyrrole) is shown in Figure 6. The non-linear character of the conductivity throughout the whole range of temperatures studied here (room temperature to liquid nitrogen) indicates that the conduction mechanism could be different in different temperature

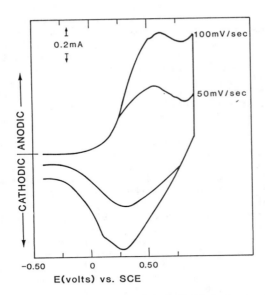

Fig. 3   Cyclic voltammograms of poly(thienylpyrrole) in 0.1M
          TBA⁺ BF₄⁻/acetonitrile.

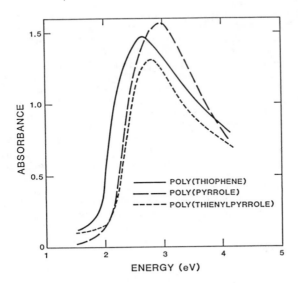

Fig. 4   Optical absorption spectra of the bleached form of
          thiophene, pyrrole, and thienylpyrrole polymers.

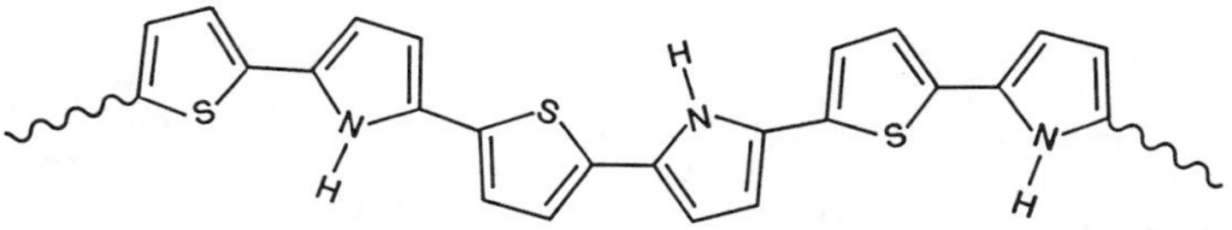

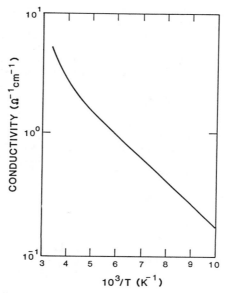

Fig. 5   Possible conformation of poly(thienylpyrrole).

Fig. 6   Temperature dependence of conductivity of as-grown
poly(thienylpyrrole).

ranges, i.e., the relative importance of the intra-chain and inter-chain conduction processes is different at different temperatures. When the conductivity is plotted vs. $T^{-1/3}$ or $T^{-1/4}$, a straight line is obtained indicating the possibility of two or three dimensional variable range hopping conduction as predicted by Mott (27). Although, the conductivity of poly(thiazole) is low, 1 X 10^{-3} S/cm, the values measured along and perpendicular to the directions of the surface indicate that a higher anisotropy than in most polyheterocycles is obtained. Such a property will be further detailed in a future publication.

CONCLUSIONS

We have shown for the first time that heterocycles such as thiazole and phthalazine, undergo electrochemical polymerization providing a new and perhaps the only synthetic route of their corresponding polymers. This work could lead to the synthesis of conducting polymers that are obtained conventionally as crystalline or perhaps liquid crystalline polymers. Such a possibility is currently under investigation. The polymerization product of thienylpyrrole may provide a material for testing predictions of particle excitations in conjugated AB polymers: solitons in the bleached material, and polarons and/or bipolarons in the oxidized material. This synthesis route will be extended to 2-heteroarylpyrroles which are precursors for the synthesis of chemotherapeutic compounds (28), such as 2-(2-furyl)pyrrole. Some of the materials synthesized in this work, e.g., poly(thienylpyrrole), could be good candidates for electro-optical switching elements with reasonable switching times, due to the fast oxidation/reduction cycles accompanied by a color change when thin films (up to 1000Å thick) ar used.

ACKNOWLEDGMENTS

The authors would like to thank Dr. S. Gottesfeld for helpful discussions, and Dr. D. Wrobleski for synthesizing the thienylpyrrole monomer. This work is supported by the Office of Basic Energy Sciences (OBES/DOE) and by the Centers for Materials Science and for Non-Linear Studies of Los Alamos National Laboratory.

LITERATURE CITED

1. Diaz, A. F.; Kanazawa, K. K.; Gardini, G. P. J. Chem. Soc. Chem. Commun. 1979, 635.
2. Tourillon, G.; Garnier, F. J. Electroanal. Chem. 1982, 135, 173.
3. Kaneto, K.; Kohno, Y.; Yoshio, K.; Inuishi, Y. J. Chem. Soc. Chem. Commun. 1983, 382.
4. Hotta, S.; Hosaka, T.; Shimotsuma, W. Synth. Met. 1983, 6, 69.
5. Diaz, A. F.; Salmon, M.; Addy J. Proc. 1st Eur. Display Research Conf., Munich, 1981, p. 111.
6. Gazard, M.; Dubois, J. C.; Champagne, M.; Garnier, F.; Tourillon, G. J. de Phys., Colloq. 1983, C3, 44, p. 595.

7. Skotheim, T.; Velazquez Rosenthal, M.; Linkous, A. J. Chem. Soc. Chem. Commun. 1985, 612.
8. Velazquez Rosenthal, M.; Skotheim, T.; Warren, J. J. Chem. Soc. Chem. Commun. 1985, 342.
9. Nazzal, A. I.; Street, G. B. J. Chem. Soc. Chem. Commun. 1985, 375.
10. Yumoto, Y.; Yoshimura, S. Synth. Met. 1986, 13, Nos. 1-3, 185.
11. Satoh, M.; Kaneto, K.; Yoshino, K. J. Chem. Soc. Chem. Commun. 1984, 1627.
12. Lazzaroni, R.; Riga, J.; Verbist, J. J.; Renson, M. J. Chem. Soc. Chem. Commun. 1985, 999.
13. Geiss, R. H.; Street, G. B.; Volksen, W.; Economy J. IBM J. Res. Develop. 1983, 27, 321.
14. Street, G. B.; Clarke, T. C.; Krounbi, M. T.; Kanazawa, K. K.; Lee, V. Y.; Pfluger, P.; Scott, J. C.; Weiser, G. Mol. Cryst. Liq. Cryst. 1982, 83, 253.
15. Pfluger, P.; Street, G. B. J. Chem. Phys. 1984, 80, 544.
16. Tourillon, G. and Garnier, F. J. Electroanal. Chem. 1982, 135, 173.
17. Rice, M. J.; Mele, E. J. Phys. Rev. Lett. 1982, 49, (19), 1455.
18. Trofimov, B. A.; Mikhaleva, A. I.; Nesterenko, R. N.; Vasil'ev, A. N.; Nakhmanovish, A. S.; Voronkov, M. G. Khim. Geterotsikl, Soedin, 1977, 8, 1136.
19. Boukou-Poba, J. P.; Farnier, M.; Guilard, R. Tetrahed. Lett. 1979, 19, 1717.
20. Wernet, W.; Monkenbusch, M.; Wegner, G. Makromol. Chem. Rapid Commun. 1984, 5, 157.
21. Diaz, A. F.; Castillo, J. I. J. Chem. Soc. Chem. Commun. 1980, 397.
22. Waltman, R. J.; Bargon, J.; Diaz, A. F. J. Phys. Chem. 1983, 87, 1459.
23. Tourillon, G.; Garnier, F. J. Phys. Chem. 1983, 87, 2289.
24. Sease, J. W.; Zechmeister, L. J. Am. Chem. Soc. 1947, 69, 270.
25. Galasso, V.; Klasinc, L.; Sabljic, A.; Trinajstic, N.; Pappalardo, G. C.; Steglich, W. J. C. S. Perkin II 1981, 127.
26. Diaz, A. F.; Vasquez Vallejo, J. M.; Martinez Duran, A. IBM J. Res. Develop. 1981, 25, 42.
27. Mott, N. F. Metal-Insulator Transition; Taylor & Francis, Ltd., London, 1974.
28. Berner, H.; Schulz, G.; Reinshagen, H. Monatsh 1977, 108, 285.

RECEIVED April 8, 1987

Chapter 48

Synthesis and Electronic Properties of Poly(8-methyl-2,3–6,7-quinolino) and Its Intermediate

J. Z. Ruan and M. H. Litt

Department of Macromolecular Science, Case Western Reserve University, Cleveland, OH 44106

2.6-Diaminotoluene reacts with formaldehyde to produce a prepolymer poly(3,5' methylane 2,6-diaminotoluene) (PM), which further condenses to poly(8-methyl, 2,3-6,7-quinolino)(PMQ). The structures of PM and PMQ were analysed by elemental analysis and NMR, UV/VIS, and IR spectroscopy. PMQ was doped with nitrosyl hexafluorophosphate (NFP) and iodine. EPR studies suggests that NO^+ oxidized the polymer to form charged spinless species. The conductivity of PMQ3 increases by 6 and 7 orders of magnitude upon doping with NFP and iodine, respectively; the highest value found at room temperature is about 10^{-2} S/cm.

Interest in the synthesis of thermally stable materials has produced much research into ladder polymers.[1,2] Both theoretical and experimental studies show that not only are aromatic ladder polymers more thermally stable but they are also more highly conducting than analogously structured nonladder systems.[3,4] In this communication, we report the synthesis and electronic properties of a ladder aromatic polymer, poly(8-methyl, 2,3-6,7-quinolino) (PMQ). The experimental procedures for preparation and characterization of PMQ are described in refs. 5 and 6.

Results and Discussion.

The expected prepolymer(PM) was obtained (Scheme 1) by the reactions between 2,6-diaminotoluene(DAT) and formaldehyde with hydrochloric acid catalysis. C, H, and N values found from element analysis: 67.15%, 7.59%, and 19.36% are in good agreement with the calculated ($C_8N_2H_{10} \cdot 0.5H_2O$) values: 66.94%, 7.67% and 19.52%.

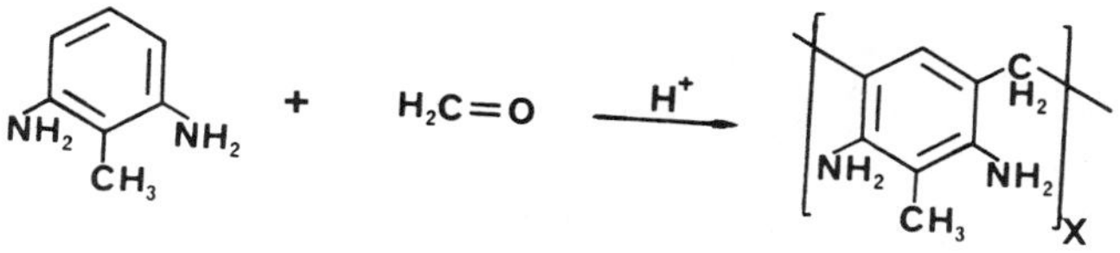

Prepolymer

Scheme 1

The prepolymer, yellow in color, dissolved in DMF or DMSO and gradually crystallized from solution as a white crystalline solid (PMC). The white color and crystallinity shows that the polymer had the proposed structure rather than some isomeric form . This was also supported by NMR and UV spectra.
^{13}C NMR spectra and peak assignment of PM and PMC are shown in Figure 1. The spectrum of PM was obtained in DMSO, while that of PMC was obtained in DMSO plus 1% deuterated trifluoroacetic acid (TFA). All peaks from TFA were carefully removed for better resolution. A normal APT method[7] was used for the simplification of the peak assignment.

The peak corresponding to singly protonated C-5 at the chain end is predicted[8] at 105.2 ppm with the peak down. No peak near 105 ppm appeared, indicating that the prepolymer has reasonably high molecular weight. In the spectrum of PMC, the peaks corresponding to C-2, C-6 and C-1 are shifted towards 128.5 ppm, the absorption of non-substituted benzene. Due to the TFA addition, the amino group is protonated and the substituent effect is reduced. The spectrum of PMC is cleaner than the spectrum of PM which also suggests that the crystallized prepolymer has a regular structure.

Figure 2 shows the proton NMR spectrum of PM. The band at 6.1 ppm is from proton d. The big peak near 4 ppm arises from the two amino groups. The resonance corresponding to the CH_2 group appears at 3.3 ppm. The sharp peak at 1.8 ppm is assigned to the methyl group. The extra peaks at 5.8 and 2.2 ppm could arise from diaminoacridine residues in the polymer[5].

IR spectra of PM and PMC are given in Figure 3; A is from PM, B is from PMC. The peaks in spectrum B are sharper than that of spectrum A because PMC is purer and crystalline. The two split sharp peaks at 3430 cm^{-1} and 3344 cm^{-1} are from asymmetrical and symmetrical N-H stretching, respectively. The sharp peaks at 1630 and 1474 cm^{-1} are due to skeletal vibration of C=C and C-N stretching. The C-H bend of CH_2 and CH_3 show as a shoulder at 1451 cm^{-1} which overlaps with the peak at 1474 cm^{-1}. All are consistent with the expected structure.

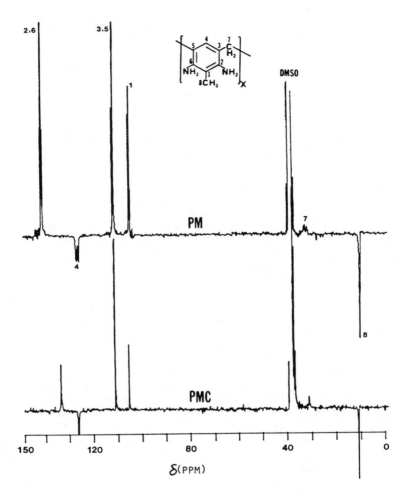

Fig. 1 ^{13}C NMR spectra of PM and PMC.

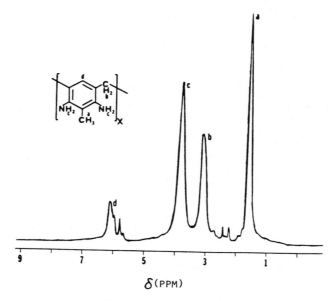

Fig. 2 ^{1}H NMR specturm of PM.

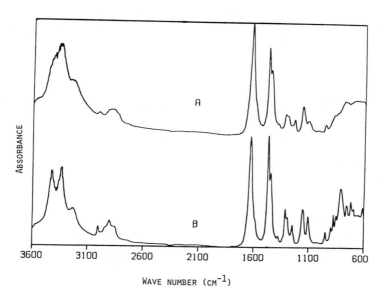

Fig. 3   IR spectra of PM and PMC.   A) from PM;
          B) from PMC.

Figure 4 shows the UV spectra of PM and PMC. The
band near 485 nm due to diaminoacridine and one near
600 nm from derivatives with five condensed rings are
considerably reduced after crystallization, which
agrees with the NMR data.

The condensation reaction was carried out at
various temperatures in polyphosphoric acid[9],[10] (Scheme
2). The final polymer is soluble in sulfuric acid if
it is condensed below 385°C. The polymer condensed at
385°C was only about 1% soluble. The insoluble part
swells in sulfuric acid. It is not certain whether
crosslinking occurred or the condensed molecular chain
has become too long to be soluble.

Four polymers: PMQ1 condensed at 250°C; PMQ2 at
310°C; PMQ3 at 345°C; and PMQ4 at 385°C were
investigated here. The polymers were purified by
dissolving them in sulfuric acid, centrifuging to
remove the inorganic residues, and precipitating them
into water. They were washed with dilute NH_3 and then
hot water until the extracts were neutral. Some one to
two percent of ash (SiO_2) remained along with some
H_2SO_4 (as salt), but the phosphorous was almost
completely removed. The problem with aromatic
heterocycles is that the analytical proceedure for
nitrogen does not work very well with them. Initial
nitrogen analyses on the materials were lower than they
should be for 1N/8C, about 0.75N/8C. Later analysis,
done after discussion with Galbraith Laboratories,
could not repeat the initial values for C, H, N, S, P
or even ash. Here the analysis gave 1.16N/8C. We
therefore base our conclusions for the polymer
structure on its solubility and x-ray patterns. All
polymers showed a large diffraction peak at 0.345nm.,
which is about that for graphite interplanar spacing.
Scherrer analysis showed that the stacks were about
2nm. thick; about six molecules were stacked in a
domain. The polymer must be planar and linear in order
to be soluble and to stack and thus the structure given
is reasonable.

PMQ3 was doped with nitrosyl hexafluorophosphate
(NFP) (0.70PF_6/monomer) and iodine (1.14 I/ monomer).

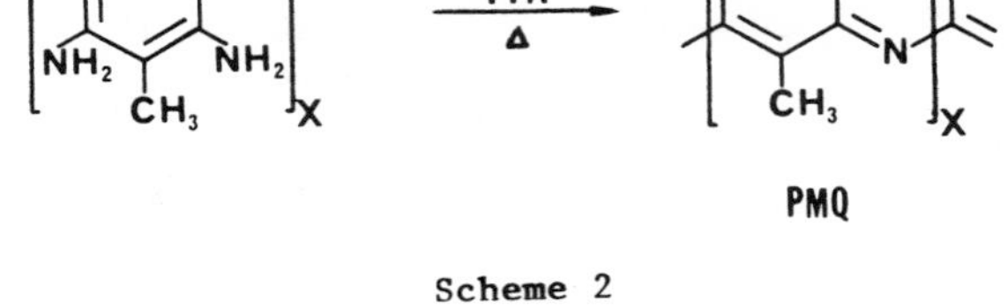

PMQ

Scheme 2

The UV/VIS spectra of the PMQ polymers are given in Figure 5. The absorption threshold of PMQ1 was found at 1.0 ev, while that for PMQ2, 0.6 ev, is about the same as that of PMQ3. This suggests that most rings must be closed at a reaction temperature region between 250°C and 310°C. The threshold of PMQ4 decreased again to 0.3 ev, indicating further ring fusion to longer conjugated chains.

Figure 6 displays IR spectra of PMQ1, PMQ2, PMQ3, and PMQ4. The sharp peak at 1612 cm^{-1} for PMQ1, 1609 cm^{-1} for PMQ2, 1595 cm^{-1} for PMQ3, and 1594 cm^{-1} for PMQ4 is assigned to aromatic C=C and C=N vibration modes. When compared with the absorption at 1630 cm^{-1} in the prepolymer, the band progressively shifted to lower wave numbers due to the extended configuration of the polymer.[8] The peak near 1110 cm^{-1} is attributed to in-plane bending of the ring C-H. The peaks look broader as reaction temperature increases, indicating more free electrons in the materials prepared at higher temperature.

Figure 7 shows the IR spectra of NFP doped PMQ3. The lower spectrum corresponds to heavier NFP doping. The aromatic C=C and C=N vibration modes shifted to higher frequency, from 1595 cm^{-1} for pristine PMQ3 to 1617 cm^{-1} for heavily doped PMQ3. The same phenomenon was also observed in iodine doped PMQ3, where the band shifted to 1610 cm^{-1} (Figure 8). This is attributed to an increase of the force constant for C=C and C=N bonds due to complexation with the acceptors.[11]

Figure 9 illustrates the temperature dependence of unpaired spin concentration for PMQ1, PMQ2, and PMQ3. At lower temperatures, the spin concentration remains constant but at higher temperatures increases exponentially as temperature increases. The data were therefore fitted using Equation 1.[6]

$$n = n_1 + n_2 \exp(-\Delta E_1/KT) \qquad (1)$$

n_1 being the Curie-like or trapped spin concentration, n_2 is the total potential thermo-excited spin concentration, and ΔE_1 is the activation energy.

At lower temperatures, the spin concentration of PMQ4 increased as temperature decreased as illustrated in Figure 10. The reason for this is presently unknown. One possibility is that a phase transition from magnetic disorder to order could occur at a critical temperature, T_c. Three determinations of spin concentration versus temperature for PMQ4 were carried out with reproducible results within experimental error. The data was fitted using Equation 2.

$$n = n_1/(1-T_c/T) + n_2 \exp(-\Delta E_1/KT) \qquad (2)$$

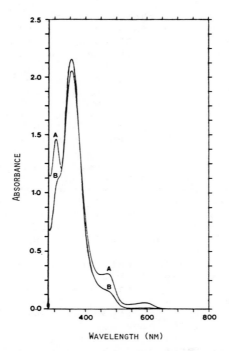

Fig. 4   UV/VIS spectra of PM and PMC.   A) from PM;
B) from PMC.

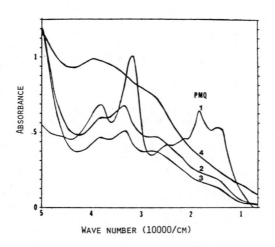

Fig. 5   UV/VIS spectra of PMQ'S.

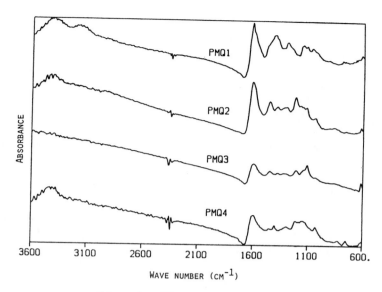

Fig. 6   IR spectra of PMQ's.

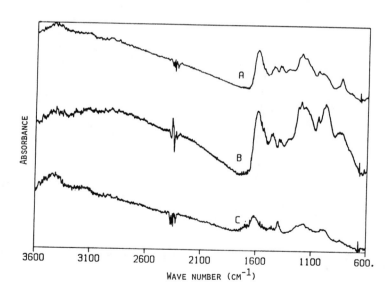

Fig. 7   IR spectra of NFP doped PMQ3.   A) from
         light doped PMQ3; B) from medium doped
         PMQ3; C) from heavy doped PMQ3.

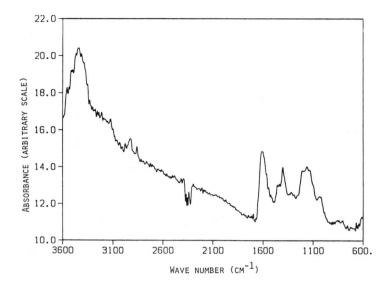

Fig. 8   IR spectrum of iodine doped PMQ3.

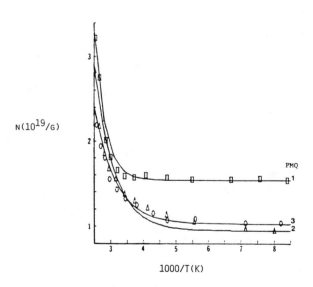

Fig. 9   Temperature dependence of unpaired spin
         density of PMQ1, PMQ2, and PMQ3.

The critical temperature, T_C=70 to 80 K, averaged from three data fittings is close to liquid nitrogen temperature. Exhaustive efforts were made to cool down to T_c for EPR measurement. Fine structure was observed in the range of 79 K (lowest temperature we could reach) to 85 K (Figure 11), but it is not certain whether the anomalous peaks are from a new phase. However, the novel spectrum has been reproduced by Dr. H. Kuska, University of Akron, so it is probably real. PMQ3 does not show this transition near 77 K.

A recent determination of magnetic susceptibility of PMQ4 by Dr. W. Hatfield (U. of N.C., Chapel Hill) at 10,000 Oersteds showed that susceptibility was independent of temperature from 4 to 300 K. At these high fields, the spins seem to be completely saturated.

The g factors, line widths, ΔHpp at room temperature, spin concentrations n_{25}, n_1, n_2, and ΔE_1, are compiled in Table 1.

The major contribution to the EPR signal at room temperature is from Curie-like spins, which are associated with localized free radicals (trapped neutral solitons).[12] The high activation energy, ΔE_1, and broad line width of PMQ1 shows the effect of short conjugated sequences on trapped spins.

Table 1: The g factor, line width, spin concentration and activation energy of pristine PMQ's

PMQ	g	$\Delta Hpp(G)$	$n_{25}(10^{19}/g)$	$n_1(10^{19}/g)$	$n_2(10^{19}/g)$	$\Delta E_1(ev)$
PMQ1	2.0028	11.8	1.58	1.54±0.02	5300±3600	0.275±0.019
PMQ2	2.0028	7.8	1.39	0.95±0.04	130± 100	0.146±0.019
PMQ3	2.0028	8.1	1.39	1.03±0.03	72± 71	0.136±0.021
PMQ4	2.0028	4.0	3.54	1.86±0.08	96± 35	0.104±0.056

After doping PMQ3 with NFP ($0.7PF_6$ per repeat unit by element analysis), there is a significant diminution of the spin concentration, 10.5 fold for the trapped spins and 5.4 fold for thermally excited spins, which may indicate that NO^+ mainly oxidized trapped spins to charged solitons.

The electronic conductivities at room temperature are on the order of 10^{-11} S/cm for pristine PMQ1, and 10^{-9} S/cm for PMQ2 and PMQ3. The conductivity of PMQ4, however is 2.1×10^{-5} S/cm, which did not change after attempted doping with NFP. After saturation doping of PMQ3 with NFP and iodine, the conductivities at 21°C increase to 1.3×10^{-3} S/cm and 2.4×10^{-2} S/cm respectively.

Temperature/conductivity plots for the pristine polymers normalized to the room temperature conductivity are given in Figure 12. They exhibit a thermally activated temperature dependence

$$\sigma = \sigma_0 \exp(-\Delta E/KT) \qquad (3)$$

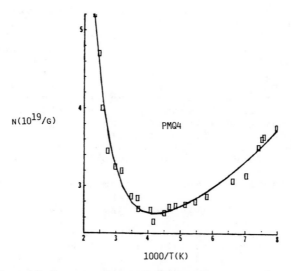

Fig. 10 Temperature dependence of unpaired spin
density of PMQ4.

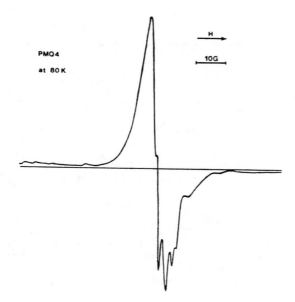

Fig. 11 EPR line shape of PMQ4 at 80 K.

where ΔE is the activation energy. The calculated σ_{25}, σ_0, and ΔE are listed in Table 2. An increase of condensation temperature results in decrease of the activation energy. The conductivity activation energy for each polymer is about equivalent to its optical absorption threshold, Eg', indicative of extrinsic conduction from the protonic acid doping.

Table 2: Electronic conductivity, activation energy, and optical threshold, Eg', of pristine polymer.

PMQ	$\sigma_{25}(S/cm)$	$\sigma_0(S/cm)$	$\Delta E(ev)$	Eg'(ev)
PMQ1	1.7×10^{-11}	$\left(4.1^{+2.1}_{-1.4}\right) \times 10^4$	0.91 ± 0.011	1.0
PMQ2	5.8×10^{-9}	$\left(80^{+43}_{-28}\right)$	0.60 ± 0.011	0.6
PMQ3	7.8×10^{-9}	$\left(3.3^{+1.2}_{-0.9}\right)$	0.51 ± 0.009	0.6
PMQ4	2.1×10^{-5}	$\left(1.7^{+0.3}_{-0.2}\right)$	0.35 ± 0.003	0.3

    Figure 13 displays the temperature dependence of conductivity for NFP doped PMQ3. The high-temperature (150°C) "bendover" might be due to decomposition of the dopant. In the temperature range -10°C to 150°C, the data have the form consistent with normal thermal excitation, Equation 3. The corresponding activation energy, 0.17 ev, is one third of that for pristine PMQ3, which could imply the formation of bipolarons[13] after doping. This is strongly suggested by the significant reduction of thermally excited spins after doping PMQ3 with NFP. At temperatures below -10°C, the conductivity data can be fitted by Equation 4, which is characterized as a three dimensional variable range hopping near the Fermi energy.[14]

$$\sigma \sim \exp(-a/T^{1/4}) \qquad (4)$$

    The temperature conductivity data of iodine doped PMQ3 (1.14 I per repeat from element analysis and weight gain.)(Figure 14) can be fitted not only with Equation 4, but also to a model of fluctuation-induced carrier tunneling (Equation 5).[15] Either of them can be expected only when the metallic domain concentration increases to above the percolation threshold from broken paths.[16]

$$\sigma = \sigma_0 \exp(-T_1/(T+T_0)) \qquad (5)$$

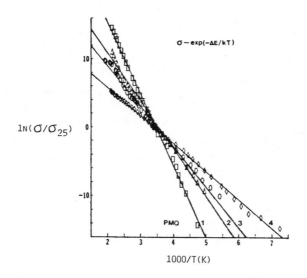

Fig. 12 Temperature dependence of electronic
         conductivity of PMQ's.

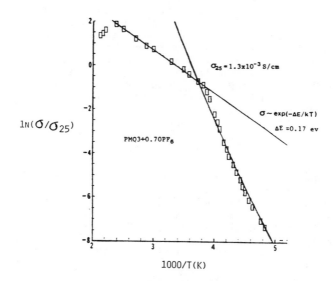

Fig. 13 Temperature dependence of electronic
         conductivity of NFP doped PMQ3.

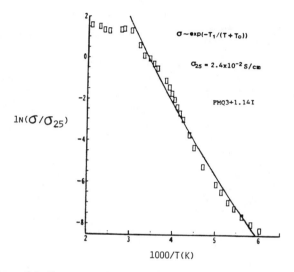

Fig. 14 Temperature dependence of electronic
conductivity of iodine doped PMQ3.

Conclusions.

A linear prepolymer, PM, can be obtained by reaction
between 2,6-diaminotoluene and formaldehyde in acidic
medium. NMR spectra show that expected prepolymer was
formed with a reasonably high molecular weight. The
final ladder polymer is very stable towards
temperature, air, acids, and base. It is soluble in
strong acid if it is polycondensed below 385°C.
Considerable differences were observed in EPR and
conductivity studies for PMQ1, PMQ2, PMQ3, and PMQ4.
Temperature dependence of spin concentration of PMQ4
indicates that a phase transition from disorder to
possible magnetic order may occur at a critical
temperature near 74 K. The mechanism is yet unknown.
Narrow EPR line width and high conductivity of pristine
PMQ4 suggests high spin mobility.[6] Significant
reduction of spin concentration of PMQ3 after NFP
doping indicates that NO^+ oxidized neutral solitons
(trapped free radicals) and/or polarons (mobile radical
cations) to form spinless charged solitons (cations)
and/or bipolarons (mobile dications). Temperature/
conductivity plots exhibit thermally activated
temperature dependence for pristine PMQ's. However,
variable range hopping and fluctuation induced
tunneling models can be applied to the NFP and iodine
doped PMQ3 respectively. Both suggest that a high
metallic domain concentration was induced by doping.

Acknowledgments.

This work was partly supported by the Dow Chemical
Company, DOE and the National Science Foundation.

Literature Cited.

1. N. R. Lerner, Polym. 24, 800, (1983).
2. Oh-Kil Kim, Mol. Cryst. Liq. Cryst. 105, 161,
   (1984).
3. M. M. Tessler, Amer. Chem. Soc., Div. of Polym.
   Chem. Preprints, 8, 165, (1967).
4. R. H. Baughman, J. Polym. Sci. Polym. Lett. Ed.,
   21, 475, (1983)
5. J. Z. Ruan and M. H. Litt, Macromolecules,
   (submitted).
6. J. Z. Ruan and M. H. Litt, Macromolecules,
   (submitted).
7. C. L. Cocq and J. V. Lallemand, J. Chem. Comm.,
   150, (1981).
8. G. C. Levy, R. L. Lichter and G. L. Nelson,
   "Carbon-13 Nuclear Magnetic Resonance
   Spectroscopy," John Wiley and Sons, New York
   Chichester Brisbane Toronto, (1980).

9. E. C. Horning, J. Koo, and G. N. Walker, J. Amer. Chem. Soc., <u>7</u>, 5826, (1951).

10. H. R. Snyder and M. S. Konecky, J. Amer. Chem. Soc., <u>80</u>, 4388, (1958).

11. O. K. Kim, J. Polym. Sci. Polym. Lett. Ed. 23, 137, (1985).

12. B. R. Weinberger, J. Kanfer, A. J. Heeger, A. Pron and A. G. MacDiarmid, Phy. Rev. B20, 223, (1979).

13. J. C. Scott, P. Pfluger, M. T. Krounbi and G. B. Street, Phys. Rev. B28, 2140, (1983).

14. A. J. Epstein, H. Rommelmann, M. Abkowitz and H. W. Gibson, Mol. Cryst. Liq. Cryst. 77, 81, (1981).

15. P. Sheng, E. K. Sichel and J. I. Gittleman, Phys. Rev. B18, 1197, (1978).

16. Y. Tomkiewiez, T. D. Schultz, H. B. Brom, and A. R. Taranko, Phys. Rev., B24, 4348, (1981).

RECEIVED April 8, 1987

Chapter 49

From Pyropolymers to Low-Dimensional Graphites

S. Yoshimura, M. Murakami, and H. Yasujima[1]

Research Development Corporation of Japan, c/o Matsushita Research
Institute Tokyo, Inc., Higashimita 3-10-1, Tama-ku, Kawasaki 214, Japan

New types of intrinsically conductive polymers which
are classified as low-dimensional graphites were prepared by pyrolysis of organic materials. Firstly, a
heat-resistant, condensation polymer, polyoxadiazole,
was converted to a highly conductive pyropolymer at
heat-treatment temperatures (HTT's) near 1000°C,
yielding a nearly-perfect graphite film for HTT's above
2800°C. Changes in electrical transport properties with
HTT favored a mechanism in which planar condensed
aromatic layers developing at HTT's above 700°C are
extended to two-dimensional graphite at higher HTT's.
Secondly, heat treatment of a perylene-tetracarboxylic
dianhydride monomer above 520°C resulted in vapor-phase
condensation of poly-peri-naphthalene (PPN), a representative one-dimensional graphite polymer, in a
whisker form. At HTT's near 2800°C, the PPN whisker was
converted to a graphite whisker with one-dimensional
graphite chains oriented normal to the whisker axis.

Since the discovery of drastic enhancement of the electrical conductivity in chemically doped polyacetylene (1), extensive synthetic
efforts have been directed at thermally stable or processible
polymers with high conductivity. Polymer pyrolysis has long been
appreciated as being a good lead for such polymers (2), polyacrylo-
nitrile (3) and polyimide (4) being extensively studied, and has
recently received an increased attention in proportion to understanding of their structural and electronic properties (5). Pyrolysis
of organic monomers or polymers possibly induces extended aromatic
rings or planar ladder polymers which, according to recent theoretical predictions (6), have a considerably smaller bandgap and

[1]Current address: Central Laboratory, Toppan Printing Company, Ltd.,
Shimotakano 1580, Sugito-cho, Kitakatsushika-gun, Saitama 345, Japan

higher mobility than polyacetylene. These polymers can be intrinsically conducting without the need of doping and also have high environmental stability as well as graphite, which will serve as useful electronic materials.

In this paper we report on the synthesis of new types of intrinsically conducting polymers using high-temperature reaction of condensation polymers and polyfunctional monomers. We shall show that the new polymers can be regarded as low-dimensional graphites both macroscopically and microscopically and discuss the change in their structural and electronic properties with heat-treatment temperature (HTT).

Two-dimensional Graphite Obtained by Pyrolytic Polyoxadiazole

Experimental. Films or fibers of several commercially-available, condensation polymers were heated at a rate of 10°C/min to a predetermined temperature (HTT) and held for 1 h. They were placed between two plates of alumina (for HTT < 1400°C) or graphite (for HTT > 1500°C) to prevent deformation due to ununiform heating. The heat treatment was made in vacuum and in an argon atmosphere for HTT's lower and higher than 2000°C, respectively. The electrical conductivity was measured along the film plane or fiber axis using an ordinary four-probe method with gold wires attached with silver paste (du Pont 4929). The temperature dependence of the electrical conductivity was measured down to 3.8 K using an Oxford helium clyostat. The data were approximated to polynomial equations to the fourth or fifth order and then fitted to various theoretical curves pertinent to conduction mechanisms.

Pyrolysis and Graphitization of Polyoxadiazole. Table 1 shows electrical conductivity at room temperature for various condensation polymers heat-treated at 1000 and 2500°C. For a HTT near 1000°C, poly(p-phenylene-1,3,4-oxadiazole) (POD) has the highest conductivity of 340 S/cm, twice as high as that of polyimide (KAP) (7). The electrical conductivity of the pyropolymers had rough correlation

POD

Table I. Electrical conductivity at room temperature for condensation polymers pyrolyzed at 1000 and 2500°C for 1 h

Materials	Conductivity (S/cm)	
	HTT = 1000°C	2500°C
Polyamideimide (HI)[1]	60	150
Polyimide (KAP)[2]	160	520
Polyamide-1 (MX)[3]	270	1000
Polyamide-2 (KEV)[4]	280	1200
Polybenzimidazole (PBI)[5]	250	2800
Polyoxadiazole (POD)[6]	340	7000

[1]Obtained from Hitachi Chemicals, [2]Kapton film from du Pont, [3]MX film from Torei Co., [4]Kevlar fiber from du Pont, [5]Obtained from Celanese Co., [6]Obtained from Furukawa Electric.

with the structural order of the starting polymer, ranging from that
(340 S/cm) of partially crystalline POD (crystallinity = 70 %) to 60
S/cm of amorphous polyamideimide (HI) (crystallinity less than 5 %).
The behavior of polyamide could not be rationalized, for both highly
crystalline KEV (> 90 %) and poorly crystalline MX (< 5 %) had
nearly the same conductivity which is less than that of POD. The
difference in the electrical conductivity was further enhanced by
heat treatment at higher HTT's; e.g., 150, 520 and 7000 S/cm for HI,
KAP and POD, respectively, at 2500°C. The highest conductivity of
POD obtained for HTT = 3000°C was 14000 - 20000 S/cm, being somewhat
lower than that of single crystal graphite (25000 S/cm) (8).

The interplanar spacing d_{002} calculated from the angle of the
x-ray (002) reflection peak was also strongly dependent on the kind
of starting polymer as shown in Fig. 1. This figure indicates that
POD is perfectly graphitized at HTT's higher than 2800°C with d_{002} =
3.354 Å and hence the degree of graphitization according to Mering's
equation (9) of 100 %. The POD films annealed at HTT's above 2800°C
consisted of highly-oriented graphite crystallites (Fig. 2(a)) and
had extremely small full width at half-maximum intensity of the
(002) reflection, within 0.17°(Fig. 2(b)) (10). Figure 3(a) shows a
transmission electron microscopy (TEM) picture of the POD film
graphitized at 2800°C. We can recognize very thin films of 20 to
100 Å in thickness constituting the graphite film. The individual
thin film exhibited a perfectly graphitic behavior as evidenced by a
selected-area electron diffraction (SAD) pattern shown in Fig. 3(b).

This result is striking because almost all polymeric materials
have been known to yield so-called hard carbons which is non-
graphitizing. This new finding may offer a promise of producing
graphite films of arbitrarily large area by polymer pyrolysis. It is
interesting to note as well that POD was perfectly graphitized at
considerably lower temperatures than HOPG (highly-oriented
pyrographite) (11), which would require a new reaction mechanism for
the pyrolytic POD.

Reaction Mechanism for the POD Films. Powder x-ray diffraction
patterns of POD pyrolyzed at HTT's below 1000°C are shown in Fig. 4.
It is recognized that the starting polymer has a high degree of
crystallinity, which was remarkably enhanced at 400°C, just before
the onset of thermal decomposition. At 1000°C, a graphitic ordering
with a lattice spacing of 3.49 Å was found. Figure 5 shows XPS
spectra in the N_{1s} region for POD films pyrolyzed at various HTT's,
which indicates that the thermolysis resulted in -C=N- and -C≡N
groups (for HTT's between 520 and 700°C) and heating at higher HTT's
gave rise to heterocyclic structures such as pyrazine or triazine.

Figure 6 shows transmitted x-ray diffraction patterns for the
POD films with progressive HTT. The characteristic (101) and (112)
reflections are absent for HTT's below 2200°C and suddenly appear at
2500°C. This result implies that the three-dimensional order of the
graphite lattice is not established at lower HTT's, namely the
condensed aromatic layers exist as individuals and have a highly
two-dimensional character. This may have come from the absence of
cross-linked network structures at lower HTT's and may have
facilitated the the recrystallization of graphite at such low
temperatures and without pressure.

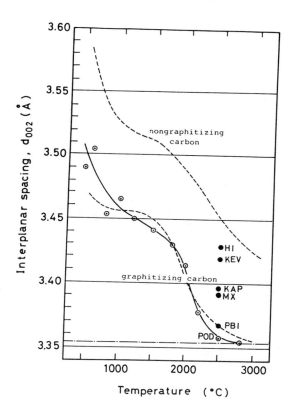

Figure 1. The interplanar spacing d_{002} as a function of HTT for polyamideimide (HI), polyamide-1 (KEV), polyimide (KAP), poly-amide-2 (MX), polybenzimidazole (PBI) and polyoxadiazole (POD).

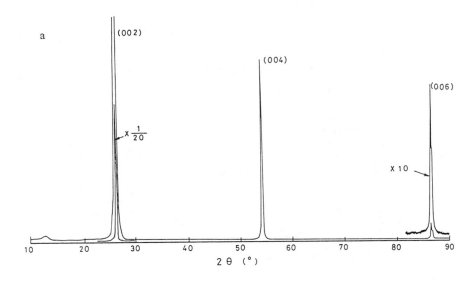

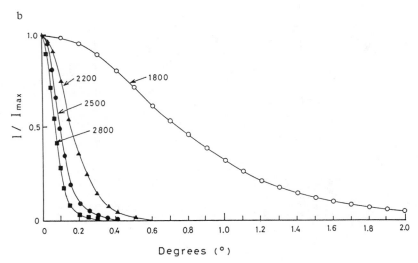

Figure 2. X-ray diffraction pattern of a POD film for HTT = 2800°C (a) and profiles of the (002) reflection peaks as a function of deviation from the Bragg angle (b). (Reproduced with permission from Ref. 10. Copyright 1986 American Institute of Physics.)

a

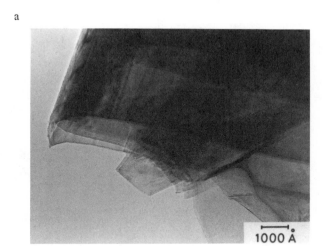

b

Figure 3.   A TEM feature (a) and a SAD pattern (b) for a thin film of POD graphitized at 2800°C.

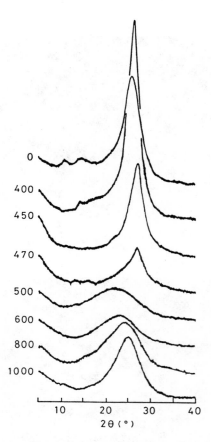

Figure 4. X-ray diffraction patterns of POD films pyrolyzed at HTT's below 1000°C.

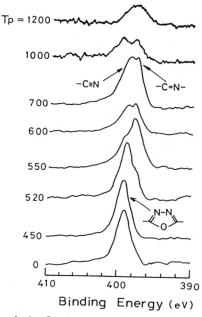

Figure 5. X-ray photoelectron spectroscopy peaks in the N_{1s} region for POD films pyrolyzed at HTT's below 1400°C.

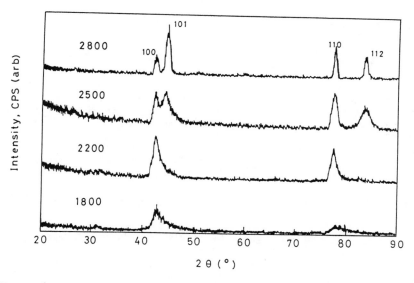

Figure 6. Change in transmitted x-ray diffraction patterns of POD films with HTT. (Reproduced with permission from Ref. 10. Copyright 1986 American Institute of Physics.)

Electrical Properties. Figure 7(a) summarizes electrical
conductivity, Hall coefficient and magnetoresistance at room
temperature as a function of HTT (12). Figure 7(b) depicts the HTT
dependence of carrier density and mobility calculated from the
results of Fig. 7(a). There are two regions of HTT where the
electronic structure drastically changes. For HTT's between 1800 and
2200°C, the electron density increases and the electron mobility
decreases with increasing HTT, while those for holes have a little
HTT dependence. This result suggests that the positive Hall
coefficient peaking at 2000°C is caused by the decrease in the
electron mobility, which is contrary to a former explanation that
holes dominate the conduction process as a result of development of
acceptor-type defects. The next feature is a critical change in
carrier densities and mobilities at 2500°C. This evidently
corresponds to the development of the graphite lattice of three-
dimensional order (see Fig. 6). The POD films heat-treated at 3000°C
had the mobilities of 5900 and 9300 cm²/Vs for electrons and holes,
respectively. The lower mobilities as compared with that of natural
single crystals (15000 cm²/Vs) (8) may be caused by the scattering
of the carriers at the surfaces of the very thin graphite
crystallites.

   Figure 8 shows the temperature dependence of electrical
conductivity (σ) plotted as $\log(\sigma\sqrt{T})$ vs. $T^{1/4}$. For higher measuring
temperatures, the conductivity is well expressed by the Mott's
variable-range hopping (VRH) model (13). In addition, in order to
explain the σ - T curves with a single equation for the whole
temperature range measured, we assumed that POD was a two-phase
composite composed of parallel VRH (σ_1) and metallic (σ_0) channels:

$$\sigma = \sigma_0 + \sigma_1$$

hence σ_0 could be best fitted to

$$\sigma_0 = 1/(\rho_0 + \rho_1 T^n)$$

The n value changed from 1.4 to 2 with increasing HTT with a kink at
2200°C as shown in Fig. 9. Note that n = 1 for ordinary metal and
graphite crystals. The characteristic feature (n > 1) can be understood if we assume that pyrolyzed POD is a two-dimensional conductor, for this T^2 dependence has been encountered with two-
dimensional organic metals (14) and graphite intercalation compounds
(15).

Polycondensation of a Polyfunctional Monomer and One-dimensional
Graphite

Synthesis of Poly-peri-naphthalene. Kaplan et al. have prepared
carbon films with relatively high electrical conductivity (250 S/cm)
by pyrolyzing a polyfunctional monomer, 3, 4, 9, 10-perylene-
tetracarboxylic dianhydride (PTCDA), at HTT's between 700 and 900°C
and claimed that the product consisted of poly-peri-naphthalene
(PPN) (16). PPN is a representative one-dimensional graphite polymer
with a small bandgap (17,18). Recently we have reported on
successful synthesis of the PPN polymer by vapor-phase condensation
of the same monomer, PTCDA, at HTT's above 520°C (19). The PPN

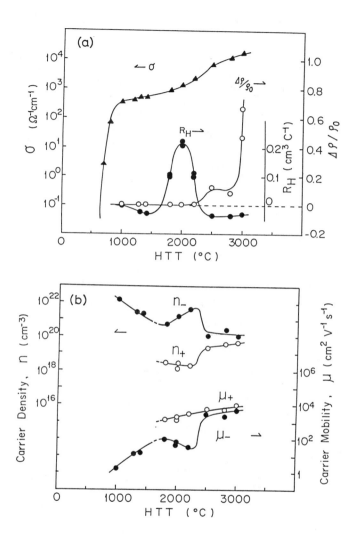

Figure 7.   Electrical conductivity (σ), Hall coefficient (R_H) and magnetoresistance ($\Delta\rho/\rho_0$) (a); carrier density (n_- and n_+ for electrons and holes, respectively) and mobility ($\bar{\mu}_-$, μ_+)(b). (Reproduced with permission from Ref. 12. Copyright 1986 American Institute of Physics.)

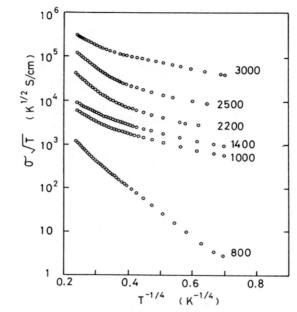

Figure 8.   Temperature dependence of electrical conductivity of
POD films pyrolyzed at various HTT's.

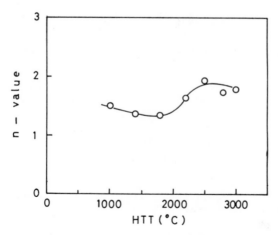

Figure 9.   Exponent (n) for electrical resistivity (in $\rho = \rho_0 + \rho_1 T^n$) as a function of HTT.

PTCDA   PPN

polymer was deposited as very fine whiskers having rectangular cross
sections of 0.1 to 0.4 μm on sides with length up to 10 mm when
synthesized in Ar, while a flat-ribbon-like morphology was obtained
for a hydrogen-containing Ar atmosphere (see Fig. 10). The polymerization reaction consisted in a simple prosess involving the thermolysis of the carboxylic groups with elimination of carbon dioxide
and monooxide gases and coupling of perylene tetraradicals.

The room-temperature conductivity of the PPN polymer was 0.2
S/cm without doping, a value almost in the middle of those of
polyacetylene and graphite. The temperature dependence of the
electrical conductivity was measured with the PPN whiskers
synthesized at various HTT's. The measured σ - T curves were fitted
to m-dimensional VRH equations:

$$\sigma = \sigma_0 \exp\{(T_0/T)^{1/m}\}.$$

Figure 11 shows the result for PPN synthesized at 530°C, for which
the best fit equation was that with m = 2. Higher values of m were
obtained for higher HTT's; for example, m = 3 for HTT = 600°C and m
= 4 for HTT > 800°C. The value of m is 3 and 4 for two- and three-
dimensional VRH, respectively (13), and m = 2 is the case where one-
dimensional VRH or charge-assisted electron tunnelling (20)
dominates. Anyway the conduction process of the PPN whiskers
obtained at HTT's below 600°C was dominated by the hopping
conduction in low-dimensional disordered systems.

Graphitization of the PPN whiskers. High-temperature heat treatment
of the whisker revealed that PPN was also a typical graphitizing
material (d_{002} = 3.357 Å for HTT = 2800°C), the graphitization
proceeding rapidly above 2000°C. Figure 12 shows a TEM picture of
the PPN whisker graphitized at 2800°C. Linear graphite chains are
oriented perpendicular to the direction of the whisker axis and
folded at the whisker edges. This result suggests a unique crystal
structure of the PPN polymer in which the one-dimensional graphite
polymer is extended normal to the whisker axis and stacked to the
direction of the whisker axis (21).

The graphitized PPN exhibited a maximum conductivity of 8000
S/cm for HTT = 2800°C, much lower than that of HOPG (25000 S/cm),
which can be explained in terms of the unique crystal structure of
graphitized PPN.

Conclusions

Pyrolysis or high-temperature polycondensation of polymers or polyfunctional monomers yielded new types of intrinsically conductive
polymers and high-quality graphite which can be classified as low-
dimensional graphites.

Pyrolysis of polyoxadiazole (POD) yielded a highly conductive
pyropolymer at heat treatment temperatures (HTT) near 1000°C and

a

b

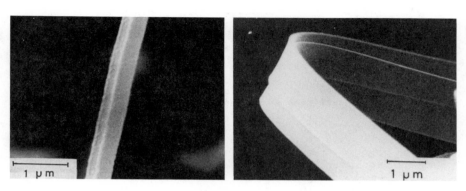

Figure 10. SEM features of PPN whiskers synthesized in Ar (a) and
in Ar/H₂ (b).

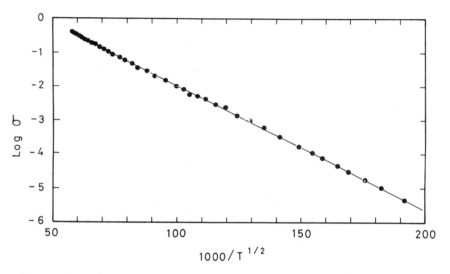

Figure 11. Temperature dependence of electrical conductivity for
a PPN whisker synthesized at 530°C.

1000 Å

Figure 12.   A TEM feature of a PPN whisker graphitized at 2800°C.

large-area, nearly-perfect graphite films were obtained at HTT's above 2800°C. Changes in electrical properties with HTT favored a mechanism in which planar condensed aromatic layers developing at HTT's near 1000°C are extended to two-dimensional graphite at higher HTT's.

Pyrolysis of a perylene-tetracarboxylic dianhydride monomer above 520°C resulted in vapor-phase condensation of poly-peri-naphthalene (PPN), a representative one-dimensional graphite polymer. PPN was obtained as fine whiskers with rectangular cross sections and exhibited electrical conduction behaviors characteristic of low-dimensional disordered systems. At HTT's above 2800°C, PPN was converted to a graphite whisker with one-dimensional graphite chains oriented normal to the axis.

The newly developed low-dimensional graphites have various technological promises, including optical materials for an X-ray monochrometer or microscope, high-energy batteries or biological materials.

Acknowledgments

This work was carried out as part of Exploratory Research For Advanced Technology (ERATO) sponsored by the Science and Technology Agency of Japan. The authors are indebted to S. Mizogami, Y. Yumoto, S. Iijima, K. Watanabe, K. Morishita and S. Naitoh for stimulating interactions in the course of the study. They also thank Professor N. Ogata of Sophia University and Professor S. Ohtani of Gunma University for valuable suggestions and encouragement.

Literature Cited

1. Shirakawa, H.; Louis, E.J.; MacDiarmid, A.G.; Chian, C.K.; Heeger, A.J. J. Chem. Soc., Chem. Commun. 1977, 1977, 578.
2. Naarmann, H. Angew. Makromol. Chem. 1982, 109/110, 295.
3. Duke, C. B.;Gibson, H. W. In Encyclopedia of Chemical Technology; Kirk-Otrhmer Ed.; John Wiley & Sons: New York, 1982; Vol. 18, p 775.
4. Chung, T. -C.; Schlesinger, Y.; Etemad, S.; MacvDiarmid, A. G.; Heeger, A. J. Polym. Sci., Polym. Phys. Ed. 1984, 22, 1239.
5. Hu, C. Z.; Andrade, J. D.; J. Appl. Polym. Sci. 1985, 30, 4409.
6. Boudreaux, D. S.; Chance, R. R.; Elsenbaumer, R. L.; Frommer, J. E.; Bredas, J. L.; Silbey, R. Phys. Rev. 1985, B31, 652.
7. Murakami, M.; Yasujima, H.; Yumoto, Y.; Mizogami, S.; Yoshimura, S. Solid State Commun. 1983, 45, 1085.
8. Spain, I. L. In Chemistry and Physics of Carbon; Walker, P. L. Ed.; Marcel Dekker: New York, 1971; Vol. 8, p 105.
9. Mering, J.; Marie, J. Les. Carbons 1965, 1, 129.
10. Murakami, M.; Watanabe, K.; Yoshimura, S. Appl. Phys. Lett. 1986, 48, 1594.
11. Moore, A. W. In Chemistry and Physics of Carbon; Walker, P. L.; Trower, P. A. Ed.; Marcel Dekker: New York, 1973; Vol. 11, p 69.
12. Yasujima, H; Murakami, M.; Yoshimura, S. Appl. Phys. Lett. 1986, 49, 499.
13. Mott, F. N. Metal-Insulator Transitions, Taylor & Francis: London, 1974.

14. Bechgaard, K.; Jacobsen, C. S.; Mortensen, K.; Pedersen, H. J.; Thorup, H. Solid State Commun. 1980, 33, 1119.
15. Dresselhause, M. S.; Dresselhause, G. Adv. Phys. 1981, 30, 139.
16. Kaplan, M. L.; Schmidt, P. H.; Chen, C.-H.; W. M. Walsh Jr., W. M. Appl. Phys. Lett. 1980, 36, 867.
17. Tanaka, K.; Ueda, K.; Koike, T.; Yamabe, T. Solid State Commun. 1984, 51, 943.
18. Bredas, J. L.; Baughman, R. H. J. Chem. Phys. 1985, 83, 1316.
19. Murakami, M.; Yoshimura, S. J. Chem. Soc., Chem. Commun. 1985, 1985, 1649.
20. Sichel, E. K.; Emma, T. Solid State Commun. 1982, 41, 747.
21. Murakami, M.; Iijima, S.; Yoshimura, S. Appl. Phys. Lett. 1986, 48, 390.

RECEIVED April 8, 1987

Chapter 50

Band-Structure Calculations on Polymeric Chains

William J. Welsh

Department of Chemistry, University of Missouri—St. Louis, St. Louis, MO 63121

Electronic band structures were calculated for several
polymeric chains structurally analogous to polyacetylene (–CH=CH)
and carbyne (–C≡C). The present calculations use the Extended Hückel
molecular orbital theory within the tight binding approximation, and
values of the calculated band gaps E_g and band widths BW were used
to assess the potential applicability of these materials as electrical semiconductors. Substitution of F or Cl atoms for H atoms in
polyacetylene tended to decrease both the E_g and BW values (relative
to that for polyacetylene). Rotation about the backbone bonds in the
chains away from the planar conformations led to sharp increases in
E_g and decreases in BW. Substitution of –SiH$_3$ or –Si(CH$_3$)$_3$ groups
for H in polyacetylene invariably led to an increase in E_g and a
decrease in BW, as was generally the case for insertion of 'Y'
(where Y = O, S, NH, CH$_2$, SiH$_2$) in carbyne to give [–C≡C–Y–C≡C]. The
degree of bondlength alternation appears to play a major role in
determining the electrical conductivity of a conjugated polymer,
with values of E_g decreasing (favoring conductivity) as bond–length
alternation decreases.

Polyacetylene, (–CH=CH), the simplest organic polymer with a fully
conjugated backbone, has generated considerable interest due to its
unusual electronic properties. Specifically, through selective
doping of the polymer its electrical conductivity can be made to
vary many orders of magnitude, from insulator to semiconductor (1–
3). The structure of the polymer chain appears to be one of the key
determinants for the electronic properties of polymer/dopant systems
(1–3). Recent structural evidence and theoretical calculations (1–3)
suggest a planar backbone structure for cis and trans forms of
(–CH=CH). However, many aspects of the structure and the configura-
tional characteristics of (–CH=CH) are not well–defined due to its
intractability, its instability to oxidation, and its insolubility
in most solvents. Many other polymeric systems have thus been
proposed or actually investigated with regard to their potential
as conductors or semi–conductors (1–3), among these the halogen-
substituted polyacetylenes (4). For example, Zeigler has recently
synthesized the perfluoropolyacetylene (–CF=CF) (4).

It appears that electrical conductivity is particularly

0097–6156/87/0346–0600$06.00/0

sensitive to the degree of conjugation and planarity along the chain backbone (1-3,5,6). In the case of substituted polyacetylenes in particular, it is crucial for effective conductivity that the substituent's steric bulk not cause appreciable deviations from planarity in an attempt to reduce steric conflicts. With regard to the halogen-substituted polyacetylenes, the fluorine atom is just small enough (r_{vdw} = 1.30 Å) to render the F...F interactions attractive even for the planar chain in which case the four-bond F...F interatomic distance is closest (2.60 Å). However, with substitution of chlorine (r_{vdw} = 1.80 Å) steric conflicts between pendant chlorine atoms will render the planar conformation highly repulsive, and this effect would become more severe in the case of Br and I substitution.

In this study, quantum mechanical methods were used to calculate the electronic band gaps E_g and band widths BW of the (-CF=CF) chain and compared with those similarly calculated for (-CH=CH) itself. Other halogen-substituted polyacetylenes considered are the chains (-CF=CH) and (-CF=CF-CH=CH) and their chlorinated analogs. Calculations were carried out as a function of rotation about the single bonds along the chain backbone in order to assess the dependence of conductivity on chain planarity. Likewise, the sensitivity of the calculated band gaps to small changes in structure (i.e., bond angles, bond lengths) has been investigated.

Depending on the size of the substituents and their sequence, the extent of steric interference between the 1- and 3-substituents (Figure 1) and between the 2- and 4-substituents in the planar trans conformation may vary considerably. In the planar cis conformation, the critical interactions in terms of possible steric conflicts are between the 2...3 and 1...4 substituents, where the latter interaction becomes particularly significant due to the decreased interatomic separation. For a given halogen-atom substituent X, steric interferences will generally increase in the order (-CH=CH-CX=CX), (-CH=CX), (-CX=CX). Comparison of the size of the substituent atoms H(r_{vdw} = 1.20 Å), F(r_{vdw} = 1.30 Å) and Cl(r_{vdw} = 1.80 Å) suggests that steric effects will be much more significant in the chlorinated derivatives.

Another structurally analogous group of substituted polyacetylenes are of the type [-CH=CX1-CH=CH-CX2=CH] with X1 and X2 chosen from among H, CH_3, SiH_3 and $Si(CH_3)_3$. The choice of $Si(CH_3)_3$ as a substituent has been suggested (4) as a means to improve the solubility and hence the processibility of the otherwise intractable polyacetylene.

Other conjugated polymeric systems have been suggested as possible electrical conductors (1-3,6,7), including (-C≡C-X-C≡C-) where X may be a group IV, V or VI element. Undoped organosilicon polyynes have been shown to possess resistivities that classify them as organic semiconductors. Consequently, band structure calculations on chains of the type (-C≡C-X-C≡C-) are included here for the cases X = O, S, NH, CH_2 and SiH_2. For comparison, calculations have also been carried out on carbyne (-C≡C) to assess the effect of the 'Y' atom or group on the otherwise fully conjugated system.

Theory

For any molecule, including polymers, the LCAO approximation and Bloch's theorem can be used to describe the delocalized crystalline orbitals $\psi_n(k)$ as a periodic combination of functions centered at the atomic nuclei. For a one-dimensional system in which N_1-1 cells (repeat units) interact with the reference cell and for a basis set of length ω describing the wave function within a given cell, the nth crystal orbital $\psi_n(k)$ is defined as (8–11)

$$\psi_n(k) = \sum_{\mu}^{\omega} C_{n\mu}(k)\phi_{\mu}(k) \tag{1}$$

where $\{\phi_{\mu}\}$ is the set of Bloch basis functions

$$\phi_{\mu}(k) = \frac{1}{N_1^{1/2}} \sum_{j_1=-(N_1-1)/2}^{(N_1-1)/2} \exp(ik \cdot R_j)\chi_{\mu}(r - R_j) \tag{2}$$

The quantity k is the wave vector. The position vector R_j, in the one-dimensional case, is given by

$$R_j = j_1 a_1 \tag{3}$$

where a_1 is the basic vector of the crystal. The crystal orbitals are, therefore,

$$\psi_n(k) = \frac{1}{N_1^{1/2}} \sum_{j_1=-(N_1-1)/2}^{(N_1-1)/2} \sum_{\mu=1}^{\omega} \exp(ik \cdot R_j)C_{n\mu}(k)\chi_{\mu}(r-R_j) \tag{4}$$

where $C_{n\mu}(k)$ is the expansion coefficient of the linear combination. The basis functions χ_{μ} are exponential functions of the Slater form. The present one-dimensional calculations included all the valence atomic orbitals of the H, C, N, and F atoms but for all other atoms only s and p orbitals could be considered.

Using the extended Hückel approximation, we obtain the corresponding eigenvalues $E_n(k)$ and coefficients $C_{n\mu}(k)$ from the eigenvalue equation

$$H(k)C_n(k) = S(k)C_n(k)E_n(k) \tag{5}$$

where $H(k)$ and $S(k)$, are respectively, the Hamiltonian and overlap matrices between Bloch orbitals defined as

$$H_{\mu\nu}(k) = \langle\phi_{\mu}(k)|H_{eff}|\phi_{\nu}(k)\rangle \tag{6}$$

and

$$S_{\mu\nu}(k) = \langle\phi_{\mu}(k)|\phi_{\nu}(k)\rangle \tag{7}$$

The distribution of the $E_n(k)$ values for a given n with respect to k (usually within the first Brillouin zone ($-0.5K \le k \le 0.5K$, where $K = 2\pi/a_1$) is the nth energy band. In the present calculations, lattice sums were extended to include second nearest neighbors. The set of all energy bands describes the band structures of the polymers. From the calculated band structures, values of the band gap E_g and the band width BW were used to predict the potential for electrical conductivity in a given polymer chain. To determine the most probable conformation in some cases, values of the total energy per unit cell $\langle E_t \rangle$ were calculated from their band structures as a function of the dihedral angle ϕ. The equation employed was (10, 12)

$$\langle E_t \rangle = \frac{1}{K} \int_{-K/2}^{K/2} E_t(k)dk \tag{8}$$

where $E_t(k)$ is the total energy at k and, according to the Extended Hückel method,

$$E_t(k) = 2 \sum_{n}^{occupied} E_n(k) \tag{9}$$

Structural Parameters

Pertinent values of the structural parameters (i.e., bond lengths and bond angles) used in the present calculations are given in Table I.

Table I. Structural Parameters Used in Calculations for Poly-acetylene and Its Halogenated and Silylated Analogues

	(–CH=CH)	Fluoro	Chloro	Silyl
Bond Lengths (Å)				
C=C	1.342	1.352	1.349	1.342
C–C	1.436	1.400	1.483	1.436
C–X[a]	1.121	1.336	1.715	1.870
Bond Angles (Degrees)				
C=C–C	127.0	128.6	123.4	127.0
C=C–X[a]	118.0	117.3	122.0	118.0

[a] X=H, F, Cl, Si, respectively.

The above values were selected, in the case of polyacetylene, from
averages taken from available experimental and theoretical structural data. For the halogenated chains, appropriate values were
selected from the results of both ab initio and CNDO/2 molecular
orbital calculations and from structural data available on small-
molecule analogues. Due to the unavailability of structural data
for the silylated analogues, values were taken from those used
for polyacetylene (except, of course, for those bonds and angles
associated with an 'X' atom) so as to deduce the effect of the
substitution alone without inclusion of concomitant structural
modifications along the polymer backbone resulting from such sub-
stitutions.

     For the calculations on carbyne (–C≡C) and the (–C≡C–X–C≡C)
chains, values of the bond lengths for the C≡C bond (1.116 Å) and
C–C bond (1.339 Å) were taken from results of ab initio studies
on carbyne (13). Associated values for the Y–C bond lengths and
C–Y–C bond angles were collected from experimental structural data
for small molecular analogs (14) or, in their absence, assigned
standard values.

Results And Discussion

     Calculated values of the band gaps E_g and band widths BW for
(–CH=CH) and some halogenated analogues, all in the trans (ϕ = 0°)
conformation, are presented in Table II.

Table II. Values of the Band Gap E_g and Band Width BW for the
           Chains in the Planar trans Conformation

Chain	$E_g{}^a$	BW^a
(–CH=CH)	1.2	6.9
(–CF=CF)	0.72	6.9
(–CH=CF)	1.0	6.9
(–CH=CH–CF=CF)	0.67	6.3
(–CCl=CCl)	0.24	6.2
(–CH=CCl)	0.25	6.1
(–CH=CH–CCl=CCl)	0.87	5.7

^{a}In units of electron-volts.

     Based on these results, it appears that substitution of F or Cl
for H has an appreciable effect on the band structure of the chain.
The band gap for (–CH=CF) is somewhat higher than expected in comparison with (–CF=CF) and (–CH=CH–CF=CF). However, such variation is
expected given the great sensitivity of the calculated E_g values to
small changes in structural geometry (6). Substitution of Cl for H
has a pronounced effect on the value of E_g. Specifically, the E_g
values for (–CCl=CCl) and (–CH=CCl) are significantly lower than
that calculated for (CH=CH). However, due to the large size of the
Cl atom relative to H and F, steric conflicts will effectively

preclude the chains from attaining the planar conformations at which these band gaps were calculated. The E_g value of (-CH=CH-CCl=CCl) is conspicuously higher than that of (-CH=CCl) and of (-CCl=CCl); whether the unusually high E_g values obtained for (-CH=CF) and (-CH=CH-CCl-CCL) are artifactual or the result of the structural differences among the chains is currently under examination.

It would appear that the chlorinated analogues have the greatest potential as possible conducting polymers. However, due to the large size of the Cl atom (r_{vdw} = 1.80 Å), severe repulsions would be encountered in the planar trans conformation. The relative magnitude of the steric interferences would decrease for the series (-CCl=CCl) > (-CH=CCl) > (-CH=CH-CCl-CCl) in which highly repulsive Cl...Cl interactions are replaced by the sterically more favorable H...H and H...Cl interactions. However, even for the latter cases the H...Cl interatomic distances (2.50 Å) in the trans conformation are not nearly large enough to accomodate the steric bulk of the Cl atom. As a result, large deviations from chain planarity are expected for all of the chlorinated derivatives.

In the case of the fluorine derivatives, the F atom is just small enough (r_{vdw} = 1.30 Å) to render F...F interactions attractive in the trans conformation where the 1...3 and 2...4 substituent interatomic distances (Figure 1) are at their closest (2.60 Å). The presence of attractive F...F interactions in the trans conformation is analogous to the attractive H...H interactions in (-CH=CH), which has been shown through structural and theoretical investigations to exist in the planar conformation in the crystalline state (1-3).

Of course, steric repulsions occurring in the planar conformations may be reduced by rotations ϕ about the C-C single bonds along the backbone. Calculated values of E_g and BW as a function of ϕ are presented in Tables III and IV. For each polymer, the value of E_g is a minimum for the planar conformations ($\phi=0°$ and $\phi=180°$) and a maximum where the planes of the chain segments on either side of the rotated bond are mutually perpendicular ($\phi=90°$). This behavior is as expected since the degree of π overlap along the chain backbone is also a maximum for the planar conformations and a minimum for the perpendicular ones. The band width BW, which is related to the delocalization of the π system along the chain backbone and to the charge carrier mobility in the band (1-3), will be larger the greater degree of π overlap. This prediction agrees with the values of BW obtained here, which are a maximum for the planar conformations of each polymer.

The preferred conformations for these chains can be predicted from the calculated band structures by computing the total unit cell energy $\langle E_t \rangle$, using Eq. 8, for selected values of ϕ, with the preferred conformation (i.e., value of ϕ) given as that associated with minimum $\langle E_t \rangle$. Using this method, the preferred conformations are $\phi=0°$ (i.e., trans) for both (-CH=CH) and all of the fluorinated analogues, $\phi=90°$ for (-CCl=CCl), and ϕ = 120° for both (-CH=CCl) and (-CH=CH-CCl=CCl). The values of E_g for each of these chains in their calculated preferred conformation are given in Table V. Inspection reveals that among the halogenated polyacetylenes, the fluorinated ones offer the greatest potential as semiconductors, and that the large values of E_g obtained for the chlorinated derivatives in their preferred, non-planar, conformations render them less attractive.

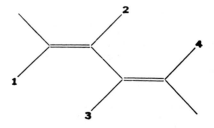

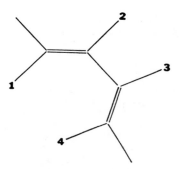

Figure 1. The trans isomeric form of polyacetylene shown in its
planar <u>trans</u> and <u>cis</u> conformations, with pertinent substituents
numbered.

Table III. Values of E_g[a] and BW[a] for (-CH=CH) and Some Fluorinated Analogues as a Function of Rotational Angle ϕ

ϕ	(-CH=CH)		(-CF=CF)		(-CH=CF)		(-CH=CH-CF=CF)	
	E_g	BW	E_g	BW	E_g	BW	E_g	BW
0°	1.2	6.9	0.7	6.9	1.0	6.9	0.7	6.3
30°	1.6	5.3	1.6	5.3	1.8	5.4	1.3	5.1
60°	3.5	3.7	3.4	3.8	3.6	4.8	3.2	4.4
90°	4.3	3.5	4.2	3.5	4.2	4.6	3.9	3.1
120°	2.7	4.7	2.5	4.6	2.6	4.7	2.4	4.3
150°	1.2	6.1	1.1	6.0	1.4	6.2	1.1	5.4
180°	1.1	6.8	0.6	6.8	1.0	6.9	0.6	6.2

[a] In units of electron-volts

Table IV. Values of E_g[a] and BW[a] for Some Chlorinated Analogues of (-CH=CH) as a Function of Rotational Angle ϕ

ϕ	(-CCl=CCl)		(-CH=CCl)		(-CH=CH-CCl=CCl)	
	E_g	BW	E_g	BW	E_g	BW
0°	0.3	6.2	0.3	6.1	0.9	5.7
30°	1.8	5.0	1.9	5.2	1.6	4.8
60°	3.4	3.8	5.5	3.8	3.2	3.5
90°	4.2	3.7	4.1	3.7	3.9	3.3
120°	2.8	4.5	2.6	4.6	2.5	4.1
150°	2.3	5.5	1.5	5.6	1.3	5.0
180°	1.7	6.6	1.2	6.3	0.9	5.7

[a] In units of electron-volts

Table V. Values of the Band Gap $E_g{}^a$ for Each Polymer Chains
at its calculated preferred value of ϕ as determined
by Equation 8

Chain	Preferred Value of ϕ	E_g
(−CH=CH)	0°	1.2
(−CF=CF)	0°	0.7
(−CH=CF)	0°	1.0
(−CH=CH−CF=CF)	0°	1.0
(−CCl=CCl)	90°	4.2
(−CH=CCl)	120°	2.6
(−CH=CH−CCl=CCl)	120°	2.5

[a]In units of electron−volts.

The value of the rotational angle ϕ will have a direct influence on the extent of steric interactions between substituent
atoms (Figure 1). Specifically, as ϕ increases from 0° (trans) to
180° (cis), the interatomic distances 1...3 and 2...4 will increase
while 1...4 and 2...3 will decrease. Of these, the 1...4 interaction
in the cis conformation is particularly critical since here contacts
are encountered with the potential for severe steric conflicts. In
fact, for (−CCl=CCl) the cis conformation is essentially precluded
since the 1...4 interactions are exclusively Cl...Cl. In (−CH=CCl)
and (−CH=CH−CCl=CCl), the Cl...Cl interactions are replaced by
H...Cl which reduce the steric conflicts encountered in the planar
conformations, but only slightly so.

The experimentally determined value of E_g for (−CH=CH) is 1.4−
1.8 eV (14). Inasmuch as these experimental values fall closer to
that given by these calculations for ϕ = 0° and 180° ($E_g \simeq$ 1.2 eV)
than for ϕ = 90° ($E_g \simeq$ 4.3 eV), the present results suggest a high
degree of planarity along the backbone.

The results listed in Table VI illustrate to what degree the
band structure of polyacetylene (corresponding to X1 = X2 = H in the
Table VI) is affected by periodic substitution of the pendant H
atoms along the chain. While substituting a pendant H by a CH_3
appears to have a negligible effect on the values of E_g and BW,
substitution instead by SiH_3 or $Si(CH_3)_3$ increases E_g and decreases
BW (suggesting poorer conductivity). A trend appears to exist for
increased E_g values (and decreased BW values) with an increase in
the steric bulk of the substituent X. Since in these immediate
calculations all chains were considered in their trans planar zig-
zag conformation, it would not seem that the steric bulk of the
substituent would have a major effect on the calculated band structures.

Table VI.   Calculated Values of the Band Gap E_g^a and Bandwidth[a]
BW for Segments of Polymer Chains with Repeat Unit
[–CH=CX1–CH=CH–CX2=CH]

X1	X2	E_g^a	BW^a
H	H	1.2	3.8
H	CH_3	1.1	3.4
CH_3	CH_3	1.4	3.6
H	SiH_3	1.4	2.2
SiH_3	SiH_3	1.8	2.4
$Si(CH_3)_3$	$Si(CH_3)_3$	2.4	1.5

[a]In units of electron-volts

The results in Table VII depict how the band structure of
carbyne [–C≡C] (corresponding to Y = '–' in Table VII) is affected
by periodic inclusion of selected atoms or groups which, by intro-
ducing a kink into the otherwise rectilinear chain of carbyne, would
provide greater conformational versatility and therefore possibly
improve processability. The chains were considered in their trans
planar zig-zag conformation. It is seen that any of the modifications to carbyne indicated in Table VII produces an increased E_g
value and a decreased BW value. That electrical conductivity would
be adversely affected is reasonable given the disruption of the
conjugated system caused by such modifications. As yet, no clear
relationship is apparent between the specific molecular nature of Y
and the effect of its inclusion on E_g and BW; the relatively minor
effect for the case Y = CH_2 is noteworthy and under continued study.

It may be surprising that, compared with (–CH=CH), (–C≡C)
yields a considerably larger E_g value and smaller BW value. Recent

Table VII. Calculated Values of the Band Gap E_g^a and
Bandwidth BW^a for Segments of Polymers Having
Repeat Unit [–C≡C–Y–C≡C]

Y	E_g^a	BW^a
'–'	1.5	2.3
O	2.8	1.9
S	3.2	0.2
NH	3.4	0.8
CH_2	1.7	1.8
SiH_2	2.1	1.8

[a]In units of electron-volts

studies suggest an explanation based on bond-length alternation (1-3,6). Specifically, in calculations on (-CH=CH) the alternating single and double bonds were assigned lengths of 1.435 Å and 1.342 Å, respectively. Correspondingly, the lengths given for the single and triple bonds in (-C≡C) are 1.339 Å and 1.116 Å, respectively. Hence, the disparity in lengths between the two types of bonds is much smaller in (-CH=CH) [0.094 Å] than in (-C≡C) [0.223 Å]. Our studies on model systems, such as (=C=C) having complete uniformity in bond lengths, indicate a correlation between bond-length uniformity along the backbone and favorable values of E_g and BW (15). Of course, bond-length uniformity and conjugation within these chains are essentially equivalent concepts, hence these results again point to a correlation between conjugation along the chain backbone and conductivity.

The choice of structural parameters (i.e., bond lengths and bond angles) in these types of calculations will certainly influence the values of E_g and BW obtained, hence it was of interest to assess the sensitivity of our calculated E_g and BW values to changes in structural geometry. Our reference polymer trans (-CH=CH) and its perfluorinated analogue trans (-CF=CF) were used for this purpose. The most spectacular effect was obtained by simultaneously increasing the lengths of the C=C bonds and decreasing those of the C-C bonds. For (-CF=CF) in the trans conformation, such a modification of only 0.02 Å reduced the value of E_g from 0.72 eV to 0.40 eV. This result is reasonable since such a modification is tantamount to increasing the extent of conjugation along the chain, and this should translate to a lower E_g value. In the other direction, decreasing the C=C bond lengths and increasing the C-C bond lengths by 0.02 Å resulted in an increase in E_g from 0.72 eV to 1.04 eV.

In contrast, calculated E_g values were largely insensitive to small (+2.0°) changes in backbone bond angles. These results confirm that conductivity is directly and strongly dependent on the degree of conjugation along the chain backbone, and that structural modifications that give rise to a decrease in bond-length alternation should provide a means for developing improved electrically conducting polymeric materials.

Acknowledgments

The author wishes to acknowledge the support provided for this research by the Plastics Institute of America and by the Air Force Office of Scientific Research (Grant AFOSR 83-0027, Chemical Structures Program, Division of Chemical Sciences).

Literature Cited

1. Baughman, R. H.; Brédas, J. L.; Chance, R. C.; Elsenbaumer, R.; Schacklette, L. W. Chem. Rev. 1982, 82, 209, and references cited therein.

2. Wegner, G. Makromol. Chem., Macromol. Symp. 1986, 1, 151, and references cited therein.

3. Roth, S. In Electronic Properties of Polymers and Related Compounds, Kuzmany, H.; Mehring, M.; Roth, S.. Eds.; Springer-Verlag, 1985.

4. Zeigler, J., Sandia National Laboratories, Albuquerque, private communications.

5. Wheland, R. C. J. Am. Chem. Soc. 1976, 98, 3926.

6. Brédas, J. L.; Silbey, R.; Boudreaux, D. S.; Chance, R. R. J. Am. Chem. Soc. 1983, 105, 6555.

7. Gourley, K. D.; Lillya, C. P.; Reynolds, J. R.; Chien, J. C. W. Macromolecules 1984, 17, 1025.

8. André, J.-M. In The Electronic Structure of Polymers and Molecular Crystals, Andre, J.-M.; Ladik, J., Eds.; Plenum: New York, 1974.

9. Hoffmann, R.; J. Chem. Phys. 1963, 39, 1397.

10. Whangbo, M.-H.; Hoffmann, R. J. Am. Chem. Soc. 1978, 100, 6093.

11. Whangbo, M.-H.; Hoffmann, R.; Woodward, R. B. Proc. R. Soc. London Ser. A 1979, A366, 23.

12. Imamura, A. J. Chem. Phys. 1970, 52, 3168.

13. Teramae, A.; Yamabe, T. Theoret. Chim. Acta 1983, 64, 1.

14. Brédas, J. L.; Chance, R. R.; Baughman, R. H. J. Chem. Phys. 1982, 76, 3673.

RECEIVED February 13, 1987

Author Index

Affiliation Index

Subject Index

A

Absorbed energy density, calculation, 87–88
Absorption of energy, inelastic
  collisions, 6
Absorption saturation, *See* Transient
  photobleaching
Acid-catalyzed thermolysis of polycarbonates
  GC–MS analysis, 144,145*f*
  imaging of two-component resist
    material, 140,142*f*
  process, 144–145
Acridine-containing polymers,
  synthesis, 227–228
Acridine/PMMA film, transmittance, 229–230
Acrylate resists, applications, 86
Adhesion
  double promoter process, 256
  of polyacrylate-based
    photoresists, 281–282
  of SiO$_2$ substrates, 256–257
Aliphatic ketones, radiolysis, 48
Alkanes, radiolysis, 18–20
Alpha particles
  advances in technology, 532,533*f*
  description, 532
  source, 532
Aminophenoxybenzene diamines,
  structure, 438,439*f*
Ancillary effect, definition, 382
Anhydride cross-linked acrylate resists,
  formation, 86–87
Anodic source, X-ray source, 149–150
Aqueous solutions, energy absorption, 8
Arene systems, radiolysis, 20–23
Aromatic ladder polymers, properties, 568
ASTRO, properties as photoresist, 548
Auger electron spectroscopy of
  poly(alkenylsilane sulfone) passivation
  atomic concentration depth
    profile, 343,344*f*,346*f*
  survey of surface, 343,344*f*

B

Becquerel, discovery of radioactivity, 5
Benzotriazole additives, effect on
  dissolution rate of resist, 241*t*

Benzotriazoles, photoresist
  applications, 238
Bilayer resist processing
  bottom layer formation, 310,311*f*
  development, 312
  inorganic resist deposition, 310,312
  pattern transfer, 312
Bilevel image transfer applications, using
  polysilane, 180–185
Bis(*p*-nitrophenyl)carbonate of
  1,4-benzenedimethanol
  synthesis, 146
  reaction with 2-cyclohexene-1,4-diol, 146–147
Block copolymers, lithographic utility, 123
Buchanan, ionic processes, 10
Bulk polarization, calculation, 382
Bulk second-order susceptibility,
  determination, 402–403
N-(*p-tert*-Butyloxycarbonyloxyphenyl)maleimide
  copolymerization with styrene, 201
  preparation, 201

C

C-beam writing, development, 66
Cage inclusion matrix, dimerization, 399
Cage processes, theory, 57,59
Cage reactions, description, 57
Cage structures, diagram, 397,398
Carreau-type non-Newtonian viscosity
  equation, effect of rheological
  properties on film thickness, 263–264
Cathode rays, definition, 32
Chalcone sulfonamide dye
  effect on dissolution rate of
    resist, 245,247*f*
  performance, 248*t*
Channel inclusion matrix, dimerization, 399
Channel structures, diagram, 397,398
Chapiro, irradiation of polymers, 12
Charlesby, irradiation of polymers, 12
Chemical amplification
  example, 200
  improvement of resist material
    sensitivity, 162

Production by Cara Aldridge Young
Indexing by Deborah H. Steiner
Jacket design by Carla L. Clemens

Elements typeset by Hot Type Ltd., Washington, DC
Printed and bound by Maple Press Co., York, PA
Dust jackets printed by Atlantic Research Corporation, Alexandria, VA

Recent ACS Books

Personal Computers for Scientists: A Byte at a Time
By Glenn I. Ouchi
288 pp; clothbound; ISBN 0–8412–1001–2

The ACS Style Guide: A Manual for Authors and Editors
Edited by Janet S. Dodd
264 pp; clothbound; ISBN 0–8412–0917–0

Silent Spring Revisited
Edited by Gino J. Marco, Robert M. Hollingworth, and William Durham
214 pp; clothbound; ISBN 0–8412–0980–4

Chemical Demonstrations: A Sourcebook for Teachers
By Lee R. Summerlin and James L. Ealy, Jr.
192 pp; spiral bound; ISBN 0–8412–0923–5

Phosphorus Chemistry in Everyday Living, Second Edition
By Arthur D. F. Toy and Edward N. Walsh
342 pp; clothbound; ISBN 0–8412–1002–0

Pharmacokinetics: Processes and Mathematics
By Peter G. Welling
ACS Monograph 185; 290 pp; ISBN 0–8412–0967–7

*Metal Complexes in Fossil Fuels: Geochemistry,
Characterization, and Processing*
Edited by Royston H. Filby and Jan F. Branthaver
ACS Symposium Series 344; 436 pp; ISBN 0–8412–1404–2

*Proteins at Interfaces: Physicochemical
and Biochemical Studies*
Edited by John L. Brash and Thomas A. Horbett
ACS Symposium Series 343; 706 pp; ISBN 0–8412–1403–4

Ordered Media in Chemical Separations
Edited by Willie L. Hinze and Daniel W. Armstrong
ACS Symposium Series 342; 293 pp; ISBN 0–8412–1402–6

Sources and Fates of Aquatic Pollutants
Edited by Ronald A. Hites and S. J. Eisenreich
Advances in Chemistry Series 216; 558 pp; ISBN 0–8412–0983–9

Nucleophilicity
Edited by J. Milton Harris and Samuel P. McManus
Advances in Chemistry Series 215; 494 pp; ISBN 0–8412–0952–9

For further information and a free catalog of ACS books, contact:
American Chemical Society
Distribution Office, Department 225
1155 16th Street, NW, Washington, DC 20036
Telephone 800–227–5558